CASE STUDIES IN IMMUNOLOGY

A CLINICAL COMPANION

6TH EDITION

CASE STUDIES IN IMMUNOLOGY

A CLINICAL COMPANION

6TH EDITION

Raif Geha • Luigi Notarangelo
Harvard Medical School

NEW YORK AND LONDON

Vice President: Denise Schanck
Senior Editor: Janet Foltin
Text Editor: Eleanor Lawrence
Assistant Editor: Janete Scobie
Senior Editorial Assistant: Sarah Wolf
Production Editor: Ioana Moldovan
Typesetter and Senior Production Editor: Georgina Lucas
Copy Editor: Bruce Goatly
Proofreader: Sally Huish
Illustrations: Matthew McClements, Blink Studio, Ltd.
Permissions Coordinator: Becky Hainz-Baxter
Indexer: Medical Indexing Ltd.

ISBN 978-0-8153-4441-4

Library of Congress Cataloging-in-Publication Data

Geha, Raif S.
Case studies in immunology : a clinical companion / Raif Geha, Luigi Notarangelo. -- 6th ed.
p. ; cm.
Includes index.
ISBN 978-0-8153-4441-4 (alk. paper)
I. Notarangelo, Luigi. II. Title.
[DNLM: 1. Immune System Diseases--Case Reports. 2. Allergy and Immunology--Case Reports. 3. Immunity--genetics--Case Reports. WD 300]
616.07'9--dc23

2011034570

Published by Garland Science, Taylor & Francis Group, LLC, an informa business
711 Third Avenue, 8th Floor, New York, NY 10017, USA and
2 Park Square, Milton Park, Abingdon, OX14 4RN, UK.

Printed in the United States of America

15 14 13 12 11 10 9 8 7 6 5 4 3 2 1

Visit our website at http://garlandscience.com

Preface

The science of immunology started as a case study. On May 15, 1796 Edward Jenner inoculated a neighbor's son, James Phipps, with vaccinia (cowpox) virus. Six weeks later, on July 1, 1796, Jenner challenged the boy with live smallpox and found that he was protected against this infection. During the past 215 years, the basic science of immunology has shed light on the pathogenesis of immune-mediated diseases. Conversely, the investigation of diseases of the immune system, particularly of genetically inherited primary immunodeficiency diseases, has provided valuable insights into the functioning of the normal immune system.

The study of immunology provides a rare opportunity in medicine to relate the findings of basic scientific investigations to clinical problems. The case histories in this book are chosen for two purposes: to illustrate in a clinical context essential points about the mechanisms of immunity and to describe and explain some of the immunological problems often seen in the clinic. For this sixth edition, we have added 10 new cases that are representative of key aspects of immune system development and function, as revealed by specific forms of primary immunodeficiencies and by common diseases with interesting underlying immunologic mechanisms. These cases include DiGeorge syndrome, familial hemophagocytic lymphohistiocytosis, Chediak–Higashi syndrome, hyper IgE syndrome, ataxia telangiectasia, WHIM syndrome, severe congenital neutropenia, recurrent herpes simplex encephalitis, juvenile arthritis, and Crohn's disease. New concepts, such as the genetic control of myeloid development, the mechanisms of lymphocyte-mediated cytotoxicity, chemokine-mediated control of leukocyte trafficking, the biology and function of T_H17 cells, and type 1 interferon-mediated control of viral infections, are also discussed in the book. We have also revised several cases to add newly acquired information about these diseases and novel developments in immunological therapeutic intervention. The cases illustrate fundamental mechanisms of immunity, as shown by genetic disorders of the immune system, immune-complex diseases, immune-mediated hypersensitivity reactions and autoimmune and alloimmune diseases. They describe real events from case histories, largely but not solely drawn from the records of the Boston Children's Hospital and the Brigham and Women's Hospital in Boston, Massachusetts. Names, places, and times have been altered to obscure the identities of the patients; other details are faithfully reproduced. The cases are intended to help medical students and pre-medical students learn about basic immunological mechanisms and understand their importance, and particularly to serve as a review aid, but we believe they will be useful and interesting to any student of immunology.

Each case is presented in the same format. The case history itself is preceded by an introduction presenting basic scientific facts needed to understand the case. The case history is followed by a brief summary of the disease under study and discussion of the clinical findings. Finally, several questions and discussion points highlight the lessons learned. These questions are not intended as a quiz but rather to shed further light on the case.

We are grateful to Dr. Peter Densen of the University of Iowa for the C8 deficiency case material, Dr. Sanjiv Chopra of Harvard Medical School for the case on mixed essential cryoglobulinemia, and Dr. Peter Schur of the Brigham and Women's Hospital for the rheumatoid arthritis case. We also thank Dr. Jane Newburger of the Boston Children's Hospital for the case on rheumatic fever and Dr. Eric Rosenberg of the Massachusetts General Hospital for the AIDS case. We thank Drs. Lisa Stutius Bartnikas, Arturo Borzutzky, Janet Chou, Ari Fried, Erin Janssen, and Andrew Shulman of Children's Hospital Boston for the cases of hyper IgE syndrome, Chediak–Higashi syndrome, ataxia telangiectasia, DiGeorge syndrome, systemic-onset juvenile idiopathic arthritis, and Crohn's disease, respectively; Dr. Jolan Walter of the Massachusetts General Hospital for the case of hemophagocytic lymphohistiocytosis; Dr. Anna Virginia Gulino of Ospedale Sant'Eugenio in Rome, for the case of WHIM syndrome; and Dr. Itai Pessach of Children's Hospital Boston for the case of severe congenital neutropenia. We are also greatly indebted to our colleagues Drs. David Dawson, Susan Berman, Lawrence Shulman, and David Hafler of the Brigham and Women's Hospital; to Dr. Razzaque Ahmed of the Harvard School of Dental Medicine; to Drs. Ernesto Gonzalez and Scott Snapper of the Massachusetts General Hospital; to Drs. Peter Newburger and Jamie Ferrara of the Departments of Pediatrics of the University of Massachusetts and the University of Michigan; to Dr. Robertson Parkman of the Los Angeles Children's Hospital; to Dr. Fabio Facchetti, Dr. Lucia Notarangelo, and Dr. Antonio Regazzoli of the Spedali Civili of Brescia, Italy; to Henri de la Salle of the Centre régional de Transfusion Sanguine in Strasbourg, France; and to Professor Michael Levin of St. Mary's Hospital, London, for supplying case materials. Our colleagues and trainees in the Immunology Division of the Children's Hospital have provided invaluable service by extracting summaries of long and complicated case histories; we are particularly indebted to Drs. Lynda Schneider, Leonard Bacharier, Francisco Antonio Bonilla, Hans Oettgen, Jonathan Spergel, Rima Rachid, Scott Turvey, Jordan Orange, Emanuela Castigli, Andrew McGinnitie, Marybeth Son, Melissa Hazen, Douglas McDonald, John Lee, and Lilit Garibyan in constructing several case histories. In the course of developing these cases, we have been indebted for expert and pedagogic advice to Fred Alt, Mark Anderson, John Atkinson, Hugh Auchincloss, Stephen Baird, Zuhair K. Ballas, Leslie Berg, Corrado Betterle, Kurt Bloch, Jean-Laurent Casanova, Talal Chatila, John J. Cohen, Michael I. Colston, Anthony DeFranco, Peter Densen, Ten Feizi, Alain Fischer, Christopher Goodnow, Edward Kaplan, George Miller, Peter Parham, Jaakko Perheentupa, Jennifer Puck, Westley Reeves, Patrick Revy, Peter Schur, Anthony Segal, Lisa Steiner, Stuart Tangye, Cox Terhorst, Emil Unanue, André Veillette, Jan Vilcek, Mark Walport, Fenella Woznarowska, and John Zabriskie.

Eleanor Lawrence has spent many hours honing the prose as well as the content of the cases and we are grateful to her for this. We would also like to acknowledge the Garland Science team for their work on the sixth edition.

A note to the reader

The main topics addressed in each case correspond as much as possible to topics that are presented in the eighth edition of *Janeway's Immunobiology* by Kenneth Murphy. To indicate which sections of *Immunobiology* contain material relevant to each case, we have listed on the first page of each case the topics covered in it. The color code follows the code used for the five main sections of *Immunobiology*: yellow for the introductory chapter and innate immunity, blue for the sections on recognition of antigen, pink for the development of lymphocytes, green for the adaptive immune response, purple for the response to infection and clinical topics, and orange for methods.

Instructor Resources Website

Accessible from www.garlandscience.com, the Instructor Site requires registration and access is available only to qualified instructors. To access the Instructor Site, please contact your local sales representative or email science@garland.com.

The images from Case Studies in Immunology, 6th Edition, are available on the Instructor Site in two convenient formats: PowerPoint® and JPEG. They have been optimized for display on a computer. The resources may be browsed by individual cases and there is a search engine. Figures are searchable by figure number, figure name, or by keywords used in the figure legend from the book.

Contributors

Lisa Bartnikas, MD, Clinical Fellow, Division of Immunology, Children's Hospital Boston

Arturo Borzutzky, MD, Fellow, Division of Immunology, Children's Hospital Boston; Research Fellow, Pediatrics, Harvard Medical School

Janet Chou, MD, Fellow, Division of Immunology, Children's Hospital Boston; Research Fellow, Pediatrics, Harvard Medical School

Ari Fried, MD, Attending Physician, Division of Immunology, Children's Hospital Boston; Instructor in Pediatrics, Harvard Medical School

Anna Virginia Gulino, MD, Staff Member, Division of Pediatric Hematology, Ospedale Sant'Eugenio, Rome, Italy

Erin Janssen, MD, PhD, Fellow, Division of Immunology, Children's Hospital Boston

Douglas McDonald, MD, PhD, Assistant in Medicine, Children's Hospital Boston; Instructor, Harvard Medical School

Itai Pessach, MD, PhD, Fellow, Division of Immunology, Children's Hospital Boston; Research Fellow, Pediatrics, Harvard Medical School

Andrew Shulman, MD, PhD, Fellow, Division of Immunology, Children's Hospital Boston; Research Fellow, Pediatrics, Harvard Medical School

Jolan Walter, MD, PhD, Clinical Instructor in Pediatrics, Massachusetts General Hospital

Contents

CASE 1 X-linked Agammaglobulinemia

An absence of B lymphocytes.

One of the most important functions of the adaptive immune system is the production of antibodies. It is estimated that a human being can make more than one million different specific antibodies. This remarkable feat is accomplished through a complex genetic program carried out by B lymphocytes and their precursors in the bone marrow (Fig. 1.1). Every day about 2.5 billion (2.5×10^9) early B-cell precursors (pro-B cells) take the first step in this genetic program and enter the body's pool of pre-B cells. From this pool of rapidly dividing pre-B cells 30 billion daily mature into B cells, which leave the bone marrow as circulating B lymphocytes, while 55 billion fail to mature successfully and undergo programmed cell death. This process continues throughout life, although the numbers gradually decline with age.

Mature circulating B cells proliferate on encounter with antigen and differentiate into plasma cells, which secrete antibody. Antibodies, which are made by the plasma cell progeny of B cells, protect by binding to and neutralizing toxins and viruses, by preventing the adhesion of microbes to cell surfaces, and, after binding to microbial surfaces, by fixing complement and thereby enhancing phagocytosis and lysis of pathogens (Fig. 1.2).

This case concerns a young man who has an inherited inability to make antibodies. His family history reveals that he has inherited this defect in antibody synthesis as an X-linked recessive abnormality. This poses an interesting puzzle because the genes encoding the structure of the immunoglobulin polypeptide chains are encoded on autosomal chromosomes and not on the X chromosome. Further inquiry reveals that he has no B cells, so that some gene on the X chromosome is critical for the normal maturation of B lymphocytes.

This case was prepared by Raif Geha, MD, in collaboration with Ari Fried, MD.

Topics bearing on this case:
Humoral versus cell-mediated immunity
Effector functions of antibodies
Effector mechanisms of humoral immunity
Actions of complement and complement receptors
B-cell maturation
Methods for measuring T-cell function

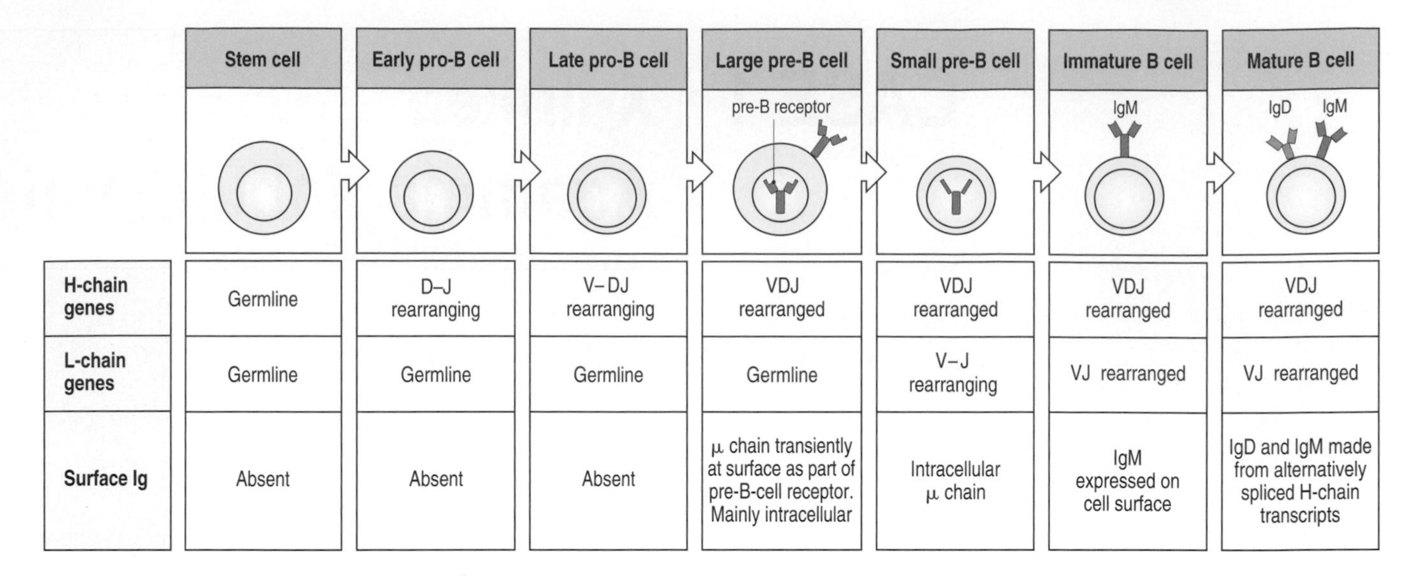

Stem cell
Early pro-B cell
Late pro-B cell
Large pre-B cell
Small pre-B cell
Immature B cell
Mature B cell
pre-B receptor
IgM
IgD
IgM
H-chain genes
Germline
D–J rearranging
V–DJ rearranging
VDJ rearranged
VDJ rearranged
VDJ rearranged
VDJ rearranged
L-chain genes
Germline
Germline
Germline
Germline
V–J rearranging
VJ rearranged
VJ rearranged
Surface Ig
Absent
Absent
Absent
μ chain transiently at surface as part of pre-B-cell receptor. Mainly intracellular
Intracellular μ chain
IgM expressed on cell surface
IgD and IgM made from alternatively spliced H-chain transcripts

Fig. 1.1 The development of B cells proceeds through several stages marked by the rearrangement of the immunoglobulin genes. The bone marrow stem cell that gives rise to the B-lymphocyte lineage has not yet begun to rearrange its immunoglobulin genes; they are in germline configuration. The first rearrangements of D gene segments to J_H gene segments occur in the early pro-B cells, generating late pro-B cells. In the late pro-B cells, a V_H gene segment becomes joined to the rearranged DJ_H, producing a pre-B cell that is expressing both low levels of surface and high levels of cytoplasmic μ heavy chain. Finally, the light-chain genes are rearranged and the cell, now an immature B cell, expresses both light chains (L chains) and μ heavy chains (H chains) as surface IgM molecules. Cells that fail to generate a functional surface immunoglobulin, or those with a rearranged receptor that binds a self antigen, die by programmed cell death. The rest leave the bone marrow and enter the bloodstream.

The case of Bill Grignard: a medical student with scarcely any antibodies.

Two-year-old boy, two maternal uncles died in infancy from infection.

Bill Grignard was well for the first 10 months of his life. In the next year he had pneumonia once, several episodes of otitis media (inflammation of the middle ear), and on one occasion developed erysipelas (streptococcal infection of the skin) on his right cheek. These infections were all treated successfully with antibiotics but it seemed to his mother, a nurse, that he was constantly on antibiotics.

His mother had two brothers who had died, 30 years prior to Bill's birth, from pneumonia in their second year of life, before antibiotics were available. She also had two sisters who were well; one had a healthy son and daughter and the other a healthy daughter.

Bill was a bright and active child who gained weight, grew, and developed normally but he continued to have repeated infections of the ears and sinuses and twice again had pneumonia. At 2 years 3 months his local pediatrician tested his serum immunoglobulins. He found 80 mg dl^{-1} IgG (normal 600–1500 mg dl^{-1}), no IgA (normal 50–125 mg dl^{-1}), and only 10 mg dl^{-1} IgM (normal 75–150 mg dl^{-1}).

Bill was started on monthly intramuscular injections of gamma globulin; his serum IgG level was maintained at 200 mg dl^{-1}. He started school at age 5 years and performed very well (he was reading at second grade level at age 5 years) despite prolonged absences because of recurrent pneumonia and other infections.

At 9 years of age he was referred to the Children's Hospital because of atelectasis (collapse of part of a lung) and a chronic cough. On physical examination he was found to be a well-developed, alert boy. He weighed 33.5 kg and was 146 cm tall (height and

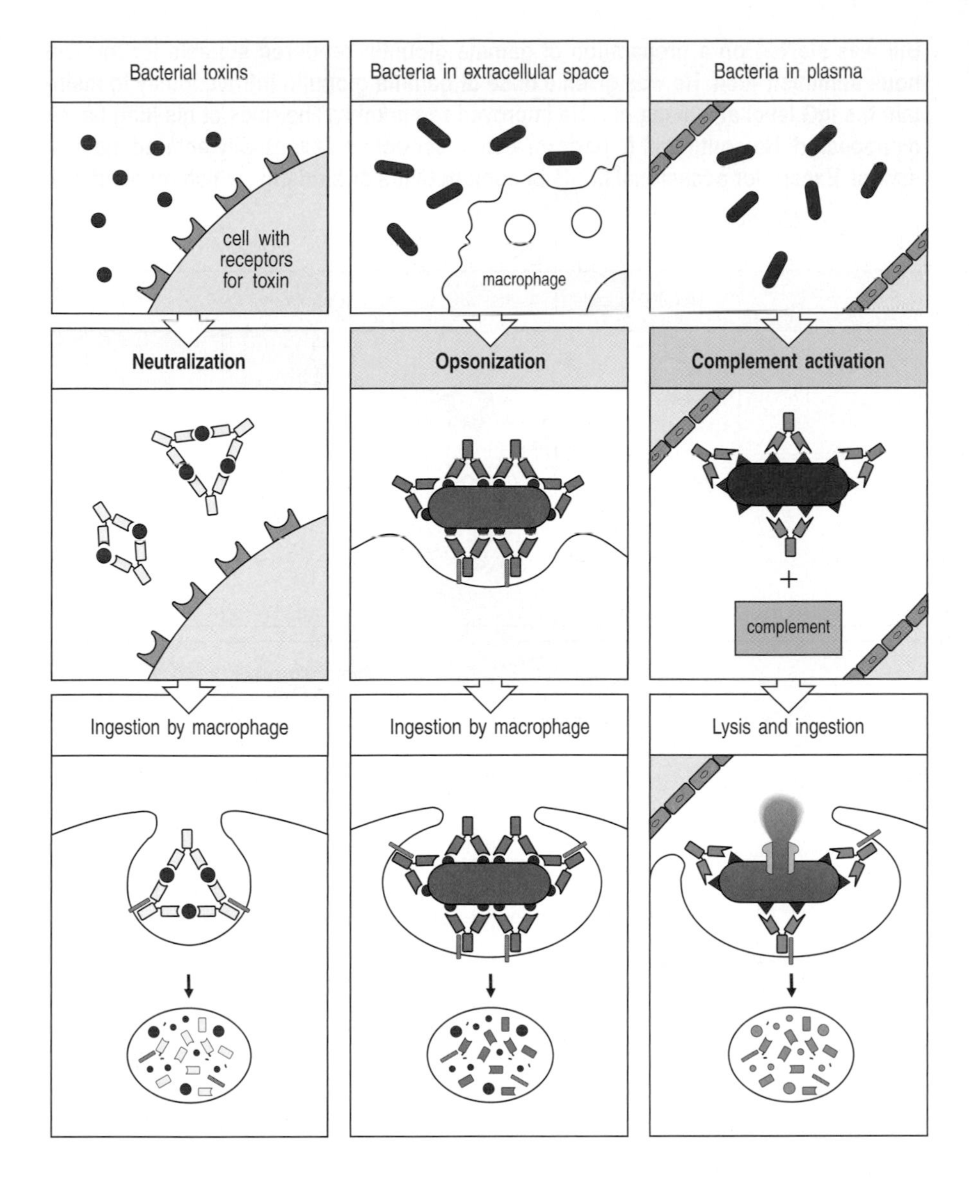

Fig. 1.2 Antibodies can participate in host defense in three main ways. The left-hand column shows antibodies binding to and neutralizing a bacterial toxin, preventing it from interacting with host cells and from causing pathology. Unbound toxin can react with receptors on the host cell, whereas the toxin:antibody complex cannot. Antibodies also neutralize complete virus particles and bacterial cells by binding to them and inactivating them. The antigen:antibody complex is eventually scavenged and degraded by macrophages. Antibodies coating an antigen render it recognizable as foreign by phagocytes (macrophages and polymorphonuclear leukocytes), which then ingest and destroy it; this is called opsonization. The central column shows the opsonization and phagocytosis of a bacterial cell. The right-hand column shows the activation of the complement system by antibodies coating a bacterial cell. Bound antibodies form a receptor for the first protein of the complement system, which eventually forms a protein complex on the surface of the bacterium that favors its uptake and destruction by phagocytes and can, in some cases, directly kill the bacterium. Thus, antibodies target pathogens and their products for disposal by phagocytes.

weight normal for his age). The doctor noted that he had no visible tonsils (he had never had a tonsillectomy). With a stethoscope the doctor also heard rales (moist crackles) at both lung bases.

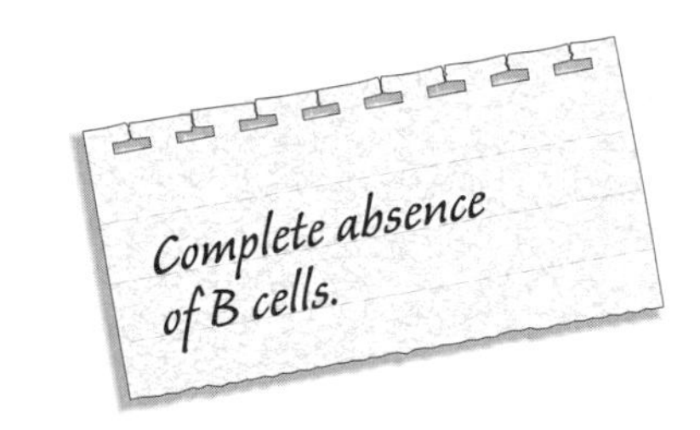

Further family history revealed that Bill had one younger sibling, John, a 7-year-old brother, who also had contracted pneumonia on three occasions. John had a serum IgG level of 150 mg dl^{-1}.

Laboratory studies at the time of Bill's visit to the Children's Hospital gave a white blood cell count of 5100 μl^{-1} (normal), of which 45% were neutrophils (normal), 43% were lymphocytes (normal), 10% were monocytes (elevated), and 2% were eosinophils (normal).

Flow cytometry (Fig. 1.3) showed that 85% of the lymphocytes bound an antibody to CD3, a T-cell marker (normal); 55% were helper T cells reacting with an anti-CD4 antibody; and 29% were cytotoxic T cells reacting with an anti-CD8 antibody (normal). However, none of Bill's peripheral blood lymphocytes bound an antibody against the B-cell marker CD19 (normal 12%) (Fig. 1.4).

T-cell proliferation indices in response to phytohemagglutinin, concanavalin A, tetanus toxoid, and diphtheria toxoid were 162, 104, 10, and 8, respectively (all normal). Serum IgG remained low at 155 mg dl^{-1}, and serum IgA and IgM were undetectable.

Bill was started on a preparation of gamma globulin rendered suitable for intravenous administration. He was given a dose of gamma globulin intravenously to maintain his IgG level at 600 mg dl^{-1}. He improved remarkably. The rales at his lung bases disappeared. He continued to perform well in school and eventually entered medical school. Except for occasional bouts of conjunctivitis or sinusitis, which respond well

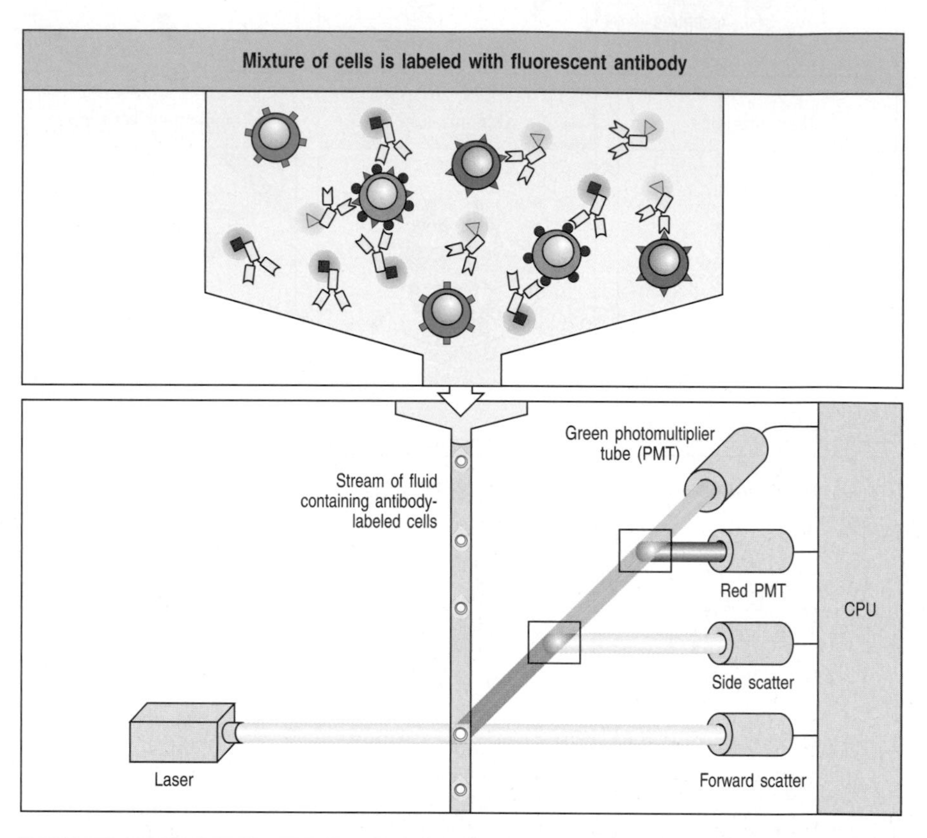

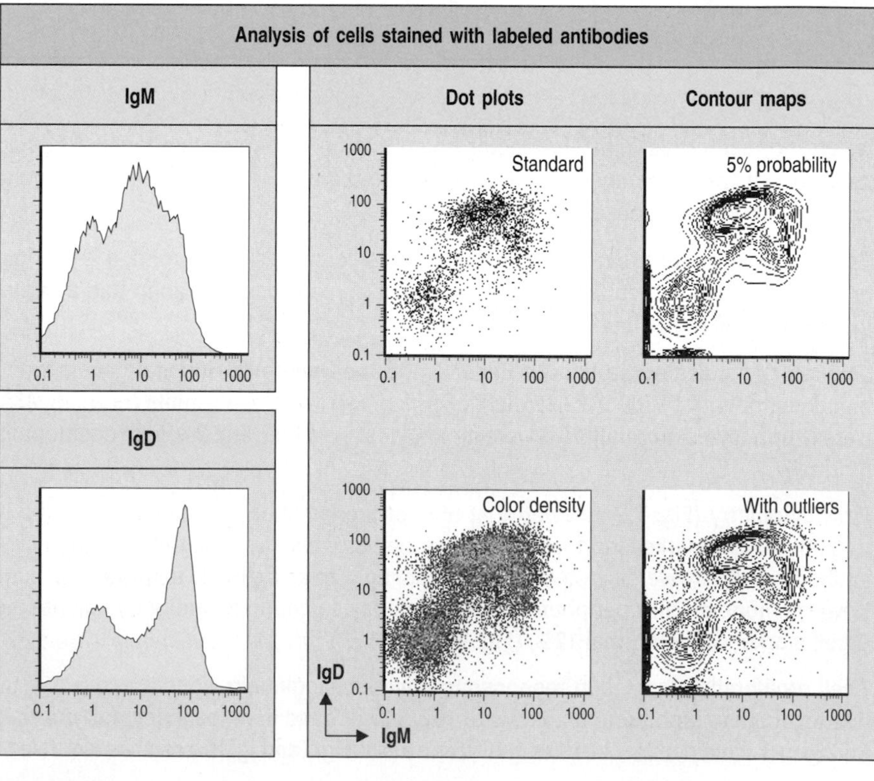

Fig. 1.3 The FACS™ allows individual cells to be identified by their cell-surface antigens and to be sorted. Cells to be analyzed by flow cytometry are first labeled with fluorescent dyes (top panel). Direct labeling uses dye-coupled antibodies specific for cell-surface antigens (as shown here), whereas indirect labeling uses a dye-coupled immunoglobulin to detect unlabeled cell-bound antibody. The cells are forced through a nozzle in a single-cell stream that passes through a laser beam (second panel). Photo-multiplier tubes (PMTs) detect the scattering of light, which is a sign of cell size and granularity, and emissions from the different fluorescent dyes. This information is analyzed by computer (CPU). By examining many cells in this way, the number of cells with a specific set of characteristics can be counted and levels of expression of various molecules on these cells can be measured. The bottom part of the figure shows how these data can be represented, using the expression of two surface immunoglobulins, IgM and IgD, on a sample of B cells from a mouse spleen. The two immunoglobulins have been labeled with different-colored dyes. When the expression of just one type of molecule is to be analyzed (IgM or IgD), the data are usually displayed as a histogram, as in the left-hand panels. Histograms display the distribution of cells expressing a single measured parameter (such as size, granularity, fluorescence color). When two or more parameters are measured for each cell (IgM and IgD), various types of two-color plot can be used to display the data, as shown in the right-hand panel. All four plots represent the same data. The horizontal axis represents the intensity of IgM fluorescence, and the vertical axis the intensity of IgD fluorescence. Two-color plots provide more information than histograms; they allow recognition, for example, of cells that are 'bright' for both colors, 'dull' for one and bright for the other, dull for both, negative for both, and so on. For example, the cluster of dots in the extreme lower left portions of the plots represents cells that do not express either immunoglobulin; these are mostly T cells. The standard dot plot (upper left) places a single dot for each cell whose fluorescence is measured. It is good for picking up cells that lie outside the main groups but tends to saturate in areas containing a large number of cells of the same type. A second method of presenting these data is the color dot plot (lower left), which uses color density to indicate high-density areas. A contour plot (upper right) draws 5% 'probability' contours, with 5% of the cells lying between each contour providing the best monochrome visualization of regions of high and low density. The lower right plot is a 5% probability contour map that also shows outlying cells as dots.

to oral antibiotic treatment, he remains in good health and leads an active life. He became skilled at inserting a needle into a vein on the back of his hand and he infuses himself with 10 g of gamma globulin every weekend.

X-linked agammaglobulinemia.

Males such as Bill with a hereditary inability to make antibodies are subject to recurrent infections. However, the infections are due almost exclusively to common extracellular bacterial pathogens—*Haemophilus influenzae*, *Streptococcus pneumoniae*, *Streptococcus pyogenes*, and *Staphylococcus aureus*. An examination of scores of histories of boys with this defect has established that they have no problems with intracellular infections, such as those caused by the common viral diseases of childhood. T-cell number and function in males with X-linked agammaglobulinemia are normal, and these individuals therefore have normal cell-mediated responses, which are able to terminate viral infections and infections with intracellular bacteria such as those causing tuberculosis.

The bacteria that are the major cause of infection in X-linked agammaglobulinemia are all so-called pyogenic bacteria. Pyogenic means pus-forming, and pus consists largely of neutrophils. The normal host response to pyogenic infections is the production of antibodies that coat the bacteria and fix complement, thereby enhancing rapid uptake of the bacteria into phagocytic cells such as neutrophils and macrophages, which destroy them. Since antibiotics came into use, it has been possible to treat pyogenic infections successfully. However, when they recur frequently, the excessive release of proteolytic enzymes (for example elastase) from the bacteria and from the host phagocytes causes anatomical damage, particularly to the airways of the lung. The bronchi lose their elasticity and become the site of chronic inflammation (this is called bronchiectasis). If affected males do not receive replacement therapy—gamma globulin—to prevent pyogenic infections, they eventually die of chronic lung disease.

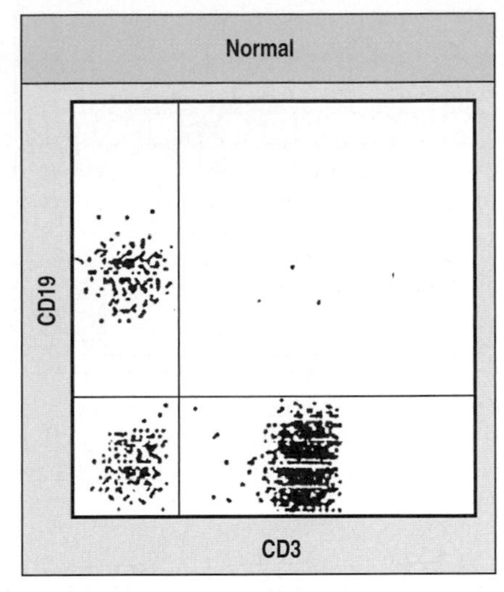

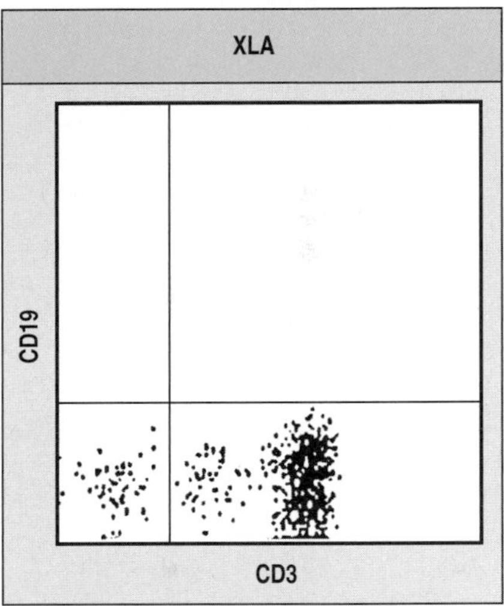

Fig. 1.4 Clinical FACS™ analysis of a normal individual (top panel) and a patient with X-linked agammaglobulinemia (XLA) (bottom panel). Blood lymphocytes from a normal individual bind labeled antibody to both the B-cell marker CD19 and the T-cell marker CD3 (see top panel). However, blood lymphocytes from an individual such as Bill with X-linked agammaglobulinemia show only binding to antibodies against the T-cell marker CD3. This indicates an absence of B cells in these patients.

Gamma globulin is prepared from human plasma. Plasma is pooled from approximately 1000 or more blood donors and is fractionated at very cold temperatures (–5°C) by adding progressively increasing amounts of ethanol. This method was developed by Professor Edwin J. Cohn at the Harvard Medical School during the Second World War. The five plasma fractions obtained are still called Cohn Fractions I, II, III, IV, and V. Cohn Fraction I is mainly composed of fibrinogen. Cohn Fraction II is almost pure IgG and is called gamma globulin. Cohn Fraction III contains the beta globulins, including IgA and IgM; Fraction IV, the alpha globulins; and Fraction V, albumin. Cohn Fraction II, or gamma globulin, is commercially available as a 16% solution of IgG. During the processing of the plasma some of the gamma globulin aggregates, and for this reason the 16% solution cannot be given intravenously. Aggregated gamma globulin acts like immune complexes and causes a reaction of shaking chills, fever, and low blood pressure when given intravenously. Gamma globulin can be disaggregated with low pH or insoluble proteolytic enzymes. It can then be safely administered intravenously as a 5% solution. In newer preparations, fractionation is followed by a further purification step using anion-exchange (DEAE) chromatography to get rid of trace contaminants. To decrease the risk of transmitting infection, the current commercially available products have several virus removal and inactivation steps incorporated into the manufacturing process.

The gene defect in X-linked agammaglobulinemia was identified when the gene was mapped to the long arm of the X chromosome at Xq22 and

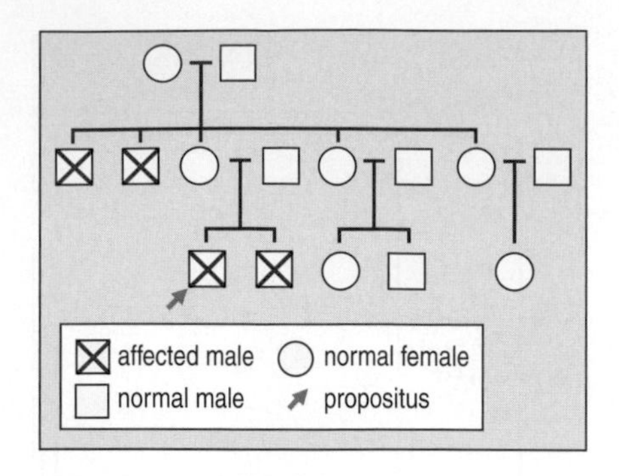

Fig. 1.5 Bill's family tree.

subsequently cloned. The gene, *BTK*, encodes a cytoplasmic protein tyrosine kinase called Bruton's tyrosine kinase (Btk), which is found in pre-B cells, B cells, and neutrophils. Btk is activated at different stages of B-cell development by the engagement of both the pre-B-cell receptor and the B-cell receptor. Btk is required to mediate the survival and further differentiation of the progenitor B cells in which successful rearrangement of their heavy-chain genes has occurred. It is also required for the survival of mature B cells.

Questions.

1 *Fig. 1.5 shows Bill's family tree. It can be seen that only males are affected and that the females who carry the defect (Bill's mother and maternal grandmother) are normal. This inheritance pattern is characteristic of an X-linked recessive trait. We do not know whether Bill's aunts are carriers of the defect because neither of them has had an affected male child. Now that the BTK gene has been mapped, it is possible in principle to detect carriers by testing for the presence of a mutant BTK gene. But there is a much simpler test that was already available at the time of Bill's diagnosis, which is still used routinely. Can you suggest how we could have determined whether Bill's aunts were carriers?*

2 *Bill was well for the first 10 months of his life. How do you explain this?*

3 *Patients with immunodeficiency diseases should never be given live viral vaccines! Several male infants with X-linked agammaglobulinemia have been given live oral polio vaccine and have developed paralytic poliomyelitis. What sequence of events led to the development of polio in these boys?*

4 *Bill has a normal number of lymphocytes in his blood (43% of a normal concentration of 5100 white blood cells per μl). Only by phenotyping these lymphocytes do we realize that they are all T cells (CD3$^+$) and that he has no B cells (CD19$^+$). What tests were performed to establish that his T cells function normally?*

5 *Bill's recurrent infections were due almost exclusively to* Streptococcus *and* Haemophilus *species. These bacteria have a slimy capsule composed primarily of polysaccharide polymers, which protect them from direct attack by phagocytes. Humans make IgG2 antibodies against these polysaccharide polymers. The IgG2 antibodies 'opsonize' the bacteria by fixing complement on their surface, thereby facilitating the rapid uptake of these bacteria by phagocytic cells (Fig. 1.6). What other genetic defect in the immune system might clinically mimic X-linked agammaglobulinemia?*

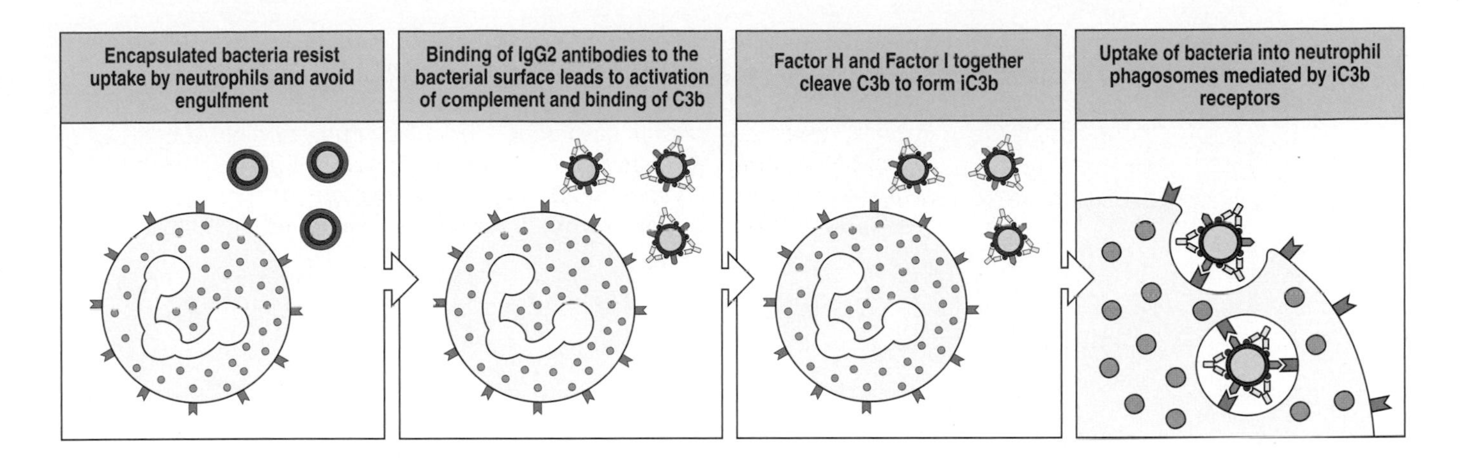
Encapsulated bacteria resist uptake by neutrophils and avoid engulfment
Binding of IgG2 antibodies to the bacterial surface leads to activation of complement and binding of C3b
Factor H and Factor I together cleave C3b to form iC3b
Uptake of bacteria into neutrophil phagosomes mediated by iC3b receptors

Fig. 1.6 Encapsulated bacteria are efficiently engulfed by phagocytes only when they are coated with complement. Encapsulated bacteria resist ingestion by phagocytes unless they are recognized by antibodies that fix complement. IgG2 antibodies are produced against these bacteria in humans, and lead to the deposition of complement component C3b on the bacterial surface, where it is cleaved by Factor H and Factor I to produce iC3b, still bound to the bacterial surface. iC3b binds a specific receptor on phagocytes and induces the engulfment and destruction of the iC3b-coated bacterium. Phagocytes also have receptors for C3b, but these are most effective when acting in concert with Fc receptors for IgG1 antibodies, whereas the iC3b receptor is potent enough to act alone, and is the most important receptor for the phagocytosis of pyogenic bacteria.

6 The doctor noted that Bill had no tonsils even though he had never had his tonsils removed surgically. How do you explain this absence of tonsils, an important diagnostic clue in suspecting X-linked agammaglobulinemia?

7 It was found by trial and error that Bill would stay healthy and have no significant infections if his IgG level were maintained at 600 mg dl^{-1} of plasma. He was told to take 10 g of gamma globulin every week to maintain that level. How was the dose calculated?

8 Females with a disease exactly mimicking X-linked agammaglobulinemia have been found. Explain how this might happen.

CASE 2 CD40 Ligand Deficiency

Failure of immunoglobulin class switching.

After exposure to an antigen, the first antibodies to appear are IgM. Later, antibodies of other classes appear: IgG predominates in the serum and extravascular space, while IgA is produced in the gut and in the respiratory tract, and IgE may also be produced in the mucosal tissues. The different effector functions of these different antibody classes are summarized in Fig. 2.1. The changes in the class of the antibody produced in the course of an immune response reflect the occurrence of heavy-chain isotype switching in the B cells that synthesize immunoglobulin, so that the heavy-chain variable (V) region, which determines the specificity of an antibody, becomes associated with

Functional activity	IgM	IgD	IgG1	IgG2	IgG3	IgG4	IgA	IgE
Neutralization	+	–	++	++	++	++	++	–
Opsonization	–	–	+++	*	++	+	+	–
Sensitization for killing by NK cells	–	–	++	–	++	–	–	–
Sensitization of mast cells	–	–	+	–	+	–	–	+++
Activates complement system	+++	–	++	+	+++	–	+	–

Distribution	IgM	IgD	IgG1	IgG2	IgG3	IgG4	IgA	IgE
Transport across epithelium	+	–	–	–	–	–	+++ (dimer)	–
Transport across placenta	–	–	+++	+	++	+/–	–	–
Diffusion into extravascular sites	+/–	–	+++	+++	+++	+++	++ (monomer)	+
Mean serum level (mg ml^{-1})	1.5	0.04	9	3	1	0.5	2.1	3×10^{-5}

Fig. 2.1 Each human immunoglobulin isotype has specialized functions and a unique distribution. The major effector functions of each isotype (+++) are shaded in dark red, while lesser functions (++) are shown in dark pink, and very minor functions (+) in pale pink. The distributions are similarly marked, with the actual average levels in serum shown in the bottom row. *IgG2 can act as an opsonin in the presence of Fc receptors of a particular allotype, found in about 50% of Caucasians.

Topics bearing on this case:

- Isotype or class switching
- Antibody isotypes and classes
- CD40 ligand and class switching
- Antibody-mediated bacterial killing

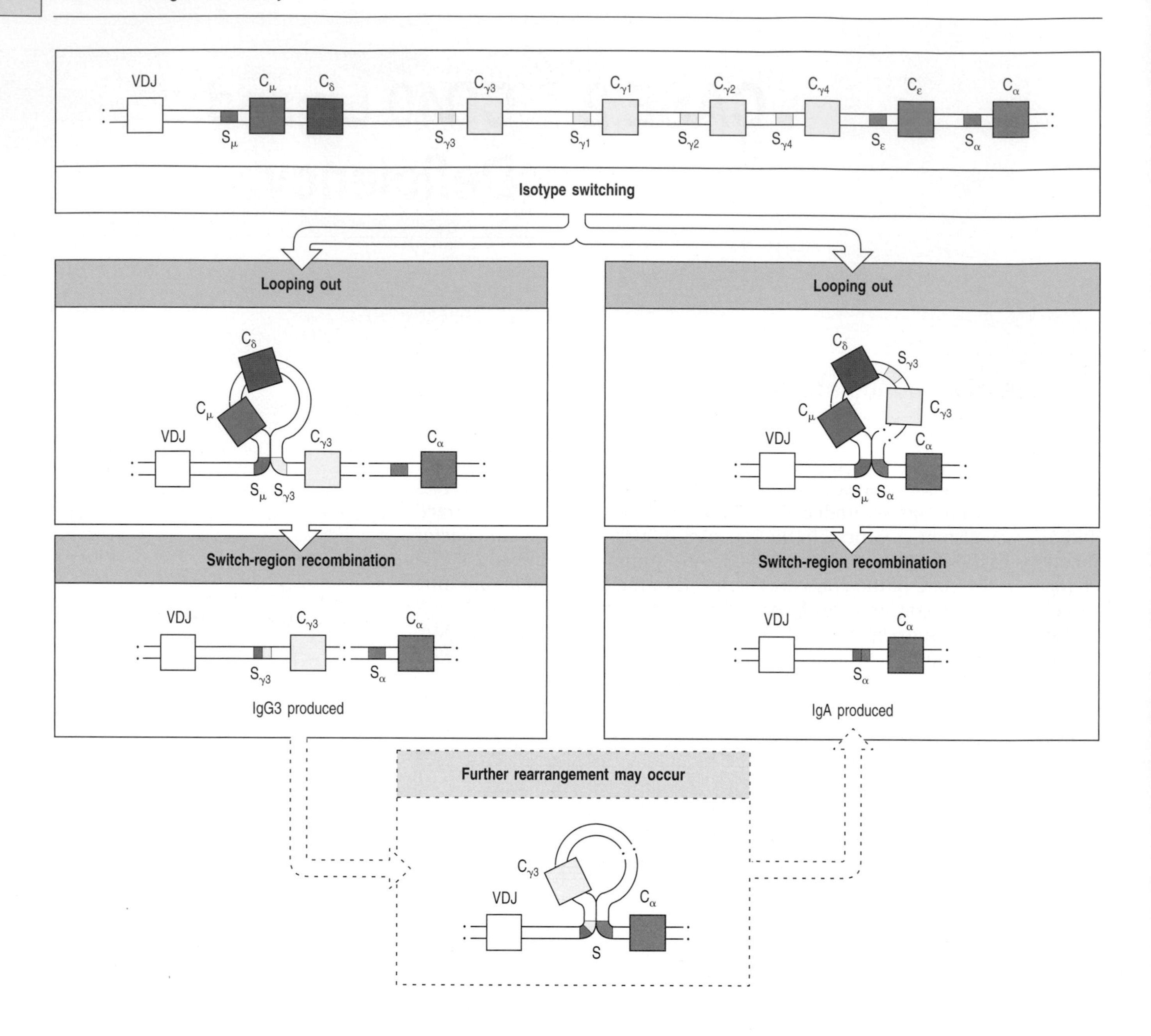

Fig. 2.2 Isotype switching involves recombination between specific switch signals. Repetitive DNA sequences that guide isotype switching are found upstream of each of the immunoglobulin C-region genes, with the exception of the C_δ gene. Switching occurs by recombination between these repetitive sequences or switch signals as a result of the repair of double-strand breaks (see Case 3), with deletion of the intervening DNA.The initial switching event takes place from the μ switch region (S_μ): switching to other isotypes can take place subsequently from the recombinant switch region formed after μ switching. S, switch region.

heavy-chain constant (C) regions of different isotypes, which determine the class of the antibody, as the immune response progresses (Fig. 2.2).

Class switching in B cells, also known as isotype switching and class-switch recombination, is induced mainly by T cells, although it can also be induced by T-cell independent Toll-like receptor (TLR)-mediated signaling. T cells are required to initiate B-cell responses to many antigens; the only exceptions are responses triggered by some microbial antigens or by certain antigens with repeating epitopes. This T-cell 'help' is delivered in the context of an antigen-specific interaction with the B cell (Fig. 2.3). The interaction activates the T cell to express the cell-surface protein CD40 ligand (CD40L, also known as CD154), which in turn delivers an activating signal to the B cell by binding CD40 on the B-cell surface. Activated T cells secrete cytokines, which are required at the initiation of the humoral immune response to drive the proliferation and differentiation of naive B cells, and are later required to induce class switching (Fig. 2.4). In humans, class switching to IgE synthesis is best understood, and is known to require interleukin-4 (IL-4) or IL-13, as well as stimulation of the B cell through CD40.

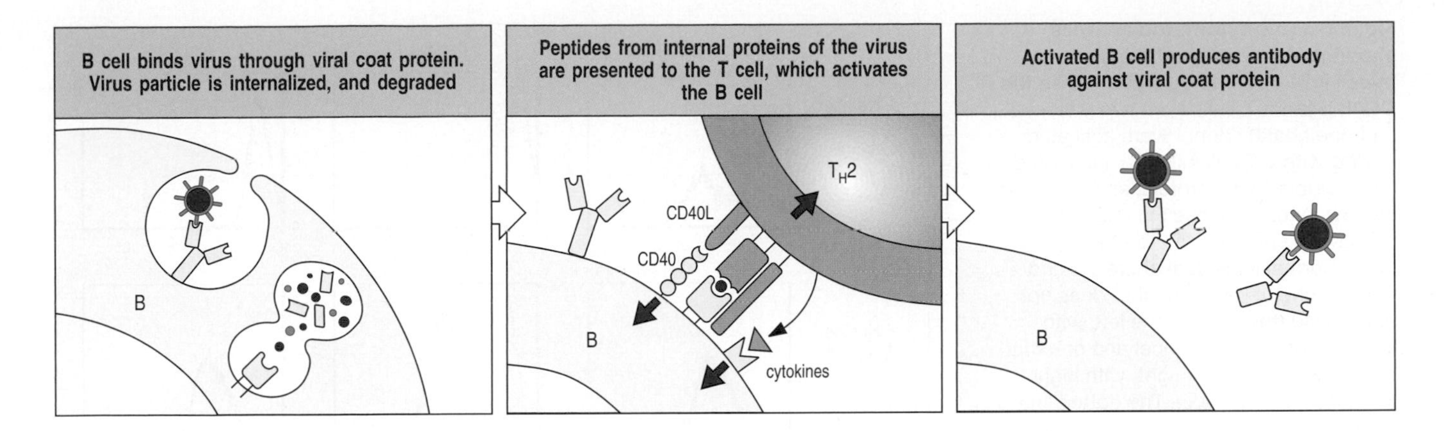

Fig. 2.3 B cells are activated by helper T cells that recognize antigenic peptide bound to class II molecules on their surface. An epitope on a viral coat (spike) protein is recognized by the surface immunoglobulin on a B cell, and the virus is internalized and degraded. Peptides derived from viral proteins are returned to the B-cell surface bound to MHC class II molecules, where they are recognized by previously activated helper T cells that activate the B cells to produce antibody against the virus.

The gene for CD40L (*CD40LG*) is located on the X chromosome at position Xq26. In males with a defect in this gene, isotype switching fails to occur; such individuals make only IgM and IgD and are severely impaired in their ability to switch to IgG, IgA, or IgE synthesis. This phenotype is known generally as 'hyper IgM syndrome,' and can also be due to defects other than the absence of CD40L (see Case 3). Similarly, defective class switching is also observed in patients with CD40 deficiency, a rare autosomal recessive condition. Defects in class switching can be mimicked in mice in which the genes for CD40 or CD40L have been disrupted by gene targeting; B cells in these animals fail to undergo switching. The underlying defect in patients with CD40L deficiency can be readily demonstrated by isolating their T cells and challenging them with soluble, fluorescently labeled CD40 (made by engineering the extracellular domain of CD40 onto the constant region (Fc) of IgG) or with monoclonal antibodies that recognize the CD40-binding epitope of CD40L. *In vitro* activated T cells from patients with CD40L deficiency fail to bind the soluble CD40–Fc (Fig. 2.5).

CD40 is expressed not only on B cells but also on the surfaces of macrophages, dendritic cells, follicular dendritic cells (FDCs), mast cells, and some epithelial and endothelial cells. Macrophages and dendritic cells are antigen-presenting

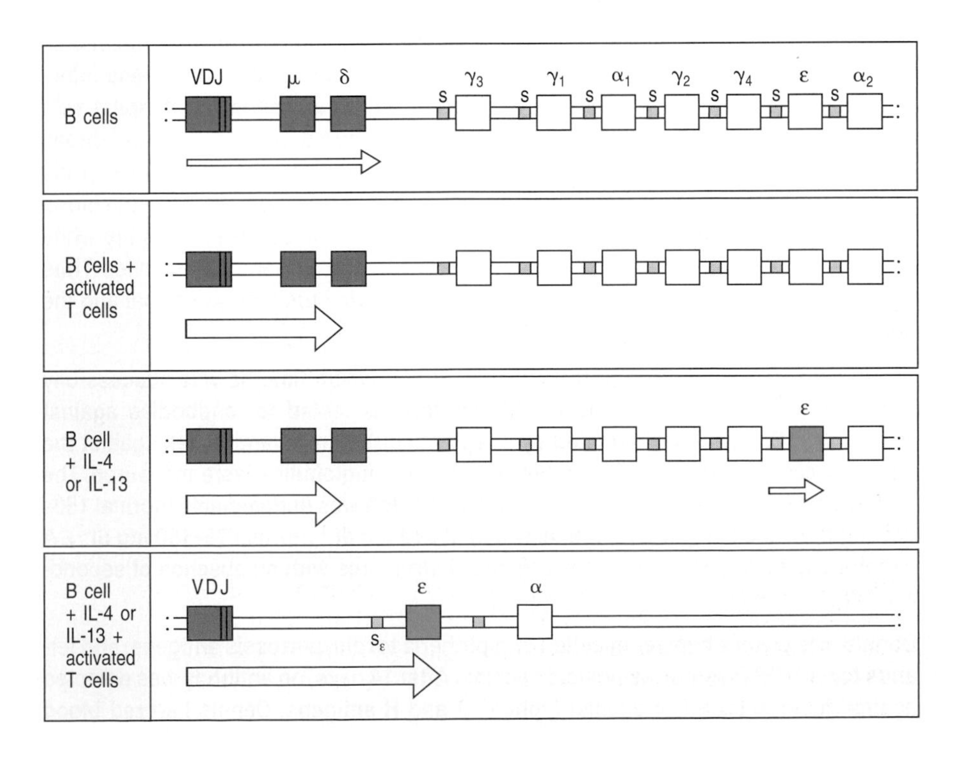

Fig. 2.4 Class switching to IgE production by human B cells. Purified human B cells in culture transcribe the μ and δ loci at a low rate, giving rise to surface IgM and IgD. On co-culture with T cells activated with ionomycin and phorbol myristate acetate (PMA), IgM is secreted. The presence of IL-4 or IL-13 stimulates an isotype switch to IgE. Purified B cells cultured alone with these cytokines transcribe the Cε gene at a low rate, but the transcripts originate in the switch region preceding the gene and do not code for protein. On co-culture with activated T cells in addition to IL-4 or IL-13, an isotype switch occurs, mature ε RNA is expressed, and IgE is synthesized.

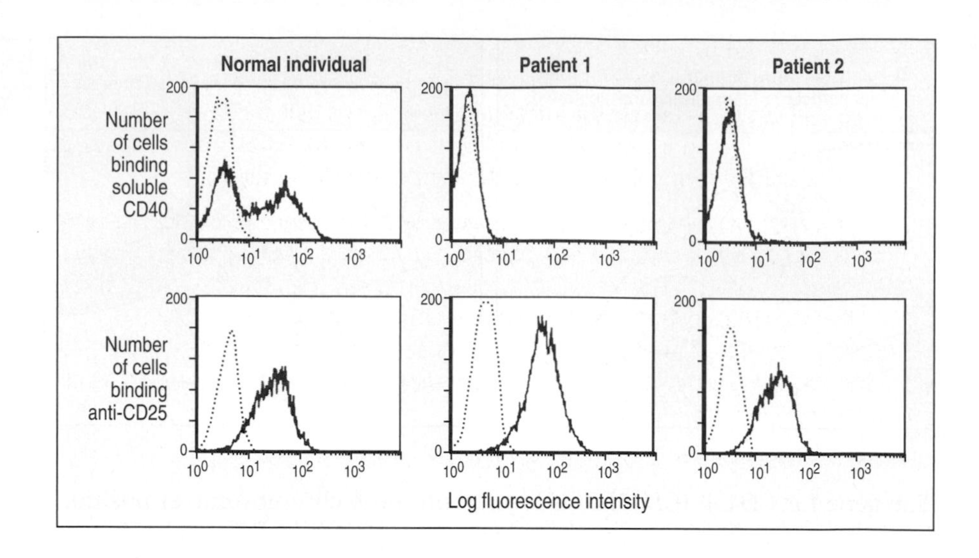

Fig. 2.5 Flow cytometric analysis showing that activated T cells from hyper IgM patients do not express the CD40 ligand. T cells from two patients and one healthy donor were activated *in vitro* with a T-cell mitogen, incubated with soluble CD40 protein, and analyzed by flow cytometry (see Fig. 1.3). The results are shown in the top three panels. In the normal individual, there are two populations of cells: one that does not bind CD40 (the peak to the left, with low-intensity fluorescence) and one that does (the peak to the right, with high-intensity fluorescence). The dotted line is the negative control, showing nonspecific binding of a fluorescently labeled protein to the same cells. In the patients with CD40L deficiency (center and right-hand panels), CD40 fluorescence exactly coincides with the nonspecific control, showing that there is no specific binding to CD40 by these cells. The bottom panels show that the T cells have been activated by the mitogen, because the T cells of both the normal individual and the two patients have increased expression of the IL-2 receptor (CD25), as expected after T-cell activation. The negative control is fluorescent goat anti-mouse immunoglobulin.

cells that can trigger the initial activation and expansion of antigen-specific T cells at the start of an immune response. Experiments in CD40L- or CD40-deficient patients and in gene-targeted CD40L-deficient mice indicate a role for the CD40–CD40L interaction in this early priming event, because in the absence of either CD40L or CD40 the initial activation and expansion of T cells in response to protein antigens is greatly reduced. The impairment of T-cell activation is the basis of some severe clinical features that distinguish CD40L and CD40 deficiency from other conditions characterized by a pure antibody deficiency.

The case of Dennis Fawcett: a failure of T-cell help.

Five-year-old boy fails to make antibody against strep infection.

Dennis Fawcett was 5 years old when he was referred to the Children's Hospital with a severe acute infection of the ethmoid sinuses (ethmoiditis). His mother reported that he had had recurrent sinus infections since he was 1 year old. Dennis had pneumonia from an infection with *Pneumocystis jirovecii* when he was 3 years old. These infections were treated successfully with antibiotics. While he was in the hospital with ethmoiditis, group A β-hemolytic streptococci were cultured from his nose and throat. The physicians caring for Dennis expected that he would have a brisk rise in his white blood cell count as a result of his severe bacterial infection, yet his white blood cell count was 4200 μl^{-1} (normal count 5000–9000 μl^{-1}). Sixteen percent of his white blood cells were neutrophils, 56% were lymphocytes, and 28% were monocytes. Thus his neutrophil number was low, whereas his lymphocyte number was normal and the number of monocytes was elevated.

Seven days after admission to the hospital, during which time he was successfully treated with intravenous antibiotics, his serum was tested for antibodies against streptolysin O, an antigen secreted by streptococci. When no antibodies against the streptococcal antigen were found, his serum immunoglobulins were measured. The IgG level was 25 mg dl^{-1} (normal 600–1500 mg dl^{-1}), IgA was undetectable (normal 150–225 mg dl^{-1}), and his IgM level was elevated at 210 mg dl^{-1} (normal 75–150 mg dl^{-1}). A lymph-node biopsy showed poorly organized structures with an absence of secondary follicles and germinal centers (Fig. 2.6).

Dennis was given a booster injection of diphtheria toxoid, pertussis antigens, and tetanus toxoid (DPT) as well as typhoid vaccine. After 14 days, no antibody was detected against tetanus toxoid or against typhoid O and H antigens. Dennis had red blood

cells of group O. People with type O red blood cells make antibodies against the A substance of type A red cells and antibodies against the B substance of type B red cells. This is because bacteria in the intestine have antigens that are closely related to A and B antigens. Dennis's anti-A titer was 1:3200 and his anti-B titer 1:800, both very elevated. His anti-A and anti-B antibodies were of the IgM class only.

His peripheral blood lymphocytes were examined by fluorescence-activated cell sorting analysis, and normal results were obtained: 11% reacted with an antibody against CD19 (a B-cell marker), 87% with anti-CD3 (a T-cell marker), and 2% with anti-CD56 (a marker for natural killer (NK) cells). However, all of his B cells ($CD19^+$) had surface IgM and IgD and none were found with surface IgG or IgA. When his T cells were activated *in vitro* with phorbol ester and ionomycin (a combination of potent polyclonal T-cell activators), they did not bind soluble CD40.

Dennis had an older brother and sister. They were both well. There was no family history of unusual susceptibility to infection.

Dennis was treated with intravenous gamma globulin, 600 mg kg^{-1} body weight each month, and subsequently remained free of infection until 15 years of age, when he developed severe, watery diarrhea. Cultures of the stools grew *Cryptosporidium parvum*. Within a few months, during which his diarrhea persisted in spite of treatment with the antibiotic azithromycin, he developed jaundice. His serum total bilirubin level was 8 mg dl^{-1}, and the serum level of conjugated bilirubin was 7 mg dl^{-1}. In addition, levels of γ-glutamyl transferase (γ-GT) and of the liver enzymes alanine aminotransferase (ALT) and aspartate aminotransferase (AST) were elevated at 93 IU l^{-1}, 120 IU ml^{-1} and 95 IU ml^{-1}, respectively, suggesting cholestasis. A liver biopsy showed abnormalities of biliary ducts (vanishing bile ducts) that progressed to sclerosing cholangitis (chronic inflammation and fibrosis of the bile ducts). In spite of supportive treatment, Dennis died of liver failure at 21 years of age.

Lymph node from a patient with CD40L deficiency (no germinal centers)

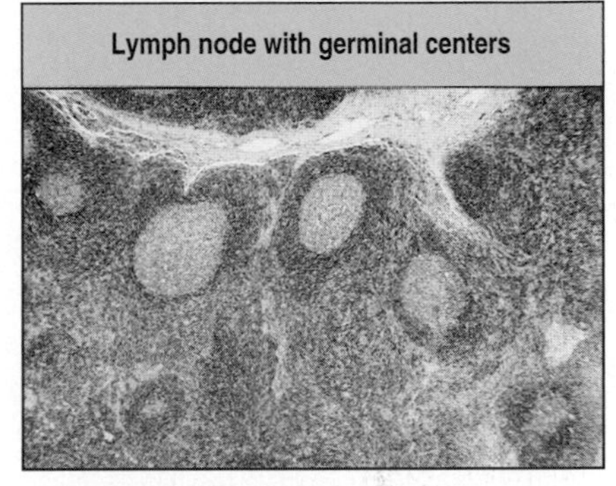

Fig. 2.6 Comparison of lymph nodes from a patient with CD40L deficiency (upper panel) and a normal individual (lower panel). Lower photograph courtesy of A. Perez-Atayde.

CD40 ligand deficiency (CD40L deficiency).

Males with a hereditary deficiency of CD40L exhibit consequences of a defect in both humoral and cell-mediated immunity. As we saw in Case 1, defects in antibody synthesis result in susceptibility to so-called pyogenic infections. These infections are caused by pyogenic (pus-forming) bacteria such as *Haemophilus influenzae, Streptococcus pneumoniae, Streptococcus pyogenes,* and *Staphylococcus aureus,* which are resistant to destruction by phagocytic cells unless they are coated (opsonized) with antibody and complement. On the other hand, defects in cellular immunity result in susceptibility to opportunistic infections. Bacteria, viruses, fungi, and protozoa that often reside in our bodies and only cause disease when cell-mediated immunity in the host is defective are said to cause opportunistic infections.

Dennis revealed susceptibility to both kinds of infection. His recurrent sinusitis, as we have seen, was caused by *Streptococcus pyogenes,* a pyogenic infection. He also had pneumonia caused by *Pneumocystis jirovecii* and diarrhea caused by *Cryptosporidium parvum,* a fungus and a protozoan, respectively, that are ubiquitous and cause opportunistic infections in individuals with defects in cell-mediated immunity.

Patients with a CD40L deficiency can make IgM in response to T-cell independent antigens but they are unable to make antibodies of any other isotype, and they cannot make antibodies against T-cell dependent antigens, which leaves the patient largely unprotected from many bacteria. They also have a defect in cell-mediated immunity that strongly suggests a role for CD40L in the T cell-mediated activation of macrophages. *Cryptosporidium* infection can cause persistent inflammation in the liver, and ultimately sclerosing cholangitis and liver failure. In addition, individuals with CD40L deficiency have severe neutropenia, with a block at the promyelocyte/myelocyte stage of differentiation in the bone marrow. Although the mechanisms underlying the neutropenia in these patients remain unclear, the lack of neutrophils accounts for the presence of severe sores and blisters in the mouth. The neutropenia and its consequences can often be overcome by administering recombinant granulocyte-colony stimulating factor (G-CSF).

Treatment of CD40L deficiency is based on immunoglobulin replacement therapy, prophylaxis with trimethoprim-sulfamethoxazole to prevent *Pneumocystis jirovecii* infection, and protective measures to reduce the risk of *Cryptosporidium* infection (such as avoiding swimming in lakes or drinking water with a high concentration of *Cryptosporidium* cysts). In spite of this, many patients with CD40L deficiency die in late childhood or adulthood of infections, liver disease, or tumors (lymphomas and neuroectodermal tumors of the gut). The disease can be cured by hematopoietic cell transplantation, and this treatment should be considered when HLA-identical donors are available and when the first signs of severe complications become manifest.

Few cases of CD40 deficiency have been reported. Its clinical and immunological features, and its treatment, are very similar to CD40L deficiency, but the disease is inherited as an autosomal recessive trait.

Questions.

1 *Dennis's B cells expressed IgD as well as IgM on their surface. Why did he not have any difficulty in isotype switching from IgM to IgD?*

2 *Normal mice are resistant to Pneumocystis jirovecii. SCID mice, which have no T or B cells but have normal macrophages and monocytes, are susceptible to this microorganism. In normal mice, Pneumocystis jirovecii organisms are taken up and destroyed by macrophages. Macrophages express CD40. When SCID mice are reconstituted with normal T cells they acquire resistance to Pneumocystis infection. This can be abrogated by antibodies against the CD40 ligand. What do these experiments tell us about this infection in Dennis?*

3 *Why did Dennis make antibodies against blood group A and B antigens but not against tetanus toxoid, typhoid O and H, and streptolysin antigens? Would he have made any antibodies in response to his Streptococcus pyogenes infection?*

4 *Most IgM is circulating in the blood, and less than 30% of IgM molecules get into the extravascular fluid. On the other hand, more than 50% of IgG molecules are in the extravascular space. Furthermore we have 30–50 times more IgG than IgM in our body. Why are IgG antibodies more important in protection against pyogenic bacteria?*

5 *Newborns have difficulty in transcribing the gene for CD40L. Does this help to explain the susceptibility of newborns to pyogenic infections? Cyclosporin A, a drug widely used for immunosuppression in graft recipients, also inhibits transcription of the gene for CD40L. What does this imply for patients taking this drug?*

CASE 3 Activation-Induced Cytidine Deaminase Deficiency

An intrinsic B-cell defect prevents immunoglobulin class switching.

Immunoglobulin class switching, also known as isotype switching, is a complex process. As we saw in Case 2, after antigen has been encountered and the mature B cell is activated, the rearranged immunoglobulin heavy-chain variable (V) region DNA sequence can become progressively associated with different constant (C)-region genes by a form of somatic recombination (see Fig. 2.2). This process of class-switch recombination requires the engagement of a specific biochemical machinery in activated B lymphocytes, which works most efficiently after the B cell has received signals from activated helper T cells, such as those delivered by the interaction of CD40 ligand (CD40L) on the T cell with CD40 on the B cell (see Fig. 2.3). In Case 2, we saw that a failure of class switching, leading to a hyper IgM syndrome characterized by production of IgM and IgD but not the other isotypes, can result from a defect in the *CD40L* gene and a consequent absence of functional CD40L protein. CD40L is encoded on the X chromosome and so this type of hyper IgM syndrome is seen only in males. Hyper IgM syndrome is, however, also encountered in females, in whom it has an inheritance pattern in many families that suggests an autosomal recessive inheritance (Fig. 3.1). As we saw in Case 2, hyper IgM due to CD40 deficiency has an autosomal recessive inheritance, and yet another cause of autosomal recessive hyper IgM syndrome has been discovered more recently—a defect in the B cell's biochemical pathways responsible for the class-switch recombination process itself.

The biochemical events underlying class switching have only been clarified relatively recently. While studying a cultured B-cell line that was being induced to undergo class switching from IgM to IgA synthesis, immunologists in Japan observed a marked upregulation of the enzyme *a*ctivation-*i*nduced cytidine *d*eaminase (AID). This enzyme converts cytidine to uridine, and it is now known to trigger a DNA breakage and repair mechanism that underlies class-switch recombination. The contribution of AID to class switching was confirmed by 'knocking out' the gene encoding AID by homologous recombination in mice; the mutant animals developed hyper IgM syndrome and were unable to make IgG, IgA, or IgE.

This case was prepared by Raif Geha, MD, in collaboration with Ari Fried, MD.

Topics bearing on this case:

- Antibody isotypes and classes
- Isotype or class switching
- Somatic hypermutation

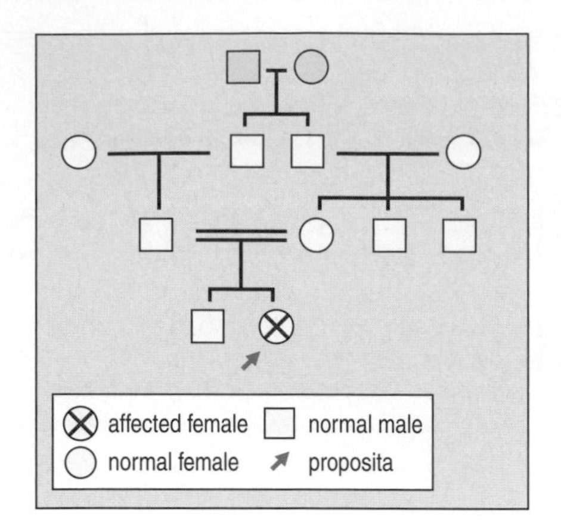

Fig. 3.1 A pedigree of a family with AID deficiency. Because the parents of the affected child (the proposita) show no signs of disease themselves, the defective *AID* gene must be recessive. The parents both carry a single copy of the defective gene and, because they are first cousins, the most likely source is their shared grandfather or grandmother (blue symbols). If one of these were heterozygous for AID deficiency, they could have transmitted the defective gene to both their sons. The fact that the affected child is a girl, and that neither her father nor any other males in her extended family show signs of disease, indicates that the gene is carried on an autosome and not the X chromosome.

The defective gene in some autosomal recessive cases of hyper IgM syndrome was mapped in several informative families to the short arm of chromosome 12, to a region corresponding with the region containing the *AID* gene in mice. This prompted the search for a link between hyper IgM syndrome and AID deficiency in humans, and several cases of the autosomal recessive form were found to have mutations in the *AID* gene.

The case of Daisy Miller: a failure of a critical B-cell enzyme.

At 3 years old, Daisy Miller was admitted to the Boston Children's Hospital with pneumonia. Her mother had taken her to Dr James, a pediatrician, because she had a fever and was breathing fast. Her temperature was high, at 40.1°C, her respiratory rate was 40 per minute (normal 20), and her blood oxygen saturation was 88% (normal >98%). Dr James also noticed that lymph nodes in Daisy's neck and armpits (axillae) were enlarged. A chest X-ray was ordered. It revealed diffuse consolidation (whitened areas of lung due to inflammation, indicating pneumonia) of the lower lobe of her left lung, and she was admitted to the hospital.

Daisy had had pneumonia once before, at 25 months of age, as well as 10 episodes of middle-ear infection (otitis media) that had required antibiotic therapy. Tubes (grommets) had been placed in her ears to provide adequate drainage and ventilation of the ear infections.

In the hospital a blood sample was taken and was found to contain 13,500 white blood cells ml^{-1}, of which 81% were neutrophils and 14% lymphocytes. A blood culture grew the bacterium *Streptococcus pneumoniae.*

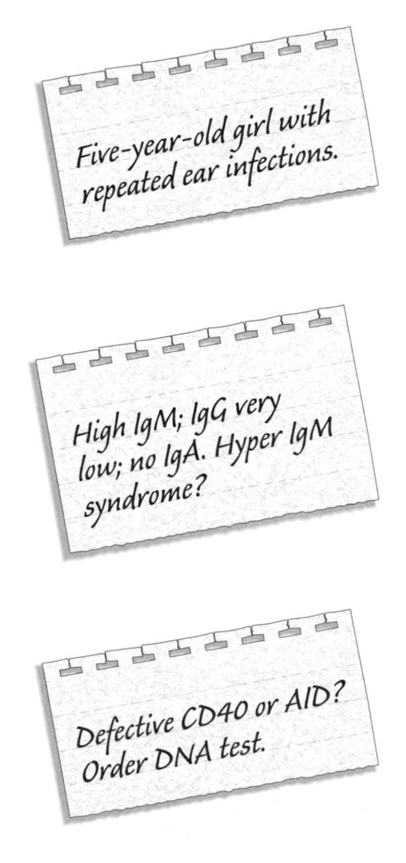

Five-year-old girl with repeated ear infections.
High IgM; IgG very low; no IgA. Hyper IgM syndrome?
Defective CD40 or AID? Order DNA test.

Because of Daisy's repeated infections, Dr James consulted an immunologist. She tested Daisy's immunoglobulin levels and found that her serum contained 470 mg dl^{-1} of IgM (normal 40–240 mg dl^{-1}), undetectable IgA (normal 70–312 mg dl^{-1}), and 40 mg dl^{-1} of IgG (normal 639–1344 mg dl^{-1}). Although Daisy had been vaccinated against tetanus and *Haemophilus influenzae*, she had no specific IgG antibodies against tetanus toxoid or to the polyribosyl phosphate (PRP) polysaccharide antigen of *H. influenzae.* Because her blood type was A, she was tested for anti-B antibodies (isohemagglutinins). Her IgM titer of anti-B antibodies was positive at 1:320 (upper limit of normal), whereas her IgG titer was undetectable.

Daisy was started on intravenous antibiotics. She improved rapidly and was sent home on a course of oral antibiotics. Intravenous immunoglobulin (IVIG) therapy was started, which resulted in a marked decrease in the frequency of infections.

Analysis of Daisy's peripheral blood lymphocytes revealed normal expression of CD40L on T cells activated by anti-CD3 antibodies, and normal expression of CD40 on B cells (Fig. 3.2). Nevertheless, her blood cells completely failed to secrete IgG and IgE after stimulation with anti-CD40 antibody (to mimic the effects of engagement of CD40L) and interleukin-4 (IL-4), a cytokine that also helps to stimulate class switching, although the blood cells proliferated normally in response to these stimuli (Fig. 3.3). cDNAs for CD40 and for the enzyme activation-induced cytidine deaminase (AID) were made and amplified by the reverse transcription–polymerase chain reaction (RT–PCR) on mRNA isolated from blood lymphocytes activated by anti-CD40 and IL-4. Sequencing of the cDNAs revealed a point mutation in the *AID* gene that introduced a stop codon into exon 5, leading to the formation of truncated and defective protein. The CD40 sequence was normal.

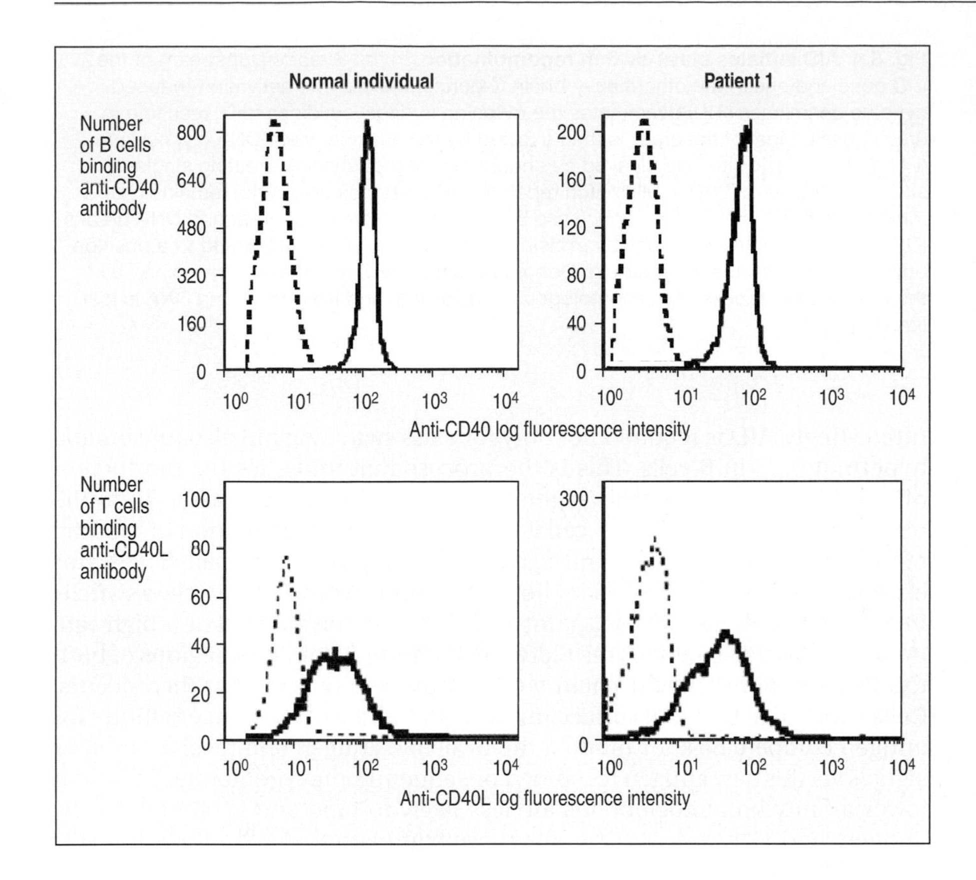

Fig. 3.2 Flow cytometric analysis showing normal expression of CD40 and CD40L in a patient with AID deficiency. Top row: CD40 measured by the binding of fluorescently tagged anti-CD40 antibodies to CD40 on B cells from (left panel) a normal individual and (right panel) a patient with AID deficiency. Bottom row: measurement of CD40L on T cells from (left panel) a normal individual and (right panel) the same patient activated *in vitro* with the mitogen phorbol ester (PMA) and ionomycin.

Activation-induced cytidine deaminase deficiency (AID deficiency).

It is now apparent that there are several distinct phenotypes of hyper IgM syndrome, resulting from different genetic defects. One phenotype, which is caused by defects in the genes encoding CD40 or CD40L (see Case 2), manifests itself as susceptibility to both pyogenic and opportunistic infections. Another phenotype, which results from defects in the *AID* gene or in the gene encoding uracil-DNA glycosylase (UNG), another DNA repair enzyme involved in class-switch recombination, resembles X-linked agammaglobulinemia in that these patients have an increased susceptibility to pyogenic infections only.

When CD40 and the IL-4 receptor on B cells are ligated by CD40L and IL-4, respectively, the *AID* gene is transcribed and translated to produce AID protein. At the same time, transcription of cytidine-rich regions at specific isotype-switch sites is induced, which involves separation of the two DNA strands at these sites. AID, which can deaminate cytidine in single-stranded DNA only, then proceeds to convert the cytidine at the switch sites to uridine. Because uridine is not normally present in DNA, it is recognized by the enzyme UNG, which removes the uracil base from the rest of the nucleotide (Fig. 3.4). The abasic sites are cleaved by a DNA endonuclease, resulting in single-strand DNA breaks. If single-strand breaks on opposite DNA strands are near to each other, a double-strand DNA break will form at each switch region. It is the repair of these double-strand breaks that brings the two switch regions together and joins a new C-region gene to the V region, resulting in class switching.

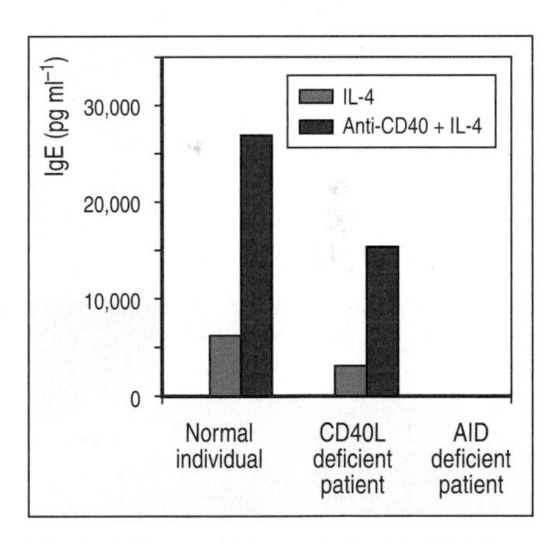

Fig. 3.3 Comparison of class switching, as judged by IgE secretion by B cells from a normal individual, a patient with CD40L deficiency, and a patient with AID deficiency. IgE production from peripheral blood mononuclear cells *in vitro* was measured after stimulation with IL-4 alone or stimulation with anti-CD40 and IL-4. The negative control was stimulation by medium alone (not shown), which produced barely detectable levels of IgE. The anti-CD40 and IL-4 together can compensate to some degree for the lack of stimulation of B cells by CD40L on T cells, but cannot compensate at all for the defect in AID.

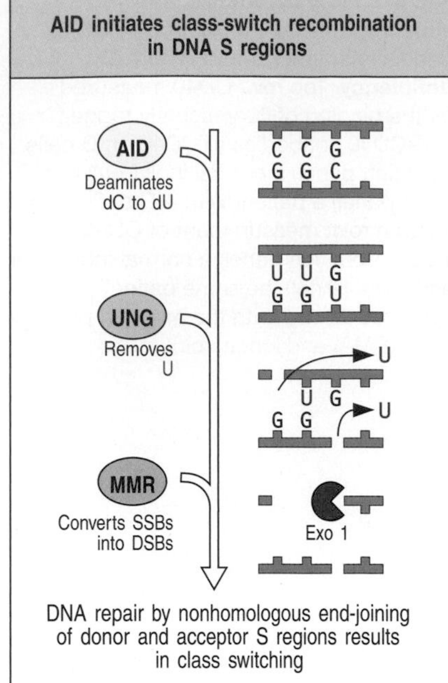
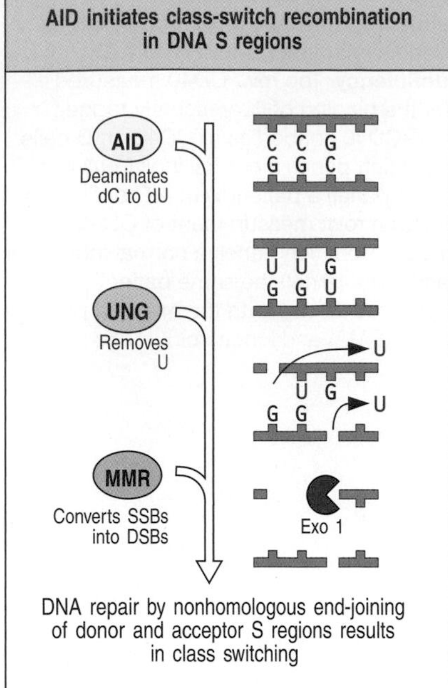

Fig. 3.4 AID initiates class-switch recombination. In the S regions upstream of the μ C gene and one of the other heavy-chain C genes, the enzyme activation-induced cytidine deaminase (AID) deaminates the cytosine in deoxycytidines (dC), resulting in uracil bases. Most of the uracil is then excised by the enzyme uracil-DNA glycosylase (UNG). The abasic sites are cleaved by endonuclease (not shown), creating single-strand DNA breaks (SSBs). Mismatch repair (MMR) enzymes detect the remaining U–G mismatches and recruit the exonuclease Exo1, which creates double-strand DNA breaks (DSBs) at each of the S regions by excising DNA from a nick on one strand to a position opposite a nick on the other strand. Donor and acceptor S regions are then joined by the DNA repair process of nonhomologous end-joining, and the intervening DNA is lost (see Fig. 2.2).

Interestingly, AID is required not only for class switching but also for somatic hypermutation in B cells. This is the process that underlies the production of antibodies of increasingly higher affinity for the antigen as an immune response proceeds. When a B cell is activated by the combination of binding of its surface IgM by antigen and signals from other cells, especially T cells (for example via CD40–CD40L interaction), not only cell division and class switching are initiated. In addition, point mutations are introduced at a high rate into the DNA that codes for the rearranged immunoglobulin V regions of both the heavy-chain and light-chain genes. A process of selection then occurs. Cells expressing mutated surface immunoglobulin with a stronger affinity for antigen compete best for binding the available antigen and receive stronger signals via this new antigen receptor; consequently, they proliferate. Cells with lower-affinity immunoglobulins are less likely to bind and be stimulated by antigen, and if they do not receive these stimulatory signals, they die. This leads to the selection of cells with mutations that result in high-affinity antibodies, a process referred to as 'affinity maturation' of the antibody response. B cells in mice and humans that lack functional AID are unable to generate these mutations after activation, and therefore cannot undergo affinity maturation. The only response these B cells can make to the activating signals is to proliferate. This results in the accumulation of IgM-positive B cells in the lymphoid organs, giving rise to an enlarged spleen (splenomegaly) and enlarged lymph nodes (lymphadenopathy) (Fig. 3.5).

Another DNA repair mechanism, called mismatch repair (MMR), has a primary role in the detection and repair of DNA mismatches during meiosis. MMR has recently been found also to contribute to class-switch recombination downstream of AID and UNG. Proteins involved in MMR, namely Msh2, Msh6, Mlh2, and Pms2, build a complex that recognizes the U–G mismatches created by AID and, with the help of the exonuclease Exo1, excises the DNA from the nearest single-strand break past the site of the mismatch, creating a double-strand break. Thus, the role of MMR in class switching is thought to be the formation of double-strand breaks if the single-strand breaks introduced by AID and UNG on opposite DNA strands are not near enough to form one. Mice deficient in any of the MMR proteins or in Exo1 have a twofold to fivefold reduction in isotype switching. Recently, a deficiency of the MMR component Pms2 has been found to cause a distinct phenotype of class-switch recombination deficiency in humans. In addition to recurrent infections, patients with Pms2 deficiency have a high rate of cancer, as a result of generally defective DNA repair.

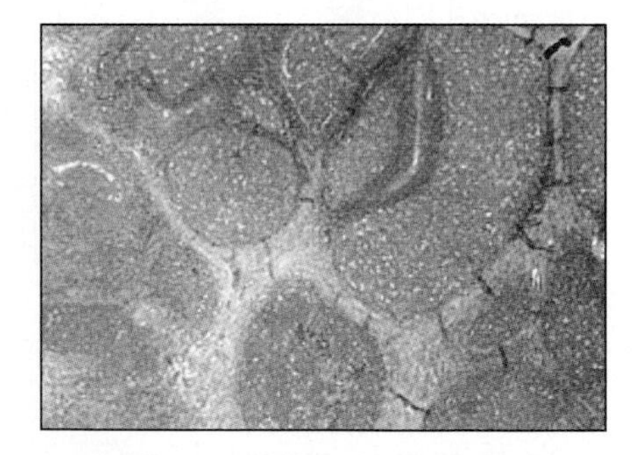

Fig. 3.5 Histology of a lymph node from a patient with AID deficiency stained with hematoxylin and eosin. Note the large follicles.

Questions.

1 *Daisy and other patients with hyper IgM syndrome caused by a mutation in AID do not seem to suffer from opportunistic infections, such as Pneumocystis jirovecii and Cryptosporidium, which are characteristic of CD40L deficiency. Why?*

2 *A 3-year-old girl presents with hyper IgM syndrome, recurrent bacterial infection and a history of Pneumocystis jirovecii pneumonia. Her CD40L and AID genes are normal. What is the potential gene defect in this patient?*

3 *A 6-year-old girl presents with hyper IgM syndrome, recurrent bacterial infection and no history of opportunistic infections. Her CD40 and AID gene sequences are normal. What is the potential gene defect in this patient?*

4 *Can you think of a simple test to distinguish hyper IgM syndrome caused by an intrinsic B-cell defect from that caused by CD40L deficiency without recourse to DNA sequencing?*

5 *Dr James noted that Daisy had enlarged lymph nodes. Patients with CD40L deficiency do not have enlarged lymph nodes. Can you explain this difference?*

CASE 4 Common Variable Immunodeficiency

A failure to produce antibodies against particular antigens.

The ability to produce antibody of all different classes after exposure to antigen is an important aspect of successful and comprehensive humoral immunity. Although antibody production and class switching in response to most protein antigens require helper T cells (see Case 2), responses to some other antigens do not. This is because the special properties of certain bacterial polysaccharides, polymeric proteins, and lipopolysaccharides enable them to stimulate naive B cells in the absence of T-cell help. These antigens are known as thymus-independent antigens (TI antigens) because they stimulate strong antibody responses in athymic individuals. They fall into two classes: TI-1 antigens, such as bacterial lipopolysaccharide, which can directly induce B-cell division; and TI-2 antigens, such as bacterial capsular polysaccharides, which do not have this property but have highly repetitive structures and probably stimulate the B cell by cross-linking a critical number of B-cell receptors. In particular, B-cell responses to thymus-independent antigens provide a prompt and specific IgM response to an important class of pathogen—encapsulated bacteria. Pyogenic bacteria such as streptococci and staphylococci are surrounded by a polysaccharide capsule that enables them to resist phagocytosis. Antibody that is produced rapidly in response to this polysaccharide capsule can coat these bacteria, promoting their ingestion and destruction by phagocytes.

Whereas TI-1 antigens are inefficient inducers of affinity maturation and memory B cells, TI-2 antigens can induce both IgM and some class-switched responses. The initiation of B-cell class switching usually depends on the interaction of B cells and helper T cells via CD40 and CD40 ligand (see Case 2). Class switching in response to TI antigens is thought to involve other members of the tumor necrosis factor (TNF)/TNF receptor (TNFR) family—the recently discovered TNF-like proteins BAFF (B-cell activating factor belonging to the TNF family) and APRIL (a proliferation-inducing TNF ligand), and their receptors on B cells. In human B cells, BAFF and APRIL induce class switching to IgA and IgG in the presence of TGF-β or IL-10 and to IgE in the presence of IL-4. In mice, in contrast, BAFF and APRIL on their own can switch B cells to IgG and IgA production.

BAFF is secreted by cells in the follicles of peripheral lymphoid tissue and can bind to several receptor proteins on B cells, which include BAFF receptor (BAFF-R), transmembrane activator and calcium-modulating cyclophilin ligand interactor (TACI), and B-cell maturation antigen (BCMA) (Fig. 4.1). In addition to its role in class switching, BAFF is involved in B-cell development

This case was prepared by Raif Geha, MD, in collaboration with Ari Fried, MD.

Topics bearing on this case:
Antibody classes
Class switching
Thymus-independent antigens

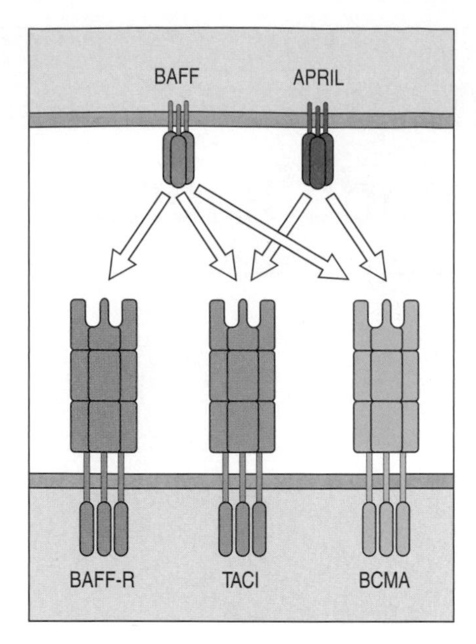

Fig. 4.1 BAFF and APRIL, and their receptors on B cells. The BAFF receptor (BAFF-R) is expressed on resting B cells and mediates the effects of BAFF on B-cell development and survival but has only a minor role in isotype switching. Studies in mice suggest that interaction of BAFF with BAFF-R has an important role in B-cell survival and is required for the development of mature B cells *in vivo*. TACI seems to be the receptor that mainly mediates isotype switching by BAFF and APRIL. TACI may also have other functions such as promoting plasma-cell differentiation and survival. BCMA is poorly expressed on resting B cells but is upregulated on germinal-center B cells and plasma cells. It has the ability to upregulate the expression of co-stimulatory molecules on the B-cell surface that enhance antigen presentation and T-cell activation, and has a role in maintaining the survival of long-lived plasma cells. BCMA has no detectable role in isotype switching by BAFF and APRIL.

and in promoting the survival of mature B cells. Mice carrying a BAFF transgene, leading to BAFF overexpression, develop high titers of autoantibodies and a systemic lupus erythematosus-like condition, whereas BAFF-deficient mice have a severe defect in B-cell development and an almost complete loss of mature B cells and of follicular and marginal zone B cells in lymph nodes and spleen. A similar phenotype is observed in mice lacking BAFF-R, or in the A/WySnJ mouse strain, which carries a naturally occurring mutation in BAFF-R.

APRIL is the closest relative of BAFF in the TNF family, sharing 33% sequence identity, and was initially reported as a protein that stimulated the proliferation of tumor cells. Unlike BAFF, it does not seem to have an effect on B-cell development and survival, because APRIL-deficient mice have normal numbers of B cells in all developmental stages. This is because APRIL binds to TACI and BCMA, but not to BAFF-R. The only detectable immune defects in APRIL homozygous null mice are diminished serum IgA levels and an impaired antibody response to immunization by the oral route. $TACI^{+/-}$ mice exhibit a marked haploinsufficiency in regard to their ability to mount an antibody response to type II dependent antigens *in vivo* and to the ability of their B cells to respond to APRIL *in vitro*.

The receptors BAFF-R, TACI, and BCMA are all members of the TNFR superfamily and have different roles in immune responses. TACI seems to be the receptor that mainly mediates isotype switching by BAFF and APRIL. BCMA, and possibly TACI, may promote plasma-cell differentiation and survival. TACI-deficient mice have low levels of serum IgA and have a deficient antibody response to immunization with TI-2 antigens in the vaccine Pneumovax, which contains polysaccharide antigens from a number of *Streptococcus pneumoniae* serotypes, and NP-Ficoll, a polysaccharide. They also have enlarged spleens and lymph nodes, with more cells than usual and an increased number of B cells. They develop autoimmunity, suggesting that TACI may have a regulatory role in B cells. Some cases of common variable immunodeficiency (CVID) in humans have recently been shown to be associated with mutations in TACI, as we shall see in this case study.

The case of Mary Johnson: impaired ability to generate all classes of antibodies leads to frequent and unusual infections.

Mrs Johnson was 40 years old when she was referred to the immunology clinic for evaluation of her immune system after suffering throughout her life from recurrent respiratory and gastrointestinal infections. As a child she was frequently diagnosed with otitis, sinusitis, and tonsillitis and had intermittent diarrhea, and from an early age she was hospitalized several times for pneumonia and gastrointestinal infections. The year before she came to the immunology clinic she had been in hospital with severe diarrhea caused by an infection with the protozoan parasite *Giardia lamblia*. When she was 25 years old, Mrs Johnson had been diagnosed with thyroid insufficiency and placed on thyroid hormone replacement therapy.

Physical examination revealed an enlarged spleen, the edge of which extended 8 cm below the left mid-costal margin. Blood tests showed that Mrs Johnson had lower than normal levels of all immunoglobulin isotypes: IgM 18 mg dl^{-1} (normal 100–200 mg dl^{-1}), IgG 260 mg dl^{-1} (normal 600–1000 mg dl^{-1}), and IgA 24 mg dl^{-1} (normal 60–200 mg dl^{-1}). Although she had been immunized several times with pneumococcal vaccines, she had been unable to respond, as shown by her lack of antibodies against

all pneumococcal serotypes present in the vaccine. Numbers of B cells and T cells were normal. No antinuclear autoantibodies or rheumatoid factor (an autoantibody against IgG) were detected, but Mrs Johnson had high levels of antithyroid antibodies.

Mrs Johnson's twin sister and her mother were both dead. They had also had hypogammaglobulinemia and an inability to make specific antibodies against polysaccharide vaccines. From her early teens onward, the sister had had recurrent viral and bacterial infections. She developed hemolytic anemia and granulomatous vasculitis, and died at the age of 31 years from gastrointestinal cancer. The mother had died of non-Hodgkin's lymphoma. Mrs Johnson's only brother had been diagnosed with common variable immune deficiency (CVID) and suffered from chronic sinusitis and recurrent chest infections. Her father was still alive.

Family history of CVID.

On the basis of the blood tests and her family history, Mrs Johnson was diagnosed with CVID and was placed on intravenous immunoglobulin 35 g every 2 weeks. This caused a remarkable improvement in her condition with a marked decrease in the frequency of infections. Sequencing of the *TACI* gene revealed that Mrs Johnson had a mutation in one of the alleles. The same mutation was also found in her brother's *TACI* gene but was absent from their father's DNA, implying that she and her brother had inherited the mutation from their mother.

Start immunoglobulin therapy.

Common variable immunodeficiency (CVID).

CVID is an immunodeficiency disorder characterized by low serum levels of all switched immunoglobulin isotypes (IgG, IgA, and IgE), an impaired ability to produce specific antibodies, even of the IgM class, after exposure to certain antigens, and increased susceptibility to infections of the respiratory and gastrointestinal tract, the latter being due to the combined decrease in IgG and IgA. It is relatively common (hence the name), being the most common primary immunodeficiency that comes to medical attention. Patients with CVID have a striking paucity of both unswitched IgM^+ $CD27^+$ memory B cells and of switched IgM^- $CD27^+$ memory B cells (Fig. 4.2). In addition they are severely deficient in plasma cells and have impaired somatic hypermutation. Consequently, they are hypogammaglobulinemic and the few antibodies they develop have diminished affinity, rendering them susceptible to chronic and recurrent infections by encapsulated bacteria such as *S. pneumoniae* (pneumococcus) or *Haemophilus influenzae*. The clinical course, as well as the degree of deficiency of serum immunoglobulins, varies from patient to patient (hence the name 'variable').

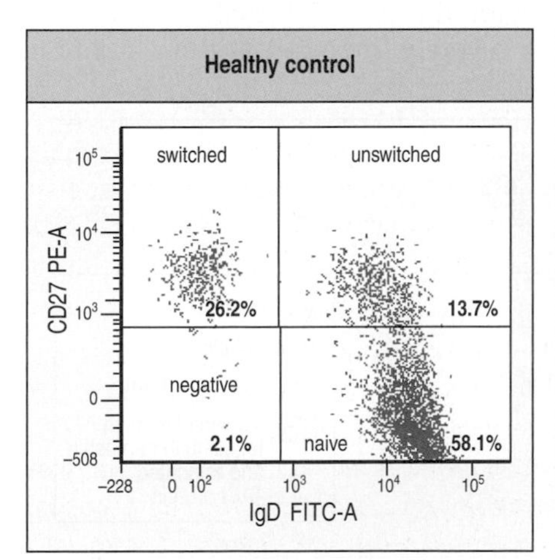

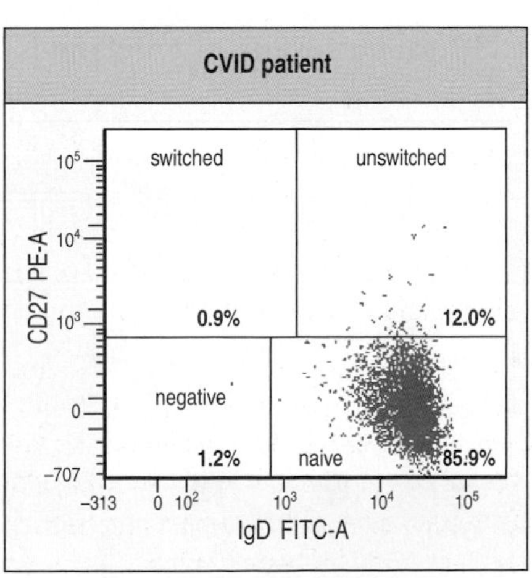

Fig. 4.2 CVID is associated with reduced numbers of switched and unswitched memory B cells. Flow-cytometric measurement of the memory B cell compartment is shown for a healthy control (left) and a patient with CVID (right). Blood lymphocytes are gated on B cells (CD19). The upper panels show $CD27^+$ memory B cells. In the left panels are B cells that underwent class switching (IgD^-) and in the right panels are unswitched B cells (IgD^+) that still carry IgD on their surface. Measurement of switched and unswitched memory B cells has become part of the routine evaluation of patients with CVID. Low percentages of switched memory B cells (<2%) predict disease-associated clinical complications such as splenomegaly and autoimmune disease.

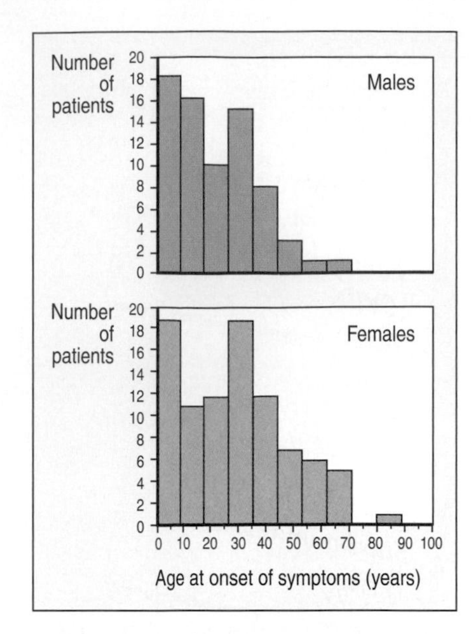

Fig. 4.3 The age at onset of clinical symptoms of immunodeficiency in CVID. The age at onset of clinical symptoms of immunodeficiency is shown in centiles. The median age at onset of symptoms is 23 years for males and 28 years for females.

Like other immunodeficiency diseases, the symptoms of CVID are frequent and unusual infections. Symptoms may appear during early childhood, adolescence, or adult life, but the age of onset of symptoms is usually in the twenties or thirties (Fig. 4.3). Patients with CVID suffer from recurrent infections of the sinuses and lungs, the ears, and the gastrointestinal tract. If therapy is delayed, the air passages in the lung may become irreversibly damaged and chronic infections may develop (bronchiectasis). Patients with CVID also are at greater risk of developing autoimmune diseases such as autoimmune thyroiditis, hemolytic anemia, autoimmune thrombocytopenia (which is due to antiplatelet antibodies), and pernicious anemia (which is caused by antibodies against the intrinsic factor that is required for the absorption of vitamin B12). Patients with CVID also have a 300-fold increased risk of lymphoma and a 50-fold increased risk of gastric carcinoma. Some patients with CVID develop granulomatous lesions (see Case 26) in the lungs and the skin that are characterized by the presence of T cells and macrophages. Human herpes virus 6 (HHV6) has been isolated from these lesions, which usually respond to immunosuppressive treatment with corticosteroids or cyclosporin.

Most cases of CVID are sporadic; that is, they are not due to an inherited defect. Studies in European populations show that around 20% of CVID cases are familial, commonly associated with autosomal dominant inheritance. Genetic analysis has shown a high degree of familial clustering of single IgA deficiency (a deficiency of IgA antibodies only, which usually does not produce any detectable symptoms) and CVID, suggesting that defects in the same genes might underlie both diseases (Fig. 4.4). Moreover, patients may first present with IgA deficiency and, after several years, may develop a full-blown picture of CVID.

The genetic and immunologic causes of CVID have been largely unknown. The heterogeneous nature of the disease is demonstrated by documentation of defects in T cells, B cells, and antigen-presenting cells, suggesting that many genes are involved. It has, however, recently been demonstrated that some familial CVID cases might be monogenic disorders. In the past few years, mutations in six genes have been described to either cause or contribute to CVID. Mutations in the gene encoding TACI have been identified in 8–10% of patients with CVID. Both homozygosity and heterozygosity for mutations in *TACI* have been shown to be associated with familial and sporadic cases of CVID (see Fig. 4.4). However, mutations in *TACI* have been detected in 1% of individuals who do not suffer recurrent infections, including family members of index cases. This reflects the variable penetrance of the gene defect, and suggests that the *TACI* mutation might synergize with defects in other genes to cause CVID. In this regard the majority of patients with CVID have poor B-cell responses to stimulation of the Toll-like receptor TLR-9, and data from mice suggest that TACI synergizes with TLR-9 and also with CD40 in promoting plasma-cell differentiation and immunoglobulin secretion. Naive B cells from CVID patients with *TACI* mutations are severely impaired in their ability to

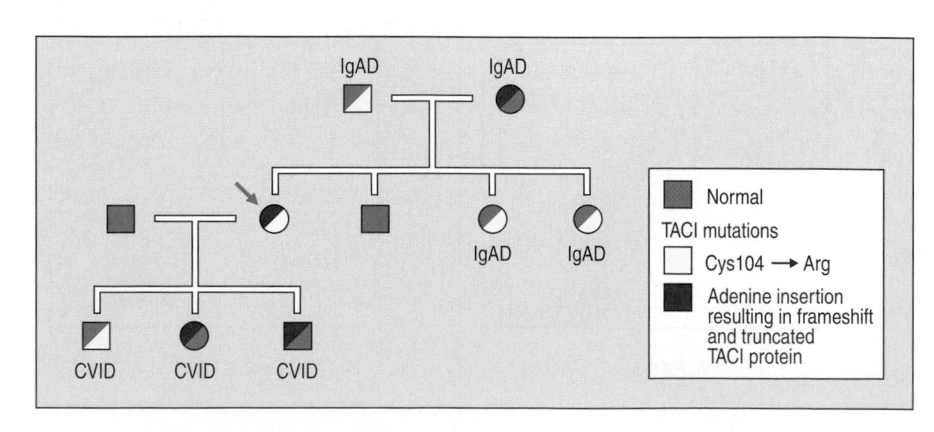

Fig. 4.4 Illustrative pedigree of a family with IgA deficiency (IgAD) and CVID. The index CVID case is marked with an arrow. Note the presence of both CVID and IgAD in family members with the same *TACI* genotype. Two different mutations in TACI have been found in this pedigree: C104R, which destroys ligand binding; and an insertion, A204bp, which results in a frameshift and premature termination of translation. Squares indicate males, circles females. Fully shaded symbols indicate homozygosity, half-shaded symbols heterozygosity. Most heterozygous TACI mutations, including C104R, cause functional haploinsufficiency with regard to the response to T-independent type II antigens (for example pneumococcal antigens). They are disease modifiers, because they are found in 10% of CVID patients but also in 1% of normal individuals.

secrete IgG and IgA in response to stimulation with APRIL *in vitro*. In addition, patients with *TACI* mutations demonstrate an impaired antibody response to the pneumococcal vaccine Pneumovax.

Other gene defects have recently been identified as very rare causes of CVID (Fig. 4.5). The phenotype of B-cell activating factor receptor (BAFF-R) deficiency is characterized by low B-cell numbers and an arrest of B cells in an immature transitional stage, which can be explained by the critical role of BAFF-R in promoting B-cell development and survival. The few patients described so far developed late-onset hypogammaglobulinemia with recurrent infections. Another B-cell intrinsic defect that results in CVID is CD19 deficiency. CD19 is a B-cell surface protein that cooperates in a complex with the complement receptor CD21 in lowering the B-cell-receptor signaling threshold after binding of immune complexes containing complement. All described patients with CD19 deficiency presented in childhood with recurrent infections and hypogammaglobulinemia. Because CD19 is a surface marker for B cells, its expression is routinely assessed by flow cytometry, when lymphocyte subsets are analyzed. A homozygous mutation in the gene encoding CD20, another B-cell surface marker, has been described as causing CVID in one patient. CD20 has a role in the regulation of Ca^{2+} transport across the plasma membrane. CD20 deficiency in this patient was associated with decreased IgG serum levels, a severe reduction of class-switched memory B cells, and an inability to mount T-independent antibody responses.

Deficiency of the inducible co-stimulator (ICOS) molecule is another very rare cause of CVID. ICOS is a co-stimulatory molecule that is upregulated on activated T cells and has a critical role in T-cell–B-cell cooperation. Patients have hypogammaglobulinemia, reduced B-cell numbers, and markedly reduced numbers of memory B cells. All identified patients have the same homozygous mutation in ICOS and are believed to be descended from a common ancestor.

Lastly, several single nucleotide polymorphisms (SNPs) in the gene encoding the mismatch repair (MMR) protein Msh5 have been found at greater frequency in patients with CVID and in patients with IgA deficiency than in healthy individuals. MMR proteins are involved in isotype switching, as discussed in Case 12, but whether Msh5 has a specific role in class switch recombination is still being debated. These polymorphisms in the *Msh5* gene are also found in healthy individuals; they are therefore more likely to cause susceptibility to CVID than to cause the disease itself. Other disease-modifying factors are yet to be determined.

Genetic defects in CVID	Relative frequency
TACI	very common
BAFF-R	extremely rare
CD19	extremely rare
CD20	extremely rare
ICOS	extremely rare
Msh5	to be determined

Fig. 4.5 Monogenetic defects associated with CVID. For most cases of CVID the gene defect remains unknown. TACI deficiency, BAFF-R deficiency, and mutations in Msh5 do not cause CVID in all affected patients, suggesting that the etiology of most cases of CVID is multifactorial. Potential disease-modifying factors are yet to be identified.

Questions.

1 *Like Mrs Johnson, most CVID patients with TACI mutations are heterozygous for the mutation. These patients express both mutant and normal TACI molecules on the surface of their B cells. Why would these heterozygous patients have the disease?*

2 *Mrs Johnson and a maternal cousin had the same TACI mutation. However, Mrs Johnson had full-blown CVID, whereas her cousin only had selective IgA deficiency. How can two individuals with the same TACI mutation present with different disorders (CVID versus IgA deficiency)?*

3 *What other monogenic defects have been described in patients with CVID?*

CASE 5 X-linked Severe Combined Immunodeficiency

The maturation of T lymphocytes.

Without T cells, life cannot be sustained. In Case 1 we learned that an absence of B cells was compatible with a normal life as long as infusions of immunoglobulin G were maintained. When children are born without T cells, they seem normal for the first few weeks or months. Then they begin to acquire infections, often with opportunistic pathogens, and in the absence of appropriate treatment they die while still in infancy. An absence of functional T cells causes severe combined immunodeficiency (SCID). It is called severe because it is fatal, and called combined because, in humans, B cells cannot function without help from T cells, so that even if the B cells are not directly affected by the defect, both humoral and cell-mediated immunity are lost. Unlike X-linked agammaglobulinemia, which results from a monogenic defect, SCID is a phenotype that can result from any one of several different genetic defects. The incidence of SCID is three times greater in males than in females and this male:female ratio of 3:1 is due to the fact that the most common form of SCID is X-linked. In the United States, about 55% of individuals with SCID have the X-linked form of the disease.

T-cell precursors migrate to the thymus to mature, at first from the yolk sac of the embryo, and subsequently from the fetal liver and bone marrow (Fig. 5.1).

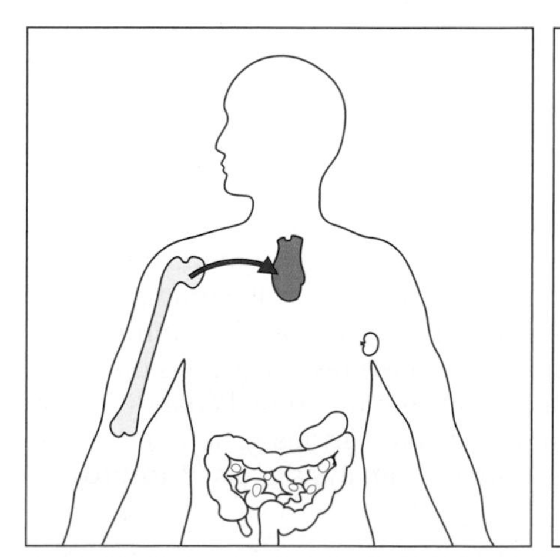

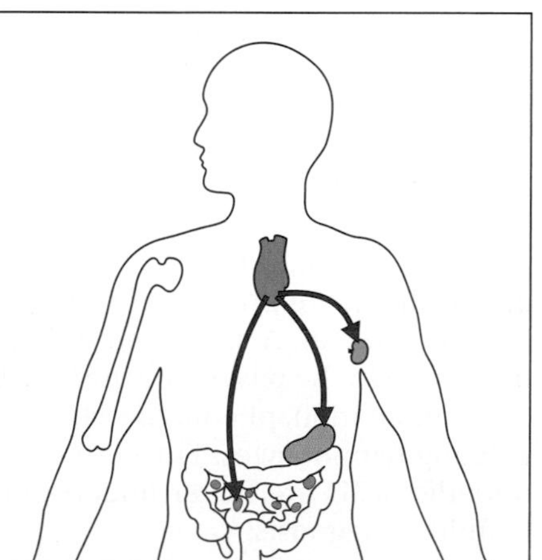

Fig. 5.1 T-cell precursors migrate to the thymus to mature. T cells derive from bone marrow stem cells, whose progeny migrate from the bone marrow to the thymus (left panel), where the development of the T cell takes place. Mature T cells leave the thymus and recirculate from the bloodstream through peripheral lymphoid tissues (right panel) such as lymph nodes, spleen, or Peyer's patches, where they may encounter antigen.

Topics bearing on this case:
The development of T cells in the thymus
Signal transduction through the T-cell receptor
Interleukin-2 receptor
Defects in T-cell function result in severe combined immunodeficiency
Testing T-cell responses with polyclonal mitogens

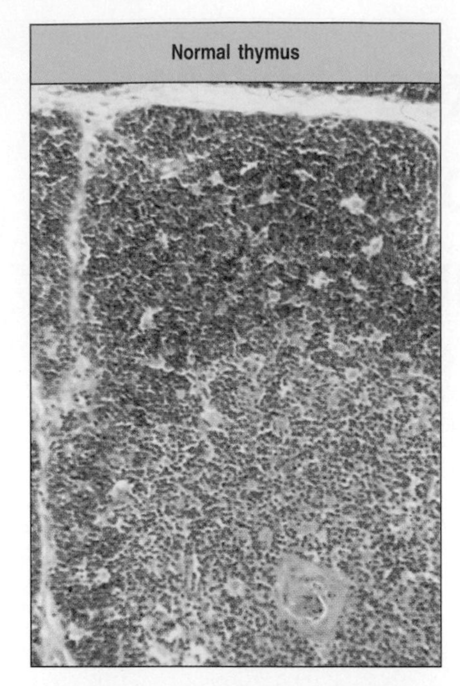

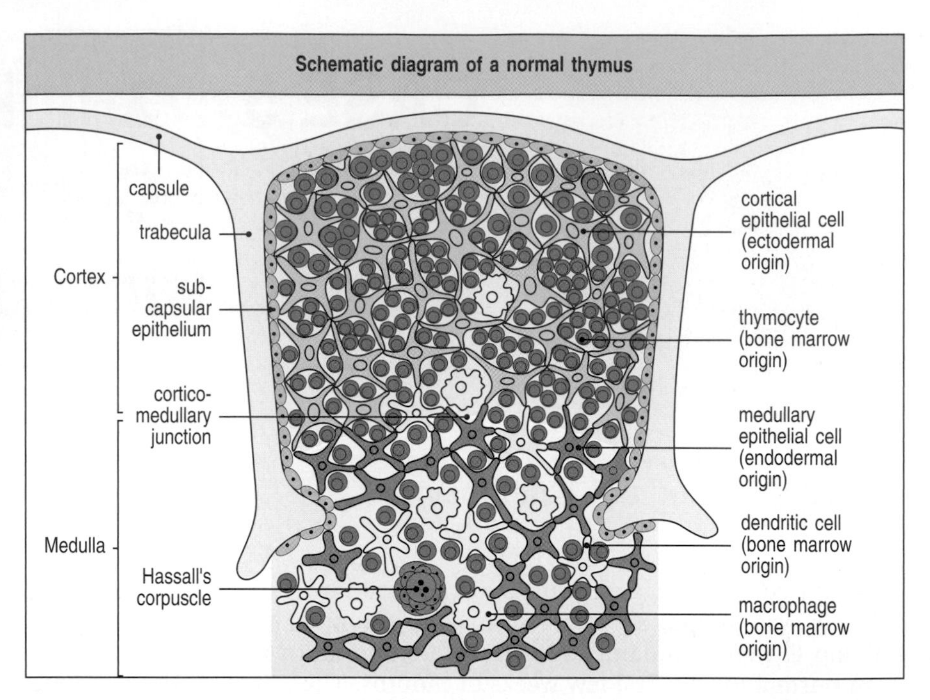

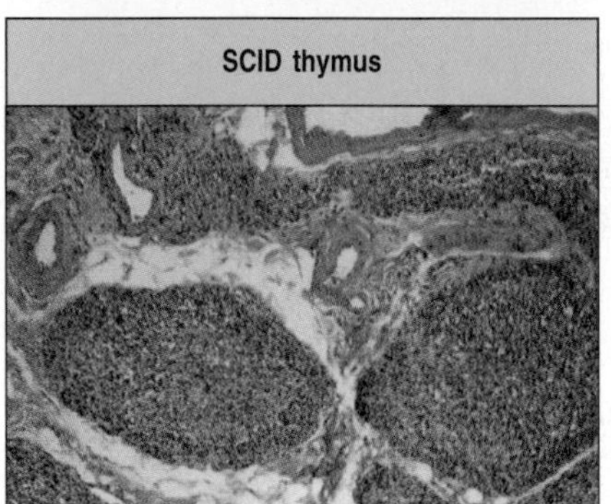

Fig. 5.2 The cellular organization of the thymus. The thymus, which lies in the midline of the body above the heart, is made up from several lobules, each of which contains discrete cortical (outer) and medullary (central) regions. The cortex consists of immature thymocytes, i.e. cells of the T lymphocyte lineage that are developing in the thymus (dark blue), branched cortical epithelial cells (pale blue), with which the immature cortical thymocytes are closely associated, and scattered macrophages (yellow) involved in clearing apoptotic thymocytes. The medulla consists of mature thymocytes (dark blue), and medullary epithelial cells (orange), along with macrophages (yellow) and dendritic cells (yellow) of bone marrow origin. Hassall's corpuscles found in the human thymus are probably also sites of cell destruction. The thymocytes in the outer cortical cell layer are proliferating immature cells; the deeper cortical thymocytes are mainly cells undergoing thymic selection. The upper photograph shows the equivalent section of a human thymus, stained with hematoxylin and eosin. The lower photograph shows a SCID thymus. The thymus is small with loss of lymphocytes and no cortico-medullary differentiation. Hassall's corpuscles are absent. Photographs courtesy of C.J. Howe (upper) and A. Perez-Atayde (lower).

The rudimentary thymus is an epithelial anlage derived from the third and fourth pharyngeal pouches. By 6 weeks of human gestation, the invasion of precursor T cells, and of dendritic cells and macrophages, has transformed the gland into a central lymphoid organ (Fig. 5.2). T-cell precursors undergo rapid maturation in the thymus gland (Fig. 5.3), which becomes the site of the greatest mitotic activity in the developing fetus. By 20 weeks of gestation, mature T cells start to emigrate from the thymus to the peripheral lymphoid organs. In all common forms of SCID the thymus fails to become a central lymphoid organ. A small and dysplastic thymus, as revealed by biopsy, has been used in the past confirm SCID. Currently, diagnosis of SCID is based on the enumeration, phenotypic characterization, and functional analysis of circulating lymphocytes. Furthermore, now that the various gene defects that underlie SCID are better understood, mutation analysis is used to provide definitive diagnosis.

Defects that result in SCID are classified into four general categories. One comprises those defects that impair lymphocyte survival; the prototype of this class is adenosine deaminase (ADA) deficiency (see Case 6), in which adenosine metabolites that are toxic to T cells and B cells accumulate.

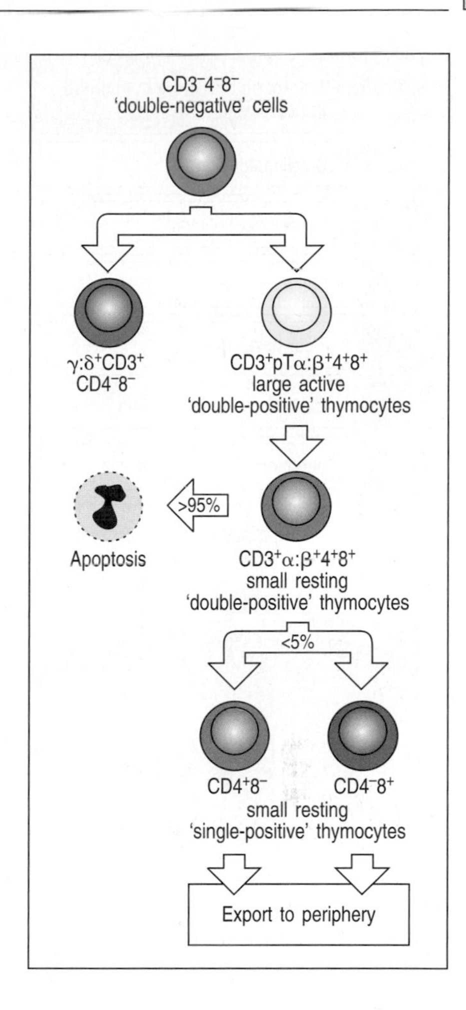

Fig. 5.3 Changes in cell-surface molecules allow thymocyte populations at different stages of maturation to be distinguished. The most important cell-surface molecules for identifying thymocyte subpopulations have been CD4, CD8, and T-cell receptor complex molecules (CD3, and α and β chains). The earliest cell population in the thymus does not express any of these. Because these cells do not express CD4 or CD8, they are called 'double-negative' thymocytes. (The γ:δ T cells found in the thymus also lack CD4 or CD8, but these are a minor population.) Maturation of α:β T cells occurs through stages in which both CD4 and CD8 are expressed by the same cell, along with the pre-T-cell receptor (pTα:β) and, later, low levels of the T-cell receptor (α:β) itself. These α:β cells are known as 'double-positive' thymocytes. Most thymocytes (about 97%) die within the thymus after becoming small double-positive cells. Those whose receptors bind self MHC molecules lose the expression of either CD4 or CD8 and increase the level of expression of the T-cell receptor. The outcome of this process is the 'single-positive' thymocytes, which, after maturation, are exported from the thymus as mature single-positive T cells.

Increased apoptosis of T-lymphocyte progenitors and of myeloid precursors is also observed in reticular dysgenesis, which is characterized by extreme lymphopenia, absence of neutrophils, and sensorineural deafness. This disease is due to mutations of the adenylate kinase 2 gene, which regulates intracellular levels of ADP.

A second category of SCID consists of defects in cytokine-mediated signals for lymphocyte maturation and proliferation, and includes the X-linked form of SCID that is the subject of this case. The *IL2RG* gene responsible for the X-linked form of SCID was mapped to the long arm of the X chromosome at Xq11 and then cloned. It encodes the common γ chain (γ_c, CD132) shared by the interleukin-2 receptor (IL-2R) and by other cytokine receptors (IL-4R, IL-7R, IL-9R, IL-15R, and IL-21R) (Figs 5.4 and 5.5). In all of these cytokine receptors, the γ_c chain associates with the intracellular tyrosine kinase JAK3, encoded by an autosomal gene. JAK3 is essential for intracellular signaling mediated by all γ_c-containing cytokine receptors upon binding of the respective cytokine.

A third category of SCID consists of defects in the machinery for somatic gene rearrangement that assembles the immunoglobulin and T-cell receptor (TCR) genes during lymphocyte development, the so-called V(D)J recombination process. This category can be divided into lymphocyte-specific defects, namely defects in the *RAG* genes that encode the lymphocyte-specific recombinase (see Case 7), and defects in ubiquitously expressed DNA repair genes that are also involved in this recombination.

Finally, another group of SCID is due to mutations in genes that encode the CD3 molecules that participate in the formation of the CD3:TCR complex. This complex allows signaling through the pre-T-cell receptor.

Other forms of combined immunodeficiency are due to defects that affect later stages in T-cell development. In these cases, there is a residual presence of circulating T lymphocytes.

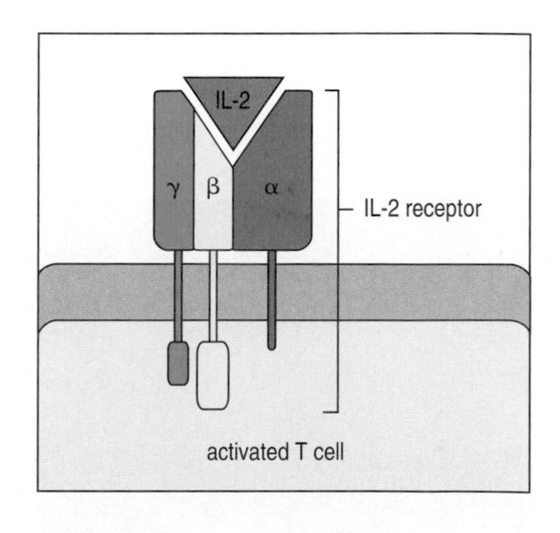

Fig. 5.4 IL-2 receptors are three-chain structures composed of α, β, and γ chains. Resting mature T cells express only the β and γ chains, which bind IL-2 with moderate affinity, allowing resting T cells to respond to very high concentrations of IL-2. Activation of T cells induces the synthesis of the α chain and the formation of the heterotrimeric receptor, which has a high affinity for IL-2 and allows the T cell to respond to very low concentrations of IL-2. The β and γ chains show amino acid similarities to cell-surface receptors for growth hormone and prolactin, all of which regulate cell growth and differentiation.

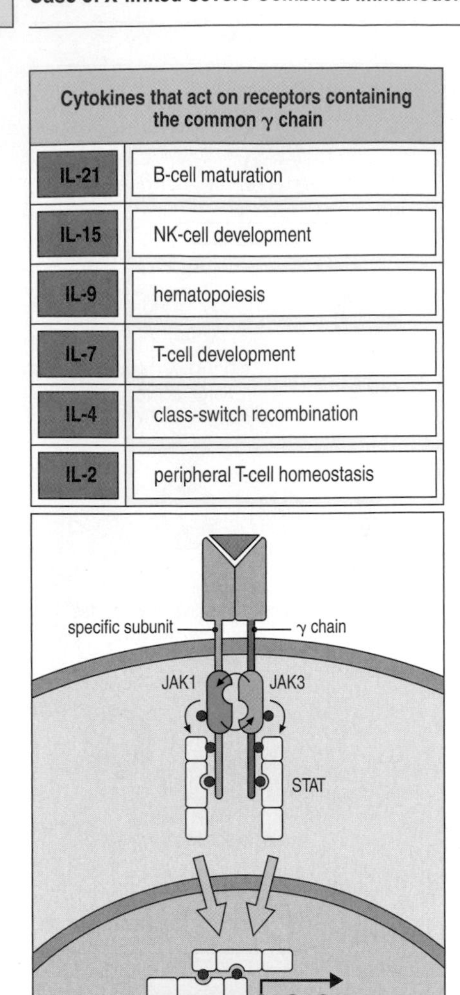

Cytokines that act on receptors containing the common γ chain
IL-21
B-cell maturation
IL-15
NK-cell development
IL-9
hematopoiesis
IL-7
T-cell development
IL-4
class-switch recombination
IL-2
peripheral T-cell homeostasis
specific subunit
γ chain
JAK1
JAK3
STAT
Nucleus
gene

Fig. 5.5 The γ_c chain is a component of multiple cytokine receptors with distinct roles in immune function. The γ_c chain is shared by cytokine receptors for IL-2, IL-4, IL-7, IL-9, IL-15, and IL-21, which have distinct roles in T-cell, B-cell, and NK-cell development and function.

The case of Martin Causubon: without T cells, life cannot be sustained.

Mr and Mrs Causubon had a normal daughter 3 years after they were married. Two years later they had a son and named him Martin. He weighed 3.5 kg at birth and seemed to be perfectly normal. At 3 months of age, Martin developed a runny nose and a persistent dry cough. One month later he had a middle ear infection (otitis media) and his pediatrician treated him with amoxicillin. At 5 months of age Martin had a recurrence of otitis media. His cough persisted and a radiological examination of his chest revealed the presence of pneumonia in both lungs. He was treated with another antibiotic, clarithromycin. Mrs Causubon noticed that Martin had thrush (*Candida* spp.) in his mouth (Fig. 5.6) and an angry red rash in the diaper area. He was not gaining weight; he had been in the 50th centile for weight at age 4 months, but by 6 months he had fallen to the 15th centile. His pediatrician had given him oral polio vaccine at ages 4 and 5 months and, at the same time, diphtheria–pertussis–tetanus (DPT) immunizations.

Martin's pediatrician referred him to the Children's Hospital for further studies. On admission to the hospital, he was fretful and had tachypnea (fast breathing). A red rash was noted in the diaper area as well as white flecks of thrush on his tongue and buccal mucosa. His tonsils were very small. He had a clear discharge from his nose, and cultures of his nasal fluid grew *Pseudomonas aeruginosa*. Coarse, harsh breath sounds were heard in both lungs. His liver was slightly enlarged.

Martin's white blood count was 4800 cells μl^{-1} (normal 5000–10,000 cells μl^{-1}) and his absolute lymphocyte count was 760 cells μl^{-1} (normal 3000 lymphocytes μl^{-1}). None of his lymphocytes reacted with anti-CD3 and it was concluded that he had no T cells. Ninety-nine percent of his lymphocytes bound antibody against the B-cell molecule CD20, and 1% were natural killer (NK) cells reacting with anti-CD16. His serum contained IgG at a concentration of 30 mg dl^{-1}, and IgM at 12 mg dl^{-1}. Serum IgA were undetectable. His blood mononuclear cells were completely unresponsive to phytohemagglutinin (PHA), concanavalin A (ConA), and pokeweed mitogen (Fig. 5.7), as well as to specific antigens to which he had been previously exposed by immunization or infection—tetanus and diphtheria toxoids, and *Candida* antigen. His red cells contained normal amounts of adenosine deaminase and purine nucleoside phosphorylase. Cultures of sputum for bacteria and viruses revealed the abundant presence of respiratory syncytial virus (RSV).

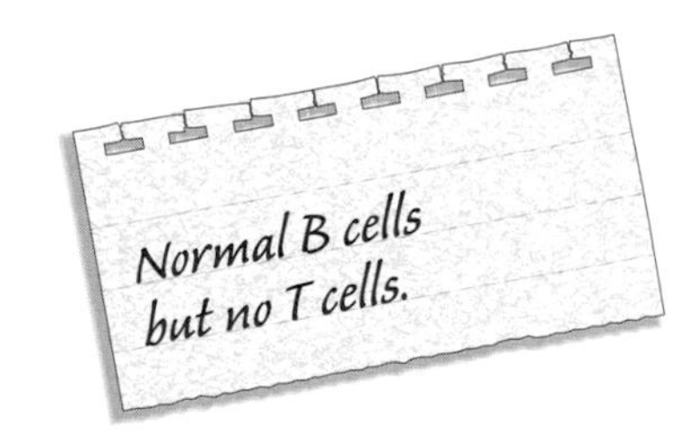

At this point a blood sample was obtained from Martin's mother to examine her T cells for random inactivation of the X chromosome (a diagnostic test explained in Case 1). It was found that her T cells exhibited complete nonrandom X-chromosome inactivation. Search for mutations in the *IL2RG* gene revealed deletion of a single A nucleotide in exon 2, leading to frameshift. HLA typing showed that Martin's sister had no matching HLA alleles. His parents, as expected, each shared one HLA haplotype with Martin.

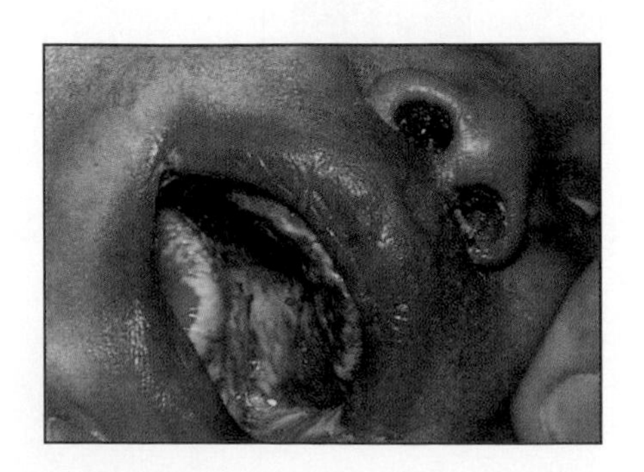

Fig. 5.6 An infant with SCID suffering from *Candida albicans* in the mouth.

Mitogen	Abbreviation	Source	Responding cells
Phytohemagglutinin	PHA	*Phaseolus vulgaris* (Red kidney beans)	T cells (human)
Concanavalin A	ConA	*Canavalia ensiformis* (Jack bean)	T cells
Pokeweed mitogen	PWM	*Phytolacca americana* (pokeweed)	T and B cells
Lipopolysaccharide	LPS	*Escherichia coli*	B cells (mouse)

Fig. 5.7 Polyclonal mitogens, many of plant origin, stimulate lymphocyte proliferation in tissue culture. Many of these mitogens are used to test the ability of lymphocytes in human peripheral blood to proliferate.

Martin was treated with intravenous gamma globulin at a dose of 0.6 g kg^{-1} body weight and his serum IgG level was maintained at 600 mg dl^{-1} by subsequent IgG infusions. He was given aerosolized ribavirin to control his RSV infection and trimethoprim-sulfamethoxazole for prophylaxis against *Pneumocystis jirovecii.* Without any chemotherapy, Martin was given 5 × 10^{6} kg^{-1} CD34^{+} cells that had been purified from his mother's bone marrow. Sixty days after receiving the maternal bone marrow, Martin's blood contained 600 maternal CD3^{+} T cells μl^{-1}, which responded to PHA. His immune system was slowly reconstituted over the ensuing 3 months, but he remained unable to produce IgA and to make specific IgG antibodies. Therefore, Martin continues to require substitution therapy with intravenous immunoglobulin (400 mg kg^{-1} every 3 weeks).

No HLA-identical sibling. Mother agrees to donate bone marrow.
Engraftment is successful.

Severe combined immunodeficiency (SCID).

Severe combined immunodeficiency, or SCID, presents the physician with a medical emergency. Unless there is a known family history, which provides the opportunity to take corrective therapeutic measures before the onset of infections, children with SCID come to medical attention only after they have been infected with a serious opportunistic infection. Because these infants die rapidly from such infections, even when treated adequately with antibiotics or antiviral agents, measures must be taken quickly to reconstitute their immune system. In most cases of SCID, the first symptoms are those of thrush in the mouth and diaper area. A persistent cough usually betrays infection with *Pneumocystis jirovecii.* Interstitial pneumonia may also be due to viral infections, and especially adenovirus, RSV, parainfluenza virus type 3, and cytomegalovirus. The third most common symptom of SCID is intractable diarrhea, often due to enteropathic coliform bacilli or to viruses.

As previously noted, SCID has many known genetic causes. Among autosomal recessive forms of SCID, deficiency of adenosine deaminase (ADA) (see Case 6) or mutations in the *RAG* genes, which initiate V(D)J recombination (see Case 7), are more common. In most patients with ADA deficiency, and in patients with severe *RAG* mutations, both T and B lymphocytes are absent. In contrast, patients with X-linked SCID lack circulating T lymphocytes but have a normal number of B cells. Natural killer lymphocytes (NK cells) are also absent.

Other cases of autosomal recessive SCID resemble the phenotype of X-linked SCID and have been ascribed to defects in the protein JAK3, which transduces the signal from γ_c-containing cytokine receptors. Therefore, patients with JAK3 deficiency also lack T and NK cells and have a normal number of B lymphocytes, similar to the situation observed in patients with X-linked SCID.

Mice lacking IL-2 do not develop SCID, which argues against an important function for IL-2 signaling in thymocyte development. The discovery that mutations in the IL-2 receptor γ chain (IL-2Rγ) caused X-linked SCID in humans seemed inconsistent with this finding in mice. This led to a search for the γ chain in other interleukin receptors that might be more important for lymphocyte development. It was found that this chain also forms part of the receptors for IL-4, IL-7, IL-9, IL-15, and IL-21, and it was renamed the common gamma (γ_c) chain. Among the cytokines that bind to γ_c-containing cytokine receptors, IL-7 has a critical role in T-cell development. In addition, IL-7 is essential for B-cell development in mice, but not in humans. *Il7*$^{-/-}$ and *Il7r*$^{-/-}$ mice lack both T and B lymphocytes; in contrast, patients with SCID due to mutations of the *IL7R* gene lack circulating T lymphocytes but have a normal number of B and NK cells. Finally, the absence of NK cells in patients with X-linked SCID and with JAK3 deficiency is due to impaired signaling through the IL-15R, as shown by impaired development of NK cells in *Il15*$^{-/-}$ and *Il15r*$^{-/-}$ mice.

Mutations of the *RAG* genes and of other genes involved in V(D)J recombination impair rearrangement of the T-cell receptor and immunoglobulin genes, and abrogate the development of both T and B lymphocytes, whereas the development of NK lymphocytes is not affected. In contrast, patients with SCID due to mutations in the genes that encode the CD3 chains (CD3γ, CD3δ, CD3ε, CD3ζ) have selective T-cell deficiency.

Later defects in T-cell development include mutations in proteins that transduce signals from the T-cell receptor, such as the tyrosine kinase ZAP-70 (Fig. 5.8). Patients with ZAP-70 deficiency have a selective lack of circulating CD8 lymphocytes. CD4 lymphocytes develop normally in these patients, but are unable to proliferate in response to mitogens and antigens. Finally, defects of the STIM1 protein, which senses levels of Ca^{2+} stores in the endoplasmic reticulum, and of ORAI1, a protein involved in the formation of calcium-release activated calcium (CRAC) channels, permit thymic T-cell development, but peripheral T cells from these patients fail to respond to activating signals. These patients therefore suffer from clinical manifestations typical of SCID. In addition, they have myopathy and may develop autoimmune manifestations.

SCID is fatal, unless reconstitution of T-cell immunity is achieved. Hematopoietic cell transplantation (HCT) is the treatment of choice in most cases of SCID. When an HLA-matched family donor is available, cure can be achieved in more than 90% of infants with SCID. In other cases, transplantation from a partially matched (haploidentical) family donor or from a matched unrelated donor can be used. These strategies allow survival in 60–80% of the patients. In some cases, HCT can also correct antibody deficiency in patients with SCID; in the remaining cases, immunoglobulin replacement therapy remains necessary. Gene therapy (using retroviral vectors that allow expression of the γ_c chain) has also been tried successfully in some patients with X-linked SCID; however, leukemic proliferation has been observed in some of these patients as the result of activation of oncogenes at the site of retroviral integration.

Questions.

1 *Martin was suspected to have X-linked SCID because he had a severe deficiency of T and NK lymphocytes, with a normal number of B lymphocytes. Even before the mutation in the* IL2RG *gene was identified, which test proved that he had X-linked SCID?*

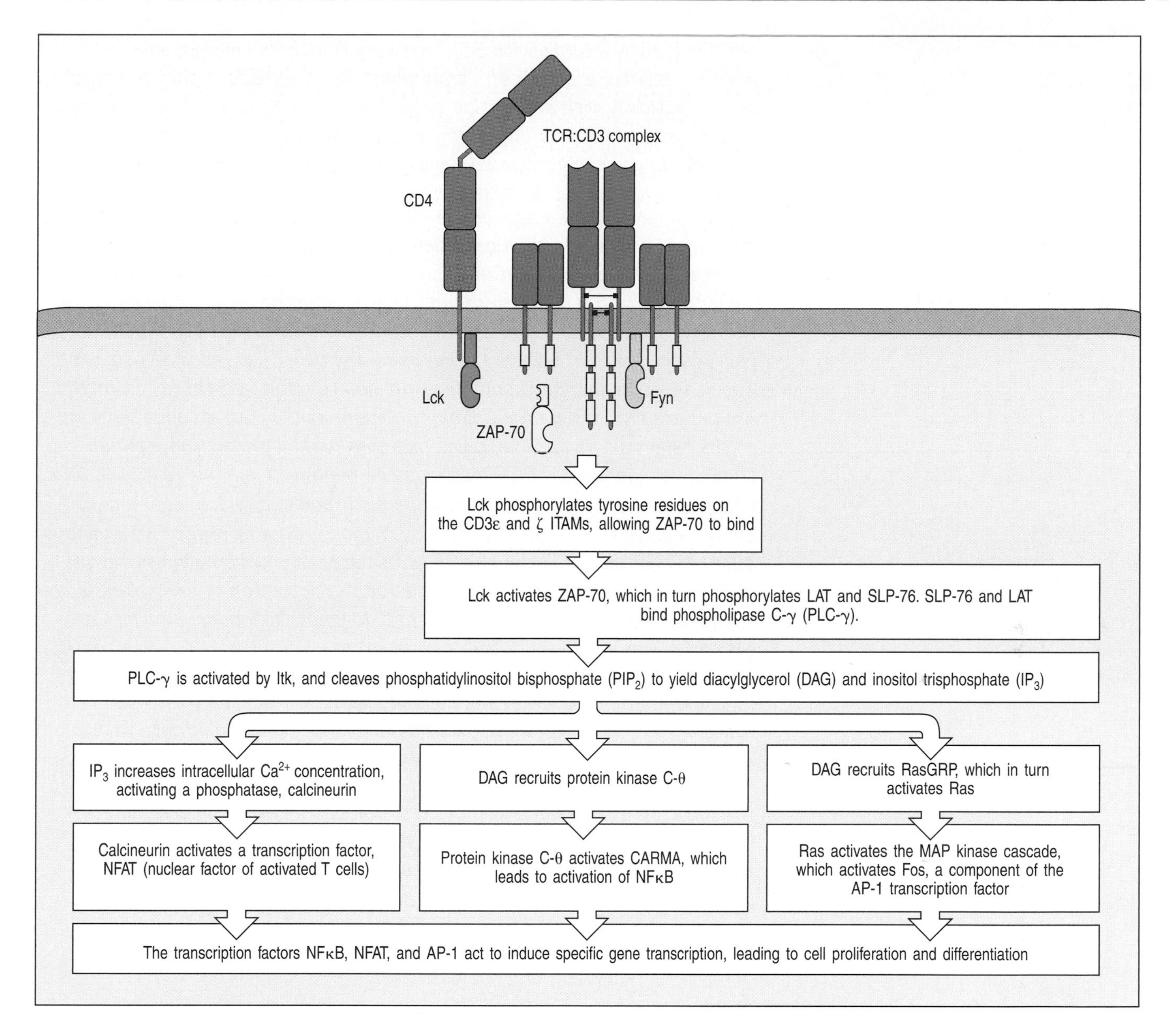

Fig. 5.8 Simplified outline of the intracellular signaling pathways initiated by the T-cell receptor and its co-receptor. The T-cell receptor and co-receptor (in this example the CD4 molecule) are associated with Src-family protein kinases Fyn and Lck, respectively. It is thought that binding of a peptide:MHC ligand to the T-cell receptor and co-receptors brings together Lck with the ITAMs in the receptor. Phosphorylation of the ITAMs in CD3ε, γ, and δ and the ζ chain enables them to bind the cytosolic tyrosine kinase ZAP-70. ZAP-70 recruited to the T-cell receptor is phosphorylated and activated by Lck. Activated ZAP-70 phosphorylates the adaptor proteins LAT and SLP-76, which in turn leads to membrane recruitment of PLC-γ and to its phosphorylation and activation by Tec kinases. Activated PLC-γ initiates three important signaling pathways that culminate in the activation of transcription factors in the nucleus. Together, NFκB, NFAT, and AP-1 act in the nucleus to initiate gene transcription that results in the differentiation, proliferation, and effector actions of T cells. This diagram is a highly simplified version of the pathways, showing the main events only.

2 *What is known about the normal functions of the receptors of which the γ_c chain forms a part, and how might these account for the phenotype of X-linked SCID?*

3 *Why was it necessary to treat the maternal bone marrow and purify CD34⁺ cells? What other strategies could have been used?*

4 *Frequently infants with SCID get very ill with* Pneumocystis jirovecii *pneumonia after a successful transplant. For this reason they are treated prophylactically with antibiotics to get rid of any* P. jirovecii *organisms that may be present in their lungs. Why is this a wise therapeutic maneuver, and how do you explain the worsening of pneumonia after a transplant?*

5 *We have already discussed the risks associated with giving live poliovirus vaccine to immunodeficient infants in the case of X-linked agammaglobulinemia (see Case 1). Martin would not have been able to clear the poliovirus he received until he was started on gamma-globulin therapy, but luckily a pathogenic variant did not arise during that time. Fortunately, Martin escaped being given any other live vaccines before he was diagnosed. In many countries (but not the United States) infants are universally given bacille Calmette–Guérin (BCG), an attenuated form of the tuberculosis bacillus, which provides partial protection against tuberculosis infection. BCG incites a cell-mediated immune response and after receiving it infants become tuberculin-positive, which means they show a delayed-type hypersensitivity response to a skin-prick with minute quantities of tuberculin. In the United States, the tuberculin test is considered so diagnostically valuable for the detection of new tuberculosis infections that BCG is not given. What do you think happens to infants with SCID who are given BCG?*

6 *Before smallpox was eradicated in the world, vaccinia virus was routinely administered to all children. What happened to infants with SCID who were vaccinated with vaccinia virus?*

7 *Although transplantation of maternal bone marrow cells resulted in successful T-cell reconstitution, Martin remained unable to produce antibodies. This failure of B-cell reconstitution is often observed in patients with X-linked SCID or with JAK3 deficiency, even when the transplant is performed from an HLA-identical donor. What could be the underlying mechanism?*

8 *What would be another strategy for reconstituting immune function in Martin?*

CASE 6 Adenosine Deaminase Deficiency

The purine degradation pathway and lymphocytes.

In Case 5 we learned about severe combined immunodeficiency (SCID) and examined the most common form—that due to an X-linked mutation in the gamma chain (γ_c) common to several interleukin receptors. In that type of SCID, T cells are virtually absent, whereas B cells are present in normal numbers although they are not functional. Further examination of this phenotype also reveals an absence of natural killer (NK) cells. Thus, X-linked SCID is classified as $T^-B^+NK^-$ SCID.

SCID patients with an almost complete absence of both B cells and T cells are also encountered; their phenotype is T^-B^-. This phenotype is exclusively associated with SCID transmitted as an autosomal recessive condition and has a quite different biochemical cause from X-linked SCID. The most common genetic defect encountered in these patients is mutation in the gene encoding the purine-degradation enzyme adenosine deaminase (ADA) that causes no enzyme to be produced. ADA is a ubiquitous housekeeping enzyme found in all mammalian cells and in blood serum; it converts the purine nucleosides adenosine and deoxyadenosine to inosine and deoxyinosine respectively, and hence to waste products that are excreted (Fig. 6.1). In its absence, cells can accumulate excessive amounts of phosphorylated adenosine and deoxyadenosine metabolites, which are toxic when present in excessive amounts. Lymphocytes are particularly susceptible to the toxic effects of these metabolites, and thus ADA deficiency leads to SCID. However, because of the ubiquitous expression of the ADA gene, other cell types are also affected, accounting for the extraimmune manifestations of ADA deficiency.

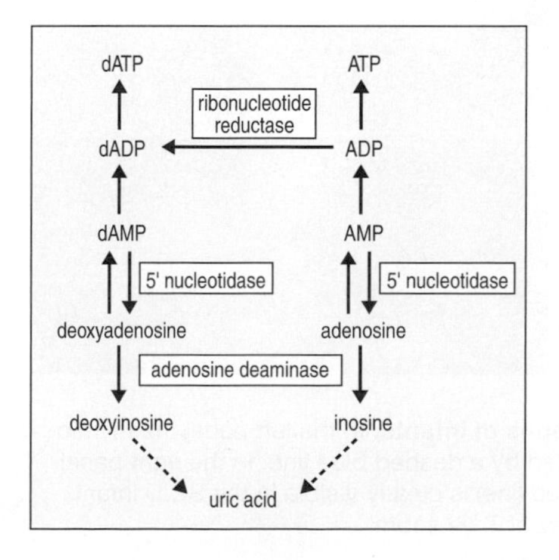

Fig. 6.1 Purine metabolic pathways. Adenosine deaminase catalyzes the conversion of deoxyadenosine and adenosine to deoxyinosine and inosine, which are eventually converted to waste products and excreted. In the absence of ADA in lymphocytes, adenosine and deoxyadenosine metabolites accumulate. In cells other than lymphocytes, the enzyme 5′ nucleotidase can convert AMP and dAMP to adenosine and deoxyadenosine, thus preventing the build-up of potentially toxic metabolites.

Topics bearing on this case:

- Defects in lymphocyte function result in severe combined immunodeficiency
- Bone marrow transplantation
- Mixed lymphocyte reaction
- Minor histocompatibility antigens
- Graft-versus-host reaction

The case of Roberta Alden: lymphocytes poisoned by toxic metabolites.

3-week-old female
infant with thrush.
Family history of SCID.

The Aldens are an African-American family from a remote rural area of the state of Georgia. Mr and Mrs Alden are probably distantly related to each other. At the time they moved to Boston they had seven healthy children—four boys and three girls. Their eighth child was a boy, who developed severe pneumonia at 3 months old and died at the City Hospital. An autopsy revealed that he had SCID. His thymus had a fetal appearance, with only rare thymocytes and no Hassall's corpuscles (see Fig. 5.2). Two years later the Aldens had a daughter, named Roberta. She appeared healthy at birth, but 6 weeks later Mrs Alden noticed thrush in Roberta's mouth (see Fig. 5.6).

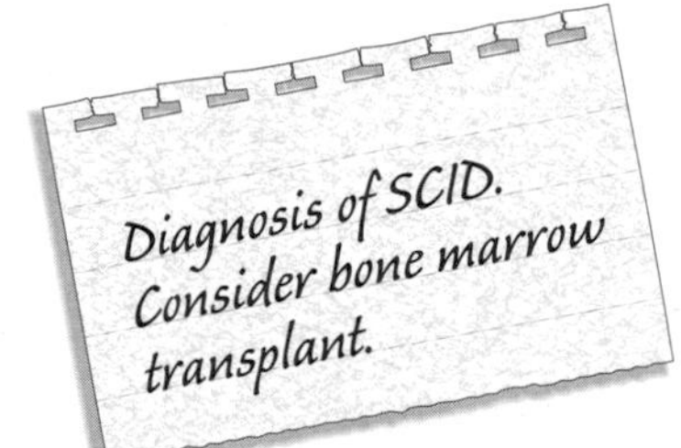

Aware of the family history, Roberta's pediatrician ordered a chest X-ray and blood studies. No thymic shadow (Fig. 6.2) was seen in the chest X-ray, and the anterior margins of the ribs were flared. Roberta's lymphocyte count was 150 cells μl^{-1} (normal for an infant is >3000 cells μl^{-1}). Her lymphocytes did not respond to the nonspecific T-cell mitogen phytohemagglutinin. A diagnosis of SCID was established.

HLA typing of Roberta, her parents, and her seven siblings revealed that her HLA type was identical with one sister, Ellen, and one brother, John. To test directly for histocompatibility, Ellen's and John's blood cells were tested separately against Roberta's in the mixed lymphocyte reaction (Fig. 6.3). No reaction was seen in either case. John and Roberta also had the same blood type (AB), so John was chosen as the bone marrow donor. Bone marrow cells (2.8×10^9 cells) were removed from John's iliac crest bone and infused into Roberta (this was calculated to be a dose of 5×10^8 cells kg^{-1} of Roberta's body weight).

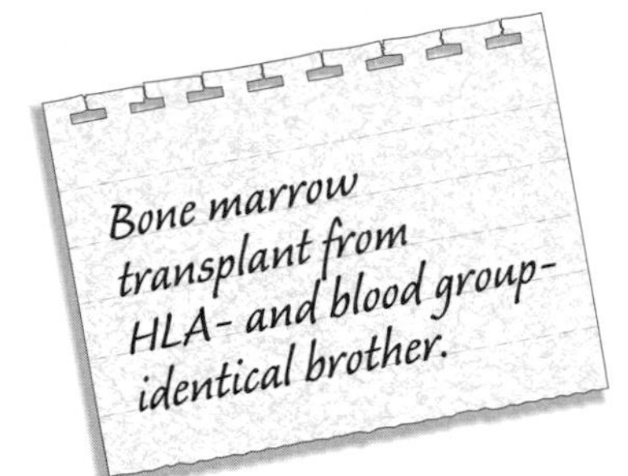

Bone marrow
transplant from
HLA- and blood group-
identical brother.

Twelve days after the bone marrow transplant, Roberta's lymphocyte count had increased to 500 μl^{-1}, and the response of her blood lymphocytes to phytohemagglutinin had risen to half normal. A karyotype of the responding cells revealed that they had XY sex chromosomes; they were thus of male origin. Roberta continued to gain weight and was discharged from the hospital one month after the transplant. Several weeks later she, her parents, and all her siblings were affected with a severe influenza-like respiratory infection. She recovered from this without problems and her lymphocyte count rose to 1575 μl^{-1}; she also now had a normal response to phytohemagglutinin. Roberta continued to grow and develop normally and remained free of infections. All her lymphocytes continue to have an XY karyotype.

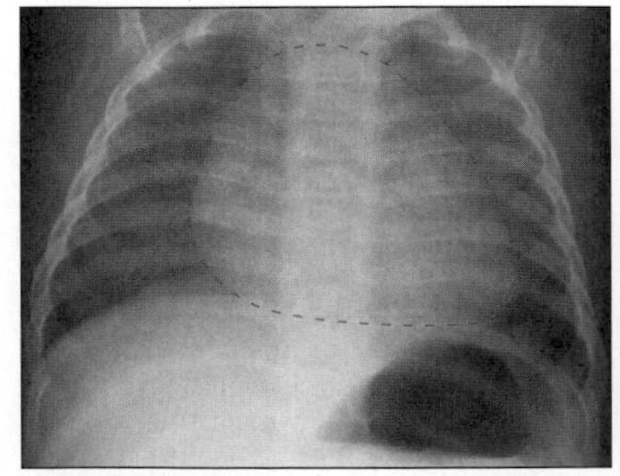
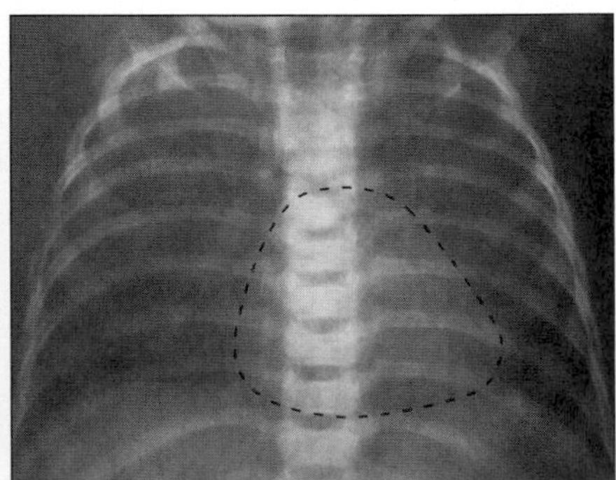

Fig. 6.2 Thymic shadow on chest radiographs of infants. In the left panel the thymic shadow in a normal healthy child is surrounded by a dashed blue line. In the right panel the heart shadow (surrounded by a dashed red line) is clearly visible in the SCID infant owing to the absence of the thymus. Courtesy of T. Griscom.

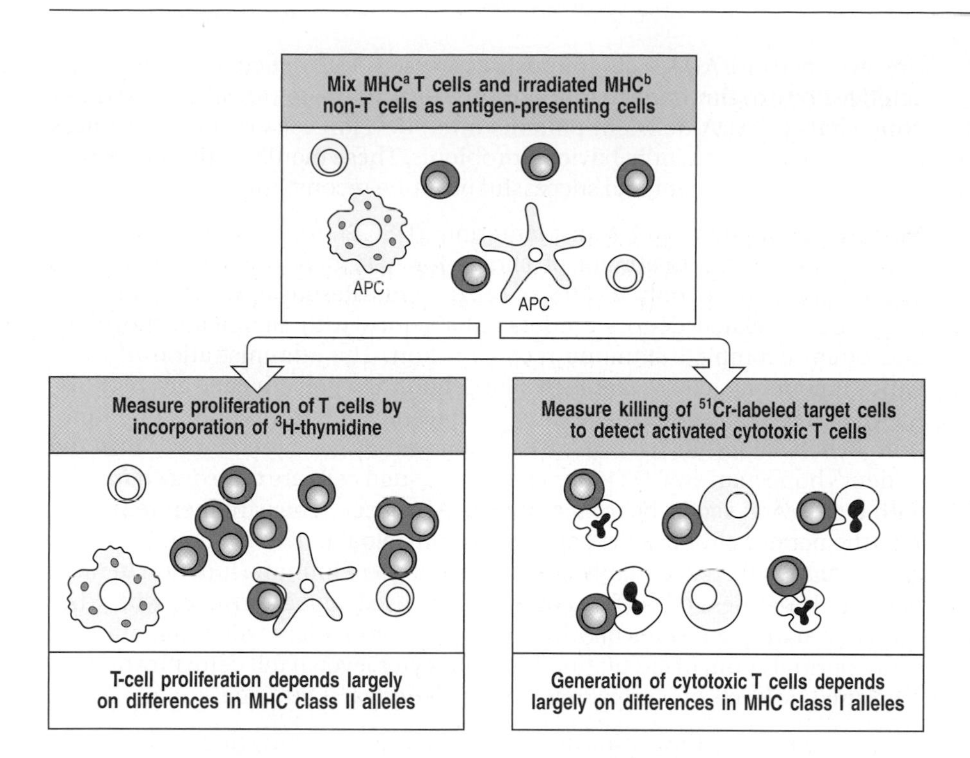

Fig. 6.3 The mixed lymphocyte reaction (MLR) can be used to detect histoincompatibility. Lymphocytes from the two individuals who are to be tested for compatibility are isolated from peripheral blood. The cells from one person (yellow), which also contain antigen-presenting cells (APCs), are either irradiated or treated with mitomycin C; they will act as stimulator cells but cannot now respond by DNA synthesis and cell division to antigenic stimulation by the other person's cells. The cells from the two individuals are then mixed (top panel). If the unirradiated lymphocytes (the responders, blue) contain alloreactive T cells, these will be stimulated to proliferate and differentiate to effector cells. Between 3 and 7 days after mixing, the culture is assessed for T-cell proliferation (bottom left panel), which is mainly the result of CD4 T cells recognizing differences in MHC class II molecules, and for the generation of activated cytotoxic T cells (bottom right panel), which respond to differences in MHC class I molecules. When the MLR is used to select a bone marrow donor, the prospective donor's cells are used as the responder cells and the prospective recipient's cells as the stimulator cells.

Adenosine deaminase deficiency.

The gene encoding adenosine deaminase (ADA) is located on chromosome 20. SCID due to ADA deficiency appears in homozygotes for defects in the ADA gene that result in an inactive enzyme or a lack of enzyme. Because production of ADA from the normal gene in heterozygotes is sufficient to compensate, the disease is inherited as an autosomal recessive condition (Fig. 6.4). Homozygotes for mutations in this gene show the most profoundly lymphopenic form of SCID. The thymus is poorly developed and there is a characteristic abnormality in the rib bones.

Normally, the adenosine and deoxyadenosine content of cells is limited by ADA, which converts these nucleotides to inosine (and deoxyinosine) and subsequently to urate, which is excreted. The limited amounts of adenosine within cells are converted to adenosine monophosphate (AMP), adenosine diphosphate (ADP), and adenosine triphosphate (ATP). Deoxyadenosine is converted to dAMP, dADP, and dATP (see Fig. 6.1). In the absence of ADA, these metabolites can accumulate in up to 1000-fold excess within cells. Excess dATP in particular inhibits the enzyme ribonucleotide reductase, which is necessary for the synthesis of all the deoxynucleotides required for DNA synthesis (Fig. 6.5); its inhibition is probably the main culprit in causing the death and nondevelopment of lymphocytes.

The thymus contains 13 times more ADA than any other tissue in the body. Because adenosine metabolites are very toxic to lymphocytes, the high level of ADA in the thymus is probably crucial for normal thymocyte development. The reason that lymphocytes are particularly vulnerable to the accumulation of these metabolic poisons is probably because they are relatively deficient in the enzyme 5′ nucleotidase. This enzyme degrades AMP and dAMP to adenosine and deoxyadenosine (see Fig. 6.1) and thereby prevents the excessive accumulation of ADP, ATP, dADP, and dATP, even in the absence of ADA.

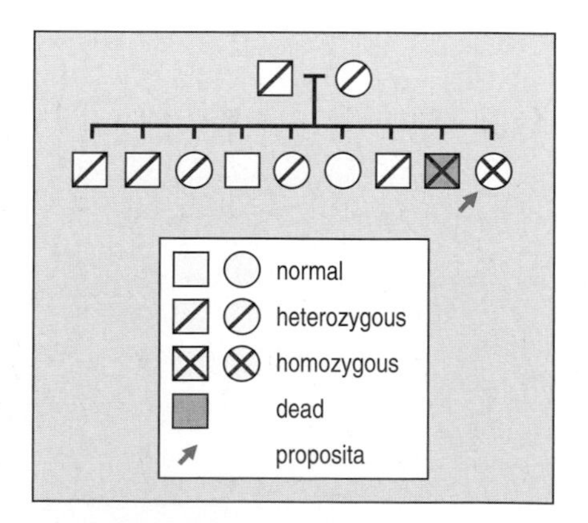

Fig. 6.4 Inheritance of ADA deficiency in Roberta's family. ADA deficiency is inherited as an autosomal recessive condition.

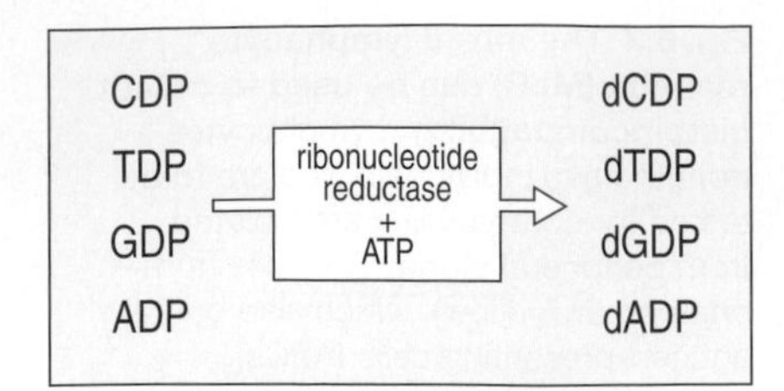

Fig. 6.5 Role of ribonucleotide reductase in generating deoxynucleotides required for DNA synthesis.

However, because ADA is ubiquitously expressed, ADA deficiency is also characterized by extraimmune manifestations. In addition to rib flaring and other bone changes, ADA-deficient patients often develop sensorineural deafness and neurological and/or behavioral problems. These manifestations may persist even after treatment and successful immune reconstitution.

Hematopoietic stem cell transplantation (HSCT) from an HLA-identical family donor is the treatment of choice for ADA deficiency. However, this option is available to only 15–20% of the patients. Results of HSCT from HLA-mismatched related donors are less satisfactory, with significant mortality, and often incomplete immune reconstitution. The administration of ADA, which is commercially available in a form bound to polyethylene glycol (PEG-ADA), clears the noxious metabolites and results in improved immune function. Beneficial results have been also achieved by gene therapy. For this, the patient's bone marrow CD34$^+$ hematopoietic stem cells are transduced *in vitro* with a retroviral vector bearing a normal ADA gene. They are then reinfused into the patient after a short course of nonmyeloablative regimen. This procedure has resulted in a significant improvement of immune function in most patients. Regardless of the form of treatment used, and at variance with what is observed in other forms of SCID, ADA-deficient patients often remain with a low-normal count of circulating T lymphocytes, even if full detoxification (as measured by levels of deoxyadenosine nucleotides) is attained.

Another enzyme in the purine degradation pathway, purine nucleoside phosphorylase (PNP), degrades guanosine to inosine. Its absence results in excessive accumulation of the guanosine metabolites GMP, GDP, and GTP, as well as the deoxyguanosine metabolites dGMP, dGDP, and dGTP, which, like the adenosine metabolites, are toxic to lymphocytes. Genetic defects in PNP resulting in SCID have also been found. In most cases, there is progressive and severe T-cell lymphopenia, whereas the number of circulating B lymphocytes can be variable. In addition to severe infections, progressive neurological deterioration and autoimmune manifestations (in particular, autoimmune hemolytic anemia), are common. HSCT is the only curative treatment.

Questions.

1 *The absence of a thymic shadow in the chest radiograph of a young infant can be helpful in the diagnosis of SCID but it is not a reliable finding. Why is this so?*

2 *In a mixed lymphocyte reaction, the cells used to stimulate the response are treated with mitomycin, a mitotic inhibitor, to prevent their responding to the responder cells. When Roberta's and John's cells were tested in a mixed lymphocyte reaction to ascertain whether his T cells would respond to Roberta's cells it was not necessary to treat Roberta's cells with mitomycin. Why was this the case?*

3 *Can you think of a reason for the fact that many patients with ADA deficiency fail to normalize their T-lymphocyte count, even after successful treatment and complete detoxification?*

CASE 7 Omenn Syndrome

A defect in V(D)J recombination results in severe immunodeficiency.

The development of B cells in the bone marrow and T cells in the thymus is initiated by the assembly of gene segments to make the variable (V) sequence that encodes the V regions of the heavy and light chains of immunoglobulins or of the α and β chains of the T-cell antigen receptors (Fig. 7.1). This process is called V(D)J recombination. A V (variable) and a J (joining) gene segment are joined to make the V-region sequences for the light chains of immunoglobulins or the α chains of T-cell receptors. An additional gene segment, D (diversity), is involved in the rearrangements that produce the V-region sequences for the immunoglobulin heavy chain and the β chain of the T-cell receptor; a D and a J gene segment are joined first, followed by joining of a V gene segment to form VDJ. In all these recombination events, the DNA between the rearranging gene segments is deleted from the chromosome. Because there are many different V, D, and J segments in the germline genome, there are several million possible combinations. This is how much of the vast diversity in the antibody and T-cell receptor repertoires is generated. Moreover, small insertions or deletions of nucleotides at the joins between V and D, and D and J segments further contribute to diversity.

The process of V(D)J recombination is initiated by enzymes encoded by the recombinase-activating genes *RAG1* and *RAG2*. The RAG-1 and RAG-2 enzymes nick double-stranded DNA. They recognize canonical DNA sequences called recombination signal sequences, which flank the coding gene segments and consist of a heptamer (CACAGTG) followed by a spacer of 12 or 23 bases and then a nonamer (ACAAAAGTG) (Fig. 7.2). RAG-1 binds to the nonamer element followed by binding of RAG-2 to the heptamer. The DNA sequence that forms the border between the heptamer and the coding segment is then nicked, and a break in the double-stranded DNA occurs. The coding ends are initially sealed by a hairpin. A series of ubiquitously expressed proteins (Ku70, Ku80, DNA-PKcs, Artemis, DNA ligase IV (LIG4), XRCC4, and Cernunnos/XLF) are then recruited and mediate DNA repair and rejoining of coding and signal ends (Fig. 7.3).

If either of the *RAG* genes is knocked out by homologous recombination in mice, the development of B cells and T cells is completely abolished and the mice have severe combined immunodeficiency. Mutations in *RAG1* and *RAG2*

Topics bearing on this case:

V(D)J recombination

RAG enzymes

Severe combined immunodeficiency

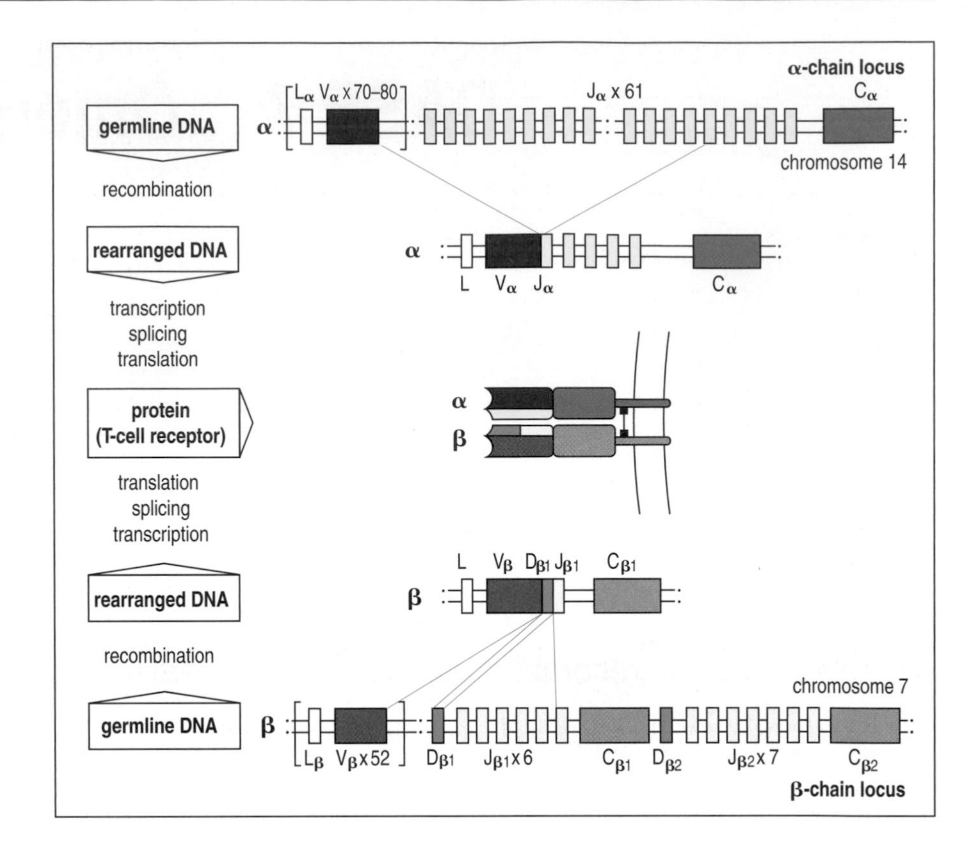

Fig. 7.1 Rearrangement of the T-cell receptor genes. The top and bottom rows of the figure show the germline arrangement of the variable (V), diversity (D), joining (J), and constant (C) gene segments at the T-cell receptor α and β loci, respectively. During T-cell development, a V-region sequence for each chain is assembled by DNA recombination. For the α chain (top), a $V_α$ gene segment rearranges to a $J_α$ gene segment to create a functional gene encoding the V domain. For the β chain (bottom), rearrangement of a $D_β$, a $J_β$, and a $V_β$ gene segment creates the functional V-domain exon. A similar array of gene segments is present at the immunoglobulin loci, and immunoglobulin gene rearrangement follows an essentially similar course.

have also been found in cases of human SCID with lack of both T and B cells (T^-B^- SCID). In addition, defects of Artemis, LIG4, and DNA-PK have been also identified in patients with T^-B^- SCID, and mutations of Cernunnos/XLF cause combined immunodeficiency with markedly reduced numbers of T and B lymphocytes. However, hypomorphic mutations in these genes may allow residual protein expression and function and may result in a different phenotype, in which autoimmune manifestations associate with severe immunodeficiency. Omenn syndrome is the prototype of these conditions, and is most often due to missense mutations in the *RAG* genes.

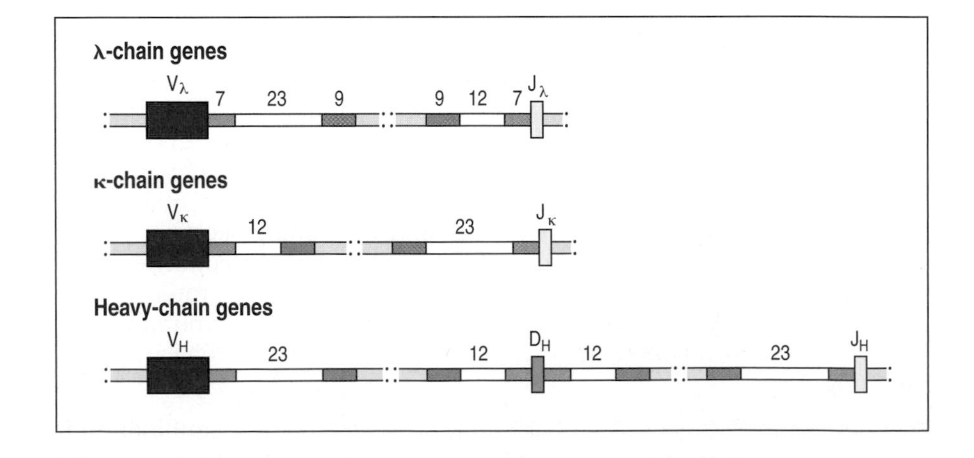

Fig. 7.2 Each V, D, or J gene segment is flanked by recombination signal sequences (RSSs). This is illustrated here with respect to the immunoglobulin genes. There are two types of RSS. One consists of a nonamer (9 nucleotides, shown in purple) and a heptamer (7 nucleotides, shown in orange) separated by a spacer of 12 nucleotides (white). The other consists of the same 9- and 7-nucleotide sequences separated by a 23-nucleotide spacer (white).

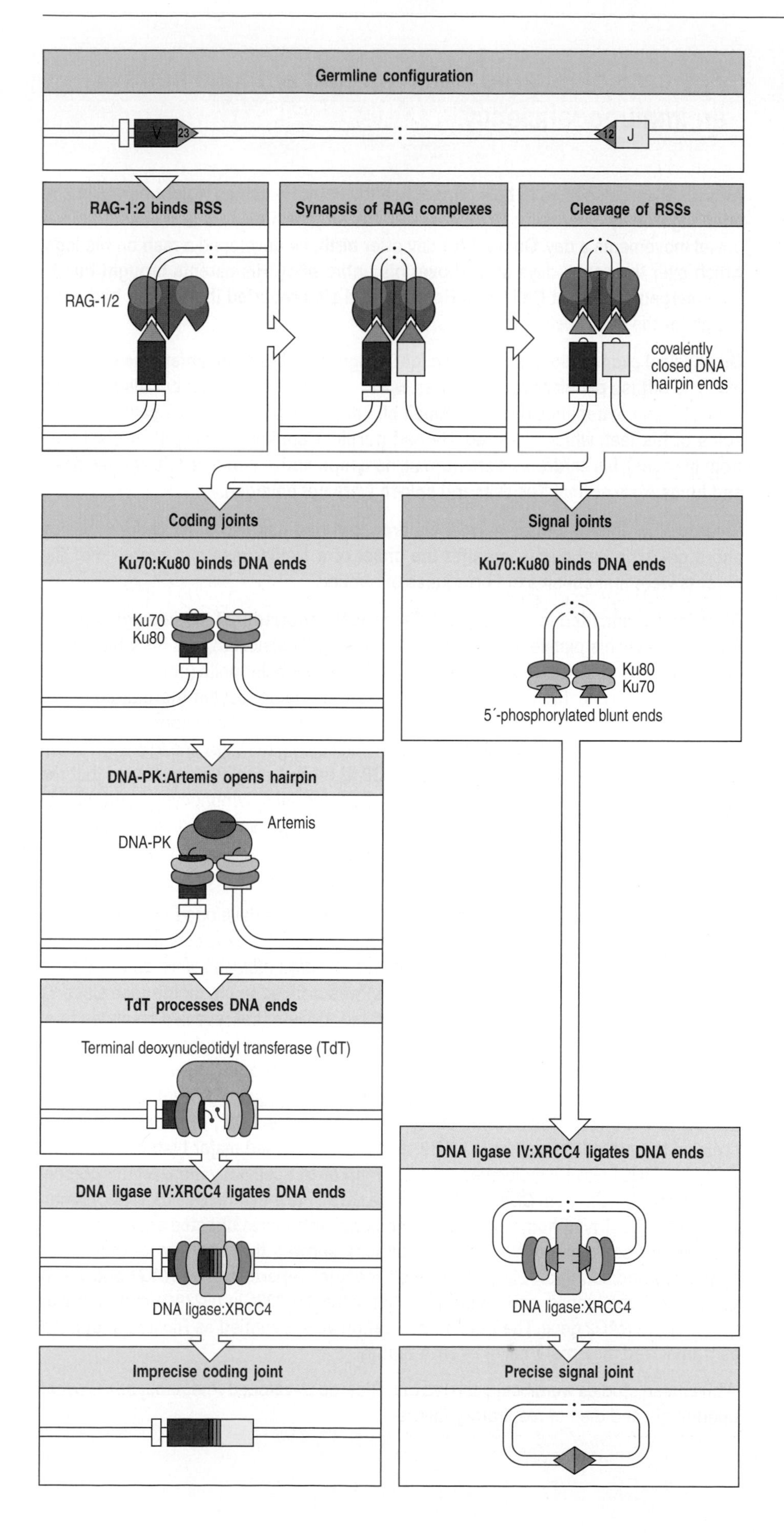

Fig. 7.3 Steps in the V(D)J recombination process. V(D)J recombination is initiated by the lymphocyte-specific RAG-1 and RAG-2 enzymes, which recognize the recombination signal sequences (RSSs) that flank the coding variable (V), diversity (D), and joining (J) elements. The DNA double-strand break at the coding ends is initially sealed by a hairpin. In the second step of the process, the ubiquitously expressed Ku70 and Ku80 are recruited both at coding ends and at signal ends. DNA-protein kinase catalytic subunit (DNA-PKcs) and Artemis are also recruited to the coding ends, and Artemis mediates opening of the coding-end hairpins. The enzyme terminal nucleotide transferase (TdT) may introduce additional nucleotides at the junction between coding elements. Finally, the enzymes DNA ligase IV and XRCC4 (involved in DNA repair and ligation) are recruited at both coding and signal ends and mediate the formation of coding and signal joints. Another enzyme, Cernunnos/XLF (not shown in the figure), also participates in the DNA repair process.

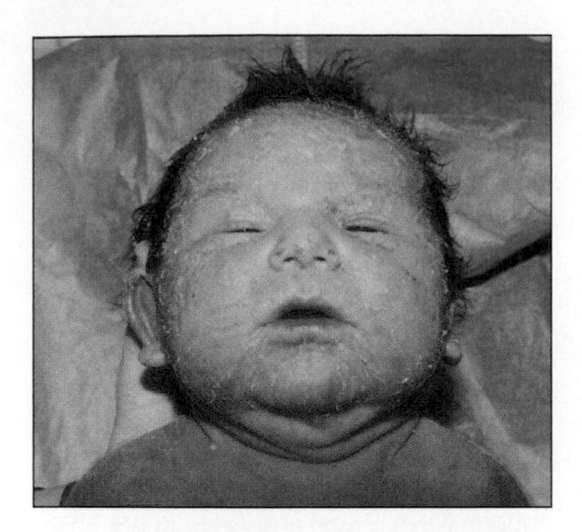

Fig. 7.4 Bright scaly red rash on the face and shoulders of an infant with Omenn syndrome.

The case of Ricardo Reis: a bright red rash betrays an immunodeficiency.

At birth, Ricardo seemed to be a normal healthy baby. He gained weight normally and cried vigorously. Soon after birth, however, his mother noticed that he had 10 loose bowel movements a day. On the 17th day after birth, he developed a rash on his legs, which over the next 7 days spread over his entire body. His parents brought him to the emergency room at Children's Hospital, and also reported that he had had a dry cough for the past week.

On physical examination, Ricardo's weight, length, and head circumference were normal. The diffuse papular scaly rash was worst on his face (Fig. 7.4) but also covered his trunk and extremities (Fig. 7.5). Small blisters were present on his palms and the soles of his feet, which were red. He had purulent conjunctivitis (yellow discharge from his eyes), his eardrums were normal, no lymph nodes could be felt, and his heart and lungs were normal. The liver and spleen were not enlarged.

One-month-old
infant with bright red
rash and purulent
conjunctivitis. Admit
to hospital.

Ricardo's parents had three normal children, but had had two other children, a boy and a girl, who had died soon after the onset of a similar rash at 1 month old. The parents were first cousins of Portuguese extraction.

Ricardo was admitted to the hospital. Blood tests showed that his hemoglobin was 8.4 g dl^{-1} (low), his platelet count was 460,000 (slightly elevated), and his white blood cell count was 8000 μl^{-1} (normal), of which 56% were eosinophils (normal <5%), 23% monocytes (normal 10%), 15% neutrophils, and 6% lymphocytes (normal 50%). An examination of his bone marrow revealed a preponderance of eosinophil precursors. Ricardo's serum IgG level was 55 mg dl^{-1} (normal 400 mg dl^{-1}), IgA and IgM were undetectable, and IgE was 7200 IU ml^{-1} (normal <50 IU ml^{-1}). A skin biopsy showed that the dermis was infiltrated with large numbers of eosinophils, lymphocytes, and macrophages. Large numbers of cells surrounded the blood vessels. An X-ray of Ricardo's chest showed clear lungs and a normal cardiac shadow; there was no thymic shadow (see Case 6).

Eosinophilia and low lymphocyte count. No thymic shadow.

In the hospital, Ricardo's condition rapidly worsened. He developed enlarged lymph nodes in the neck and groin, and pus accumulated in the skin behind his ear. This was drained, and *Staphylococcus aureus* and *Candida albicans* were cultured from the drainage fluid. Thrush (*Candida albicans*) was noticed in his mouth (see Case 5). An immunologist was consulted. He ordered blood tests that revealed an absence of B cells and a paucity of T cells. Ricardo's peripheral blood lymphocytes responded poorly to stimulation with phytohemagglutinin and with anti-CD3 monoclonal antibody. On FACS analysis, no cells were found that reacted with anti-CD19, which detects B cells (see Case 1). All the lymphocytes were $CD3^+$, of which 90% coexpressed the activation marker CD45R0, and 65% expressed major histocompatibility complex (MHC) class II molecules, another marker of T-cell activation. Eighty percent of the lymphocytes were $CD4^+$, and 15% were $CD8^+$. Flow-cytometry analysis of Ricardo's peripheral T lymphocytes, using monoclonal antibodies directed against various families of T-cell receptor V_α and V_β sequences showed that only few of them were expressed, indicating an oligoclonal T-cell receptor repertoire. The *RAG1* and *RAG2* genes were sequenced, and homozygosity for the Arg222Gln (R229Q) mutation was found in the *RAG2* gene. The T cells were definitively identified as Ricardo's (and not as transferred maternal T cells) by HLA typing.

Lymph nodes enlarged. Opportunistic infections noted. Immunodeficiency?

While these studies were being carried out, Ricardo developed *Pneumocystis jirovecii* pneumonia and died of respiratory failure.

Omenn syndrome.

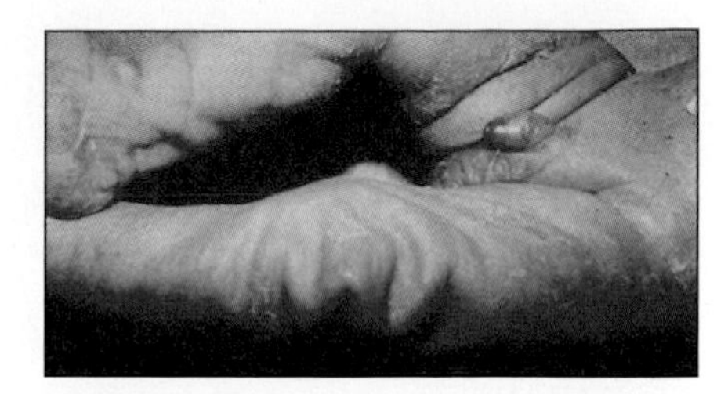

Fig. 7.5 Legs and groin of an infant with Omenn syndrome. The skin is bright red and wrinkled from edema and the infiltration of inflammatory cells.

The RAG enzymes essential for V(D)J recombination were first discovered in mice and later identified in humans. Infants with the autosomal recessive form of severe combined immunodeficiency (SCID) were screened for mutations in these genes, and several cases were identified in which RAG-1 or RAG-2 was deficient. These infants lacked T and B lymphocytes, but had a normal number of NK cells; hence they had $T^-B^-NK^+$ SCID.

Some patients were found with missense mutations in the *RAG* genes such that only partial enzyme activity was expressed. An examination of patients with a form of SCID called Omenn syndrome revealed further missense mutations in *RAG* genes. This syndrome is characterized by early onset of a generalized red rash (erythroderma), failure to thrive, protracted diarrhea, and enlargement of the liver, spleen, and lymph nodes. A high eosinophil count (eosinophilia) is usually encountered, together with a lack of B lymphocytes and a marked decrease in T cells. Immunoglobulins are also markedly decreased, but IgE levels are raised. As only partial ability to execute V(D)J recombination is retained by the mutated enzyme, in most cases no mature circulating B cells are detected and the few T cells that are found are oligoclonal; that is, they are the products of a limited number of different clones. These oligoclonal T cells infiltrate and cause significant damage in target organs.

As illustrated by this case, Omenn syndrome is usually rapidly fatal unless it is treated by bone marrow transplantation, which may result in full correction of the disease.

Genetic defects that result in a severe, but incomplete, impairment of T-cell development by interfering with mechanisms other than V(D)J recombination can also result in Omenn syndrome. These include IL-7Rα chain deficiency (IL-7 is required for lymphocyte development), γ_c deficiency (X-linked SCID; see Case 5), and mutations of the *RMRP* gene. The last of these causes cartilage hair hypoplasia, a condition characterized by dwarfism, sparse hair, a variable degree of immunodeficiency, and hematological abnormalities. As with the *RAG* genes, Omenn syndrome occurs when the defect is 'leaky'; that is, due to a missense mutation that severely impairs but does not abolish function, allowing a few T cells to develop. It is likely that in Omenn syndrome the autoimmune manifestations, with infiltration of target organs by oligoclonal T cells, reflect several mechanisms, as demonstrated by studies in patients and in animal models of the disease. Poor generation of T lymphocytes in the thymus results in impaired maturation of medullary thymic epithelial cells and reduced expression of *AIRE*, thus impinging on the deletion of self-reactive T cells (see Case 17). Furthermore, generation of regulatory T cells in the thymus is also impaired, affecting peripheral tolerance (see Case 18). Finally, the few T cells that are generated in the thymus of patients with Omenn syndrome undergo extensive peripheral expansion (homeostatic proliferation) and secrete increased amounts of cytokines, including inflammatory (IFN-γ) and TH_2 (IL-4, IL-5) cytokines.

Apart from the lymphocyte-specific RAG proteins, V(D)J recombination also involves proteins of the nonhomologous end-joining pathway that are universally used for DNA repair and recombination in human cells. In addition to $T^-B^-NK^+$ SCID, patients with defects in these genes (Artemis, DNA-PK, LIG4, and Cernunnos/XLF) present increased cellular sensitivity to ionizing radiation, because they are unable to repair radiation-induced DNA damage. These radiosensitive forms of $T^-B^-NK^+$ SCID are often associated with extraimmune clinical manifestations, such as microcephaly, neurodevelopmental problems, and growth and development defects.

Questions.

1 *How do you explain the high IgE level and eosinophilia in this patient?*

2 *How do you explain the enlargement of the lymph nodes in this patient?*

3 *How does Ricardo's family history help you determine the mode of inheritance of Omenn syndrome?*

4 *A bright red rash (erythroderma) is characteristic of Omenn syndrome. What causes this rash?*

CASE 8 MHC Class II Deficiency

An inherited failure of gene regulation.

The class II molecules of the major histocompatibility complex (MHC) are involved in presenting antigens to $CD4^+$ T cells. The peptide antigens that they present are derived from extracellular pathogens and proteins taken up into intracellular vesicles, or from pathogens such as *Mycobacterium* that persist intracellularly inside vesicles. MHC class II molecules are expressed constitutively on antigen-presenting cells, including B lymphocytes, macrophages, and dendritic cells. In humans, together with the MHC class I molecules (see Case 12), they are known as the HLA antigens. They are also expressed on the epithelial cells of the thymus and their expression can be induced on other cells, principally by the cytokine interferon-γ. T cells also express MHC class II molecules when they are activated.

MHC class II molecules are heterodimers consisting of an α chain and a β chain (Fig. 8.1). The genes encoding both chains are located in the MHC on the short arm of chromosome 6 in humans (Fig. 8.2). The principal MHC class II molecules are designated DP, DQ, and DR and, like the MHC class I molecules, they are highly polymorphic. Peptides bound to MHC class II molecules can be recognized only by the T-cell receptors of CD4 T cells and not by those of CD8 T cells (Fig. 8.3). MHC class II molecules expressed in the thymus also have a vital role in the intrathymic maturation of CD4 T cells.

Expression of the genes encoding the α and β chains of MHC class II molecules must be strictly coordinated and it is under complex regulatory control by a series of transcription factors. The existence of these transcription factors and a means of identifying them were first suggested by the study of patients with MHC class II deficiency.

Topics bearing on this case:
Role of MHC class II molecules in antigen presentation to CD4 T cells
Role of co-receptor molecule CD4 in antigen recognition by T cells
Intrathymic maturation of CD4 T cells
Mixed lymphocyte reaction
Lymphocyte stimulation by polyclonal mitogens
FACS analysis

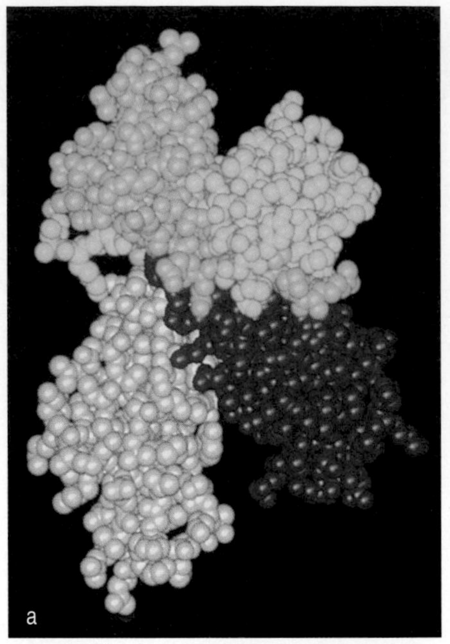

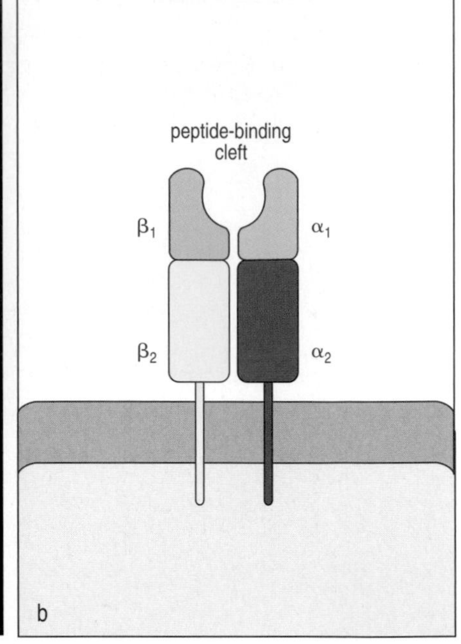

Fig. 8.1 Structure of an MHC class II molecule. Panel a shows a computer graphic representation of the MHC class II molecule HLA-DR1. Panel b is a schematic representation of the molecule. It is composed of two transmembrane glycoprotein chains, α and β, each folded into two protein domains. The antigenic peptide binds in a cleft between the two chains. Photograph courtesy of C. Thorpe.

The case of Helen Burns: a 6-month-old child with a mild form of combined immunodeficiency.

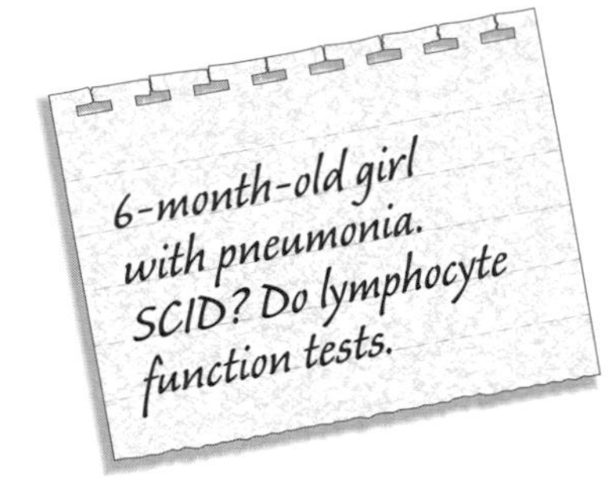

Helen Burns was the second child born to her parents. She thrived until 6 months of age when she developed pneumonia in both lungs, accompanied by a severe cough and fever. Blood and sputum cultures for bacteria were negative, but a tracheal aspirate revealed the presence of abundant *Pneumocystis jirovecii*. She was treated successfully with the anti-*Pneumocystis* drug pentamidine and seemed to recover fully.

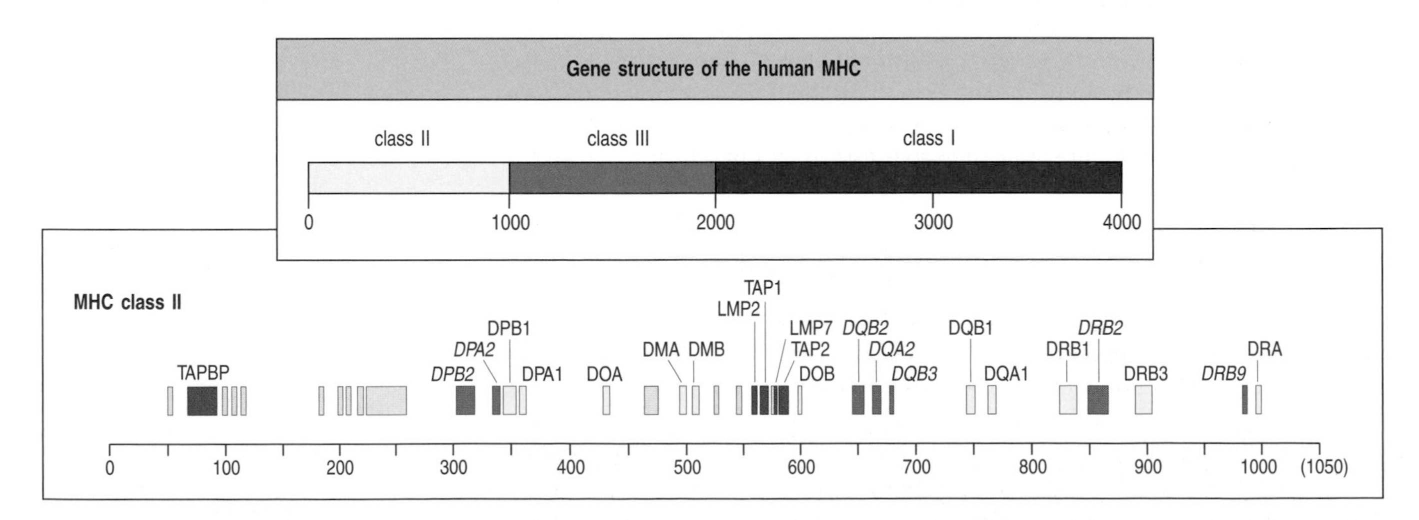

Fig. 8.2 Detailed map of the MHC class II region. The genes for the α and β chains of the HLA-DP, HLA-DR, and HLA-DQ molecules are shown as *DPA*, *DPB*, etc. The situation is complicated because there are two *DPA* genes, two *DPB* genes, and several *DRB* genes. Genes shown in gray and named in italic are pseudogenes. MHC class II genes are shown in yellow. Genes in the MHC region that have immune functions but are not related to the MHC class I and class II genes are shown in purple. Approximate genetic distances given in thousands of base pairs.

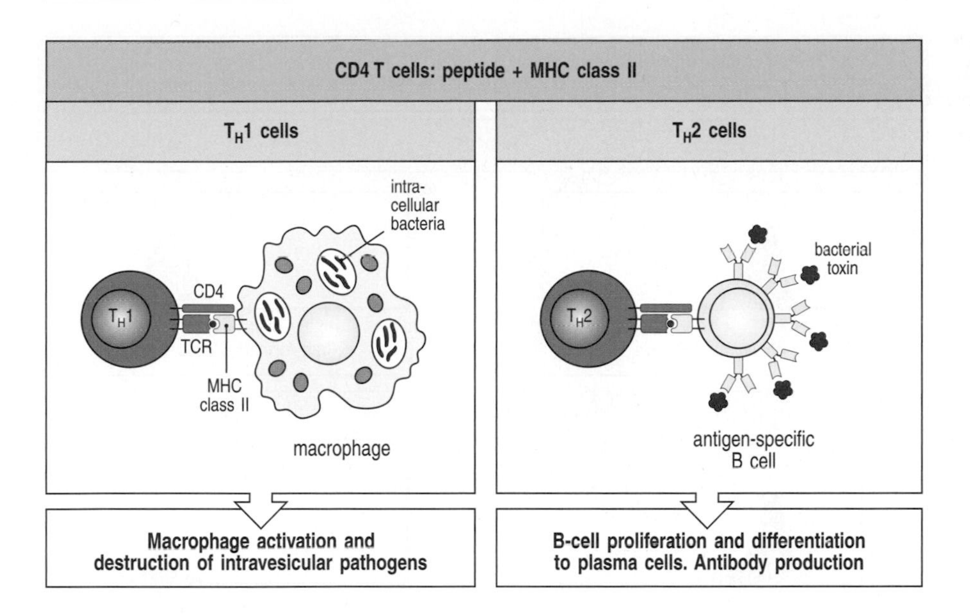

Fig. 8.3 Effector CD4 cells recognize antigens bound to MHC class II molecules. CD4 T cells carry the co-receptor molecule CD4, which binds to MHC class II molecules on the antigen-presenting cell and helps to stabilize the binding of T-cell receptor to antigen. Effector CD4 T cells fall into several different subclasses: T_H1 and T_H2 are shown here. T_H1 cells are involved mainly in responding to antigens presented by macrophages, whereas T_H2 cells respond to antigen presented by B cells, stimulating the differentiation of B cells to plasma cells and the production of antibodies.

As her pneumonia was caused by the opportunistic pathogen *P. jirovecii*, Helen was suspected to have severe combined immunodeficiency. A blood sample was taken and her peripheral blood mononuclear cells were stimulated with phytohemagglutinin (PHA) to test for T-cell function by ^{3}H-thymidine incorporation into DNA. A normal T-cell proliferative response was obtained, with her T cells incorporating 114,050 counts min^{-1} of ^{3}H-thymidine (normal control 75,000 counts min^{-1}). Helen had received routine immunizations with orally administered polio vaccine and DPT (diphtheria, pertussis, and tetanus) vaccine at 2 months old. However, in further tests her T cells failed to respond to tetanus toxoid *in vitro*, although they responded normally in the ^{3}H-thymidine incorporation assay when stimulated with allogeneic B cells (6730 counts min^{-1} incorporated, in contrast with 783 counts min^{-1} for unstimulated cells).

When it was found that Helen's T cells could not respond to a specific antigenic stimulus, her serum immunoglobulins were measured and found to be very low. IgG levels were 96 mg dl^{-1} (normal 600–1400 mg dl^{-1}), IgA was 6 mg dl^{-1} (normal 60–380 mg dl^{-1}), and IgM 30 mg dl^{-1} (normal 40–345 mg dl^{-1}).

Helen's white blood cell count was elevated at 20,000 cells μl^{-1} (normal range 4000–7000 μl^{-1}). Of these, 82% were neutrophils, 10% lymphocytes, 6% monocytes, and 2% eosinophils. The calculated number of 2000 lymphocytes μl^{-1} was low for her age (normal >3000 μl^{-1}). Of her lymphocytes, 27% were B cells as determined by an antibody against CD20 (normal 10–12%), and 47% reacted with antibody to the T-cell marker CD3. In particular, 34% of Helen's lymphocytes were positive for CD8, and 10% were positive for CD4. Thus, at 680 cells μl^{-1} her number of CD8 T cells was within the normal range, but the number of CD4 T cells (200 μl^{-1}) was much lower than normal (her CD4 T-cell count would be expected to be twice her CD8 T-cell count). The presence of substantial numbers of T cells, and thus a normal response to PHA, ruled out a diagnosis of severe combined immunodeficiency (see Case 5).

Helen's pediatrician referred her to the Children's Hospital for consideration for a bone marrow transplant, despite the lack of a diagnosis. When an attempt was made to HLA-type Helen, her parents, and her healthy 4-year-old brother by serology, a DR type could not be obtained from Helen's white blood cells. Her circulating B lymphocytes were transformed with the Epstein–Barr virus (EBV) to establish a B-cell line, which was then analyzed by flow cytometry. The EBV-transformed B lymphocytes did not express HLA-DQ or HLA-DR molecules. Hence, a diagnosis of MHC class II deficiency was established (Fig. 8.4).

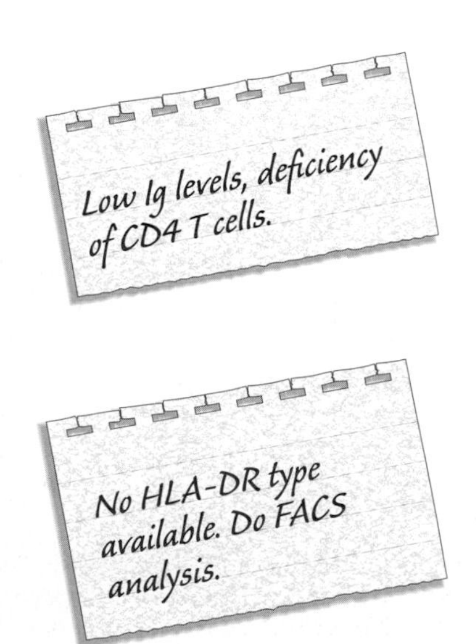

Low Ig levels, deficiency of CD4 T cells.
No HLA-DR type available. Do FACS analysis.

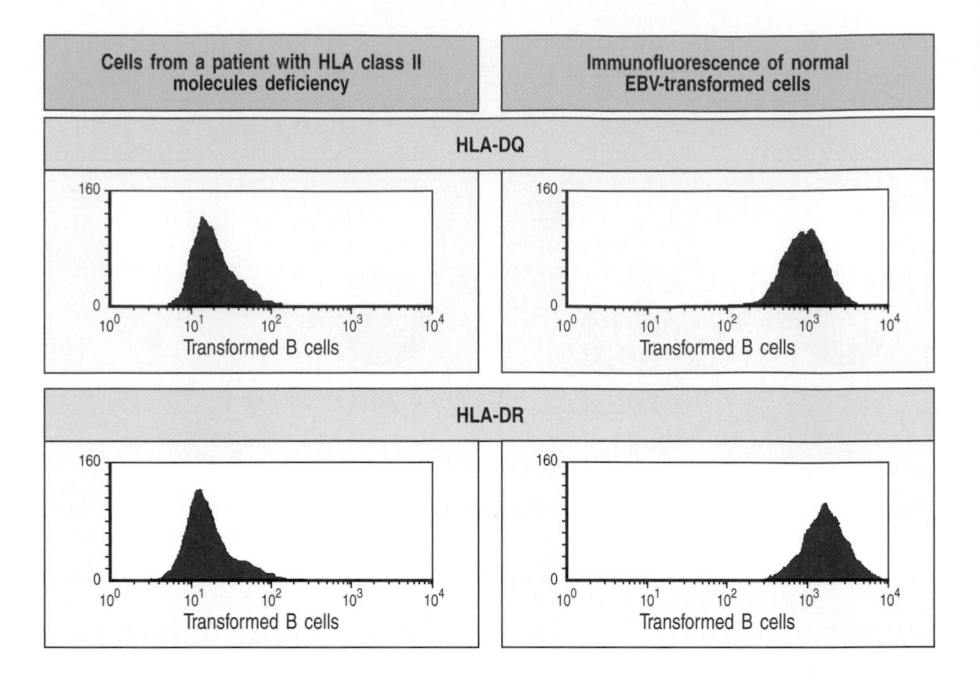

Fig. 8.4 Detection of MHC class II molecules by fluorescent antibody. Helen's transformed B-cell line was examined by using a fluorescent antibody against HLA-DQ and HLA-DR. Helen (left panels) expressed approximately 1% of the amount of MHC class II molecules compared with a transformed B-cell line from a normal control (right panels).

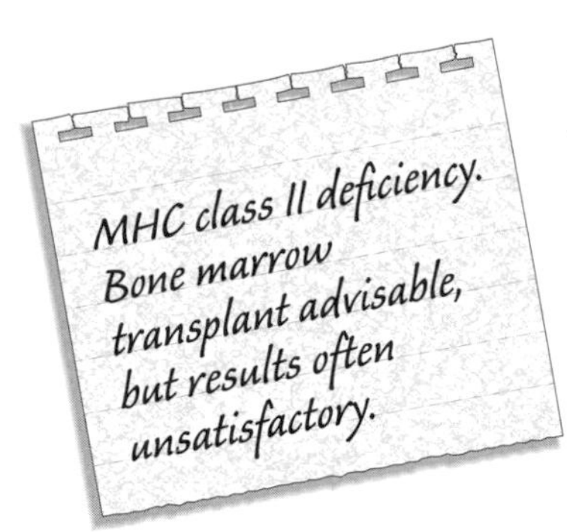

MHC class II deficiency.
Bone marrow transplant advisable, but results often unsatisfactory.

Her brother was found to have the same HLA type as Helen, and therefore was chosen as a bone marrow donor. Helen was given 1 mg kg^{-1} of body weight of the cytotoxic drug busulfan every 6 hours for 4 days and then 50 mg kg^{-1} cyclophosphamide each day for 4 days to ablate her bone marrow. The brother's bone marrow was administered to Helen by transfusion without any *in vitro* manipulation. The graft was successful and immune function was restored.

MHC class II deficiency.

MHC class II deficiency is inherited as an autosomal recessive trait. Health problems show up early in infancy. Affected babies present the physician with a mild form of combined immunodeficiency as they have increased susceptibility to pyogenic and opportunistic infections. However, they differ from infants with severe combined immunodeficiency (SCID; see Case 5) in that they have T cells, which can respond to nonspecific T-cell mitogens such as PHA and to allogeneic stimuli. Unlike in some other types of immunodeficiency, progressive infection with the attenuated live vaccine strain BCG has not been observed in MHC class II-deficient patients after BCG vaccination against tuberculosis (most cases of MHC class II deficiency have been observed in North African migrants in Europe, where BCG vaccination is routine). This is because mycobacterial antigens derived from BCG can be presented on MHC class I molecules and infected cells can be destroyed by cytotoxic T cells. In contrast, and for reasons that are unclear so far, patients with MHC class II deficiency are highly prone to severe viral infections.

Patients with MHC class II deficiency are deficient in CD4 T cells, in contrast with MHC class I deficiency, in which CD8 T-cell numbers are very low and the levels of CD4 T cells are normal (see Case 12). Typically, patients with MHC class II deficiency also have moderate to severe hypogammaglobulinemia.

Hematopoietic stem cell transplantation (HSCT) is the treatment of choice for patients with MHC class II deficiency. Helen Burns was cured after a bone marrow transplant from her HLA-identical brother. However, the results of HSCT in patients with MHC class II deficiency, even when transplanted from HLA-identical donors, are often not satisfactory, and the number of circulating CD4 T cells frequently remains low. This is likely to be because positive selection of donor-derived CD4 thymocytes is compromised, owing to a lack of MHC class II molecules on the surface of the patient's thymic epithelial cells.

Genetic linkage analysis in large extended families with MHC class II deficiency has shown that this condition is not linked to the MHC locus on the short arm of chromosome 6 and that the genes encoding the MHC class II molecules at this locus are normal. Interferon-γ induces the expression of MHC class II molecules on antigen-presenting cells from normal people but fails to induce their expression on the antigen-presenting cells of patients with MHC class II deficiency. This suggested that the defect might lie in the regulation of expression of the MHC class II genes.

The search for the cause of the defect was complicated further by the discovery that MHC class II deficiency in different patients seems to have different causes. B-cell lines isolated from class II-deficient patients do not express MHC class II molecules. However, when B cells from two different patients are fused, MHC class II expression is often observed. The fusion of the two cell lines has corrected the defect. This means that one cell must be able to replace whatever is lacking in the other, and thus the two cells must carry different genetic defects causing the MHC class II deficiency. Pairwise fusions were performed on a large number of cell lines from different patients, and four complementation groups were found (Fig. 8.5).

These experiments provided clues that led eventually to the identification of the defect. The lack of MHC class II molecules turns out to result from defects in the transcription factors required to regulate their coordinated expression. All four of these transcription factors, which bind to the 5′ regulatory region of the MHC class II genes, have been identified.

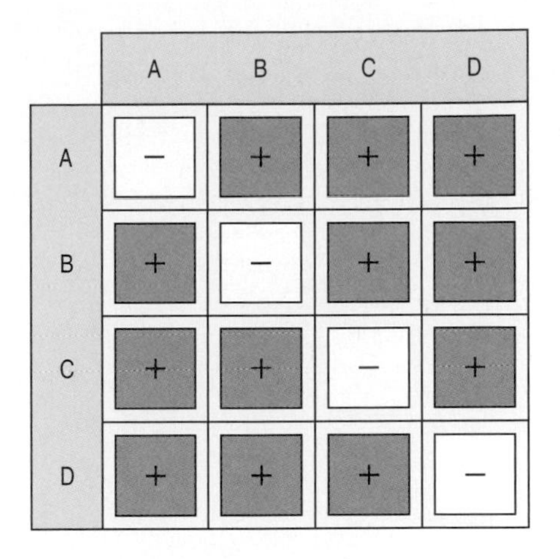

Fig. 8.5 Complementation groups of MHC class II deficiency. B-cell lines isolated from different patients were fused in all pairwise combinations to determine whether they could correct each other's defect. If two cell lines do not correct each other (–), they are in the same complementation group and have the same genetic defect. However, if the defect is corrected (+), the two cell lines belong to two different complementation groups and have two different defects. Four complementation groups, A, B, C, and D, were discovered by this technique.

Questions.

1 *Why did Helen lack CD4 T cells in her blood?*

2 *Why did Helen have a low level of immunoglobulins in her blood?*

3 *In SCID, lymphocytes fail to respond to mitogenic stimuli. Although Helen was first thought to have SCID, this diagnosis was eliminated by her normal response to PHA and an allogeneic stimulus. How do you explain these findings?*

4 *If a skin graft were to be placed on Helen's forearm do you think she would reject the graft?*

CASE 9 DiGeorge Syndrome

The embryonic development of the thymus.

The thymus is the central lymphoid organ in which T cells develop and mature (see Cases 5 and 6). It is composed of an epithelial stroma that becomes populated with precursor T cells and other cells of hematopoietic origin such as macrophages and dendritic cells. The thymic stromal cells provide a microenvironment that is essential for the attraction, survival, expansion, and differentiation of the T-cell precursors. The thymus initially develops in the embryo as an epithelial anlage that gives rise to the thymic stroma. The thymic epithelium derives from the endoderm of the third pharyngeal pouch (Fig. 9.1). The pharyngeal arches and pouches are the embryologic segmental structures that develop into organs of the face and upper thorax. Part of the third and the fourth pharyngeal pouches give rise to the parathyroid glands, to which early thymic development is closely linked. Mesenchymal cells from the pharyngeal arches are also essential for thymic development, giving rise to the connective tissue of the thymus as well as to the smooth muscle of the heart and the major arteries.

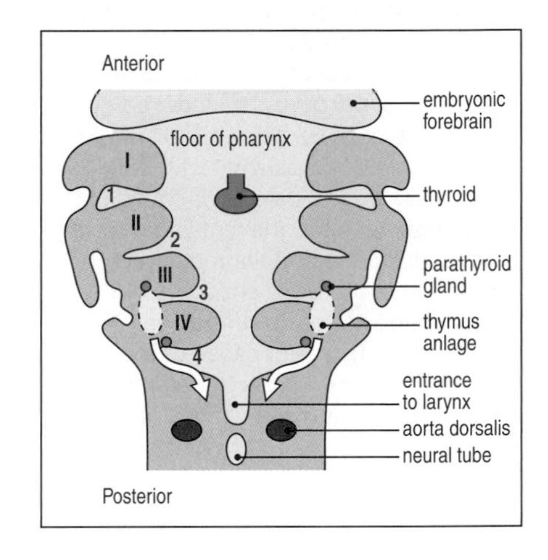

Fig. 9.1 The thymus originates from the pharyngeal pouches. This cutaway view of the embryonic pharynx shows the segmented structure of the region comprising the four pharyngeal arches (I–IV, shaded pink) on either side of the ventral midline, separated internally by pharyngeal pouches (1–4). A thymus anlage (yellow) arises from the third pharyngeal pouch on each side, in close proximity to the parathyroid glands. As the embryo develops, the two anlagen come together to form a single organ. The correct segmentation of the embryonic pharynx, which is under the control of the transcription factor Tbx1, is required for the correct development of the thymus and other organs and structures that develop from this region.

The transcription factor Tbx1 has a central role in the development of the pharyngeal apparatus and its derivates, including the thymus, the parathyroid glands and some tissues of the developing heart. Tbx1 belongs to a family of transcription factors that have a common DNA-binding sequence, designated the T-box. These T-box factors have a role in early embryonic cell fate decisions and in the regulation of the development of many embryonic and extraembryonic structures. Tbx1 is expressed in the endoderm of the third pharyngeal pouch and in the adjacent mesenchyme. It regulates the expression of several growth factors and transcription factors important for development of the thymus and the parathyroid glands. As a regulator of embryonic patterning, Tbx1 not only controls the segmentation of the embryonic pharynx but is also required for the growth, proper alignment, and septation of the cardiac outflow tract.

As one might expect, therefore, deletion or mutation of the *TBX1* gene leads to a wide range of congenital defects, including defects in thymus development. *TBX1* is one of the more than 35 genes located at chromosome 22q11.2. Deletion of this region in one of the two chromosomes 22 is the most common cytogenetic abnormality associated with DiGeorge syndrome (Fig. 9.2), a condition characterized by congenital heart defects, hypoparathyroidism and hypocalcemia, and a variable degree of immunodeficiency. A minority of patients with DiGeorge syndrome carry mutations of the *TBX1* gene without 22q11.2 deletion.

Topics bearing on this case:

- Thymic development
- Thymic epithelial cells
- Immunodeficiency
- Thymus transplantation

This case was prepared by Luigi Notarangelo, MD, in collaboration with Ari Fried, MD.

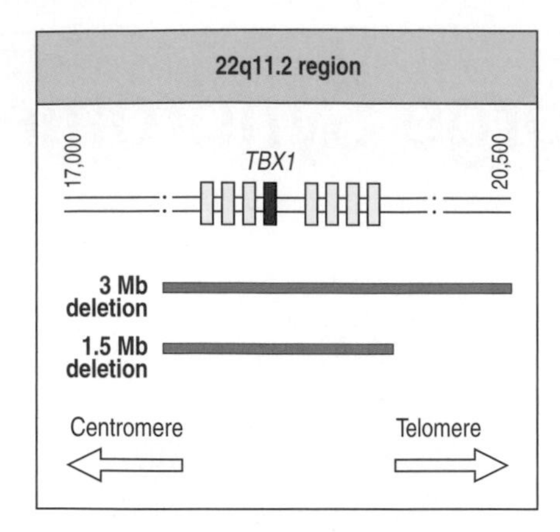

Fig. 9.2 Map of the chromosome region that harbors the 22q11.2 deletion commonly observed in patients with DiGeorge syndrome. The most common abnormality is a deletion of 3 Mb, which is found in 90% of patients with 22q11.2 deletion. A 1.5 Mb deletion is found in 8% of patients, and the remainder have smaller deletions or point mutations. Disruption of the gene encoding the transcription factor Tbx1 has been identified as the major cause of the heart abnormalities, as well as of defects in the thymus and parathyroid glands that are associated with DiGeorge syndrome.

The case of Elizabeth Bennet: severe immunodeficiency as a result of disrupted development of the thymus.

Elizabeth Bennet was born at term after an uncomplicated pregnancy. She had a low birth weight of 2.1 kg, and dysmorphic facial features were noted at birth, including low-set ears as well as a relatively small mouth with an undersized lower jaw (micrognathia) (Fig. 9.3). At 2 days old, Elizabeth developed feeding difficulties, rapid breathing, increased fatigue, and a bluish discoloration of the skin. She was diagnosed with truncus arteriosus, a severe congenital heart defect, characterized by a single common outflow tract leaving the heart instead of the normal two separate blood vessels—the aorta and the pulmonary artery. At 4 days of age, Elizabeth developed seizures. She was found to have very low blood levels of calcium (6.2 mg dl^{-1}, normal 8.5–10.2 mg dl^{-1}) and was treated with calcium and vitamin D.

Her hypocalcemia resulted from very low levels of parathormone in her blood, a hormone that is made by the parathyroid glands and is critical for regulating calcium and phosphorus homeostasis in the body.

Elizabeth underwent cardiac surgery for repair of her heart defect. No thymic tissue was identified intraoperatively. After successful surgery, genetic studies were done; these revealed a normal karyotype, which excluded major chromosomal rearrangements. However, with the help of fluorescence *in situ* hybridization (FISH), she was found to have a deletion of chromosome 22q11.2, consistent with a diagnosis of DiGeorge Syndrome (Fig. 9.4).

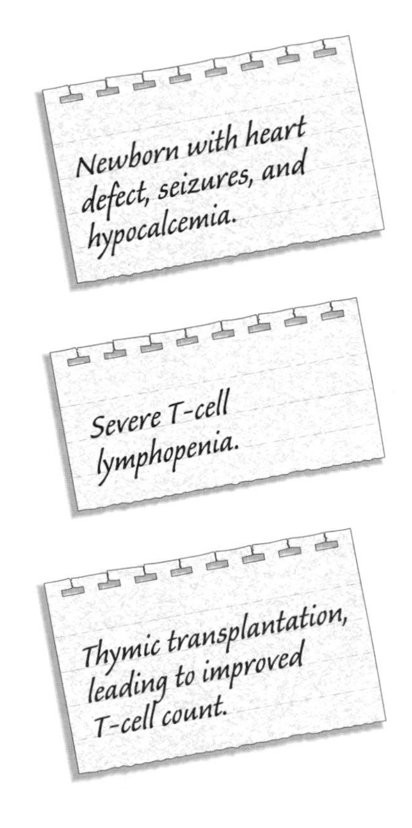

Newborn with heart defect, seizures, and hypocalcemia.
Severe T-cell lymphopenia.
Thymic transplantation, leading to improved T-cell count.

At an immune evaluation at 2 weeks old, Elizabeth's absolute lymphocyte count was low for her age, with 560 cells μl^{-1} (normal 3000 lymphocytes μl^{-1}). She had almost no $CD3^{+}$ T cells, with a count of 11 cells μl^{-1}, which was less than 1% of her total lymphocyte count, while numbers of $CD19^{+}$ B-cell numbers and $CD16^{+}/CD56^{+}$ NK cells were in the normal range for her age. Her peripheral blood mononuclear cells (PBMCs) responded poorly to the mitogens phytohemagglutinin (PHA) and concanavalin A (ConA), which is indicative of poor T-cell function (see Case 5 and Fig. 5.7). Her significant T-cell defect led to the diagnosis of complete DiGeorge syndrome, a rare variant of DiGeorge syndrome (less than 1% of all cases of DiGeorge syndrome) that is associated with severe immunodeficiency and death if not treated early in life.

Elizabeth was started on prophylactic antibiotics to prevent infection with the opportunistic pathogen *Pneumocystis jirovecii.* At 6 months of age she received a thymic transplant into her right leg quadriceps muscle. A biopsy of the thymic graft, performed a few months after transplantation, showed significant presence of thymocytes within the transplanted thymic tissue. One year after transplantation, Elizabeth had developed a significant number of T cells (860 $CD3^{+}$ cells μl^{-1}), although she did not reach normal T-cell counts. Her T cells responded normally to mitogens and antigens *in vitro.* After immunization, Elizabeth mounted protective antibody responses to tetanus toxoid, *Haemophilus influenzae* type b, and *Streptococcus pneumoniae* (pneumococcus). She required calcium supplementation for several months and needed repeat cardiac surgery at the age of 3 years, which she tolerated well.

At 6 years old Elizabeth developed purple bruises (purpura) and pinpoint red lesions (petechiae) on her skin. She was found to have a low platelet count of 15,000 μl^{-1} (normal 150,000–300,000 μl^{-1}). This was due to destruction of her platelets by autoantibodies, resulting in insufficient blood clotting and therefore bleeding into the skin, a condition called immune thrombocytopenia. She was successfully treated with high-dose intravenous gamma globulin (1 g per kg body weight), which resulted in a

rapid increase in the platelet count to 115,000 μl^{-1}. The treatment was repeated every 3 weeks for a further 2 months, with full normalization of the platelet count.

DiGeorge syndrome.

A microdeletion of the 22q11.2 region on one chromosome 22 is the most commonly diagnosed cytogenetic deletion in humans, with an estimated prevalence at birth of 1:4000. It results in a variable clinical phenotype, which includes DiGeorge syndrome, velocardiofacial syndrome (VCFS), or conotruncal anomaly face syndrome. Most 22q11.2 deletions are spontaneous and arise as a result of an aberrant meiotic exchange event. Patients with both DiGeorge syndrome and VCFS have numerous, overlapping clinical features, including the absence (aplasia) or underdevelopment (hypoplasia) of the thymus, hypoparathyroidism, cardiovascular defects, and structural defects of the face and pharynx (see Fig. 9.3). Ninety percent of patients with a 22q11.2 deletion share the same 1.5 Mb or 3 Mb monoallelic microdeletion (see Fig. 9.2); a few patients have smaller deletions that result in the same phenotype.

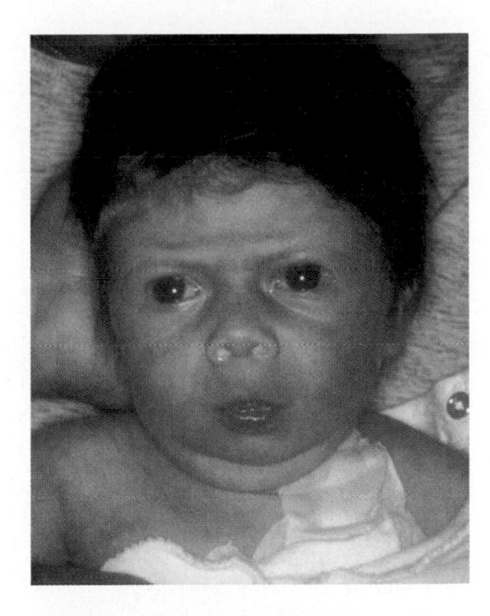

Fig. 9.3 Typical features of DiGeorge syndrome. These include low-set ears, hypertelorism (an increased distance between the eyes), and a small mouth and an underdeveloped jaw (micrognathia).

Patients with the hemizygous 22q11.2 deletion can be easily diagnosed by FISH (see Fig. 9.4). However, nearly half of patients with DiGeorge syndrome do not carry 22q11.2 deletions. Point mutations of the *TBX1* gene have been identified in a few of these patients, and deletions on chromosome 10 in others. A phenotype similar to that of DiGeorge syndrome can also be observed in patients with CHARGE syndrome, which is due to mutations in the gene for chromodomain helicase DNA-binding protein 7 (*CHD7*). CHARGE stands for *c*oloboma (small structural defects) of the eye, *h*eart defects, *a*tresia of the choanae (a blockage of the nasal passages), *r*etardation of growth and/or development, *g*enital and/or urinary abnormalities, and *e*ar abnormalities. Patients with CHARGE syndrome may also present with immunodeficiency. Because of the similarities of clinical phenotype, CHARGE syndrome must be considered in the differential diagnosis with DiGeorge syndrome.

DiGeorge syndrome can encompass a broad range of clinical features, but most infants present with a congenital cardiac defect, mild to moderate immunodeficiency, facial dysmorphisms, developmental delay, palatal dysfunction, feeding difficulties, and hypocalcemia due to absent or low function of the parathyroid glands. Neurobehavioral and psychiatric abnormalities (schizophrenia) may be observed in a significant fraction of patients, especially during adolescence or adulthood.

The degree of immunodeficiency in DiGeorge syndrome is highly variable and does not correlate with the presence or severity of other clinical features of the syndrome. Most patients have a small thymus and a milder immune defect, characterized by a mild to moderate decrease in T-cell counts, but intact T-cell function. These patients are classified as having 'incomplete DiGeorge syndrome.'

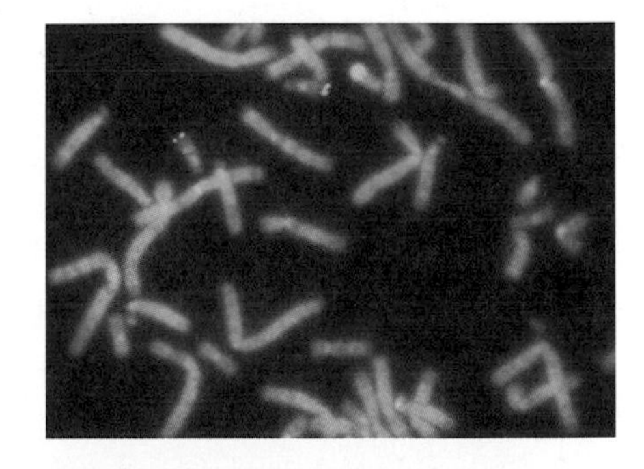

Fig. 9.4 Hemizygous deletion of 22q11.2 detected by fluorescence *in situ* hybridization (FISH). Chromosomes were stained with a green probe that recognizes the subtelomeric region of chromosome 22, and a red probe that stains the 22q11.2 region. The absence of this region is apparent in one of the two chromosomes 22. Micrograph courtesy of Sergio Barlati.

At the severe end of the spectrum are patients with complete absence of a functional thymus and profound T-cell lymphopenia. These patients represent less than 1% of all patients with DiGeorge anomaly, and are given the diagnosis of 'complete DiGeorge syndrome.' They have severe combined immunodeficiency with increased susceptibility to opportunistic infections and tend to die by the age of 1–2 years if not treated adequately. To restore T-cell function, infants with complete DiGeorge syndrome have to undergo thymic transplantation.

In some cases, patients with complete DiGeorge syndrome may develop expansion of a small number of clones of T cells and display severe skin rash and lymphadenopathy, leading to a phenotype that resembles Omenn syndrome (see Case 7). In this subgroup of patients (known as 'atypical complete DiGeorge syndrome'), variable (or even increased) numbers of circulating T cells (expressing the activation marker CD45R0) are detected, and T lymphocytes infiltrate the skin and other organs. Autoimmunity, especially leading to a reduction in blood cells (cytopenia), is another sign of immune dysregulation that is frequently observed in patients with DiGeorge syndrome, even after successful thymic transplant, as seen in Elizabeth's platelet deficiency.

Mutations of the *FOXN1* gene, which encodes a transcription factor essential for thymic epithelial cell development, account for another rare immunodeficiency with very severe T-cell lymphopenia. These patients also present with generalized alopecia and lack of hair follicles. This condition represents the human equivalent of the 'nude' mouse phenotype.

Questions.

1 *Bone marrow transplantation is the treatment of choice in other cases of severe immunodeficiency. Would it work in patients with DiGeorge syndrome and in patients with* FOXN1 *deficiency?*

2 *Thymic transplantation in patients with DiGeorge syndrome is performed using thymic tissue from an unrelated donor, who typically is not HLA-matched to the patient. How can the transplant be successful?*

3 *How could you explain why most patients with DiGeorge syndrome do not have a very severe T-cell deficiency, and that they tend to normalize calcium levels with time?*

4 *What could be the reason that patients with DiGeorge syndrome are at higher risk of autoimmune manifestations?*

CASE 10 Acquired Immune Deficiency Syndrome (AIDS)

Infection can suppress adaptive immunity.

Certain infectious microorganisms can suppress or subvert the immune system. For example, in lepromatous leprosy, *Mycobacterium leprae* induces T cells to produce lymphokines that stimulate a humoral response but suppress the development of a successful inflammatory response to contain the leprosy bacillus. The leprosy bacillus multiplies and there is a persistent depression of cell-mediated immune responses to a wide range of antigens (see Case 48). Another example of immunosuppression is provided by bacterial superantigens, such as toxic shock syndrome toxin-1. Superantigens bind and stimulate large numbers of T cells by binding to certain V_β chains of the T-cell receptor, inducing massive production of cytokines by the responding T cells (see Case 47). This, in turn, causes a temporary suppression of adaptive immunity.

At the beginning of the 20th century, when tuberculosis was the leading cause of death and fully half the population was tuberculin-positive, it was well known that an intercurrent measles infection would cause a well-contained tuberculosis infection to run rampant and result in death. The mechanism responsible is now known to be the suppression of IL-2 synthesis after binding of measles virus to CD46 on macrophages.

Some of the microorganisms that suppress immunity act by infecting lymphocytes. Infectious mononucleosis or glandular fever is caused by a virus (Epstein–Barr virus) that infects B lymphocytes. The infection activates cytotoxic CD8 T cells, which destroy the B cells in which the Epstein–Barr virus is replicating. In the third week of infection, at the height of activation of CD8 T cells, all adaptive immunity is suppressed. The cytokines responsible for the immunosuppression are not well defined but probably include IL-10 and TGF-β (see Case 45).

Topics bearing on this case:
Failure of cell-mediated immunity
Infection with the human immunodeficiency virus (HIV)
Control of HIV infection
Drug therapy for HIV infection
ELISA test

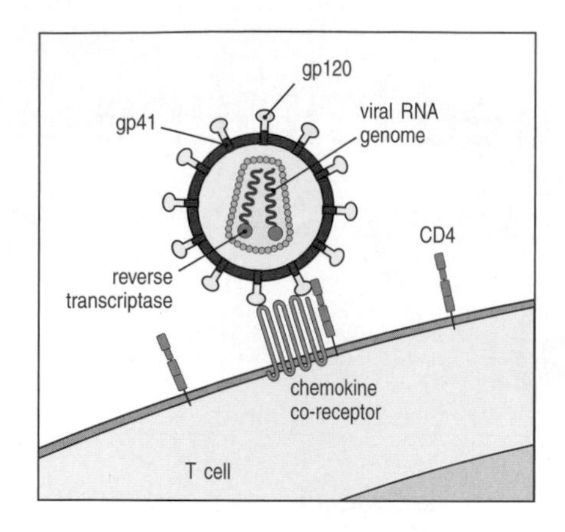

Fig. 10.1 HIV binds to CD4 T cells through its coat glycoprotein gp120. The gp120 molecule on the surface of the virus binds CD4 on T cells and macrophages; the viral protein gp41 then mediates fusion of the enveloped virus with the target cells, allowing the viral genome to enter the cell. The chemokine receptors CCR5 and CXCR4 act as co-receptors of HIV.

The human immunodeficiency virus (HIV) presents a chilling example of the consequences of infection and destruction of immune cells by a microorganism. CD4 molecules on the T-cell surface act as the receptors for HIV (Fig. 10.1). CD4 is also expressed on the surface of cells of the macrophage lineage and they, too, can be infected by this virus. The chemokine receptors CCR5 and CXCR4 act as obligatory co-receptors for HIV. As we shall see, the primary infection with HIV may go unnoticed, and the virus may replicate in the host for many years before symptoms of immunodeficiency are seen. During this period of clinical latency, the level of virus in the blood and the number of circulating CD4 cells remain fairly steady, but in fact both virus particles and CD4 cells are being rapidly destroyed and replenished, as rounds of virus replication take place in newly infected cells. When the rate at which CD4 cells are being destroyed exceeds the capacity of the host to replenish them, their number decreases to a point at which cell-mediated immunity falters. As we have seen in other cases, such as severe combined immunodeficiency (see Case 5), the failure of cell-mediated immunity renders the host susceptible to fatal opportunistic infections.

The case of Martin Thomas: a police officer whose past comes back to haunt him.

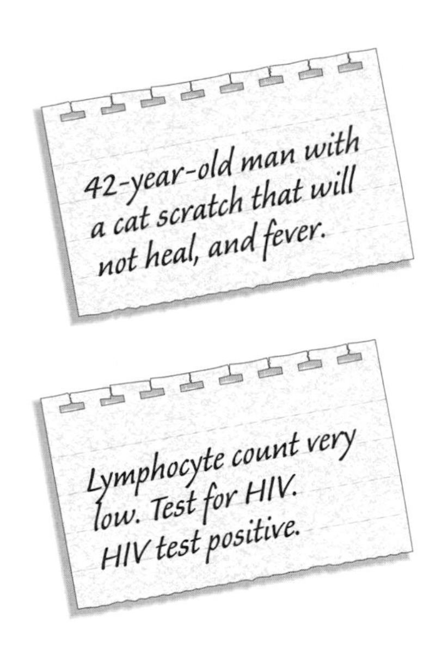

42-year-old man with a cat scratch that will not heal, and fever.
Lymphocyte count very low. Test for HIV. HIV test positive.

Martin Thomas is a 42-year-old African-American police officer who has always been in good health. He has been married for 10 years and has one child, an 8-year-old daughter. Six months ago he went to the emergency room at the local hospital complaining of a fever and a swollen right hand. He was admitted to hospital for the hand infection, which was assumed to be the result of a cat scratch. His blood lymphocyte count was found to be very low, so a blood sample was sent to be tested for antibodies against the human immunodeficiency virus (HIV). Both an ELISA (enzyme-linked immunosorbent assay) (Fig. 10.2) and a Western blot (Fig. 10.3) revealed the presence of anti-HIV antibodies. Officer Thomas was referred to Dr Wright, an AIDS specialist, at the Massachusetts General Hospital.

Martin Thomas told Dr Wright that he had had several homosexual encounters before his marriage 10 years ago. He had always been in good health until 6 months before the present consultation, when he began to have drenching night sweats several times a week. Over this period his body weight had gone down from 94.5 kg to 90 kg. He could not remember having any infections other than the one in his hand, nor any rashes, gastrointestinal problems, cough, shortness of breath, or any other symptoms. His mother had been 84 years old when she died of a heart attack, and his father had died at age 87 from cirrhosis of the liver, cause unknown. His wife and child were both in good health and his wife had recently tested negative for anti-HIV antibodies. Mr Thomas told Dr Wright that he did not smoke or use intravenous drugs. He drank large amounts of beer at weekends. A cat and a dog were the only pets in the house.

On physical examination his blood pressure was 130/90, his pulse rate 92, and temperature 37.5°C (all normal). Nothing abnormal was found during the physical examination. His white blood cell count was 5800 μl^{-1} (normal), his hematocrit was 31.3, and his platelet count was 278,000 μl^{-1} (both normal). His CD4 T-cell count was very low at 170 μl^{-1} (normal 500–1500 μl^{-1}) and his load of HIV-1 RNA was 67,000 copies ml^{-1}.

Mr Thomas was prescribed trimethoprim sulfamethoxazole for prophylaxis against *Pneumocystis jirovecii* pneumonia (see Case 5). He was also given a combination antiretroviral therapy consisting of zidovudine (Retrovir, AZT), lamivudine (Epivir, 3TC), and efavirenz (Sustiva). He was counseled about safe sex with his wife.

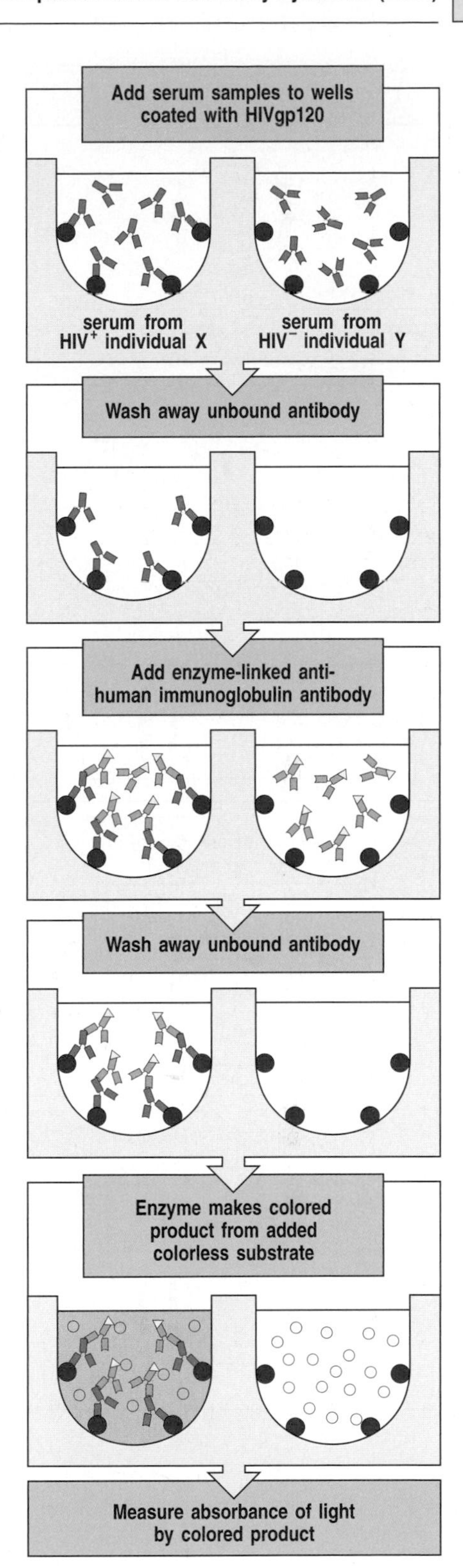

Add serum samples to wells coated with HIVgp120
serum from HIV+ individual X
serum from HIV− individual Y
Wash away unbound antibody
Add enzyme-linked anti-human immunoglobulin antibody
Wash away unbound antibody
Enzyme makes colored product from added colorless substrate
Measure absorbance of light by colored product

Fig. 10.2 Use of the enzyme-linked immunosorbent assay (ELISA) to detect the presence of antibodies against the HIV coat protein gp120. Purified recombinant gp120 is coated onto the surface of plastic wells to which the protein binds nonspecifically; residual sticky sites on the plastic are blocked by adding irrelevant proteins (not shown). Serum samples from the individuals being tested are then added to the wells under conditions where nonspecific binding is prevented, so that only binding to gp120 causes antibodies to be retained on the surface. Unbound antibody is removed from all wells by washing, and anti-human immunoglobulin that has been chemically linked to an enzyme is added, again under conditions that favor specific binding alone. After further washing, the colorless substrate of the enzyme is added, and colored material is deposited in the wells in which the enzyme-linked anti-human immunoglobulin is found. This assay allows arrays of wells known as microtiter plates to be read in fiberoptic multichannel spectrometers, greatly speeding the assay.

After 5 weeks of this therapy his HIV-1 viral load declined to 400 copies of RNA ml^{-1} and after 8 weeks to <50 copies of RNA ml^{-1}, in other words to undetectable levels. In the meantime his CD4 T-cell count rose to 416 μl^{-1}. The prophylaxis for *Pneumocystis* was discontinued. Mr Thomas remains well and active and works full time.

Acquired immune deficiency syndrome (AIDS).

AIDS is caused by the human immunodeficiency virus (HIV), of which there are two known types, HIV-1 and HIV-2. HIV-2 was largely confined to West Africa but now seems to be spreading into Southeast Asia. HIV infections in North and South America and in Europe are exclusively from HIV-1. HIV can be transmitted by homosexual and heterosexual intercourse, by infusion of contaminated blood or blood products, or by contaminated needles, which are the major source of infection among drug addicts. The infection can also be passed from mother to child during pregnancy, during delivery or, more uncommonly, by breastfeeding. In the past, between 25% and 35% of infants born to HIV-positive mothers were infected, but the rate of vertical transmission in industrialized countries has more recently dropped to 3–10% by giving HIV-positive pregnant women antiretroviral drugs such as zidovudine.

Contact with the virus does not necessarily result in infection. The standard indicator of infection is the presence of antibodies against the virus coat protein gp120. The initial infection, as in Mr Thomas's case, may pass unnoticed and without symptoms. More often, a mild viral illness within 6 weeks of infection is sustained, with fever, swollen lymph nodes, and a rash. It subsides at about the time that seroconversion (the appearance of anti-HIV antibodies) occurs, and although virus and antibody persist, the patient feels well. A period of clinical latency lasting years, and perhaps even decades, may ensue during which the infected person feels perfectly well. Then they begin to experience low-grade fever and night sweats, excessive fatigue, and perhaps candidiasis (thrush) in the mouth. Lymph nodes in the neck or axillae (armpits) or groin may swell. Weight loss may become very marked. These are the prodromal symptoms of impending AIDS. (A prodrome is a concatenation of signs and symptoms that predict the onset of a syndrome.) The number of CD4 T cells in the blood may have been normal up to this time but, with the onset of the prodrome, the CD4 T-cell count begins to fall (Fig. 10.4). When the number of CD4 T cells decreases to the range of 200–400 cells μl^{-1}, the final phase of the illness, which is called AIDS, starts. At this time serious, eventually fatal, opportunistic infections as well as certain unusual malignancies occur (Fig. 10.5).

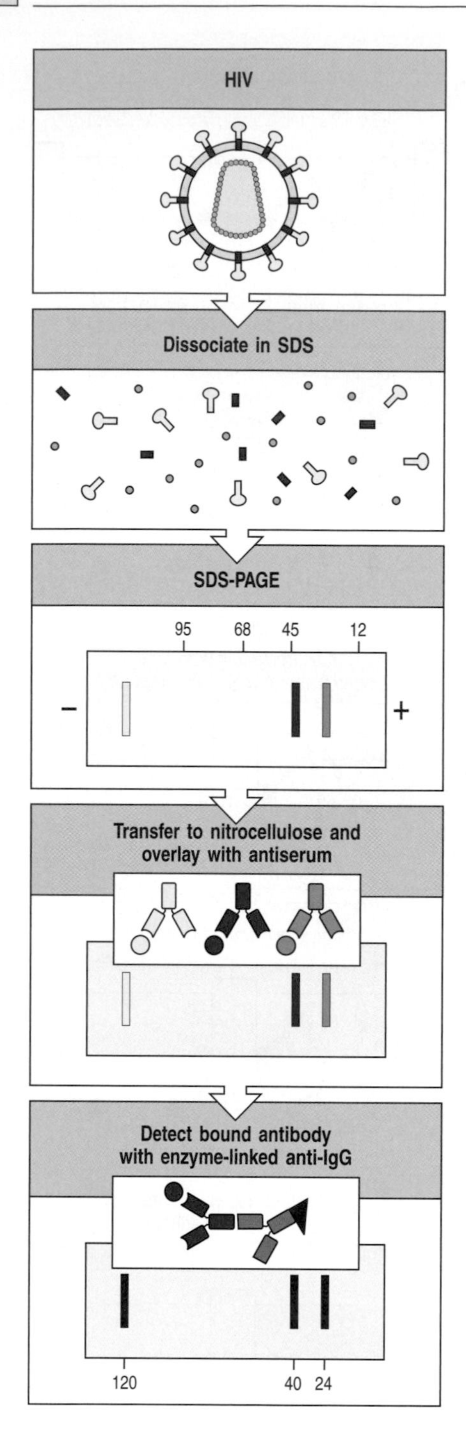

Fig. 10.3 Western blotting is used to identify antibodies against the human immunodeficiency virus (HIV) in serum from infected individuals. The virus is dissociated into its constituent proteins by treatment with the detergent SDS, and its proteins are separated by SDS-PAGE. The separated proteins are transferred to a nitrocellulose sheet and reacted with the test serum. Anti-HIV antibodies in the serum bind to the various HIV proteins and are detected by using enzyme-linked anti-human immunoglobulin, which deposits colored material from a colorless substrate. This general methodology will detect any combination of antibody and antigen and is used widely, although the denaturing effect of SDS means that the technique works most reliably with antibodies that recognize the antigen when it is denatured.

At any time after the infection, HIV may infect megakaryocytes, which have some surface CD4. Because megakaryocytes are the bone marrow progenitors of blood platelets, extensive infection of megakaryocytes causes the platelet count to fall (thrombocytopenia) and bleeding to occur. HIV may also infect the glial cells of the brain. Glial cells are of the monocyte–macrophage lineage and have some CD4 on their surface. The infection of glial cells may cause dementia and other neurological symptoms.

Questions.

1 *When Mr Thomas was first seen by Dr Wright, a chest X-ray revealed some enlarged hilar lymph nodes. If a lymph-node biopsy had been obtained, in what way would its histopathology have differed from that of lymph nodes from patients with severe combined immunodeficiency (SCID; see Case 5) or X-linked agammaglobulinemia (XLA; see Case 1)?*

2 *The course of an HIV infection in adults is very different from that in an infant infected* in utero *or intrapartum (during birth). What are the major differences between pediatric and adult AIDS and how do you account for them?*

3 *What are the mechanisms of resistance to the progression of HIV infection?*

4 *A few individuals, mostly hemophiliacs, are known to have been infected with HIV as long as 20 years ago and yet they remain asymptomatic. What factors may contribute to long-term survival with this infection?*

5 *Dr Wright told Mr Thomas to take zidovudine, lamivudine, and efavirenz. What are these drugs?*

6 *What do HIV protease inhibitors do? Does Mr Thomas need one of these drugs?*

7 *What is the mechanism of CD4 T-cell depletion in HIV infection?*

8 *What is the most important known determinant of the progression of HIV infection?*

9 *Which cytokine released during an HIV infection causes weight loss?*

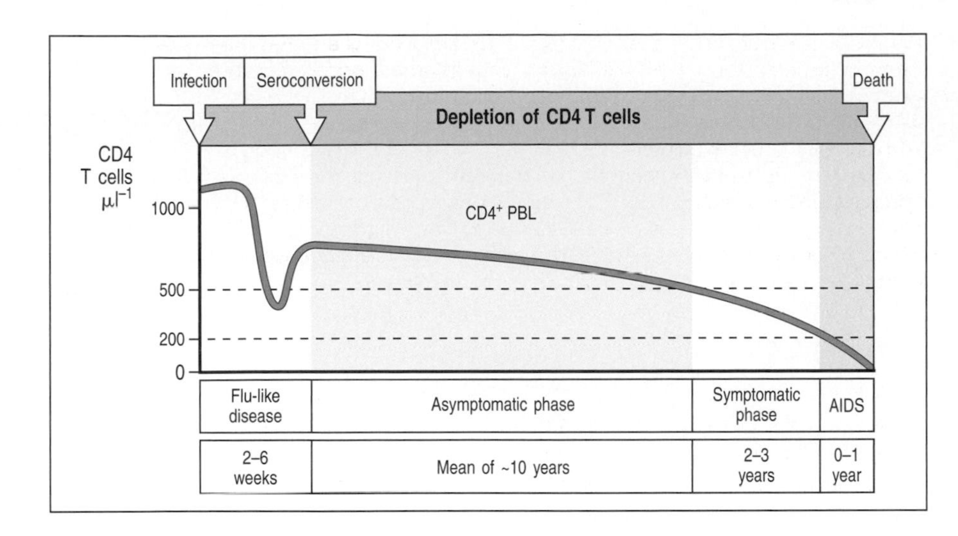

Fig. 10.4 The typical course of infection with HIV. The first few weeks are typified by an acute influenza-like viral illness, with high titers of virus in the blood. An adaptive immune response follows, which controls the acute illness and largely restores CD4 T cell levels but does not eradicate the virus. Opportunistic infections and other symptoms become more frequent as the CD4 T-cell count falls, starting at around 500 cells ml^{-1}. The disease then enters the symptomatic phase.

Infections	
Parasites	*Toxoplasma* spp. *Cryptosporidium* spp. *Leishmania* spp. *Microsporidium* spp.
Bacteria	*Mycobacterium tuberculosis* *Mycobacterium avium intracellulare* *Salmonella* spp.
Fungi	*Pneumocystis jirovecii* *Cryptococcus neoformans* *Candida* spp. *Histoplasma capsulatum* *Coccidioides immitis*
Viruses	Herpes simplex Cytomegalovirus Herpes zoster
Malignancies	
Kaposi's sarcoma (invasive) Non-Hodgkin's lymphoma, including EBV-positive Burkitt's lymphoma Primary lymphoma of the brain	

Fig. 10.5 A variety of opportunistic pathogens and cancers can kill AIDS patients. Infections are the major cause of death in AIDS; of these, respiratory infection with *Pneumocystis jirovecii* is the most prominent. Most of these pathogens require effective macrophage activation by CD4 T cells or effective cytotoxic T cells for host defense. Opportunistic pathogens are present in the normal environment but cause severe disease primarily in immunocompromised hosts, such as AIDS patients and cancer patients. AIDS patients are also susceptible to several rare cancers, such as Kaposi sarcoma and lymphomas, suggesting that immune surveillance by T cells may normally prevent such tumors. EBV, Epstein–Barr virus.

Fig. 10.1 The typical course of infection with HIV.

CASE 11 Graft-Versus-Host Disease

Alien T cells react against their new host.

Bone marrow transplantation has proved to be useful therapy for some forms of leukemia, bone marrow failure (aplastic anemia), and primary immunodeficiency diseases. More recently, other sources of hematopoietic stem cells, such as peripheral blood stem cells and cord blood, have also been used for these purposes. Bone marrow and most other sources of hematopoietic stem cells contain mature T lymphocytes, which may recognize the tissues of their new host as foreign and cause a severe inflammatory disease in the recipient. This is known as graft-versus-host disease (GVHD) and is characterized by a rash, which often starts on the face (Fig. 11.1), diarrhea, pneumonitis (inflammation in the lung), and liver damage.

To achieve successful engraftment of bone marrow and avoid rejection of the transplant by the host, the immune system of the recipient must be destroyed and the recipient rendered immunoincompetent. This is usually accomplished with lethal doses of radiation or the injection of radiomimetic drugs such as busulfan, and the use of immunosuppressive drugs (cyclophosphamide, fludarabine). In children with severe combined immunodeficiency, who cannot produce T lymphocytes (see Case 5), this preparative treatment is not needed.

GVHD occurs not only when there is a mismatch of classical MHC class I or class II molecules but also in the context of disparities in minor histocompatibility antigens; such minor differences are likely to be present in all donor–recipient pairs other than identical twins, even when HLA-matched. Mature

Topics bearing on this case:
T-cell recognition of nonself MHC molecules
Reactions of T cells against foreign antigens
Transplantation
Minor histocompatibility antigens
Mixed lymphocyte reaction

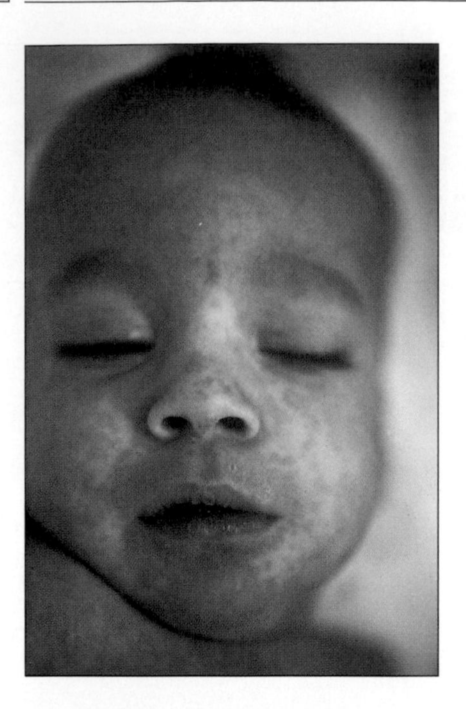

Fig. 11.1 The rash characteristic of GVHD often starts on the face.

CD4 T cells in the graft that are activated by allogeneic molecules produce a 'cytokine storm' that recruits other T cells, macrophages, and natural killer (NK) cells to create the inflammation characteristic of GVHD. Although B cells may also be present in GVHD inflammation, they do not have a significant role in causing or sustaining GVHD.

GVHD is arbitrarily called 'acute' if it occurs less than 100 days after the transplant, and 'chronic' if it develops after 100 days. Chronic GVHD differs from acute GVHD in other respects and is a more severe and difficult-to-treat problem. The presence of alloreactive T cells in the donor bone marrow is usually detected in routine laboratory testing by the mixed lymphocyte reaction (MLR) (see Fig. 6.3), in which lymphocytes from the potential donor are mixed with irradiated lymphocytes from the potential recipient. If the donor lymphocytes contain alloreactive T cells, these will respond by cell division. Although the MLR is routinely used for the selection of donors it does not accurately quantify alloreactive T cells. Although the limiting-dilution assay more precisely counts the frequency of alloreactive T cells, it is too cumbersome for routine clinical use.

The case of John W. Wells: a curative therapy becomes a problem.

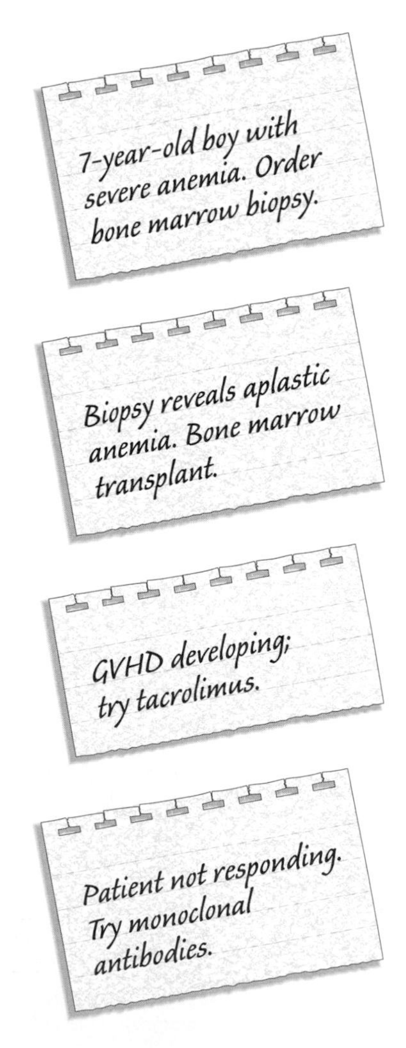

7-year-old boy with severe anemia. Order bone marrow biopsy.
Biopsy reveals aplastic anemia. Bone marrow transplant.
GVHD developing; try tacrolimus.
Patient not responding. Try monoclonal antibodies.

John was healthy until he was 7 years old, when his mother noticed that he had become very pale. She also noticed small hemorrhages (petechiae) on the skin of his arms and legs and took John to the pediatrician. Apart from the pallor and skin petechiae, a physical examination showed nothing unusual. The pediatrician ordered blood tests, which revealed that John was indeed very anemic. His hemoglobin was 7 g dl^{-1} (normal 10–15 g dl^{-1}) and platelet count was 20,000 μl^{-1} (normal 150,000–300,000 μl^{-1}). His white blood cell count was also lower than normal. The pediatrician sent John to a hematology consultant for a bone marrow biopsy.

The biopsy showed that John's bone marrow had very few cells and that red cell, platelet, and white cell precursors were almost completely absent. Aplastic anemia (bone marrow failure) of unknown cause was diagnosed. Aplastic anemia is ultimately fatal but can be cured by a successful bone marrow transplant. Fortunately, John had an HLA-identical 11-year-old brother who could be the bone marrow donor. John was admitted to the Children's Hospital and given a course of fludarabine and cyclophosphamide to eradicate his own lymphocytes. He was then given 2×10^8 nucleated bone marrow cells per kg body weight obtained from his brother's iliac crests. He was also started on cyclosporin A (CsA) to prevent GVHD.

John did well for 3 weeks after the bone marrow transplant and was then sent home to recover. However, in spite of GVHD prophylaxis with CsA, on the 24th day after the transplant he was readmitted to hospital with a skin rash and watery diarrhea consistent with acute GVHD. On admission he had a patchy red rash on palms and soles, scalp, and neck. He had no fever and was not jaundiced. His lungs were clear and the heartbeat normal. The liver and spleen were not enlarged.

John was treated with corticosteroids. His skin rash faded, but the intestinal symptoms did not abate and the diarrhea became more profuse. He was given rabbit antithymocyte globulin (ATG) for two consecutive days. This brought about a 90% decrease in the volume of his stool and the intestinal bleeding stopped. Two weeks later, John was sent home, with continuing treatment with low doses of corticosteroid, and his GVHD remained under control.

Graft-versus-host disease (GVHD).

Graft-versus-host disease was first described more than 30 years ago by Billingham, Brent, and Medawar, who gave allogeneic lymphocytes to newborn mice. The mice became runted (their growth was retarded), lymphoid tissue was destroyed, and they developed diarrhea and necrosis of the liver. GVHD was first recognized in human infants with severe combined immunodeficiency disease who inadvertently received allogeneic lymphocytes contained in a blood transfusion (Fig. 11.2). For a recipient to develop GVHD, the graft must contain immunocompetent cells, the recipient must express major or minor histocompatibility molecules that are lacking in the graft donor, and the recipient must be incapable of rejecting the graft.

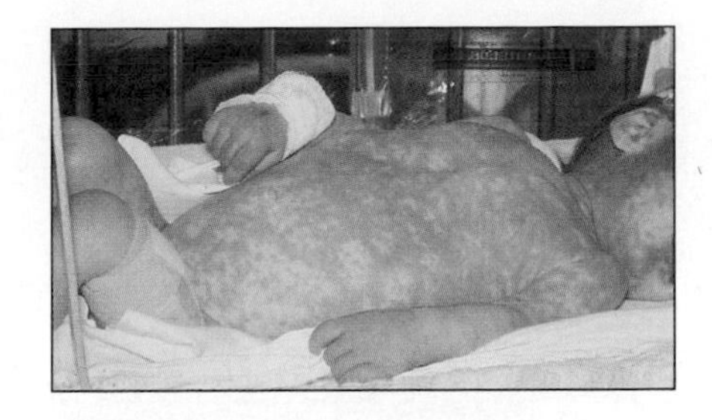

Fig. 11.2 GVHD on trunk and limbs of affected infant.

The first clinical manifestation of GVHD is a bright red rash that characteristically involves the palms and soles. The rash usually begins on the face and neck and progresses to involve the trunk and limbs, particularly the palms and soles. The rash may itch a great deal and its onset may be accompanied by fever. After the skin manifestations appear (Fig. 11.3), the gastrointestinal tract becomes involved (Fig. 11.4) and profuse watery diarrhea is produced.

Liver function tests may become abnormal and reveal destruction of hepatic tissue (Fig. 11.5). Eventually, other tissues such as the lungs and bone marrow become sites of GVHD inflammation. The only satisfactory therapy at present for GVHD is elimination of the T cells that initiate the reaction, either by immunosuppressive drugs or, as in John's case, by T-cell-depleting agents.

Hematopoietic cell transplantation from a haploidentical (that is, half-matched) donor (typically, one of the two parents) carries a very high risk of GVHD. For this reason, the bone marrow from haploidentical donors is manipulated to deplete the mature T lymphocytes (or to purify the stem cells only) before attempting the transplant. Transplantation from Matched Unrelated Donors (MUDs) has also become current practice at many centers and there is also a significant risk of GVHD with this type of transplant. Some bone marrow transplant centers use T-cell depletion also for transplantation from MUDs, but this approach is not followed by the majority of centers. Transplantation with unmanipulated bone marrow from MUDs is associated with a higher rate of engraftment and may provide a 'graft-versus-leukemia' reaction, in which donor-derived mature T cells may kill the residual leukemic cells of the recipient, making relapse less likely in the case of bone marrow transplants for the treatment of leukemia.

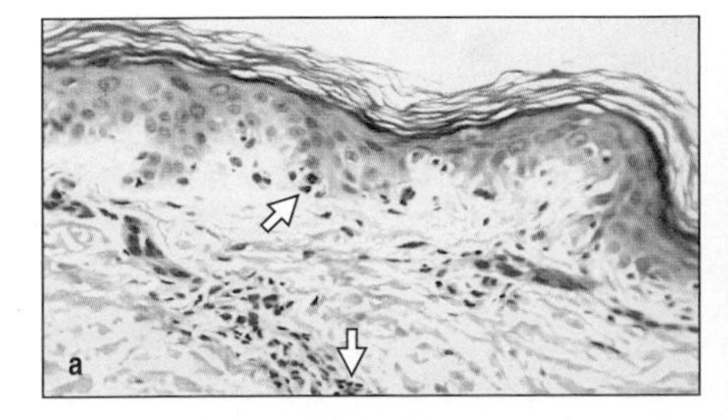

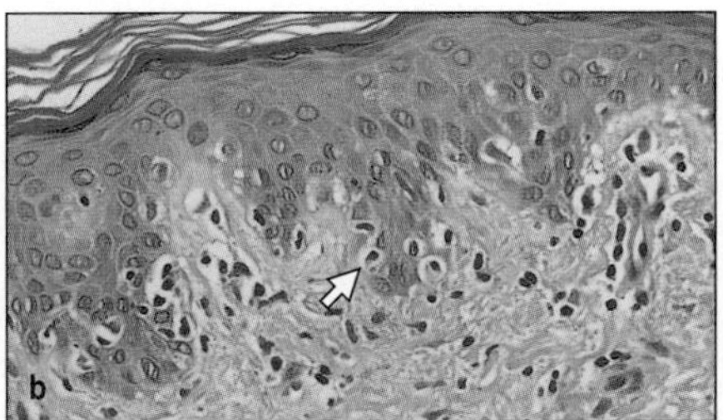

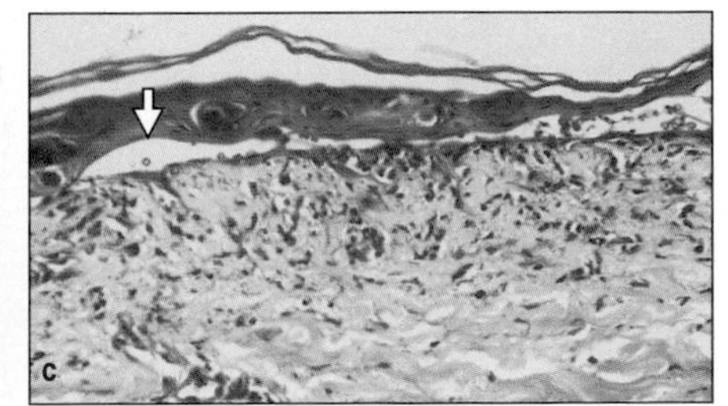

Fig. 11.3 GVHD in the skin. Panel a: early GVHD in the skin. Lymphocytes are emerging from blood vessels (lower arrow) and adhering to the basal layer of the epidermis (upper arrow). Panel b: the basal cells of the epidermis begin to swell and vacuolate. Their nuclei become condensed (dark staining) as these cells die (arrow). Panel c: advanced destruction of the skin by GVHD, with sloughing of the epidermis (arrow). Photographs kindly provided by Robert Sackstein.

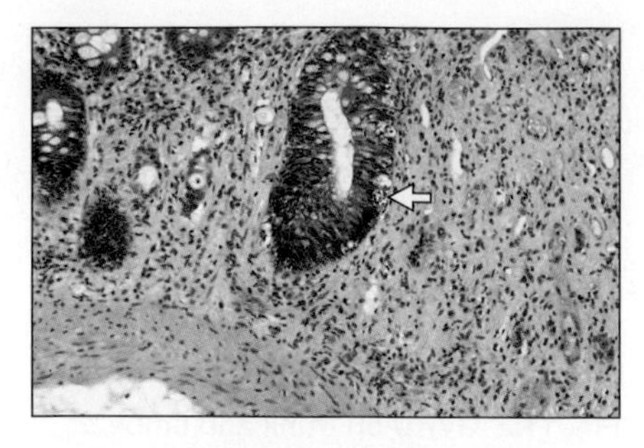

Fig. 11.4 GVHD in the colon. Inflammatory cells have invaded the crypts of the intestine and destroyed the normal architecture (arrow). Photograph kindly provided by Mark Shlomchik.

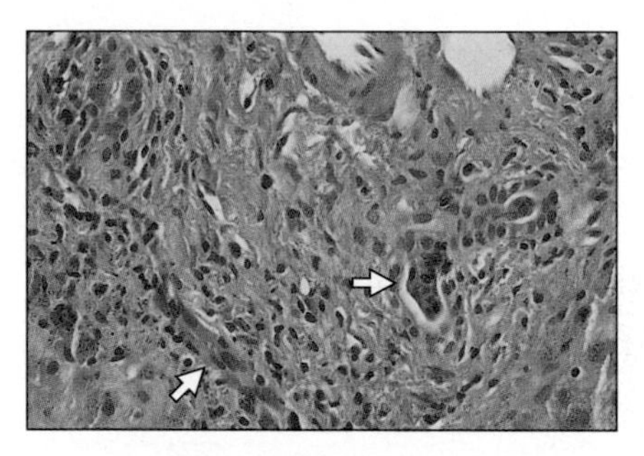

Fig. 11.5 Liver damage as a result of GVHD. Inflammatory cells have invaded the liver and destroyed the hepatic ducts (arrows). Photograph kindly provided by Robert Sackstein.

Questions.

1 *Bone marrow is often depleted of T cells before transplantation, to try and avoid GVHD. However, in the treatment of leukemia by bone marrow transplantation, T cells in the graft can have a beneficial effect. How do you explain this?*

2 *CD4 T cells in the graft that recognize foreign histocompatibility molecules become activated and produce the cytokine interferon (IFN)-γ. This helps sustain and increase GVHD. Why?*

3 *John was given ATG (in addition to steroids) to control acute GVHD. Do you know any other ways to achieve* in vivo *depletion of T lymphocytes?*

4 *Why are the skin and intestinal tract the major sites of GVHD?*

CASE 12 MHC Class I Deficiency

A failure of antigen processing.

The class I molecules encoded by the major histocompatibility complex (MHC) are expressed to a greater or lesser extent on the surface of all the cells of the body except the red blood cells. MHC class I molecules bind peptides derived from proteins synthesized in the cytoplasm, and carry them to the cell surface, where they form a complex of peptide and MHC molecule on the cell surface. This complex can then be recognized by antigen-specific CD8 T cells. T cells as a class recognize only peptides presented to them as a complex with an MHC molecule; the T-cell receptors of CD8 T cells recognize only peptides presented by MHC class I molecules, whereas those of CD4 T cells recognize only peptides presented by MHC class II molecules (see Case 8).

MHC class I molecules are involved principally in immune reactions against virus infections. Cytotoxic CD8 T cells specific for viral antigens terminate viral infections by recognizing viral peptides carried by MHC class I molecules on the surface of virus-infected cells, and then killing these cells (Fig. 12.1). They release the pore-forming protein perforin and cytotoxic granzymes, as well as the inflammatory cytokines tumor necrosis factor α (TNF-α) and lymphotoxin. In addition, cytotoxic T cells express the Fas ligand, which engages the cell-surface molecule Fas on target cells. Both processes induce the target cells to undergo programmed cell death (apoptosis).

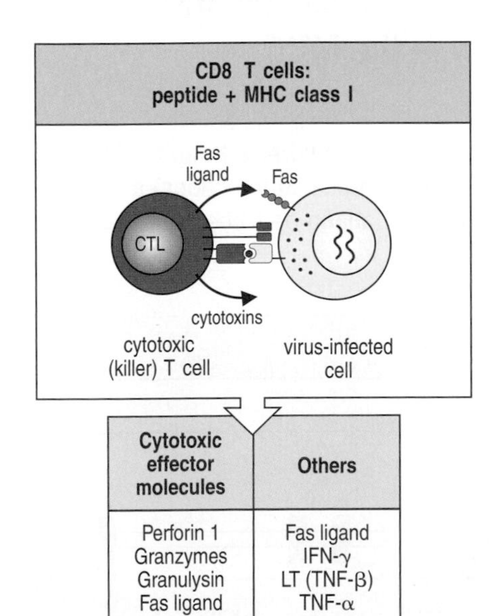

12.1 Cytotoxic CD8 T cells recognize and kill virus-infected cells. CD8 T cells recognize virally derived peptides presented by MHC class I antigens at the surface of virus-infected cells. They kill target cells by releasing cytotoxins and cytotoxic cytokines and by binding to Fas on the target cell. LT, lymphotoxin.

Topics bearing on this case:

- Genetic organization of the MHC
- Processing of intracytoplasmic protein antigens
- Transport of MHC class I molecules to the cell surface
- MHC class I molecules and CD8 T-cell intrathymic maturation
- Effector CD8 T-cell function in virus infection
- FACS analysis

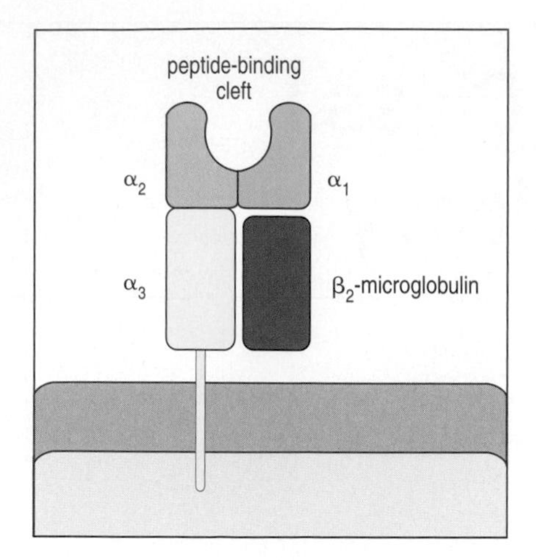

Fig. 12.2 A schematic representation of a human MHC class I molecule. It is a heterodimeric glycoprotein, composed of one transmembrane α chain bound noncovalently to β_2-microglobulin. The α chain is folded into three protein domains, two of which form a cleft into which the peptide antigen binds.

In humans, the class I and class II MHC molecules are known as the HLA antigens, and together they determine the tissue type of an individual. MHC class I molecules are particularly abundant on T and B lymphocytes, and also on macrophages and other cells of the monocyte lineage as well as on neutrophils. Other cells express smaller amounts. Each individual expresses three principal types of class I molecule—HLA-A, HLA-B, and HLA-C. These are heterodimeric glycoproteins, composed of an α chain and a β chain, the latter known also as β_2-microglobulin (Fig. 12.2).

The genes encoding the α chains of the human MHC class I molecules are located close together in the MHC on the short arm of chromosome 6 (Fig. 12.3). In humans, the gene encoding β_2-microglobulin, the polypeptide chain common to all class I molecules, is located not in the MHC but on the long arm of chromosome 15. The genes encoding the α chains are highly polymorphic, and so there are numerous variants of HLA-A, HLA-B, and HLA-C within the population.

Viral proteins, like cellular proteins, are made in the cytoplasm of the infected cell, and some are soon degraded into peptide fragments by large enzyme complexes called proteasomes. The peptides are then transported from the cytosol into the endoplasmic reticulum by a complex of two transporter proteins called TAP1 and TAP2, which is located in the endoplasmic reticulum membrane (Fig. 12.4). The genes encoding TAP1 and TAP2 are also located in the MHC, in the region containing the class II genes (see Fig. 12.3).

The endoplasmic reticulum contains MHC class I molecules, which enter as separate α and β chains as soon as they have been synthesized and are retained there. After the antigenic peptides enter the endoplasmic reticulum they are loaded onto the complex of α chains and β_2-microglobulin. The TAP-binding protein (also known as TAPBP or tapasin) facilitates the interaction of MHC class I molecules with TAP1 and TAP2, and promotes loading of antigenic peptides into this complex. The peptide:MHC class I complex is then released and transported onward to the cell surface (Fig. 12.5). In humans, the *TAPBP* gene is also located within the MHC cluster on the short arm of chromosome 6.

This case describes a rare inherited immune deficiency accompanied by the absence of MHC class I molecules on the patients' cells.

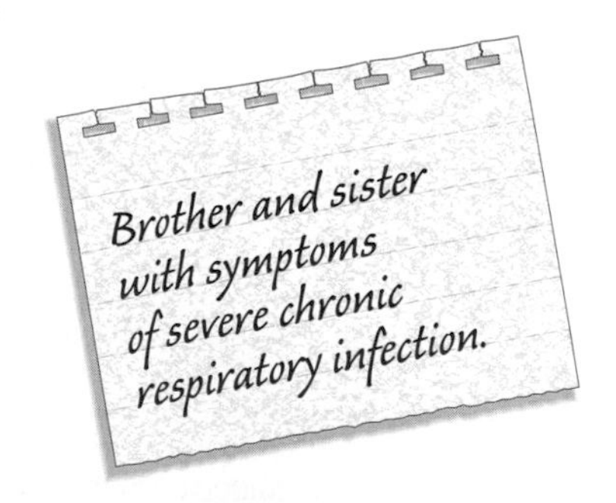

Brother and sister with symptoms of severe chronic respiratory infection.

The children of Sergei and Natasha Islayev: the consequences of a small flaw in the MHC.

Tatiana Islayev was 17 years old when first seen at the Children's Hospital. She had severe bronchiectasis (dilatation of the bronchi from repeated infections) and a persistent cough that produced yellow-green sputum. Tatiana had been chronically ill

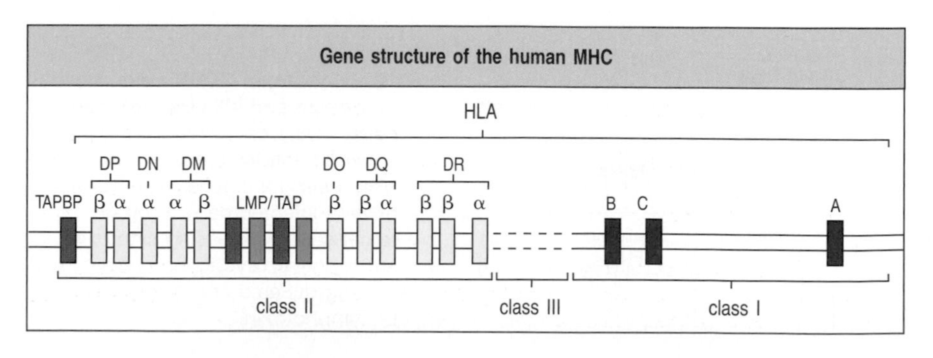

Fig. 12.3 The organization of the major histocompatibility complex (MHC) on chromosome 6 in humans. There are separate regions of class I and class II genes. The class I genes are called HLA-A, HLA-B, and HLA-C. The gene for β_2-microglobulin is located on chromosome 15. The genes for the TAP1:TAP2 transporter are located in the class II region of the MHC.

from the age of 4 years, when she started to get repeated infections of the sinuses, middle ears, and lungs, apparently due to a variety of respiratory viruses. The bacteria *Haemophilus influenzae* and *Streptococcus pneumoniae* could be cultured from her sputum, and she had been prescribed frequent antibiotic treatment to control her persistent fevers and cough. Her brother Alexander, aged 7 years, also suffered from chronic respiratory infections. Like his sister, he had begun to suffer severe repeated viral infections of the upper and lower respiratory tracts at an early age. He also had severe bronchiectasis, and *H. influenzae* could be cultured from his sputum.

Owing to the chronic illness of Tatiana and Alexander, the Islayevs had emigrated recently from Russia to the United States, where they hoped to get better medical treatment. When they came to America they had three other children, aged 5, 10, and 13 years, who were all healthy and showed no increased susceptibility to infection. As infants in Moscow, both Tatiana and Alexander had received routine immunizations with oral poliovirus as well as diphtheria, pertussis, and tetanus (DPT) vaccinations. They had also been given BCG as newborn babies for protection against tuberculosis, and had tolerated all these immunizations well.

When they were examined, Tatiana and Alexander both had elevated levels of IgG, at more than 1500 mg dl^{-1} (normal levels 600–1400 mg dl^{-1}). They had white blood cell counts of 7000 and 6600 cells μl^{-1}, respectively. Of these white cells, 25% (1750 and

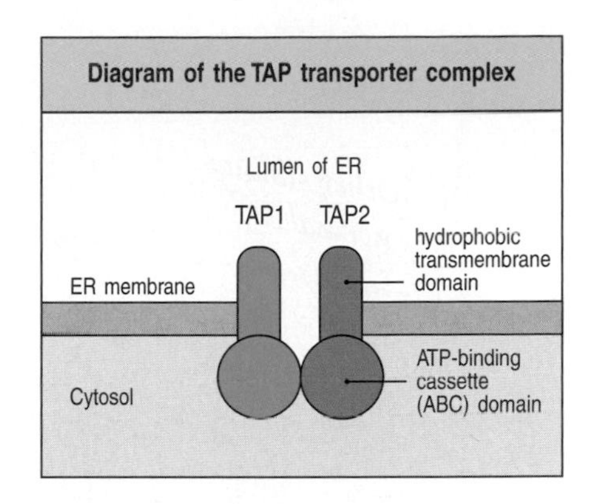

Fig. 12.4 The TAP1 and TAP2 transporter proteins form a heterodimer in the endoplasmic reticulum membrane.

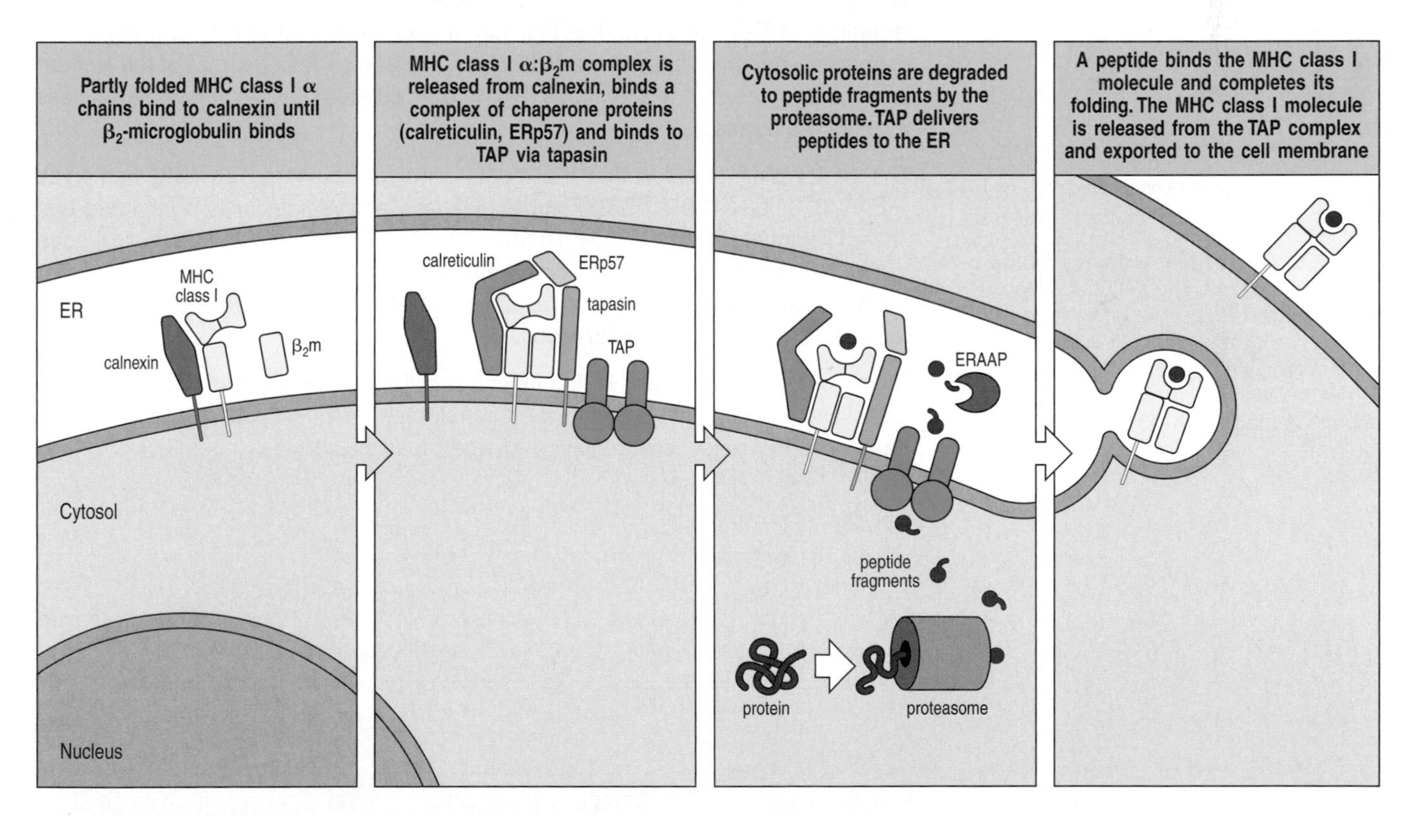

Fig. 12.5 MHC class I molecules do not leave the endoplasmic reticulum unless they bind peptides. MHC class I α chains assemble in the endoplasmic reticulum with the membrane-bound protein, calnexin. When this complex binds β_2-microglobulin (β_2m), the MHC class I α:β_2m dimer is released from calnexin, and the partly folded MHC class I molecule then binds to the TAP1 subunit of the TAP transporter by interacting with one molecule of the TAP-associated protein, tapasin. The chaperone molecules ERp57, which forms a heterodimer with tapasin, and calreticulin also bind to form the MHC class I peptide-loading complex. The MHC class I molecule is retained within the endoplasmic reticulum until it binds a peptide, which completes the folding of the MHC class I molecule. Peptides generated by the degradation of proteins in the cytoplasm are transported into the lumen of the endoplasmic reticulum by the TAP transporter. Peptides too long to bind MHC class I molecules are trimmed by the peptidase ERAAP. Once peptide has bound to the MHC molecule, the peptide:MHC complex is transported through the Golgi complex to the cell surface.

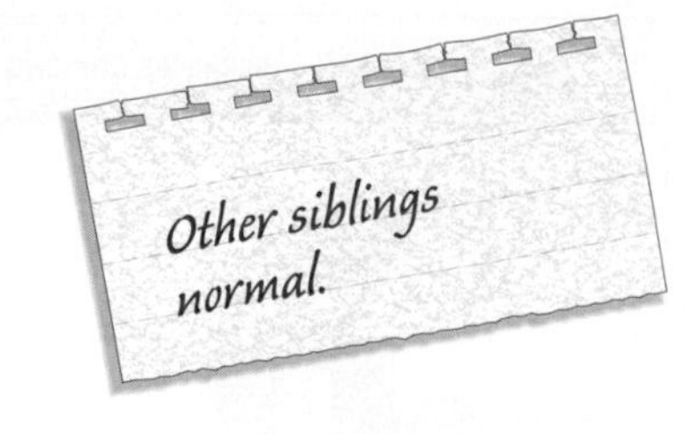

1650 μl^{-1}, respectively) were lymphocytes. Ten per cent of the lymphocytes reacted with an antibody against B cells (anti-CD19) (a normal result) and 4% with an antibody against natural killer (NK) cells (anti-CD16) (normal). The remainder of the lymphocytes reacted with an anti-T-cell antibody (anti-CD3). More than 90% of the T cells were CD4-positive, and 10% were CD8-positive. This represented a profound deficiency of CD8 T cells. Blood tests on their siblings and parents showed no deficiency of CD8 T cells. Both Tatiana and Alexander had normal neutrophil function and complement titers.

Furthermore, their cell-mediated immunity also seemed normal when tested by delayed hypersensitivity skin tests to tuberculin and antigen from *Candida*, a fungal component of the normal body flora (see Case 53); they developed the normal delayed-type hypersensitivity response of a hard, raised, red swelling some 50 mm in diameter at the site of intradermal injection of these antigens. Both children were found to have high titers of antibodies against herpesvirus and cytomegalovirus as well as against mumps, chickenpox, and measles viruses. When asked, the parents recalled that the children had been immunized against influenza several times, and antibodies against five different strains of influenza were found. However, the anti-influenza antibodies were present in very low titers. They also had low titers of antibody against Epstein–Barr virus.

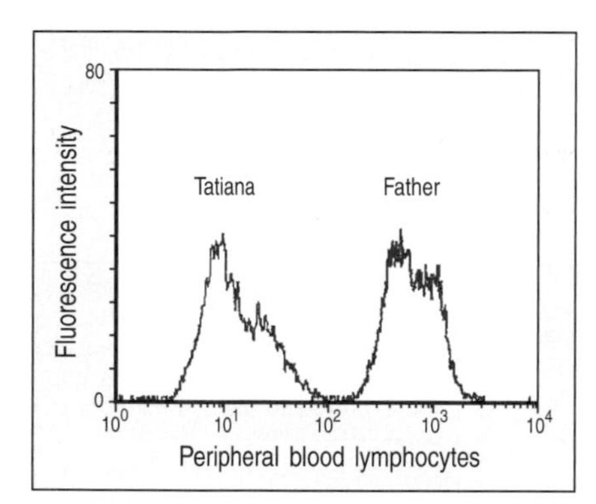

Fig. 12.6 Fluorescent antibody typing for HLA antigens. The peripheral blood lymphocytes of Tatiana and her father were typed by a fluorescent antibody against the MHC class I molecule HLA-A3. Tatiana's lymphocytes express about 1% of the HLA-A3 expressed on the lymphocytes of her father.

When white blood cells from all family members were typed for HLA antigens by standard typing procedures, no MHC class I molecules at all could be found on Tatiana's and Alexander's cells. When their blood cells were examined using the more sensitive technique of FACS analysis (see Fig. 1.3), it was found that Tatiana and Alexander expressed very small amounts of MHC class I molecules, less than 1% of the amount expressed on the cells of their father (Fig. 12.6). In contrast, they expressed MHC class II molecules normally.

The HLA typing revealed that the mother and father shared an MHC haplotype (HLA-A3, -B63, HLA-DR4, -DQ3) and that Tatiana and Alexander had inherited this shared haplotype from both their parents (Fig. 12.7). They were therefore homozygous for the MHC region. The other children were heterozygous. It was thus concluded that the two children's susceptibility to respiratory infections was linked to the MHC locus, for which only they were homozygous.

To try to determine the underlying defect, B cells from Tatiana and Alexander were established as cell lines in culture by transformation with Epstein–Barr virus. The transformed B cells were examined for the presence of messenger RNA for MHC

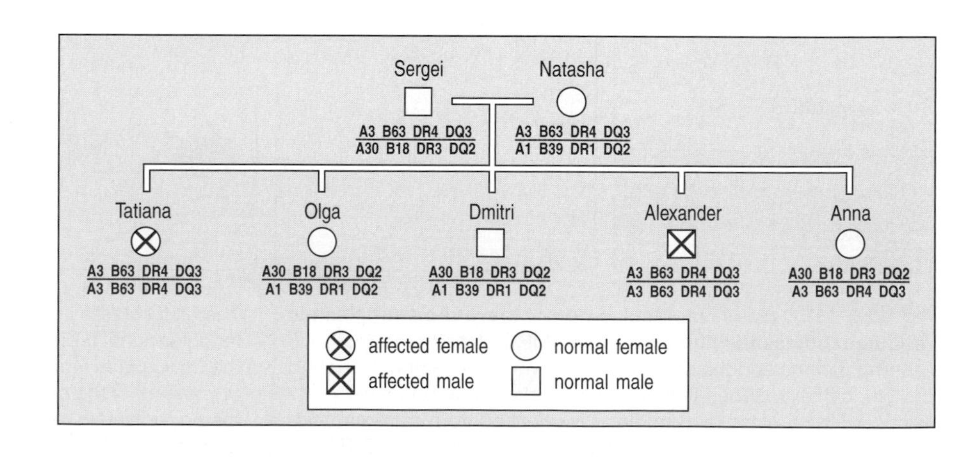

Fig. 12.7 Inheritance of MHC haplotypes in the Islayev family.

class I molecules: normal levels were found. This eliminated the possibility that they had a structural or regulatory defect in genes encoding the α chain of the MHC class I molecules. Because the gene for β_2-microglobulin maps to chromosome 15, it was highly unlikely that their MHC-linked condition resulted from a defect in that gene. Only one possibility remained to be explored—a mutation in the *TAP1*, *TAP2*, or *TAPBP* genes. When the DNA sequences of these genes were determined, both Tatiana and Alexander were found to have the same nonsense mutation in their *TAP2* genes. Their parents were found to be heterozygous for this mutation.

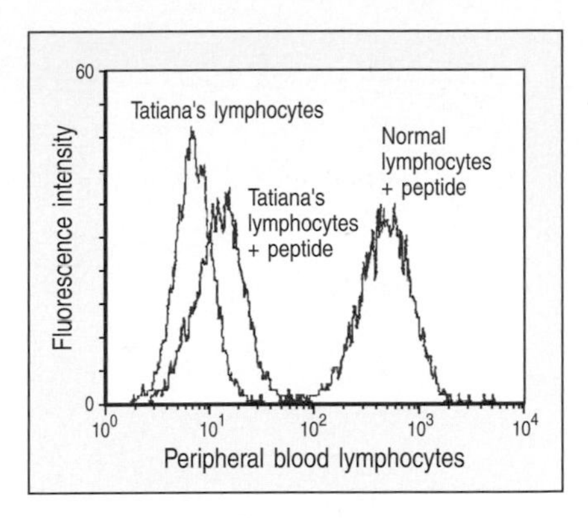

Fig. 12.8 Increased expression of MHC class I molecules in response to virus infection. Peripheral blood lymphocytes from Tatiana were examined with a fluorescent antibody against HLA-A3 before and after loading the cells with an antigenic peptide from influenza virus. There is a small increase in the number of MHC class I molecules expressed on the lymphocytes to approximately 5% of the amount expressed by normal cells loaded with the same peptide.

MHC class I deficiency.

A recessively inherited immunodeficiency was first suggested by the similarity of Tatiana's and Alexander's condition and its complete absence from other members of the family.

The clinical phenotype observed in the Islayev children and associated with MHC class I deficiency has some unexpected features. Despite a profound deficiency in the number of CD8 T cells and the inability to present viral antigens to CD8 T cells because of the absence of MHC class I antigens, the Islayev children were apparently capable to fight some viral infections. The high levels of antibodies against chickenpox, measles, and mumps viruses in their blood showed that they had been exposed to and successfully overcome these infections.

They had sustained innumerable respiratory viral infections, however. And their poor antibody responses to a variety of influenza strains showed that they might have had problems responding to respiratory viruses in general. It might have been that some viruses were better able than others to stimulate an increased expression of MHC class I molecules on their cells. Tatiana and Alexander expressed MHC class I molecules at very low levels, and when their isolated B cells were loaded with an antigenic peptide from influenza virus, this stimulated a small increase in the number of MHC class I molecules on these cells (Fig. 12.8). It is therefore possible that other virus infections that they sustained, such as chickenpox, were able to induce sufficient expression of MHC class I molecules to terminate the infection properly.

The repeated respiratory infections caused anatomic damage to their airways, resulting in the bronchiectasis. The abundant *Haemophilus* and pneumococci in their sputum is characteristic of patients with bronchiectasis, and in their case was not due directly to any deficiency of immunity against these capsulated bacteria (compare with Case 1). Patients with MHC class I deficiency often suffer from midline granulomatous disease, resembling Wegener's granulomatosis. Vasculitis is also common, especially on the extremities.

The profound reduction of CD8 T cells in patients with MHC class I deficiency is a direct consequence of the lack of MHC class I molecules on the surface of epithelial and dendritic cells in the thymus, the organ in which all T cells bearing α:β T-cell receptors mature. The interaction of thymocytes with MHC class I molecules expressed by thymic epithelial and dendritic cells is crucial for the intrathymic maturation of CD8 T cells.

This family reveals that we have much to learn about the role of MHC class I molecules in protection against intracellular pathogens.

Questions.

1 *Why was the number of CD8 T cells in Tatiana and Alexander decreased despite normal levels of CD4 T cells?*

2 *When the T-cell antigen receptors on their CD8 T cells were examined, it turned out that they were all γ:δ receptors and none were α:β receptors. How do you explain this?*

3 *Tatiana and Alexander had normal delayed-type hypersensitivity responses to tuberculin and* Candida. *Is this surprising in view of their deficiency in CD8 T cells?*

4 *Why did Tatiana and Alexander have high levels of serum IgG?*

5 *Genetic defects have also been found in the genes encoding TAP1 and TAPBP. Do you think that the clinical course in these patients would differ from that observed in Tatiana and Alexander?*

CASE 13 X-linked Lymphoproliferative Syndrome

A defect in the immune response to a virus.

Viruses pose a constant challenge to our immune system. Unable to reproduce on their own, they have evolved as parasites, capable of residing within living cells whose biosynthetic machinery they subvert for their own reproduction. An effective two-pronged immune response to these hidden invaders is to kill the host cells within which they reside by means of the cytotoxic cells of the immune system, and to reduce the number of extracellular virus particles by means of antibodies.

Both innate and adaptive immune responses control viral infections. The natural killer (NK) cells of innate immunity (Fig. 13.1) are constantly on surveillance for cells with telltale markers of viral infection. NK cells are large granular cells of the lymphocyte lineage, which differ from cytotoxic T lymphocytes in not expressing antigen-specific receptors. Instead they carry a number of receptors that recognize virus-infected cells in other ways and, like cytotoxic T cells, release cytotoxic proteins that induce apoptosis and death of the virus-infected cell (see also Case 14). These proteins include perforin and granzyme, which are also released by cytotoxic CD8 T cells. This response does not require any previous immunological experience with the virus and is particularly important when an individual first encounters a virus.

In the adaptive immune response, virus-specific cytotoxic T lymphocytes are generated during the primary immune response to the virus and specifically kill infected cells through the release of cytotoxic granules similar to those of NK cells. Naive virus-specific CD8 T cells are activated to effector status as cytotoxic T cells through engagement of the T-cell antigen receptor with a complex of virus-derived peptide and major histocompatibility complex (MHC) class I molecule on the surface of an antigen-presenting dendritic cell. Naive CD8 T cells are not easily activated to effector status. Strong co-stimulation by the dendritic cell together with co-stimulatory help from activated CD4 T cells is usually required to activate a naive cytotoxic T cell.

During the primary immune response to viruses, memory cytotoxic T cells are also produced. In the event of reexposure to the same virus, either by reinfection from the environment or by reactivation of virus latent in the body, these cytotoxic T cells rapidly recognize and kill infected host cells displaying viral

Topics bearing on this case:
NK-cell activating receptors and killer function
NK T cells
Cell signaling from tyrosine kinase-associated receptors
Activation and function of cytotoxic T cells

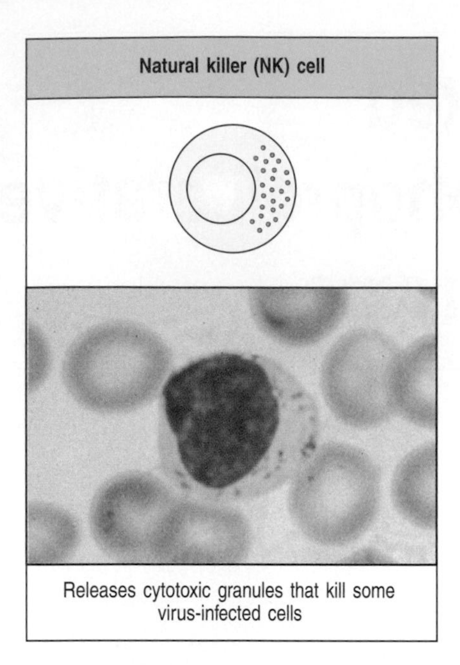

Fig. 13.1 Natural killer (NK) cells. These are large granular lymphocyte-like cells with important functions in innate immunity. Although lacking antigen-specific receptors, they can detect and attack certain virus-infected cells. Photograph courtesy of N. Rooney and B. Smith.

antigens. Equally important to antiviral immunity are virus-specific primary and secondary antibody responses, especially for viruses causing significant viremia.

In this case we look at the effects of a deficiency in an intracellular signaling molecule found in both NK cells and T cells on the ability of the immune system to control infection by the Epstein–Barr virus (EBV). EBV infects epithelial cells and B cells and is present in small numbers in most humans over the age of 15 years. Its expansion is usually well controlled by cytotoxic T cells and NK cells. A primary EBV infection triggers the activation and cell division of B cells infected by the virus. The infected B cell expresses a number of viral antigens that are targets for specific cytotoxic responses by NK cells and CD8 T cells that keep the proliferation of infected B cells under control. In most people, EBV infection remains asymptomatic or may cause infectious mononucleosis, which eventually subsides within a span of 6–10 weeks. The normal course of acute infectious mononucleosis is described in Case 45, and you may find it helpful to read that case before embarking on this one. After resolution of the acute infection, the virus persists in a latent form in B cells, salivary glands, and epithelial cells of the nose and throat, and can be shed in saliva. Occasional reactivation of the virus later in life is rapidly brought under control by EBV-specific memory cytotoxic T cells. This cellular immune surveillance is critical in maintaining the balance between host and virus. Primary and acquired deficiencies of T-cell function are associated with a marked susceptibility to lethal EBV infection.

For example, in very rare instances acute EBV infection in boys is not contained, and results in a failure to eliminate the virus that is accompanied by massive overproliferation of lymphocytes (lymphoproliferation), overproduction of cytokines, destruction of liver and bone marrow, B-cell lymphoma, and/or dysgammaglobulinemia (selective deficiencies of one or more, but not all, classes of immunoglobulins), in most cases resulting in death. Such susceptibility to overwhelming infection can be inherited through unaffected females, an indication that the susceptibility trait is due to a gene on the X chromosome, and the condition is therefore called X-linked lymphoproliferative syndrome (XLP). Two distinct gene defects have been found to cause XLP. More rarely, severe susceptibility to EBV infection may be inherited as an autosomal recessive trait, so that females may also develop the disease. Here we will review the case of a boy with XLP due to the loss of an intracellular signaling protein called SAP.

The case of Nicholas Nickleby: inefficient killing of EBV-infected B cells by cytotoxic lymphocytes.

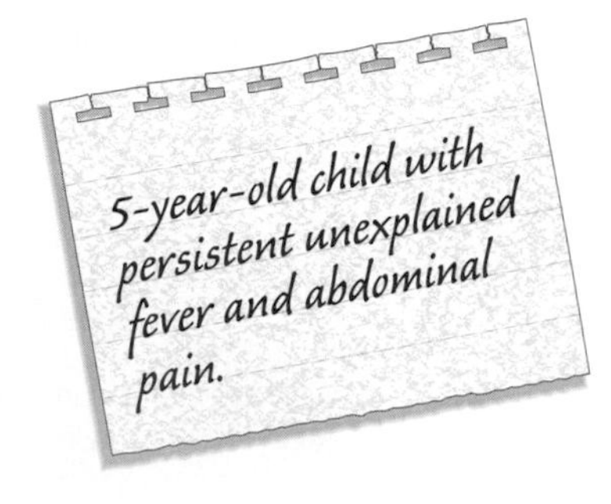

5-year-old child with persistent unexplained fever and abdominal pain.

Nicholas was brought to the pediatrician at 5 years old because of several days of fever (38–39°C). He had no cough, runny nose, rash, diarrhea, or any other symptoms of infection. The physical examination revealed only some mildly enlarged lymph nodes in his neck, and his parents were advised to treat the fever with acetaminophen. Over the following weeks, the fevers persisted and Nicholas seemed less energetic than usual. He was brought to the doctor several times but the only consistent finding was persistent enlarged, nontender lymph nodes. Finally, after 6 weeks of illness, Nicholas complained of abdominal pain and was referred to the Children's Hospital.

The past medical history was significant, revealing problems with persistent and recurrent middle ear infections (otitis media) as well as several episodes of bacterial pneumonia between the ages of 2 and 3 years. Immunological evaluation at the time

had revealed decreased blood levels of IgG of 314 mg dl^{-1} (normal 600–1500 mg dl^{-1}), and normal IgA and IgM. Nicholas was briefly treated with prophylactic antibiotics with a good response. These were discontinued at the age of 4 years and he had no further infections or follow-up tests before his admission to Children's Hospital. The family history was notable for the presence of a maternal uncle who died of aplastic anemia following an acute febrile episode. Furthermore, Nicholas's maternal grandfather had died with a history of recurrent lymphomas.

On admission to hospital, Nicholas appeared tired but not acutely ill. His temperature was 38.5°C and his heart rate, respiration, and blood pressure were all normal. His height and weight were both in the 25th centile for age. A few scattered small skin hemorrhages (petechiae) were noted on his legs and feet. Several lymph nodes were palpable in his neck, and these appeared significantly larger than on previous examinations. Supraclavicular, axillary, or inguinal lymph nodes were not enlarged. The tonsils were moderately enlarged but were not red, and there was no evidence of inflammation. The heart sounds were normal. The abdomen was moderately distended but soft, and slightly tender in the right upper quadrant. The liver was enlarged and its edge was palpable 4 cm below the right costal margin.

Laboratory evaluation showed a mild anemia with a hematocrit of 28% (normal 35–40%). The white blood cell count was 6400 μl^{-1} (normal 5000–10,000 μl^{-1}) and the platelet count was decreased at 47,000 μl^{-1} (normal 150,000–300,000 μl^{-1}). The count of different types of white blood cell was remarkable for the very high proportion (22%) of atypical lymphocytes (normal less than 2%) (see Case 45). Liver function tests indicated significant liver damage, with elevated ALT (alanine aminotransferase) at 1250 U l^{-1} (normal 5–45 U l^{-1}) and AST (aspartate aminotransferase) at 824 U l^{-1} (normal 5–45 U l^{-1}). Tests for antibodies against hepatitis A, B, and C viruses were negative. The titer of IgM antibody against EBV viral capsid antigen (VCA) was positive at more than 1:40. Anti-VCA IgG antibody was 1:320 and antibodies against Epstein–Barr nuclear antigen (EBNA) and early antigen (EA) were undetectable, consistent with an acute EBV infection (see Fig. 45.4). EBV viremia was detected in Nicholas's blood by the polymerase chain reaction (PCR).

High proportion of
atypical lymphocytes
in blood count. Test for
EBV infection.

A chest X-ray showed clear lungs and a normal-sized heart, but the lymph nodes in the mediastinum were enlarged. Ultrasound examination of the abdomen revealed a significant amount of free fluid in the abdominal cavity (ascites) and an enlarged liver. An abdominal CT scan revealed marked enlargement of lymph nodes in the retroperitoneum.

In the light of the family history and laboratory evidence of acute EBV infection, a diagnosis was made of X-linked lymphoproliferative syndrome (XLP) with fulminant infectious mononucleosis. Nicholas was kept in hospital and initially treated with an antiviral agent, ganciclovir, and intravenous immune globulin (IVIG) in an attempt to control the EBV infection. However, the fever persisted and his liver dysfunction and ascites rapidly worsened. The glucocorticoid dexamethasone was added to his therapy. Despite these aggressive interventions, Nicholas developed severe shock symptoms, resembling those after bloodstream infection (sepsis), with diffuse vascular leakage, a fall in blood pressure (hypotension), poor circulation, and multiorgan failure. All cultures for bacterial pathogens were negative. He died 10 days after admission to hospital.

Diagnosis of XLP in
view of family history.

A post-mortem examination showed that the liver was markedly enlarged. Fluid had accumulated in the abdomen and around the lungs (pleural effusions). EBV was identified by culture and PCR in the liver and bone marrow. There was a striking infiltration of the liver, spleen, and lymph nodes by a mixed population of mononuclear cells including small lymphocytes, plasma cells, and lymphoblasts. In the liver, these infiltrates were associated with extensive tissue death (necrosis). Examination of the bone marrow revealed a decreased number of erythroid, megakaryocytic, and myeloid cells, along with increased numbers of histiocytic cells, lymphocytes, and plasma cells.

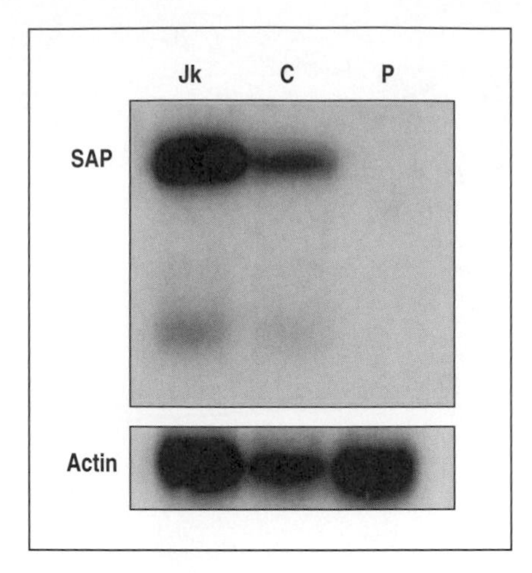

Fig. 13.2 Northern blot analysis of SAP mRNA in peripheral blood mononuclear cells obtained from Nicholas. RNA was prepared from the Jurkat T-cell line (Jk) and from the peripheral blood mononuclear cells of a control (C) and the patient (P). A ^{32}P-labeled probe specific for SAP was used to probe the Northern blot, and a β-actin probe was used as a loading control. Courtesy of Hans Oettgen.

To confirm a diagnosis of X-linked lymphoproliferative syndrome, expression of SAP (*SH2D1A*) mRNA was analyzed by Northern blotting, revealing a complete absence of the product (Fig. 13.2). None of the four exons encoding SAP could be amplified by PCR, consistent with a complete deletion of the gene.

X-linked lymphoproliferative disease (XLP).

Most cases of familial XLP are due to defects of the *SH2D1A* gene, which has been mapped to the X chromosome at position Xq25 and encodes the intracellular signaling protein SLAM-associated protein (SAP). Patients with this defect show uncontrolled T-cell activation, especially in response to an EBV infection, but a reduced capacity to kill EBV-infected B cells. In a minority of cases, symptoms of XLP occur without evident past or current EBV infection. Boys presenting with EBV-induced fulminant infectious mononucleosis, and who have a family history of affected male relatives, will have XLP as a result of mutations in the gene encoding SAP or, less frequently, in the *BIRC4* gene, encoding the X-linked inhibitor of apoptosis (XIAP) protein. The fulminant infectious mononucleosis that develops in these patients after their initial encounter with Epstein–Barr virus commonly proves lethal; among 161 boys with XLP, 57% died of this disease. Of those who survived, half developed lymphomas, as did Nicholas's maternal grandfather, and the other half became agammaglobulinemic as a result of the destruction of their B cells. In rare instances, the bone marrow of affected males may be destroyed, resulting in the fatal disease of aplastic anemia, as in Nicholas's maternal uncle.

In normal individuals, EBV-infected B cells are attractive targets for killing by both NK cells and virus-specific effector cytotoxic T cells. NK cells express receptors that activate or inhibit the cell's killer activity. The fate of potential target cells is determined by the balance of activating and inhibitory signals they deliver to NK cells through these receptors (Fig. 13.3). Inhibitory receptors

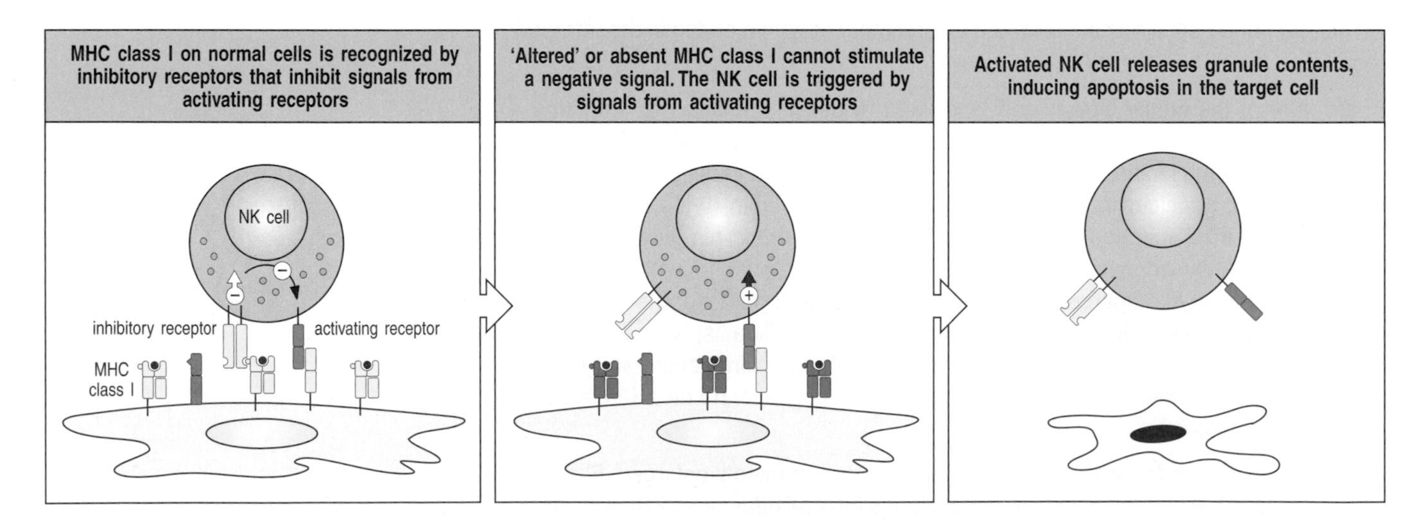

Fig. 13.3 Receptor–ligand interactions in the activation of NK cells. NK cells carry both activating and inhibitory receptors on their surface. Left panel: when an inhibitory receptor recognizes an MHC class I molecule, it sends an inhibitory signal that prevents activation via the activating receptors (which recognize a variety of carbohydrate ligands on cell surfaces). Center panel: altered or absent MHC does not engage the inhibitory receptors, however, and in those circumstances the activating signal dominates. Right panel: as a result, cells expressing low levels of MHC molecules are susceptible to killing by NK cells. The SLAM-family member CD244 interacts with CD48, a protein that is upregulated on B cells after EBV infection. This interaction may enhance NK cell activity by signaling via SAP and Fyn.

interact with MHC class I molecules and help prevent NK cells from attacking healthy, uninfected cells. NK cells also carry a receptor called CD244 (2B4), a member of the SLAM (signaling lymphocytic activation molecule) family, which can act as an activating or an inhibitory receptor, depending on which intracellular signaling proteins it is associated with.

The cytoplasmic portions of SLAM receptors have tyrosine-containing motifs that provide potential docking sites for the intracellular adaptors SAP (SLAM-associated protein; encoded by *SH2D1A*) or an alternative adaptor protein EAT-2. SAP is composed of a single SH2 domain and is found in all T cells, germinal center B cells, and NK cells. SAP is also expressed by NKT cells, which are a subpopulation of lymphocytes with features of both NK and T cells. NKT cells express invariant T-cell receptors recognizing glycolipids. SAP is unusual among intracellular adaptor proteins in that its single SH2 domain can both bind to the tails of activated SLAM receptors and recruit the cytoplasmic Src kinase Fyn to the receptor complex by interacting with the SH3 domain of Fyn.

B cells infected with the Epstein–Barr virus (EBV) increase expression of the SLAM family member CD48 on their surface, and in normal individuals CD48 interacts with CD244 on NK cells, providing an activating signal that enables the NK cells to kill the infected B cells (see Fig. 13.3). Signaling via CD244 is critical in driving NK killing of EBV-infected target cells. CD244 contains several tyrosines in its cytoplasmic tail, which after the receptor is activated become phosphorylated and constitute docking sites for cytosolic proteins containing SH2 domains. Under normal conditions, SAP binds to these cytoplasmic tyrosines, and recruits the Src-family tyrosine kinase Fyn to the receptor complex, allowing propagation of the activating signal onward (Fig. 13.4). In the absence of SAP, active Fyn is not recruited to CD244, and alternative signaling pathways are activated that result in inhibition of killing by the NK cells.

The SLAM family is also implicated in the function of CD8 cytotoxic T cells, with receptors such as SLAM itself (CD150) and CD244 on T cells interacting with SLAM-family molecules on antigen-presenting dendritic cells, B cells, and monocytes. With the exception of the CD48/CD244 pair, SLAM-family molecules are homotypic, interacting with an identical molecule on another cell. In SAP-knockout mice and in patients with XLP, the CD8 T-cell response is exaggerated, suggesting that signaling via SAP normally downregulates this response. In the absence of SAP, SLAM-family receptor signaling is dysregulated, leading to unchecked proliferation of cytokine-producing CD8 cells but impaired cytotoxic function.

The overproduction of cytokines as a result of uncontrolled T-cell proliferation seems to be important in causing the life-threatening tissue damage that

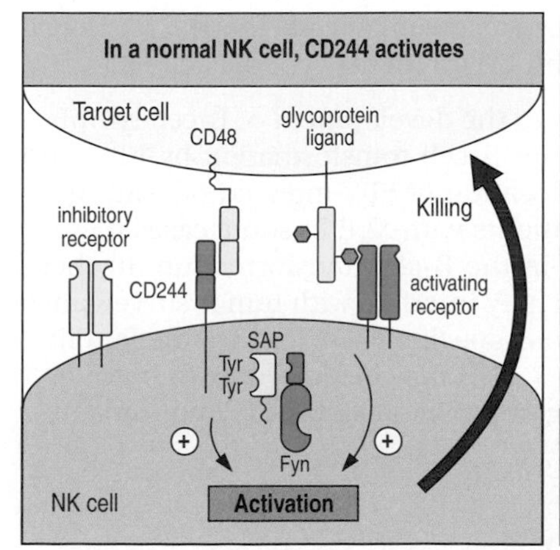

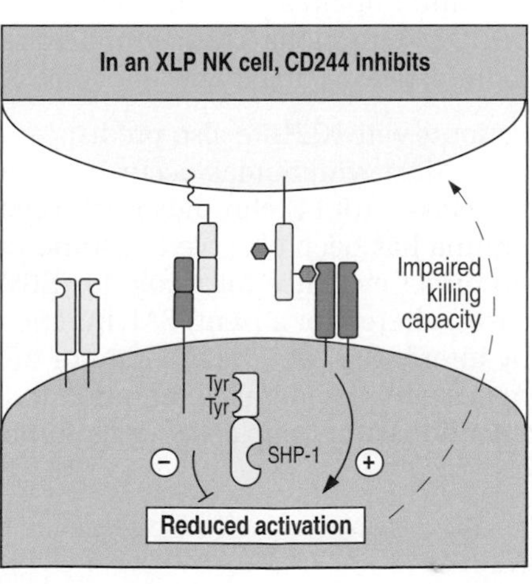

Fig. 13.4 The effect of defects in SAP on NK-cell killer function. In normal NK cells (left panel), interaction with EBV-infected B cells provides positive signals that lead to activation of the NK cytotoxic mechanism. These positive signals are provided by ligands that bind activating NK receptors and by the binding of NK-cell CD244 to CD48 on the B cell. The cytosolic tail of CD244 contains tyrosine residues, which interact with the SH2 domain of SAP. SAP couples CD244 to activation of the tyrosine kinase Fyn, thus enhancing NK-cell activation. In the absence of SAP (right panel), this activating signal is interrupted by the binding of the phosphatase SHP-1 to the receptor, and the killing capacity of the NK cell is impaired.

occurs in XLP. On infection with EBV, patients with XLP suffer a destruction of the liver that is most probably due to uncontrolled cytokine-mediated injury. Indeed, plasma levels of a number of T-cell-derived cytokines, including interferon (IFN)-γ, interleukin (IL)-2, and tumor necrosis factor (TNF)-α, are all markedly elevated. The same cytokines are present at much lower (or undetectable) levels in normal EBV-infected individuals. In patients with fulminant infectious mononucleosis, the uncontrolled lymphocyte proliferation and cytokine secretion lead to a syndrome of severe inflammation of the liver (hepatitis), destruction of bone marrow cells, and systemic shock. T-cell-derived cytokines, particularly TNF-α, as well as monocyte-derived IL-1, result in enhanced vascular permeability and loss of intravascular volume. This is analogous to the clinical scenario observed in toxic shock syndrome (see Case 47) in which uncontrolled T-cell activation and cytokine secretion can also lead to multiorgan damage and a clinical picture resembling sepsis.

In a second potential mechanism of tissue injury, particularly in the liver, T-cell-derived cytokines may promote the expression of Fas on hepatocytes. The ligand, FasL, is expressed on the surface of activated T cells and can induce hepatocyte apoptosis (see Case 19). T-cell-derived cytokines, particularly IFN-γ, also trigger the activation of monocytes/macrophages, which engage in indiscriminate phagocytosis of surrounding cells. Histologic analysis of tissues from patients with fulminant infectious mononucleosis often reveals hemophagocytosis, a phenomenon in which some macrophages seem to engulf entire red blood cells and others are heavily laden with cellular debris. Hemophagocytosis is further enhanced because the polyclonal B-cell activation induced by EBV infection (see Case 45) produces complement-fixing antibodies that interact with red-cell antigens. Finally, numbers of NKT cells are severely diminished in XLP, possibly because activation of the kinase Fyn is essential for their development.

Although more than half of the patients with XLP are extremely susceptible to EBV, one-third of the patients develop dysgammaglobulinemia without an episode of severe mononucleosis. Experiments with SAP-deficient mice show that SAP is essential for the functional integrity of both T-cell and B-cell responses to soluble T-dependent antigens, and pivotal for IgM and IgG responses to viruses, well-defined protein antigens, and haptens, and for immunoglobulin class switching and germinal center formation. Thus, when SAP is absent, both helper T cells and B cells are defective in function, which may explain the progressive dysgammaglobulinemia in a subset of patients with XLP without involvement of EBV. It has been shown that SAP has a critical role in development and function of follicular helper T cells (T_{FH}), a subpopulation of T cells that secrete IL-21 and govern recruitment of antigen-specific B lymphocytes to the germinal centers, allowing maturation of antibody responses. Thus, the SAP-dependent deficiency of T_{FH} cells accounts for the dysgammaglobulinemia and impaired antibody responses observed in patients with XLP due to *SH2D1A* mutations. These impaired responses can precede the severe clinical manifestations that follow infection with EBV, as in the case of Nicholas.

Patients with XLP are also predisposed to the development of B-cell lymphomas. Most lymphomas occur because of B-cell transformation by EBV, and the classic t(8;14) chromosomal translocation of EBV-induced Burkitt's lymphoma has been observed in some patients with XLP. In some cases there is no direct evidence for a role for EBV in the B-cell transformation. In these cases, altered (or absent) SAP function may interfere with tumor surveillance by interfering with the function of tumor-specific NK cells. In some families, males with the identical mutation in *SH2D1A* may be present with fulminant infectious mononucleosis, dysgammaglobulinemia, or B-cell lymphoma.

A minority of male patients with XLP do not carry mutations in SAP but in the *BIRC4* gene, encoding the XIAP protein, whose function is to inhibit caspase-mediated apoptosis. Accordingly, lymphocytes from patients with XIAP deficiency show increased susceptibility to apoptotic stimuli. However, the pathophysiology of the XLP phenotype in these patients is still unclear. The terms XLP-1 and XLP-2 are used to distinguish patients with XLP due to SAP or to XIAP mutations, respectively. Finally, EBV-induced lymphoproliferative disease may occur also in patients with mutations of the IL-2-inducible tyrosine kinase (ITK), which is activated in T cells in response to engagement of the antigen receptor. ITK deficiency is inherited as an autosomal recessive trait, and so both males and females may be affected.

Questions.

1 *Did Nicholas's history provide any clues to his diagnosis before his presentation with fulminant infectious mononucleosis?*

2 *How can the diagnosis of X-linked lymphoproliferative syndrome be made?*

3 *If a girl presents with fulminant infectious mononucleosis and evidence of acute EBV infection, what might be the basis of her disease?*

4 *Nicholas has a 3-year-old brother, Alexander, who has SAP deficiency. What can be done to prevent the occurrence of disease in this boy?*

CASE 14 Hemophagocytic Lymphohistiocytosis

The killing machinery of NK cells and cytotoxic T cells.

Virus-infected cells are eliminated by natural killer (NK) cells and cytotoxic CD8 T cells. NK cells are lymphocytes of innate immunity that are activated to kill virus-infected cells by recognizing alterations in the MHC molecules on their surface. NK cells carry numerous activating and inhibitory receptors that between them control the cells' cytotoxicity. In contrast, the T-cell receptors on cytotoxic CD8 T lymphocytes specifically recognize viral antigens presented on the infected cell's surface by MHC class I molecules. Antigen binding by the T-cell receptor activates the cytotoxic T cell. Despite the differences in recognition mechanisms, both these cell types kill their targets in the same way—by release of the cytotoxic proteins perforin, granzyme B, and granulysin onto the target cell. Perforin forms multimeric pores in the cell membranes that enable the delivery of the other cytotoxic proteins from the killer cell into the target.

Cytotoxic proteins are preformed and stored in endosomal 'lytic granules' within the lymphocyte. Once activated, cytotoxic lymphocytes reorient the microtubule-organizing center of the cell toward the point of contact with the target, and this cytoskeletal rearrangement guides the lytic granules to the contact point. After the lytic granules dock at the cell membrane, perforin polymerizes and inserts into the membrane of the target cell, thus forming pores that connect the killer and target cells. Granzyme B and granulysin are released through the pore into the target cell, where they induce apoptosis (Fig. 14.1).

Topics bearing on this case:

Lymphocyte cytotoxic activity

Macrophage activation

This case was prepared by Luigi Notarangelo, MD, in collaboration with Jolan Walter, MD, PhD.

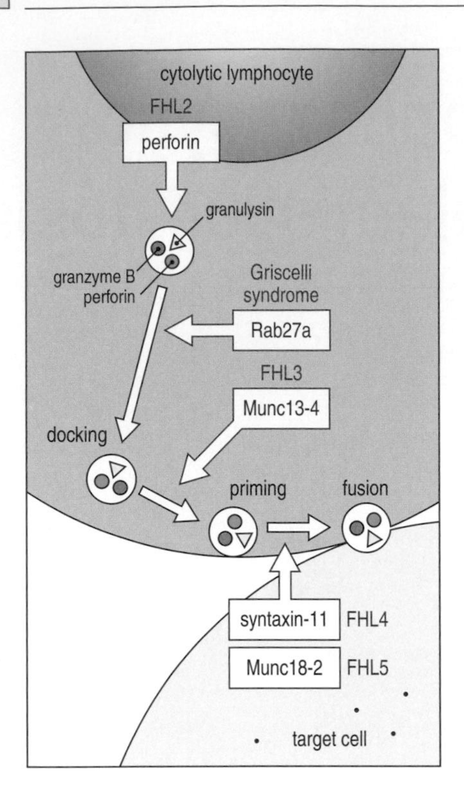

Fig. 14.1 Intracellular trafficking of lytic granules and genetic defects that can lead to HLH. On activation, cytotoxic CD8 T cells and NK cells mobilize lytic granules filled with cytotoxic proteins that are transported toward the point of contact between the lymphocyte and the target cell, where they dock with the membrane. This is followed by the exocytosis of granule contents and the release of cytotoxic proteins through pores formed by perforin. Different steps of this process are impaired in the forms of FHL for which the genetic defect is known (FHL 2, 3, 4, and 5) and in another disease in which HLH is a symptom (Griscelli syndrome); the protein deficient in each disease is shown next to each form of FHL. A deficiency in perforin does not affect the formation of the cytolytic granules but impairs the release of the cytotoxins and their entry into the target cell.

Several other proteins apart from the cytotoxins are crucial to successful cell-mediated cytotoxicity. The small GTPase Rab27a promotes the docking of mature cytotoxic granules to the cell membrane. The Munc13-4 protein promotes the priming of the cytolytic granules; finally, syntaxin-11 and Munc18-2 (both of which are part of the docking complex) enable fusion of the secretory granules with the cell membrane (see Fig. 14.1). Cytotoxic proteins may then be released by exocytosis through the pores formed by perforin.

The process of degranulation can be detected by the appearance of lysosomal-membrane-associated glycoproteins (CD107a, CD107b, and CD63) in the cell membrane. Under resting conditions, these proteins are located on the inner surface of the lytic granule membrane; on degranulation, they become exposed on the lymphocyte surface.

Several genetic defects that affect the cytotoxic machinery and cause disease in humans have been identified. This case concerns one of these, the disease hemophagocytic lymphohistiocytosis. Other defects cause disorders of cytotoxicity and pigmentation such as Griscelli syndrome type 2, Chediak–Higashi syndrome (see Case 15), and Hermansky–Pudlak syndrome type 2 (Fig. 14.2). All these disorders are characterized by increased susceptibility to viral disease and an overwhelming inflammatory response, with increased production of the cytokine interferon-γ (IFN-γ).

Fig. 14.2 Characteristics of FHL and other diseases causing HLH.

	FHL1	FHL2	FHL3	FHL4	FHL5	Griscelli syndrome type 2	Chediak–Higashi syndrome	Hermansky–Pudlak syndrome type 2
Gene affected	Unknown	*PRF1*	*UNC13D*	*STX11*	*STXBP2*	*RAB27A*	*CHS1*	*AP3BP1*
Protein	Unknown	Perforin	Munc13-4	Syntaxin-11	Munc18-2	Rab27a	LYST	AP3B1
Function	Unknown	Pore formation	Priming	Granule fusion with cell membrane	Granule fusion with cell membrane	Docking of granules	Protein sorting	Protein sorting
HLH	+	+	+	+	+	+	+	+
Hypopigmentation	–	–	–	–	–	+	+	+
Other features	–	–	–	–	–	–	Giant lysosomes, peripheral neuropathy	Neutropenia, tendency to bleeding

The case of Jude Fawley: a febrile illness with altered mental status.

Fever and hepatosplenomegaly.
Coagulopathy and markedly increased ferritin.

Jude was born after an uneventful pregnancy; his parents were distantly related. At 2 months of age he developed rhinorrhea ('running nose') and fever. Seven days later he was referred to the emergency room with a high fever (39.5°C) and difficulty in feeding. His physical examination showed significant enlargement of the liver and the spleen, which were both palpable 4 cm below the costal edge.

A complete blood count showed marked lymphocytosis (29,000 μl^{-1}), thrombocytopenia (platelet count 73,000 μl^{-1}; normal 150,000–300,000 μl^{-1}), and anemia (hemoglobin (Hb) 6.2 g dl^{-1}; normal 9.0–14.0 g dl^{-1}). Spinal fluid sampled by lumbar puncture showed mild pleocytosis (increased white blood cell count). Laboratory investigations revealed increased levels of liver enzymes (alanine aminotransferase (ALT) 375 U l^{-1}; aspartate aminotransferase (AST) 220 U l^{-1}), markedly high ferritin (7500 ng ml^{-1}; normal 50–200 ng ml^{-1}), positive C-reactive protein, and elevated fibrin degradation products.

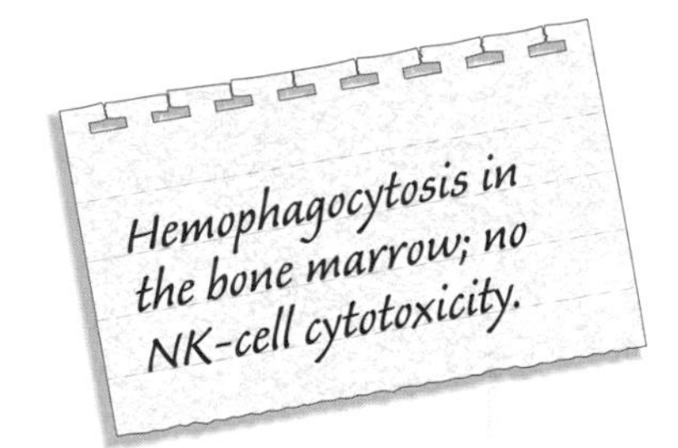

Hemophagocytosis in the bone marrow; no NK-cell cytotoxicity.

Serological tests for Epstein–Barr virus, cytomegalovirus, and adenovirus were all negative. Although blood cultures were negative, Jude was treated with intravenous antibiotics (ampicillin and amikacin), but his condition continued to worsen, and his consciousness rapidly deteriorated. Brain magnetic resonance imaging (MRI) with T2 intensity revealed spotty high-density lesions in the white matter of the cerebrum, basal nuclei, and cerebellum. Laboratory investigations showed a worsening of anemia (Hb 5.2 g dl^{-1}) and thrombocytopenia (31,000 μl^{-1}), but also high serum levels of triglycerides (750 mg dl^{-1}; normal value less than 150 mg dl^{-1}) and low fibrinogen (40 mg dl^{-1}; normal value 200–400 mg dl^{-1}). Bone marrow aspiration was performed and showed hypocellularity with an increased number of large granular lymphocytes and macrophages with hemophagocytic activity. Jude was diagnosed with hemophagocytic lymphohistiocytosis (HLH). Flow cytometry analysis revealed an absence of perforin. Complete absence of NK cell-mediated cytotoxicity was demonstrated using a suitable target (K562) cell line. Sequence analysis of the perforin gene revealed a single nucleotide deletion in exon 2, resulting in a frameshift and the production of an amino-terminally truncated perforin protein that could not be detected by flow cytometry analysis. Both parents were heterozygous for the same mutation.

Jude received chemotherapy according to the HLH-94 protocol, which consisted of dexamethazone, etoposide, and cyclosporin A (CsA). He subsequently underwent matched unrelated hematopoietic stem-cell transplantation (HSCT) with a conditioning regimen based on busulfan, cyclophosphamide, and etoposide, and CsA for graft-versus-host disease prophylaxis. Three years after HSCT, Jude is doing well, with mild mental and motor delay in his development.

Hemophagocytic lymphohistiocytosis.

The case of Jude Fawley illustrates the typical features of hemophagocytic lymphohistiocytosis (HLH). Both congenital and acquired forms of HLH are known; Jude had the congenital form, which is often known as familial hemophagocytic lymphohistiocytosis (FHL). Acquired HLH may be secondary to infections, malignancies, or autoimmune diseases. In particular, patients with juvenile arthritis or systemic lupus erythematosus can develop a disease similar to HLH, called macrophage activation syndrome. HLH is an aggressive and potentially life-threatening disease.

FHL is a rare disorder, with an estimated frequency of 1 in 50,000 births in the United States. The disease typically appears in infancy or early childhood; both sexes are affected, which is consistent with autosomal recessive inheritance. The clinical presentation is characterized by high and persistent fever, and spleen and liver enlargement. Neurological manifestations (ranging from seizures and confusion to coma) are common and are associated with pleocytosis in the cerebrospinal fluid, also known as lymphocytic meningitis. Patients develop severe anemia and thrombocytopenia, abnormal liver function, and coagulopathy (decreased fibrinogen and increased levels of fibrin degradation products, which make the blood less able to clot). Increased levels of serum triglycerides and of inflammatory markers (ferritin and C-reactive protein) are also observed.

These manifestations are the consequence of the defect in lymphocyte cytotoxicity that makes patients with HLH unable to kill virus-infected cells. In this situation, there is continuous activation of NK cells and CD8-positive T lymphocytes, which infiltrate the liver, spleen, bone marrow, and central nervous system and secrete high amounts of IFN-γ. This cytokine is a potent activator of macrophages, which are induced to secrete pro-inflammatory cytokines such as IL-6 and TNF-α. The hemophagocytosis—the phagocytic destruction of red blood cells—is also a consequence of macrophage activation (Fig. 14.3). An increase in serum levels of the soluble IL-2 receptor (sIL-2R) is also characteristic of HLH during active phases of the disease, and is a marker of T-cell activation.

In FHL, the clinical symptoms of fever, immune activation, and increased inflammation are precipitated especially by infections with herpesviruses, such as cytomegalovirus, Epstein–Barr virus, and varicella-zoster virus. These episodes are known as the 'accelerated phase' of the disease. They can lead to multiple organ failure and death, and can occur multiple times in a patient's life.

At least five variants of FHL are known (see Fig. 14.2), and the underlying genetic defect is known in four of them. Jude had FHL2, and was unable to produce a functional perforin. Some forms of HLH (Griscelli syndrome type 2, Chediak–Higashi syndrome, and Hermansky–Pudlak syndrome type 2) are also characterized by hypopigmentation, reflecting the role of the affected proteins in melanogenesis (melanin production). Patients with Chediak–Higashi syndrome also exhibit peripheral neuropathy and their leukocytes have characteristically large lysosomes (see Case 15), whereas patients with Hermansky–Pudlak syndrome type 2 show neutropenia (low levels of circulating neutrophils) and increased bleeding due to functionally abnormal platelets.

The diagnosis of HLH is based on a combination of clinical and laboratory features. In FHL, there is a genetically determined defect in lymphocyte-mediated cytotoxicity. NK cell-mediated cytotoxicity is usually assessed by culturing the patient's peripheral blood mononuclear cells with ^{51}Cr-labeled

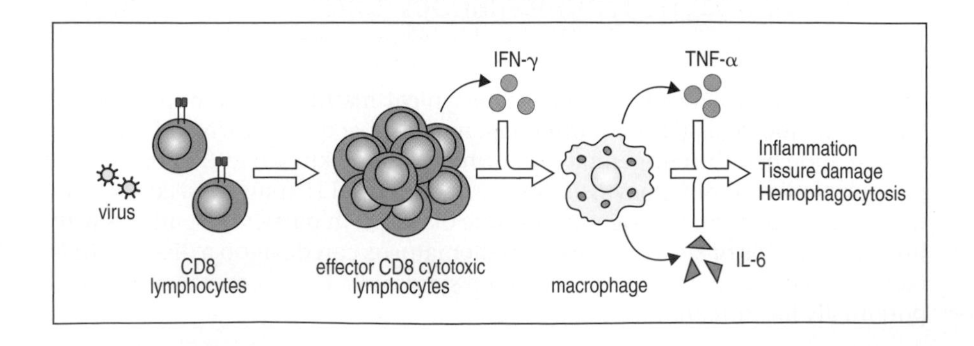

Fig. 14.3 Cytokine storm in HLH. In patients with HLH, viral infections trigger the activation of CD8 T cells that undergo clonal expansion and activation, but cannot clear the viral infection. The activated CD8 T cells secrete large amounts of IFN-γ, which promotes the activation of macrophages and the release of TNF-α, IL-6, and other pro-inflammatory molecules. This series of events ultimately causes tissue damage and hemophagocytosis.

K562 target cells. If the patient's NK cells have intact cytotoxic activity, ^{51}Cr is released into the supernatant. Any defect in activation, docking, or priming of the lytic granules can be detected by flow-cytometric analysis for the appearance of CD107a on the surface of the patient's lymphocytes activated *in vitro*. This test gives a normal result in patients such as Jude, whose specific deficiency is in perforin. Flow cytometry may also be used to diagnose FHL2, by means of intracellular staining for perforin.

Treatment of FHL is based on aggressive immunosuppression to stop the ongoing immune activation. However, even if this is sufficient to achieve clinical remission, patients remain highly prone to other episodes of accelerated phase, and the disease has a very high mortality rate. The only cure is through HSCT. Experience in animal models suggests that administration of anti-IFN-γ monoclonal antibody might be effective to achieve remission, without the risk of side effects related to the use of chemotherapy, steroids, or other immunosuppressive drugs. For the nonfamilial, secondary forms of HLH, treatment is based on elimination of the trigger, in addition to immune suppression.

Questions.

1 *Why do patients with FHL develop hyperinflammatory responses in spite of their immune deficiency?*

2 *Why did Jude develop enlarged liver and spleen?*

3 *Why did Jude show hypocellularity in his bone marrow?*

CASE 15 Chediak–Higashi Syndrome

A defect of intracellular vesicle trafficking.

Cytoplasmic vesicles are essential components of all cells in the human body. These small membrane-enclosed sacs participate in diverse cellular functions, including transport within the cell and with the exterior, storage of nutrients, digestion of cellular waste and foreign products, and processing of proteins. They also serve as closed chambers for multiple chemical reactions that could otherwise damage the cell if they were to occur outside the vesicle. The types of vesicles and their specific functions vary between different cells, although most of them would appear as undistinguishable intracellular granules under the microscope.

Correct trafficking of vesicles is as important as the function of the vesicle itself. The process through which cells absorb molecules or structures from the exterior by invaginating the cell wall and forming a vesicle is called endocytosis. In contrast, release of the content of intracellular vesicles to the exterior is called exocytosis. Vesicles and their trafficking in the cell are particularly important for normal functioning of the immune system. They are essential in the first line of defense, in which cells of the innate immune system, such as neutrophils and macrophages, engulf invading microorganisms through a form of endocytosis called phagocytosis (Fig. 15.1). This newly created vesicle is called a phagosome. The phagosome then fuses with a lysosome, a different type of vesicle that contains the machinery to kill the invading organism and digest its components. This fused vesicle, called a phagolysosome, degrades its contents by several mechanisms, including the production of reactive oxygen species, such as superoxide, and the action of proteolytic enzymes, such as lysozyme, and of defensins and other antimicrobial peptides.

Another group of immune cells that engage in endocytosis are antigen-presenting cells such as dendritic cells. In this case, the cell captures a foreign macromolecule and degrades it, but the main function in this instance is the loading of peptide onto antigen-presenting proteins (MHC proteins), which will expose the newly acquired antigen to examination by T cells. T cells that carry receptors recognizing the foreign antigen in association with MHC proteins get activated and trigger the adaptive immune response.

Secretory vesicles, whose function is to release their contents to the extracellular space or into other cells through exocytosis, also have an essential function in the immune response. Cytotoxic T lymphocytes (CD8 T cells) and natural killer cells (NK cells) recognize other host cells that have been infected by viruses or intracellular bacteria and kill them by transporting secretory vesicles to the cell surface, where they fuse with the cell membrane and release their contents of enzymes and other mediators, such as perforin and granzymes, thus inducing apoptosis of the target cell.

This case was prepared by Raif Geha, MD, in collaboration with Arturo Borzutzky, MD.

Topics associated with this case:
Phagocytosis
Endocytosis and processing of antigen
Killing by cytotoxic T cells and NK cells

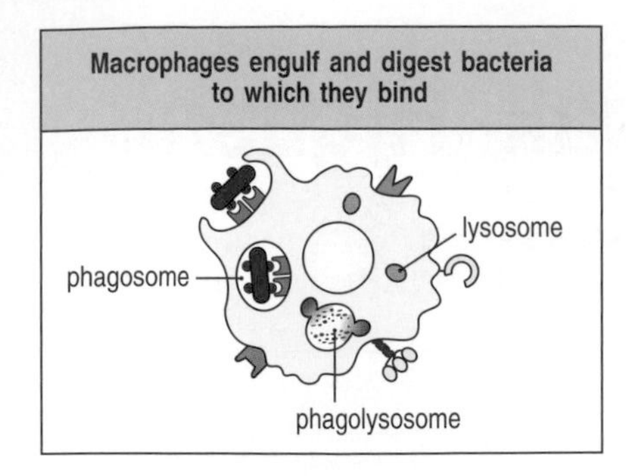

Fig. 15.1 Cell vesicles involved in phagocytosis. A phagocyte, such as a macrophage, binds to bacteria and engulfs them, creating a phagosome. The phagosome fuses with a lysosome, which contains hydrolytic enzymes, creating a vesicle called the phagolysosome, in which the bacteria are digested and killed.

In addition to their role in the functioning of the immune system, vesicles have an important role in pigmentation of the skin and eye, synapses between neurons, and platelet activation. As we shall illustrate, a defect in the trafficking of vesicles can lead to a devastating disease affecting many different functions of the immune system as well as other organs and systems.

The case of Shweta Amdra Devi: the cause of recurrent infections is found in a peripheral blood smear.

Recurrent infections.
Fairer complexion and
lighter hair color than
other family members.

Shweta was seen by the immunologists at the Children's Hospital at the age of 4 years, with a history of recurrent ear infections and two episodes of pneumonia. At 2 years of age she had also had cellulitis (bacterial infection of the skin) and lymphadenitis (lymph node infection) due to *Staphylococcus aureus*, which required intravenous antibiotics. In addition, Shweta had a history of easy bruising and multiple episodes of mild nosebleeds.

Shweta was born at full term and with a normal birth weight. Her umbilical cord separated normally at 10 days. Her parents reported that ever since Shweta was a baby, she seemed to be bothered by bright lights. She also had fair skin and hair, which was much lighter than the skin and hair of her parents and siblings and was unusual for her Indian descent. This was the reason she had been named Shweta, which means "fair complexioned" in the Hindi language.

Her parents were second cousins, who had recently emigrated to the United States from a small town near Bangalore, India. Shweta had two healthy siblings, a 10-year-old boy and a 17-year-old girl. There were no family members with a history of recurrent infections.

On physical examination, Shweta was relatively thin and small for her age. She had a mild fever of 37.7°C and slightly fast breathing. Her skin was fair and she had light brown hair with a silvery sheen. She had an atrophic scar on her left leg, the consequence of an old skin infection. The irises of her eyes were gray. An ear examination revealed a perforation and pus drainage in the right tympanic membrane, and she had moderate cervical lymphadenopathy (enlarged lymph nodes). Examination of her lungs revealed very poor aeration and crackles in the right lower lobe. Her abdomen had mild hepatosplenomegaly (enlarged liver and spleen).

A chest radiograph confirmed that Shweta had a consolidation in the right lower lobe and a moderate pleural effusion, thus confirming a new episode of pneumonia. Aspiration of the pleural fluid grew *S. aureus,* which was broadly sensitive to antibiotics. She was admitted to the hospital and started on intravenous nafcillin (an antistaphylococcal antibiotic), after which she slowly recovered.

Laboratory testing revealed a white blood cell count of 5200 μl^{-1} (normal), of which 21% (1100 cells μl^{-1}) were neutrophils (mildly decreased), 68% were lymphocytes (normal) and 8% were monocytes (normal). Serum immunoglobulins and complement titers were normal. Because of her history of infections and current staphylococcal pneumonia, neutrophil function was measured with a nitro blue tetrazolium (NBT) test, which assays the capacity of the NADPH oxidase in neutrophils to produce the superoxide that reduces the NBT dye from yellow to blue (see Case 26). The results showed that Shweta's neutrophils could reduce NBT normally. The laboratory technician examining the peripheral blood smear observed giant cytoplasmic granules in her leukocytes, which stained positive with myeloperoxidase. She called the physicians in charge of Shweta's care and told them, "this patient has Chediak–Higashi syndrome."

After Shweta recovered from the acute infection, she was started on prophylactic antibiotics, and plans were made for a bone marrow transplant. Shweta's brother proved to be a full HLA match and donated the bone marrow. Shweta was successfully transplanted with full resolution of infections. Years later, during adolescence, Shweta developed a progressive neurological disease characterized by weakness, tremors, and ataxia, which confined her to a wheelchair.

Giant cytoplasmic granules in leukocytes.

Chediak–Higashi syndrome.

Patients such as Shweta have a very rare and severe disease called Chediak–Higashi Syndrome (CHS). This is an autosomal recessive disorder clinically characterized by recurrent bacterial infections and partial absence of pigmentation of skin, hair, and eyes (oculocutaneous albinism) (Fig. 15.2). Affected patients also have a tendency to bleeding, due to platelet dysfunction, which is usually mild to moderate. If patients survive into adolescence or early adulthood, most will develop progressive neurological defects, including cerebellar ataxia, central nervous system atrophy, seizures, peripheral neuropathy, and cognitive defects. In addition, most patients with CHS undergo at some point an accelerated phase of uncontrolled lymphocyte proliferation and lymphohistiocytic infiltration characterized by fever, lymphadenopathy, hepatosplenomegaly, and pancytopenia (reduction of platelets and of white and red blood cells), which is usually lethal.

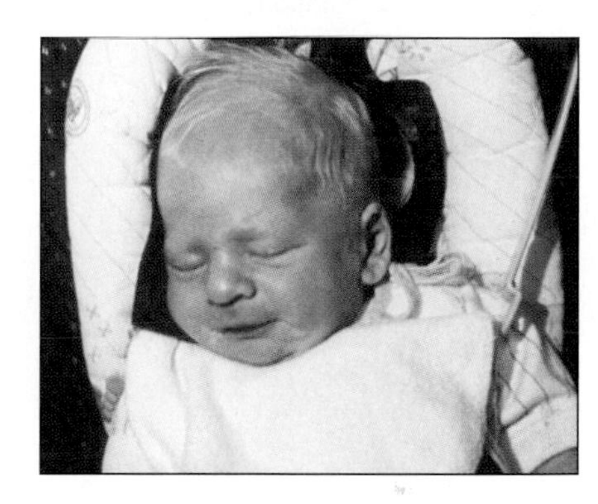

Fig. 15.2 A child with Chediak–Higashi syndrome (CHS). Note the white hair (albino) and the silvery sheen of the hair.

The diagnosis of CHS can be made by examination of a peripheral blood smear for the distinctive giant cytoplasmic granules (vesicles) in leukocytes and platelets (Fig. 15.3). A similar phenotype of giant intracellular granules was known in the *beige* mouse strain, which has hypopigmentation of the coat (Fig. 15.4). The *beige* mouse was crucial in the identification of the human gene responsible for CHS, which was named *CHS1* (*Chediak Higashi Syndrome 1*). This gene, which is also known as *LYST* (*LYSosomal Trafficking regulator*), is part of the BEACH family of proteins involved in vesicle formation and trafficking. In addition to humans and mice, other mammals with mutations in *CHS1* have been identified, including the Aleutian mink, whose coat has a blue tinge.

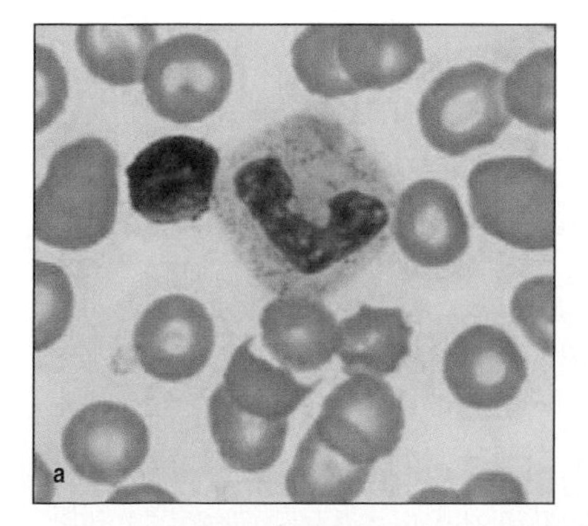
a

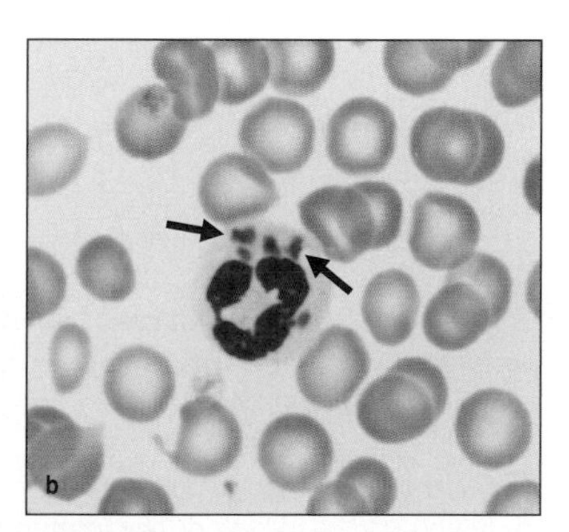
b

Fig. 15.3 Large intracellular granules in leukocytes are typical of CHS. Compare the peripheral blood smear from a healthy person (panel a) with that from a patient with Chediak–Higashi syndrome (panel b). The large cell in the center of both photographs is a neutrophil, distinguishable by the irregular lobed nucleus (stained purple). Note the abnormally large granules in the cytoplasm of the neutrophil from the patient with CHS. Preparations stained with hematoxylin and eosin.

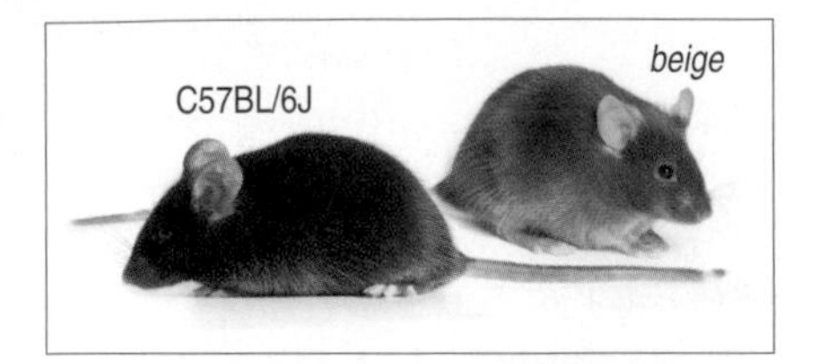

Fig. 15.4 The *beige* mouse is an animal model for CHS. The color of the fur of the *beige* mouse (on the right) is lighter than the C57BL/6J wild-type mouse because of abnormalities in pigmentation associated with mutations in the *CHS1* gene.

All cells from patients with CHS have an abnormal clustering of giant lysosome-like vesicles around the nucleus. It is not yet clear how mutations in *CHS1/LYST* cause the vesicle defect, because the precise biological function of the protein is unknown. It is suspected that abnormal organellar protein trafficking may lead to aberrant fusion of vesicles and a failure to transport lysosomes to the appropriate location in the cell. Most mutations in *CHS1* are nonsense or null mutations that lead to an absence of the protein, although some cases with milder phenotypes due to missense mutations have been described.

The main immunological defect of patients with CHS is in innate immunity, affecting the first line of defense to pyogenic infections of the skin and respiratory tract. Infections are frequent and usually severe, beginning shortly after birth. Commonly implicated microorganisms include the bacteria *Staphylococcus aureus*, *Streptococcus pyogenes*, and *Streptococcus pneumoniae* and occasionally fungi such as *Candida* or *Aspergillus*. Neutrophil counts are mildly to moderately decreased, probably as a result of the destruction of neutrophils in the bone marrow. These cells also have decreased intracellular microbicidal activity. Affected neutrophils and monocytes have a chemotactic and migratory capacity that is about 40% that of normal cells. This may be due in part to the fact that the large fused granules impair the cells' ability to move. In addition, cytotoxic T lymphocytes and NK cells have severely impaired cytotoxicity because of their inability to secrete the granules that contain lytic proteins such as granzymes and perforin. Studies of B cells of affected patients also suggest that these are less able to load peptide onto MHC class II molecules and have decreased antigen presentation compared with normal B cells.

The pigmentation defect of CHS is due to the inability of melanocytes, the cells that produce the pigment, to transfer pigment-containing secretory granules (melanosomes) to keratinocytes and other epithelial cells (Fig. 15.5). Platelet dysfunction is due to a reduction in a particular group of vesicles called platelet dense bodies that participate in sustaining platelet aggregation. The pathogenesis of the neurological disease that affects patients with CHS is believed to be due to problems of vesicular trafficking in neurons and glia.

CHS is treated with prophylactic antibiotics and aggressive management of infections. Bone marrow transplantation corrects the immunological and hematological defects, and prevents the accelerated phase of CHS. However, manifestations of CHS in nonhematopoietic organs, particularly the oculocutaneous albinism and the progressive neurological disease, are not modified by bone marrow transplantation.

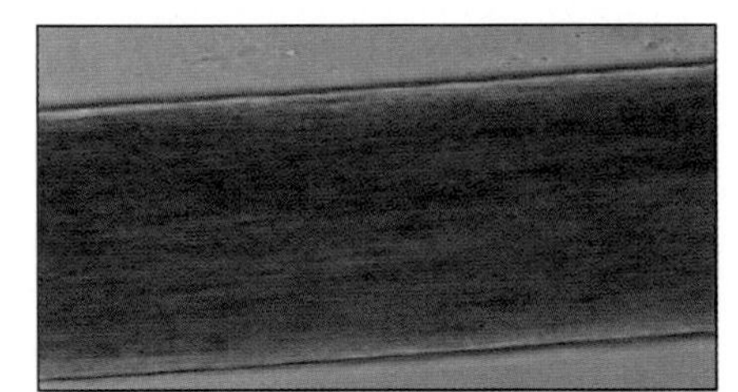

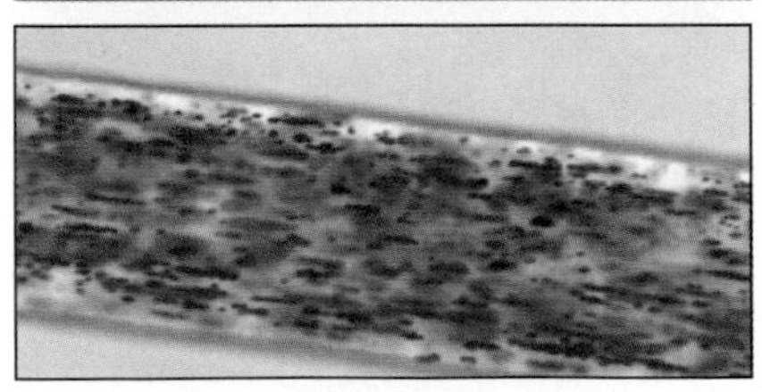

Fig. 15.5 Abnormalities of hair pigmentation are diagnostic of CHS. Photomicrograph of a hair shaft from an unaffected individual (panel a) and a patient with CHS (panel b). Note the uniform distribution of the pigment melanin in the normal hair shaft and the abnormal speckles of pigment in the hair from the patient with CHS.

Questions.

1 *Why do you think Shweta's NBT test was normal?*

2 *Besides the observation of giant granules in the leukocytes of a peripheral blood smear, what other non-invasive test can be helpful to confirm the diagnosis of CHS?*

3 *The accelerated phase of CHS consists of massive organ infiltration by lymphocytes. Which of the immunological defects in this disease would you think is involved?*

4 *Why does bone marrow transplantation not affect the oculocutaneous albinism and neurological disease?*

CASE 16 Wiskott–Aldrich Syndrome

Role of the actin cytoskeleton in T-cell function.

Many functions of T cells require the directed reorganization of the cell's cytoskeleton, in particular the actin cytoskeleton. The eukaryotic cell cytoskeleton as a whole consists of actin filaments, microtubules, and intermediate filaments. It provides a framework for the internal structural organization of the cell and is also essential for cell movement, cell division, and many other cell functions. In T cells, as in other animal cells, the actin cytoskeleton is found mainly as a meshwork of actin filaments immediately underlying the plasma membrane (Fig. 16.1). The actin cytoskeleton is a dynamic structure and can undergo rapid reorganization by the depolymerization and repolymerization of actin filaments. As we shall see for Wiskott–Aldrich syndrome, an inability of T cells to reorganize their actin cytoskeleton when required has profound effects on their function and thus on immune function as a whole.

The functions of T cells in immune defense involve interactions with effector cells such as B cells, dendritic cells, and infected target cells that are initiated by direct cell–cell contact via cell-surface receptors. For example, helper T cells interact with B cells through cell-surface receptors to stimulate B-cell proliferation and the subsequent differentiation into antibody-producing plasma cells. T-cell–B-cell interactions are also involved in isotype switching and the generation of memory cells, whereas cytotoxic T-cell killing of virus-infected target cells also involves direct contact with the target cell. These interactions are accompanied by reorganization within the T-cell cortical actin cytoskeleton that, for example, focuses secreted T-cell products onto the target cell.

The cytoskeleton is linked to cell-surface receptors in the plasma membrane so that events occurring at the membrane can affect cytoskeleton reorganization. For example, cross-linking of T-cell antigen receptors and co-receptors by antigen:MHC complexes leads to their aggregation at one pole of the T cell, with an accompanying concentration of the actin cytoskeleton at that point. Binding of a helper T cell to a B cell through the binding of its T-cell receptors to antigen:MHC complexes on the B-cell surface also leads to a reorganization of the actin cytoskeleton locally in the zone of contact, which in turn causes a microtubule-dependent mechanism to focus the secretory apparatus of the T cell on the point of contact with the B cell (Fig. 16.2); the release of cytokines

Topics bearing on this case:
Isotype switching
T-cell help in antibody responses
Interaction of cytotoxic T cells with their targets
Methods for measuring T-cell function
Gene knockouts

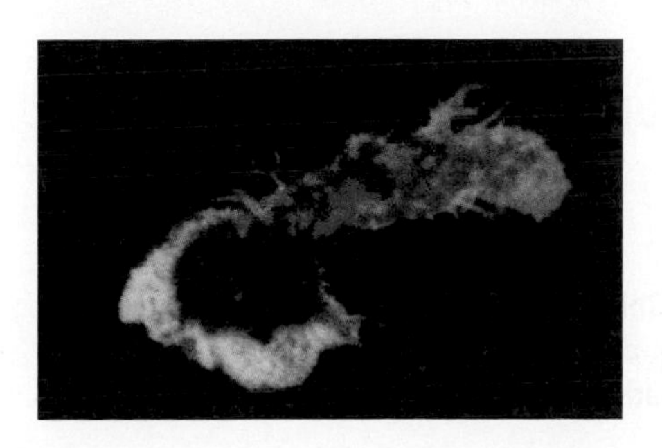

Fig. 16.1 An activated cytotoxic T cell moving over a surface. The actin cytoskeleton is stained green. The red staining indicates the lytic granules containing cytotoxic proteins. Photograph courtesy of Gillian Griffiths.

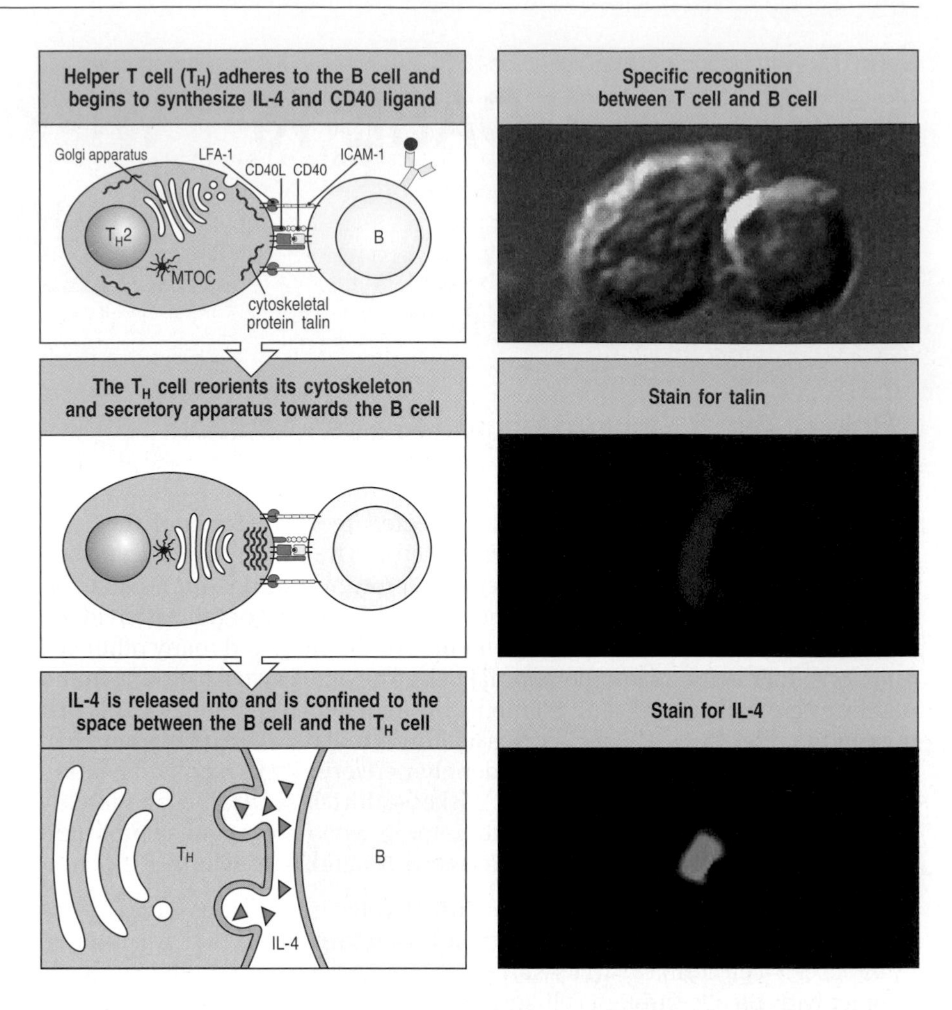

Fig. 16.2 Binding of a helper T cell to an antigen-binding B cell causes a reorganization of the cytoskeleton in the T cell. Engagement of the T-cell receptor causes the T cell to express the CD40 ligand (CD40L), which binds to CD40 on the surface of the B cell (top panel). The cross-linking of the receptors at the point of contact leads to a reorganization of the cortical actin cytoskeleton, shown here by the redistribution of the protein talin (shown in red in top left and both middle panels), which is associated with the actin cytoskeleton. Subsequent reorientation of the secretory apparatus of the T cell leads to cytokine release at the point of contact with the antigen-binding B cell, as shown by staining for IL-4 (bottom right panel), which shows the IL-4 (green) confined to the space between the B cell and the T cell. Photographs courtesy of A. Kupfer.

from the T cell is thus directed to the contact point. Similar cytoskeletal reorganizations occur when a cytotoxic T cell contacts its target cell.

Many other T-cell functions depend on the actin cytoskeleton. Like many other animal cells, T cells move in a crawling 'ameboid' fashion. The movement of T cells as they emigrate from the thymus into the blood vessels and subsequently 'home' from the bloodstream into lymphoid tissue requires the active participation of the actin cytoskeleton. Cell division induced by the activation of T cells by antigen or by nonspecific mitogens also involves the actin cytoskeleton in that the cell is divided into two by the action of a contractile ring formed of actin filaments and myosin.

T cells from patients with Wiskott–Aldrich syndrome are deficient in all these normal cellular abilities and, in particular, seem unable to interact successfully with B cells and other target cells. In addition, because the Wiskott–Aldrich syndrome protein is broadly expressed within the hematopoietic system, other blood cell lineages are also affected.

The case of Jonathan Stilton: the consequences of a failure of T-cell–mediated immune responses.

Jonathan Stilton was first referred to the Children's Hospital at 20 months of age with a history of recurrent infections, eczema, asthma, and episodes of bloody diarrhea and autoimmune hemolytic anemia. He had been a full-term baby and seemed quite

normal at birth. Routine immunizations of DPT (diphtheria, pertussis, and tetanus), conjugated *Haemophilus influenzae* type b (Hib) and pneumococcus vaccines, and inactivated polio vaccine had been given at 2, 4, and 6 months of age without any untoward consequences. At 6 months Jonathan's mother noticed eczema developing on his arms and legs; this was treated with 1% cortisone ointment. The eczema became infected with *Staphylococcus aureus*, and petechiae (small skin hemorrhages) appeared in the eczematous areas as well as on unaffected areas of skin. By 9 months of age, Jonathan began to have frequent infections of the middle ear and upper respiratory tract. At 1 year old, he developed pneumonia, which was confirmed by a chest radiograph. Between respiratory infections he started to wheeze and was found to have asthma.

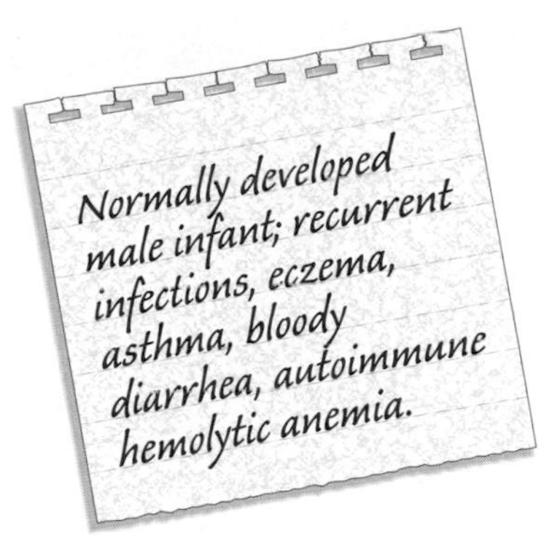

Normally developed male infant; recurrent infections, eczema, asthma, bloody diarrhea, autoimmune hemolytic anemia.

At 16 months of age, Jonathan had an episode of bloody diarrhea. A blood analysis showed low levels of hemoglobin of 9.5 g dl^{-1} and a normal white blood cell count of 6750 μl^{-1}. The proportions of granulocytes and lymphocytes were also normal. The platelet count was, however, very low (thrombocytopenia), at 10,000 μl^{-1} (normal 150,000–350,000 μl^{-1}), and the platelets were abnormally small (Fig. 16.3). He received two transfusions of platelets, after which the bloody diarrhea stopped, although the platelet count remained low at 40,000 μl^{-1}.

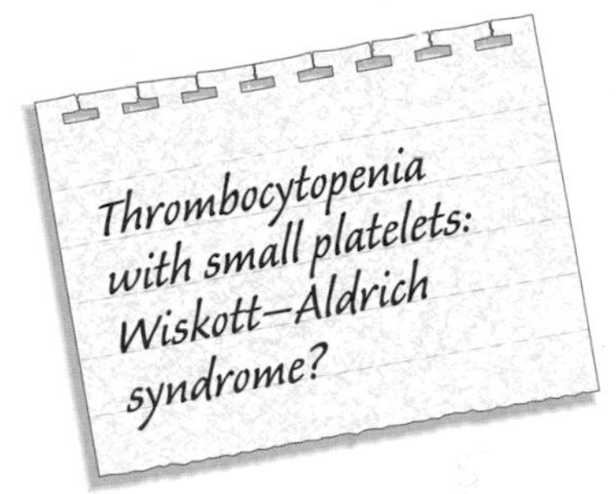

Two months later, Jonathan's mother noticed that he had become very pale, and his urine was darker than usual. He was brought in to the Emergency Room. At that time, his hemoglobin was as low as 5 g dl^{-1} and his platelet count remained low at 15,000 μl^{-1}. A positive Coombs direct antibody test was demonstrated, indicating that his red blood cells were coated with IgG antibody, and prompting the diagnosis of autoimmune hemolytic anemia. He was treated with corticosteroids (methylprednisone, 2 mg kg^{-1} body weight intravenously for 5 days). This resulted in an increase in hemoglobin levels to 8.5 g dl^{-1}. Jonathan was referred to the Children's Hospital for additional investigations.

Further immunological investigations at the time revealed IgG levels of 750 mg dl^{-1} (normal), IgM 25 mg dl^{-1} (decreased), IgA 475 mg dl^{-1} (increased), and IgE 1,750 ng l^{-1} (increased). Jonathan's red blood cells were type O, but no anti-A or anti-B isohemagglutinins were present in his serum. His titer of antibodies against tetanus toxoid (a protein) was 0.15 unit ml^{-1} (a borderline normal result). In addition, his antibody response to pneumococcus was protective to only two of the seven strains contained in the conjugated vaccine he had received. He was given a boosting immunization with polysaccharide vaccine to 23 pneumococcal strains, but 4 weeks later antibody titers remained unchanged.

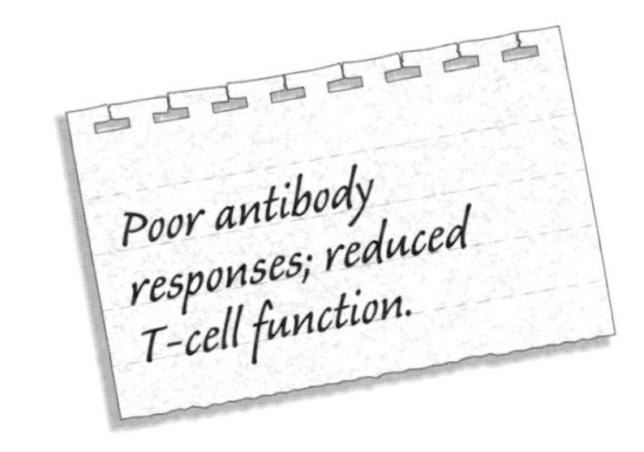

Jonathan had normal numbers of circulating B cells (11% of peripheral blood lymphocytes) and T cells (85%) with a normal distribution of CD4 T cells and CD8 T cells. However, the *in vitro* proliferative response of his T cells to the mitogen phytohemagglutinin was slightly diminished, and response to the mitogenic effect of anti-CD3 antibodies was severely decreased.

Because of the combination of recurrent infections, eczema, thrombocytopenia with small platelets, bloody diarrhea, autoimmune hemolytic anemia, and antibody deficiency, Wiskott–Aldrich syndrome (WAS) was suspected. Flow-cytometric and Western blot analysis showed lack of expression of the Wiskott–Aldrich syndrome protein (WASP) in the cytoplasm of Jonathan's lymphocytes, and a C$\rightarrow$T change at position

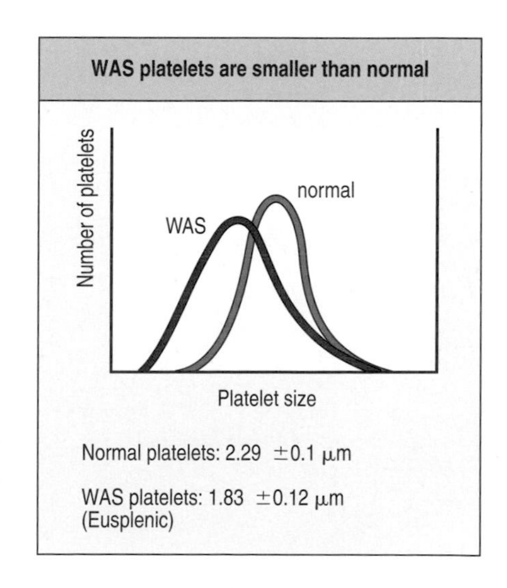

Fig. 16.3 Platelet sizing. Blood from a normal subject and from a patient with WAS was drawn into sodium citrate, an anticoagulant. The blood was then centrifuged and allowed to sediment at 1*g*; the platelet-rich plasma was removed. This was diluted in sterile buffer and scanned in a particle sizer. The mean diameter of normal platelets is 2.29 ± 0.1 μm. The mean diameter of platelets from patients with Wiskott–Aldrich syndrome is 1.83 ± 0.12 μm. Data kindly supplied by Dianne Kenney.

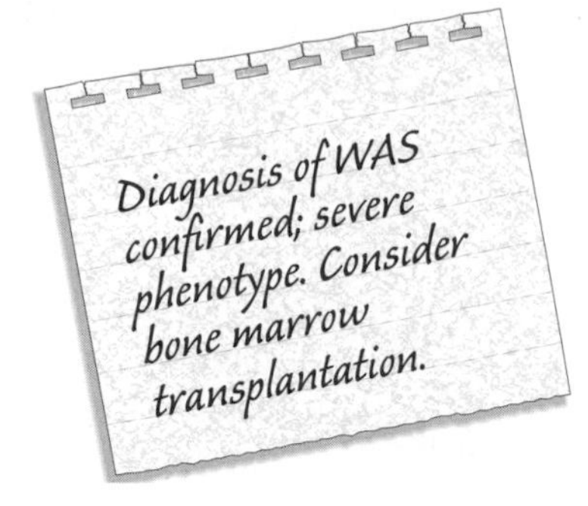

Diagnosis of WAS confirmed; severe phenotype. Consider bone marrow transplantation.

71 in the *WAS* gene, resulting in a premature stop at codon 13, was demonstrated, confirming the diagnosis of WAS.

Treatment with intravenous immunoglobulins (400 mg kg^{-1} every 3 weeks) and antibiotic prophylaxis with trimethoprim–sulfamethoxazole were started. Because of the severity of the disease, a bone marrow transplant was considered. Jonathan had two siblings, a 7-year-old brother and a 5-year-old sister. However, neither of them was HLA-identical. The search for a matched unrelated donor was initiated, and a suitable donor (matched at eight out of eight HLA loci tested) was identified 2 months later. At 26 months of age, Jonathan received a bone marrow transplant, following myeloablative chemotherapy. In addition, he received cyclosporin A for prophylaxis against graft-versus-host disease (GVHD). However, three weeks after transplant, he developed fever (at 39°C), skin rash, and slightly elevated liver enzymes (ALT 55 IU ml^{-1}; AST 47 IU ml^{-1}). Tests for pathogens were negative, and acute GVHD was diagnosed. Jonathan was treated with methylprednisone, resulting in rapid disappearance of the fever and rash, and normalization of liver enzymes. Now, 3 years after the bone marrow transplant, Jonathan lives at home. He has experienced no significant infections, his eczema has completely resolved and his platelet count has fully normalized at 220,000 ml^{-1}. He has received another round of immunization and has developed protective antibody titers to tetanus, Hib and to six out of the seven pneumococcal strains in conjugated anti-pneumococcal vaccine. His blood type has changed to B+ (the blood type of the bone marrow donor) and he has a normal titer of anti-A isohemagglutinin at 1:64.

Analysis of the family history revealed that Jonathan's mother had two older brothers. Of these, Mark was alive and well at 42 years of age, whereas Raymond had died at the age of 37 years. Raymond had developed eczema early in life that had persisted during the years. He had also suffered from recurrent bronchitis and had three episodes of pneumonia, which were treated with antibiotics. He also had petechiae and had been diagnosed with chronic idiopathic thrombocytopenic purpura because of his persistent thrombocytopenia. When he was 35 years old, Raymond developed a B-cell lymphoma positive for the Epstein–Barr virus. Although he was successfully treated with chemotherapy, Raymond died 2 years later of sepsis caused by a Gram-negative enterobacillus.

Wiskott–Aldrich syndrome.

Jonathan Stilton and his maternal uncle Raymond illustrate between them all the principal features of Wiskott–Aldrich syndrome (WAS). The syndrome was first described by Wiskott in 1937 in Munich when he observed three male infants with bloody diarrhea, eczema, and thrombocytopenia with small platelets. In 1947, Aldrich and his colleagues studied a large Dutch-American family in Minnesota and established the X-linked inheritance of the disease in five generations of affected males. The immunodeficiency that accompanies WAS was not appreciated until 1968, when it was described by Blaese and Waldmann and by Cooper and Good.

The *WAS* gene, mutations in which are responsible for WAS, has been mapped to the short arm of the X chromosome and cloned. It encodes a protein named the Wiskott–Aldrich syndrome protein (WASP), which has homology with actin-binding cytoskeletal proteins involved in the reorganization of the actin cytoskeleton in white blood cells and platelets. WASP is expressed only in white blood cells and megakaryocytes (from which platelets are derived), which explains the restriction of its effects to immune-system and blood clotting functions. In patients with WAS, T cells and platelets are defective in number and function. T-cell movement, capacity for cell division, capping of antigen receptors, and reorientation of the cytoskeleton on engagement with other cells are all impaired. In addition, the function of monocytes, macrophages, and dendritic cells is also affected, with significant defects in directional motility and phagocytosis. The cytotoxic function of natural killer cells (NK cells) is impaired. Finally, distinct abnormalities of B lymphocytes have also been described.

Abnormally small platelets are a distinct and characteristic feature of WAS. The destruction of blood platelets by the spleen is greatly increased, but it is not known what the spleen recognizes as abnormal on WAS platelets. The platelets in the circulation become spontaneously activated and they extrude their granules. For this reason they appear small and in fact prove to be so when their volume or diameter is measured in a particle sizer (see Fig. 16.3). Splenectomy may increase the platelet number; however, it also increases the risk of sepsis. T cells are also morphologically abnormal in patients with WAS because the cells lose their surface microvilli and assume a characteristically bald appearance (Fig. 16.4).

Patients with WAS have increased susceptibility to both pyogenic bacterial infections and opportunistic infections. Among the latter, severe chickenpox (varicella), herpes simplex and molluscum contagiosum (a viral infection of the skin, particularly eczematous skin) are frequently encountered. The increased susceptibility to such viral infections may be at least partly due to the impaired cytotoxic function of CD8 T cells and NK cells in WAS; the impairment seems to be in their inability to attach to target cells. Antibody formation, particularly against carbohydrate antigens, is defective, as we saw for Jonathan, who could not respond to anti-pneumococcal vaccine. He also did not make antibodies against the blood group antigens, which are complex carbohydrates.

Antibody responses to carbohydrate antigens in humans are mostly restricted to the IgM and IgG2 classes. 'Natural' antibodies of the IgM class include isohemagglutinins and antibodies directed against common blood-borne pathogens. The production of these natural antibodies is dependent on normal development of the marginal zone in the spleen, which usually takes about 2 years to mature in normal individuals. Patients with WAS, however, have severe abnormalities of the splenic architecture, with virtual lack of the marginal zone (Fig. 16.5), and this explains the lack of IgM antibodies against carbohydrate moieties. However, failure to produce IgG2 antibodies against capsular polysaccharides and blood group antigens in WAS indicates that, in

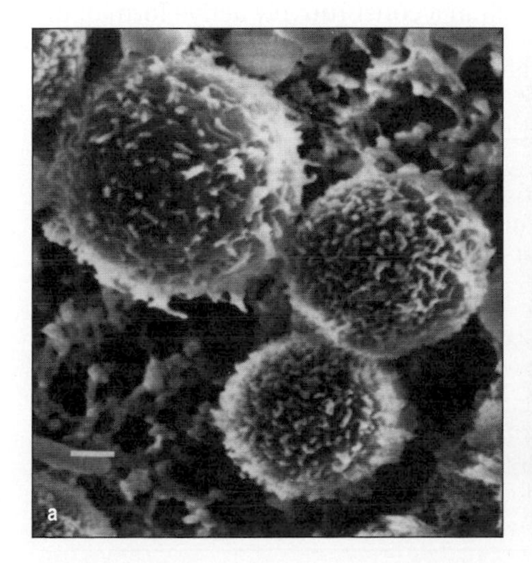

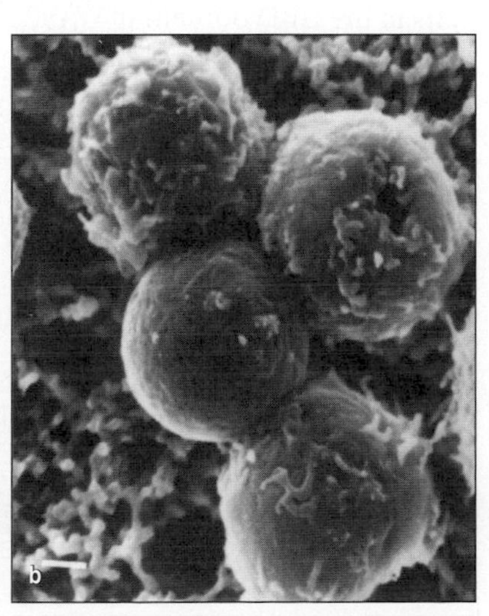

Fig. 16.4 Loss of microvilli on T cells in Wiskott–Aldrich syndrome. Scanning electron micrographs of normal lymphocytes (panel a) and lymphocytes from a patient with Wiskott–Aldrich syndrome (panel b). Note that the normal lymphocyte surface is covered with abundant microvilli, which are sparse or absent from the patient's lymphocytes. Photographs courtesy of Dianne Kenney.

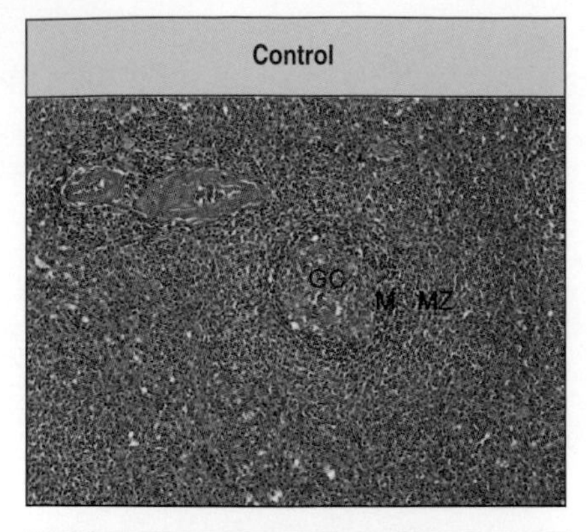

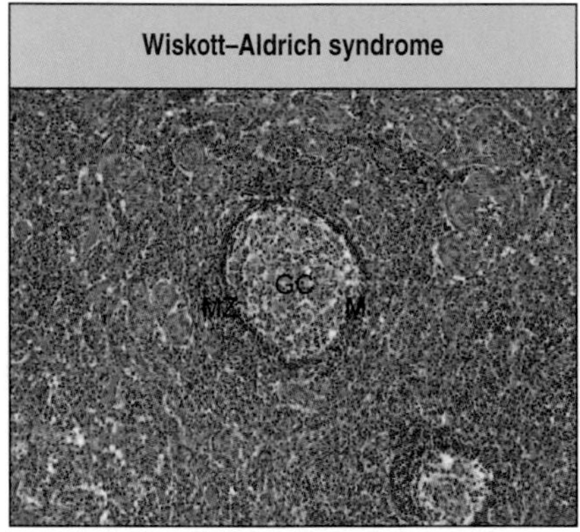

Fig. 16.5 Severe reduction of the marginal zone in the spleen from Wiskott–Aldrich syndrome patients. Hematoxylin and eosin staining of sections of spleen from a control subject (upper panel) and a patient with Wiskott–Aldrich syndrome (WAS) (lower panel). The germinal centers (GC) are surrounded by the mantle (M) and by the marginal zone (MZ), which contains B lymphocytes and specialized macrophages. In patients with WAS, the MZ is severely reduced in size. Magnification: ×40. Photographs courtesy of Fabio Facchetti and William Vermi.

humans, the production of antibodies against complex linear polysaccharides is not independent of T-cell help.

The importance of anti-polysaccharide antibodies is highlighted by the problems with infection encountered in individuals with X-linked agammaglobulinemia (see Case 1), and the importance of being able to switch to IgG isotypes is illustrated by the case of CD40 ligand deficiency (see Case 2). The fact that the lack of production of polysaccharide antibodies in patients with WAS is also due to T-cell defects was confirmed by early attempts at bone marrow transplantation to correct WAS. A few patients were converted into partial T-cell chimeras (the B lymphocytes remaining of recipient origin), which was sufficient to clear their eczema and enable them to make antibodies against carbohydrate antigens. Subsequently, complete success at replacing the T-cell population has been achieved by bone marrow transplantation after irradiation to wipe out the abnormal cells in the host.

Although Jonathan's T-cell number seemed normal when tested at 2 years old, patients with WAS typically show a progressive decline in T-cell count, often associated with poor control of viral infections, and especially of the Epstein–Barr virus (EBV), which can infect B lymphocytes. When this happens, the B cells undergo polyclonal expansion, which may eventually become oligoclonal or even monoclonal, leading to the development of B-cell lymphoma, as happened in one of Jonathan's uncles.

The structure and function of WASP have been studied intensively during the past few years (Fig. 16.6). The carboxy-terminal portion of the protein, the so-called VCA domain, binds to the actin-related protein complex (Arp2/3), and this binding initiates the nucleation of actin. Arp2/3 guides the branching of the actin filaments as they polymerize to make new filaments. WASP resides in the cytoplasm in an inactive form because the carboxy-terminal acidic domain (A) is bound to the GTPase-binding domain (GBD) so that inactive WASP appears as a hairpin. When the small G protein Cdc42 (in its GTP-bound form) binds to the GBD, WASP springs open into a linear activated form and the VCA domain can bind the Arp2/3 complex. Interestingly, missense mutations in the GBD domain of WASP result in a constitutively active form of the

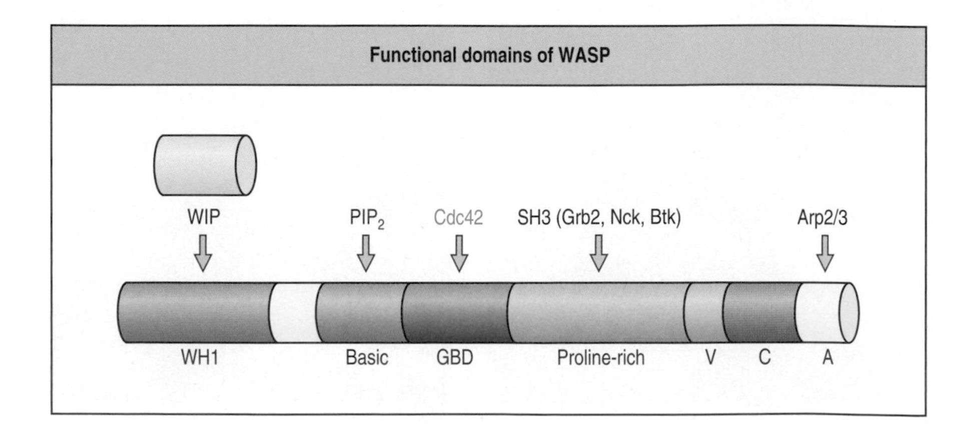

Fig. 16.6 A stick diagram of WASP (502 amino acids) shows the various domains of the molecule. The amino-terminal domain (WH1) binds WIP (the WASP interactive protein, which keeps WASP inactive until WIP is released). It is followed by a stretch of basic amino acids that bind the carboxy-terminal acid residues (A) of WASP, and this also keeps WASP in an inactive state. Phosphatidylinositol bisphosphate (PIP_2) also binds to this region and moves WASP from the cytoplasm to the immunologic synapse. Then comes a domain that binds the G protein Cdc42, followed by a proline-rich region, which is an SH3 domain that binds many adaptor proteins. Finally, the carboxy-terminal region of WASP is composed of verprolin (V)-like, cofilin (C)-like, and acidic (A) regions. These regions interact with the Arp2/3 complex and promote actin polymerization.

protein; this is associated with severe cellular abnormalities, especially in myeloid cells, that lead to severe congenital neutropenia and an increased risk of myelodysplasia, a pre-leukemic condition of the bone marrow. Finally, the amino-terminal WH1 domain of WASP binds to the WASP-interacting protein (WIP), and this binding stabilizes WASP. Amino acid substitutions in this domain result in decreased levels of WASP, and are typically responsible for a milder form of WAS, known as X-linked thrombocytopenia (XLT). Patients with XLT suffer from chronic thrombocytopenia, but they are less prone to severe infections, autoimmunity, or malignancies.

Mice with knockouts of the *WAS* gene show defective homing of lymphocytes and defective capping of the T-cell antigen receptor (Fig. 16.7), as well as defective responses of T cells to mitogens and anti-CD3 antibodies. B-cell function seems to be normal, with the exception of severe abnormalities of marginal zone B cells.

The high frequency of immune-complex disease and allergy in WAS remains to be explained. It may relate to defective function of regulatory T cells. That may also account for the high frequency of autoimmune manifestations in patients with WAS, especially autoimmune cytopenias.

WAS is a severe disease. Although regular administration of immunoglobulins, antibiotic prophylaxis and measures to avoid severe trauma-related bleeding are all important components of the treatment plan, severe manifestations and a failure to express WASP are an indication for hematopoietic cell

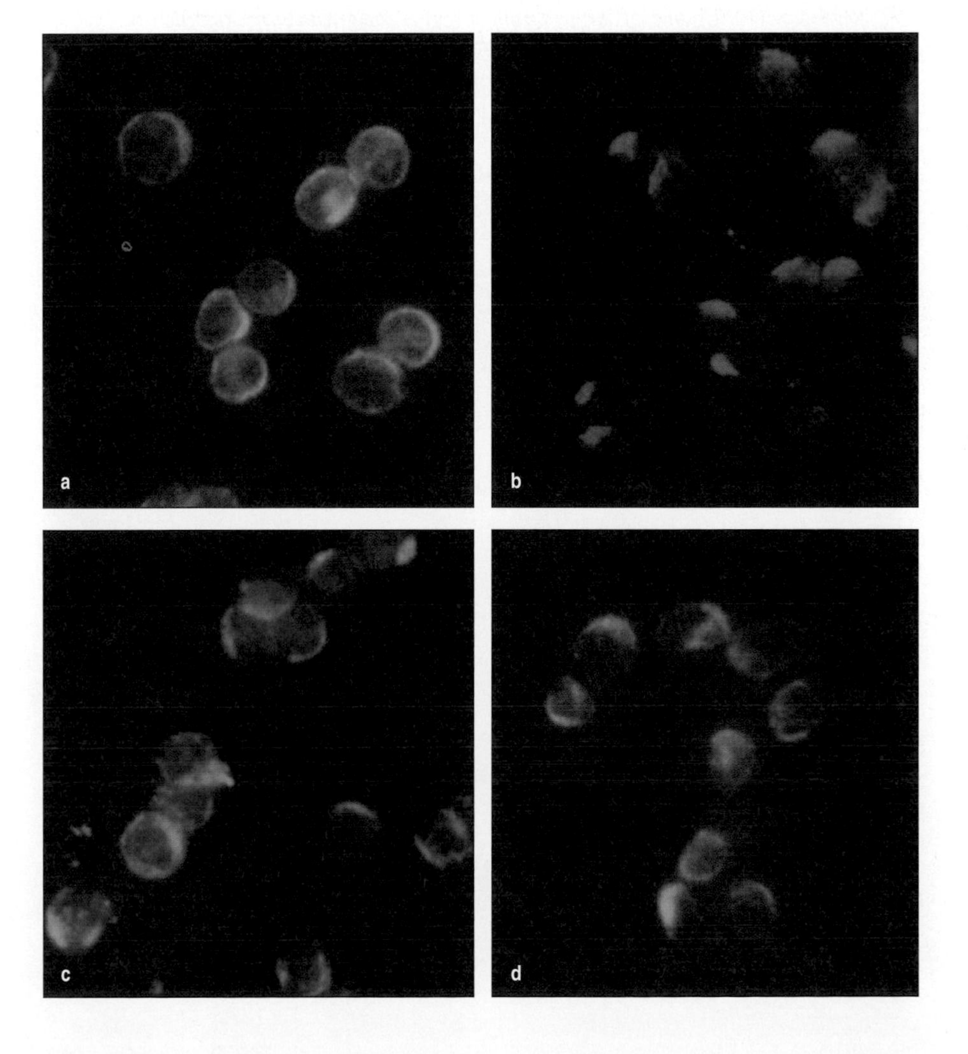

Fig. 16.7 Failure of T-cell capping in *Was* knockout mice. T cells were obtained from $Was^{-/-}$ mice in which the *Was* gene encoding the protein WASP had been knocked out by homologous recombination. The T cells were treated with anti-CD3ε antibody, which binds to the T-cell receptor:CD3 complex, and were then examined with a fluorescent antibody against CD3. Normal resting T cells (panel a) show capping of the T-cell receptor after anti-CD3ε treatment (panel b). Panel c shows resting T cells from $Was^{-/-}$ mice, and panel d shows the failure of cap formation after stimulation of the T-cell receptor. Photographs courtesy of Scott Snapper.

transplantation. Excellent results have been achieved with transplantation from matched related or unrelated donors; however, prior myeloablative treatment is required to allow the robust engraftment of donor cells. Splenectomy is no longer considered a treatment of choice, because of the increased risk of sepsis. Initial trials of gene therapy to treat WAS are being conducted. Patients with XLT are typically treated with conservative management.

Questions.

1 *When Mrs Stilton and other female heterozygous carriers of WAS are examined for randomness of X-chromosome inactivation, nonrandom inactivation of the WAS-bearing X chromosome is found in all the blood cell lineages—monocytes, eosinophils, basophils, neutrophils, B lymphocytes, and CD4 and CD8 T lymphocytes. When hematopoietic stem cells are isolated from these women, they also exhibit nonrandom X inactivation. How might this be explained?*

2 *Can you devise a strategy that might induce isotype switching in WAS B cells to overcome the lack of T-cell–B-cell collaboration?*

3 *Males with XLT are often initially given a diagnosis of chronic idiopathic thrombocytopenia purpura. Can you think of any laboratory test that would make you suspect XLT?*

CASE 17 Autoimmune Polyendocrinopathy–Candidiasis–Ectodermal Dystrophy (APECED)

Role of the thymus in the negative selection of T lymphocytes.

Millions of different T-cell antigen receptors are generated during the development of thymocytes in the thymus gland. This stochastic process inevitably leads to the formation of some T-cell antigen receptors that can bind to self antigens. When a T-cell receptor of a developing thymocyte is ligated by antigen on stromal cells in the thymus, the thymocyte dies by apoptosis. This response of immature T lymphocytes to stimulation by antigen is the basis of negative selection (Fig. 17.1). Elimination of these T cells in the thymus aborts their subsequent potentially harmful activation.

Experiments in mice have shown that self-MHC:self-peptide complexes encountered in the thymus purge the T-cell repertoire of immature T cells bearing self-reactive receptors. The bone marrow derived dendritic cells and macrophages of the thymus are thought to have the most significant role in negative selection (Fig. 17.2), but thymic epithelial cells may also be involved.

Most of what we know about negative selection has come from studying transgenic mice. How the myriad of self antigens, many of them organ specific, can be 'seen' in the developing thymus gland by developing thymocytes remained largely undefined until the genetic defect in the human autosomal recessive disease APECED (autoimmune polyendocrinopathy–candidiasis–ectodermal dystrophy) was identified. The gene that is mutated in this disease is called *AIRE* (for *a*uto*i*mmune *re*gulator), and its discovery has given new insight into the process of negative selection of thymocytes.

Topics bearing on this case:
Lymphocyte development
Negative selection of T lymphocytes
Autoimmune disease

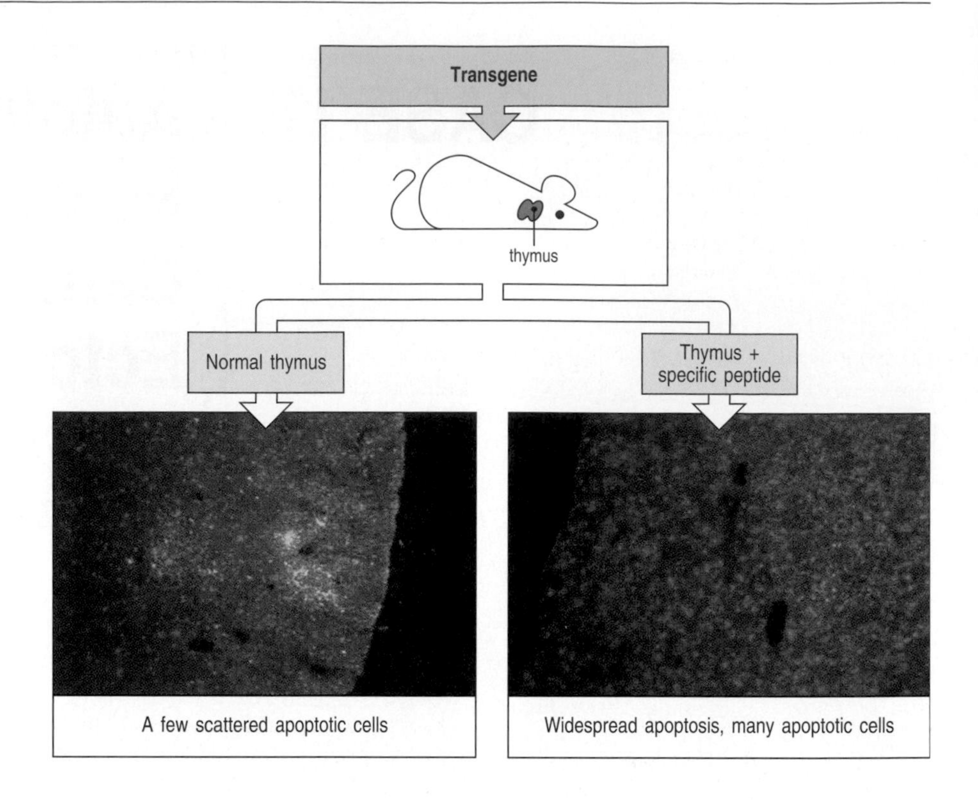

Fig. 17.1 T cells specific for self antigens are deleted in the thymus. In mice transgenic for a T-cell receptor that recognizes a known peptide antigen complexed with self MHC, all the T cells have the same specificity. In the absence of the peptide, most thymocytes mature and emigrate to the periphery. This can be seen in the lower left panel, in which a normal thymus has been stained with antibody to identify the medulla (in green), and by the TUNEL technique to identify apoptotic cells (in red). If the mice are injected with the peptide that is recognized by the transgenic T-cell receptor, then massive cell death occurs in the thymus, as shown by the increased numbers of apoptotic cells in the lower right-hand panel. Photographs courtesy of A. Wack and D. Kioussis.

The case of Robert Jordan: a young man with a lifelong history of autoimmune diseases of various endocrine glands.

When Robert was 18 months old his mother took him to the pediatrician, complaining that his skin was very dry and that his movements seemed sluggish compared with those of a normal active infant. Robert's pediatrician made the diagnosis of hypothyroidism. He prescribed Synthroid (thyroid hormone) and in a few weeks the child was much improved. But by the time Robert was almost 6 years old his mother felt that he was not growing at a normal rate. The pediatrician referred Robert to Dr Hemingway, an endocrinologist at the Children's Hospital. On examination he found that Robert's height and weight were below the third centile for his age. An X-ray of Robert's wrist revealed a bone age of 3 years 6 months when he was, in fact, 5 years 10 months old.

A blood test showed that Robert's level of thyroid-stimulating hormone (TSH) was more than 60 ng dl^{-1}. This is several times the upper limit of normal TSH levels and indicated that Robert was receiving inadequate thyroid hormone replacement. The dose of Synthroid was increased from 0.05 to 0.075 mg per day. Robert resumed normal growth and his bone age subsequently caught up with his actual age.

As Dr Hemingway pondered the cause of Robert's hypothyroidism, he learned from Mrs Jordan that she had another child, a daughter 2 years older than Robert. This girl had nail dystrophy, perioral candidiasis, hypoparathyroidism, and serum antibodies against islet cells of the pancreas (but no clinically apparent diabetes). This led Dr Hemingway to suspect that Robert had the inherited disease APECED, an autosomal recessively inherited abnormality. He tested Robert's serum calcium levels to make sure that he had not developed hypoparathyroidism like his sister; they were normal. Neither did Robert's serum contain antibodies against any other endocrine glands.

Age 6, growth less than normal. Increase thyroid hormone.

18-month-old boy with lethargic movements and dry skin.

Older sibling with various autoimmune conditions. Inherited autoimmune disease? APECED?

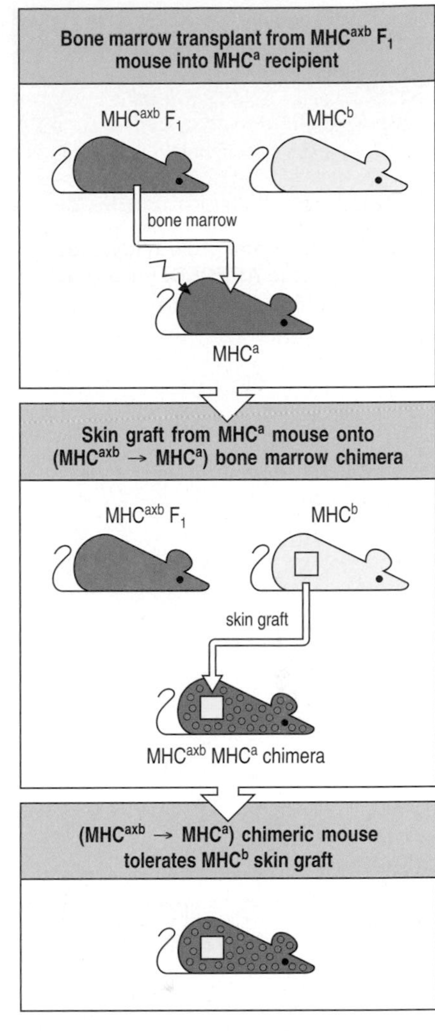

Fig. 17.2 Bone marrow derived cells mediate negative selection in the thymus. When MHC$^{a×b}$ F_1 bone marrow is injected into an irradiated MHC^a mouse, the T cells mature on thymic epithelium expressing only MHC^a molecules. Nevertheless, the chimeric mice are tolerant to skin grafts expressing MHC^b molecules (provided that these grafts do not present skin-specific peptides that differ between strains a and b). This implies that the T cells whose receptors recognize self antigens presented by MHC^b have been eliminated in the thymus. As the transplanted MHC$^{a×b}$ F_1 bone marrow cells are the only source of MHC^b molecules in the thymus, bone marrow derived cells must be able to induce negative selection.

Dr Hemingway had noticed that Robert's fingernails were thickened, with longitudinal notches and ridging (Fig. 17.3). He referred Robert to a dermatologist, who agreed that this abnormality was consistent with APECED. Nail scrapings cultured for *Candida albicans* were negative. The dermatologist also noticed two patches of hair loss at the top and back of Robert's scalp. He could easily pull out the hair, whose roots looked atrophied under the microscope, and made the diagnosis of alopecia areata (patchy hair loss).

Robert continued to grow satisfactorily, and at 8 years old his TSH level was 2.5 ng dl^{-1}, a normal level which indicated adequate thyroid hormone replacement. But he continued to lose hair in patches and lost his eyebrows. He developed fissures at the angle of his mouth as a result of infection with *C. albicans*. He was taunted at school about his bizarre appearance, and his schoolwork deteriorated. With his parents' divorce pending, Robert became depressed and took an overdose of Synthroid in a suicide attempt. He received intensive psychotherapy for his depression.

As Robert approached puberty, his scrotum became darkly pigmented, as did the areolae around his nipples. Dr Hemingway suspected that Robert was developing adrenal insufficiency (Addison's disease). A blood test revealed a level of adrenocorticotropic hormone (ACTH) three times the upper limit of normal. The doctor prescribed the steroid prednisone at 5 mg per day and Fluorinef (which conserves sodium and potassium excretion) at 0.1 mg a day.

When he was 18 years old Robert noticed that he had started to bruise easily and that his gums bled after brushing his teeth. Dr Hemingway sent him to a hematologist, who found a low platelet count of 34,000 μl^{-1} The hematologist decided that Robert had developed yet another autoimmune disease, idiopathic thrombocytopenic purpura (ITP).

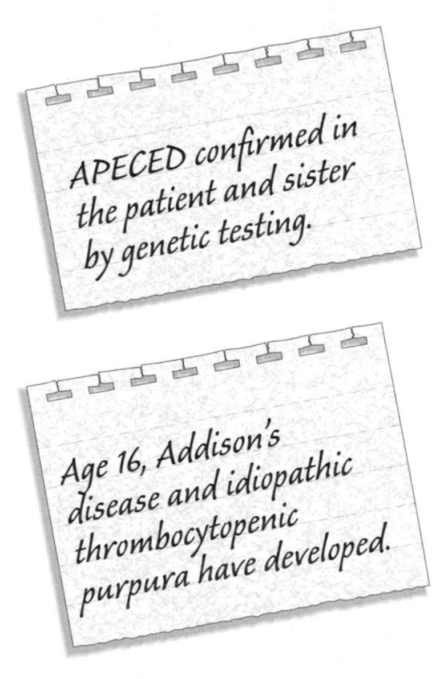

APECED confirmed in the patient and sister by genetic testing.
Age 16, Addison's disease and idiopathic thrombocytopenic purpura have developed.

Autoimmune polyendocrinopathy–candidiasis–ectodermal dystrophy (APECED).

APECED is also known as autoimmune polyglandular syndrome (APS) type I. Although there are several polyglandular diseases of unknown origin, APECED is the only one known to have a pattern of autosomal recessive inheritance. Affected individuals such as Robert Jordan develop a wide range of autoantibodies not only against organ-specific antigens of the endocrine glands but also against antigens in the liver and skin, and, as we have seen in this case, against blood cells such as platelets. In addition to these autoimmune diseases, affected individuals have abnormalities of various ectodermal elements such as fingernails, teeth, and skin. They also have an increased susceptibility to infection with the yeast *Candida albicans*.

APECED has a high incidence among Finns, Sardinians, and Iranian Jews, as high as 1 in 5000 to 1 in 9000 births in some populations. This made it possible to map the gene responsible to chromosome 21 in affected Finnish families.

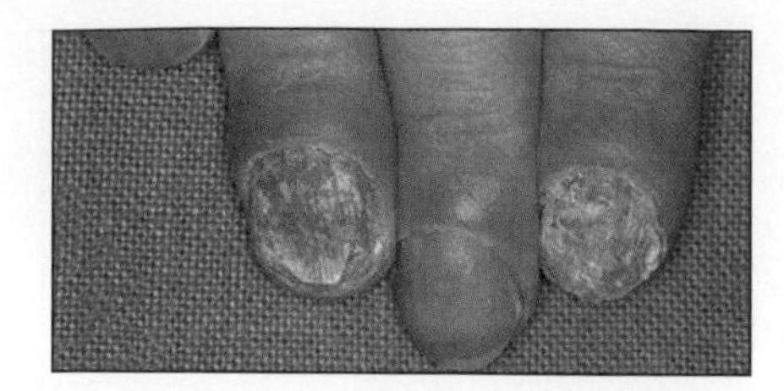

Fig. 17.3 Abnormal growth (dysplasia) of fingernails in APECED. Photograph courtesy of Mark S. Anderson.

The gene was cloned and named *AIRE* for autoimmune regulator. It was the first identified gene outside the major histocompatibility complex to be associated with autoimmune disease, and seems to encode a transcriptional regulator. Robert and his sister were found to have a deletion of 13 base pairs in exon 8 of the gene. Their parents were both heterozygous for this mutation.

When the equivalent gene *Aire* was 'knocked out' in mice by homologous recombination, they developed autoimmune diseases just like the human patients; the extent and severity of these autoimmune diseases progressed as the mice aged. The protein AIRE was found normally to be expressed predominantly in the epithelial cells of the medulla of the thymus and only weakly in peripheral lymphoid tissue. AIRE induces the expression of 200–1200 genes in the thymic medulla, including genes that encode organ-specific antigens of the salivary glands and the zona pellucida of ova (Fig. 17.4), among others. In normal circumstances, the expression of these self peripheral tissue antigens in the thymus permits negative selection of those developing T cells that react to them. When AIRE is lacking, these antigens are also not present. The potentially self-reactive T cells are not removed from the repertoire in the thymus and leave to cause autoimmune reactions in peripheral organs. The molecular mechanisms by which AIRE controls the expression of peripheral tissue antigens in the thymic medulla are complex, and involve recognition of unmethylated histone 3, transcriptional activation, and control of mRNA splicing.

It is has recently been shown that autoimmunity accounts also for the selective susceptibility to candidiasis (*C. albicans* infection) in patients with APECED. In particular, these patients have autoantibodies against IL-17A, IL-17F, and IL-22—that is, cytokines with a critical role in the control of fungal infections.

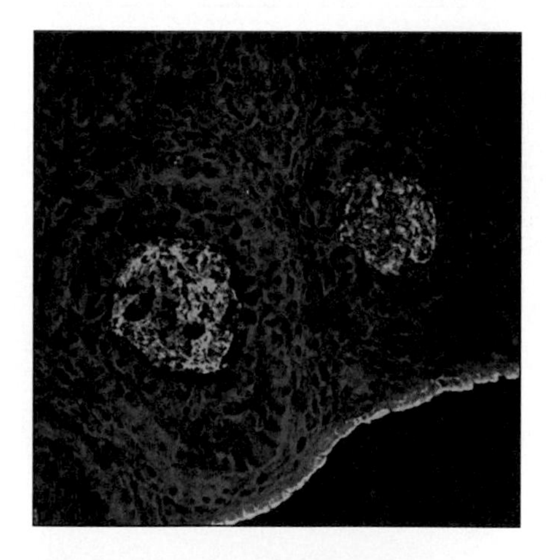

Fig. 17.4 Ovarian follicles stained with serum from a patient with APECED. Serum was reacted with a microscopic section of ovary and then with fluorescent antibody against human IgG, which reveals serum IgG bound to antigens in the zona pellucida around the oocyte (green). Photograph courtesy of Mark S. Anderson.

Questions.

1 *What caused the deep pigmentation of Robert's scrotum and the areolae around his nipples that led to the suspicion that he had developed an autoantibody against his adrenal cortical cells (Addison's disease)?*

2 *What abnormality might be found in the lymph nodes of the mice that lacked the Aire gene?*

3 *When lymphocytes from Aire-deficient mice were transferred into Rag-deficient mice, which have no mature lymphocytes of their own, autoimmune disease developed in the recipient. This did not occur when lymphocytes from normal mice of the same strain were transferred into Rag-deficient mice. What can be deduced from this experiment?*

4 *Primary ovarian failure has frequently been observed in females with APECED. What is the underlying mechanism?*

CASE 18 Immune Dysregulation, Polyendocrinopathy, Enteropathy X-linked Disease

A failure of peripheral tolerance due to defective regulatory T cells.

The primary role of the immune system is to recognize pathogens and eliminate them from the body. However, an equally important task is to distinguish potentially dangerous antigens from those that are harmless. Countless innocuous foreign antigens are encountered every day by the lungs, gut, and skin, the interfaces between the body and the environment. Similarly, the body contains numerous self antigens that might bind to the specific antigen receptors on B and T cells. Activation of the immune system by such innocuous antigens is unnecessary and may lead to unwanted inflammation. Allergic and autoimmune diseases are well-known examples of such unwanted and potentially destructive responses.

Fortunately, unwanted immune responses are normally prevented or regulated by the phenomenon of immunologic tolerance. This is defined as nonresponsiveness of the lymphocyte population to the specific antigen, and arises at two stages of lymphocyte development. Central tolerance is the result of the removal of self-reactive lymphocytes in the central organs; an autoimmune disease due to a defect in central tolerance is described in Case 17. Peripheral tolerance, in contrast, inactivates those T and B cells that escape central tolerance and exit to the periphery. Defects in either central or peripheral tolerance can result in unwanted or excessive immune responses.

Several mechanisms of peripheral tolerance exist (Fig. 18.1). One whose importance is increasingly being recognized is the network of regulatory cells that prevent or limit the activation of T cells, including self-reactive T cells,

Topics bearing on this case:
Peripheral tolerance
Regulatory T cells
Central tolerance
Autoimmunity

This case was prepared by Raif Geha, MD, in collaboration with Itai Pessach, MD.

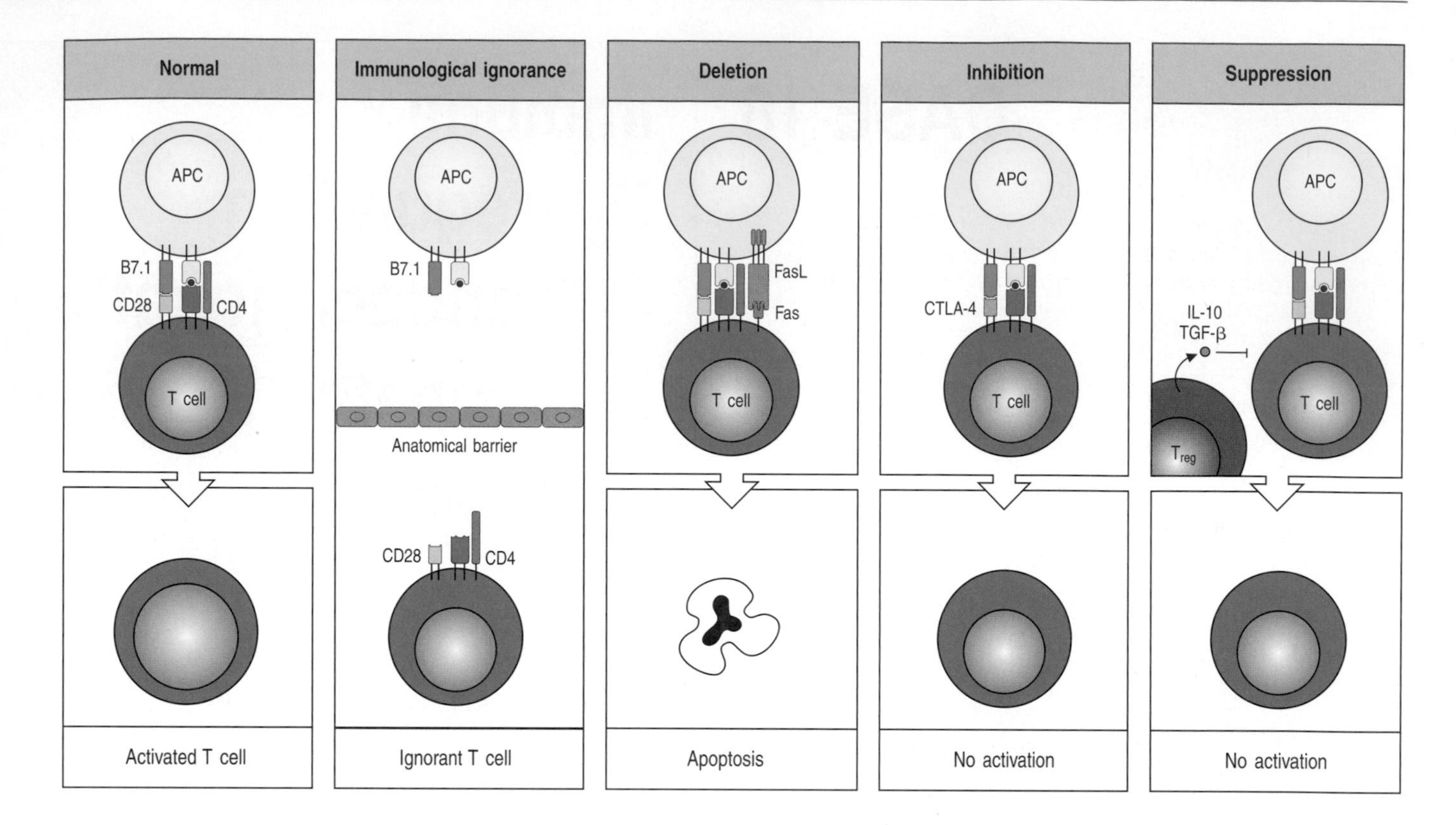

Fig. 18.1 Mechanisms of peripheral immunologic tolerance. T cells that are physically separated from their specific antigen—for example, by the blood–brain barrier—cannot become activated, a circumstance referred to as immunologic ignorance. T cells that express Fas (CD95) on their surface can receive signals from cells that express Fas ligand, leading to their deletion. The activation of naive T cells can be inhibited if the cell-surface protein CTLA-4 (CD152) binds B7.1 (CD80) on antigen-presenting cells (APCs). Regulatory T cells (mainly CD4 CD25 Foxp3-expressing) can inhibit, or suppress, other T cells, most probably through the production of inhibitory cytokines such as IL-10 and TGF-β.

and the consequent destructive inflammatory processes. When these regulatory cells do not function properly, problems can arise. A key cell type responsible for the maintenance of peripheral tolerance is the CD4 CD25 regulatory T cell (T_{reg}), also known as the natural regulatory T cell, which seems to become committed to a regulatory fate while still in the thymus and represents a small subset of circulating T cells (5–10%).

Although these cells were first characterized by their cell-surface CD25 (the α chain of the IL-2 receptor), this protein also appears on other T cells after activation. Natural T_{reg} cells are better characterized by their expression of the transcription factor Foxp3, which is essential for their specification and function as regulatory cells. Over the past decade they have emerged as crucial to the maintenance of peripheral tolerance. Neonatal thymectomy in mice and thymic hypoplasia in humans (DiGeorge syndrome; see Case 9) result in impaired generation of natural T_{reg} cells and the development of organ-specific autoimmune disease. The generation of natural T_{reg} cells in the thymus requires interaction with self-peptide:MHC class II complexes on cortical epithelial cells.

A second group of regulatory T cells seems to be induced from naive CD4 T cells in the periphery. These cells are $CD4^+$ $CD25^-$ and are heterogeneous, including subsets known as T_H3, T_R1, and a $CD4^+$ $CD25^-$ $Foxp3^+$ subset. Recently, a novel and rare population of T_{reg} cells that are CD8-positive has been described. Whereas CD4 T_{reg} subsets have been extensively studied, less is known about CD8 T_{reg} cells, their subsets, and their modes of action. NK cells and NKT cells have also been shown to be able to regulate immune responses. As a group, regulatory cells represent just one mechanism in a complex system of immunologic tolerance, acting to prevent or rein in unwanted immune responses.

The following case illustrates how a breakdown in peripheral tolerance as a result of a defect in regulatory T cells leads to a constellation of allergic symptoms, gastrointestinal symptoms, and autoimmune disease in infancy.

The case of Billy Shepherd: a defect in peripheral tolerance leading to dermatitis, diarrhea, and diabetes.

Billy was born at full term and developed atopic dermatitis (see Case 51) shortly after birth. This was treated by skin hydration and by the local application of hydrocortisone and antihistamines to control itching, the treatment being only partly successful. At 4 months of age, Billy developed an intractable watery diarrhea. Although he had initially gained weight well, by now his weight had fallen below the third centile for his age. At 6 months old, Billy started to develop high blood glucose levels and glucose in the urine. He was diagnosed with type 1 diabetes (insulin-dependent diabetes mellitus) and was referred by his pediatrician to the endocrine clinic at the Children's Hospital.

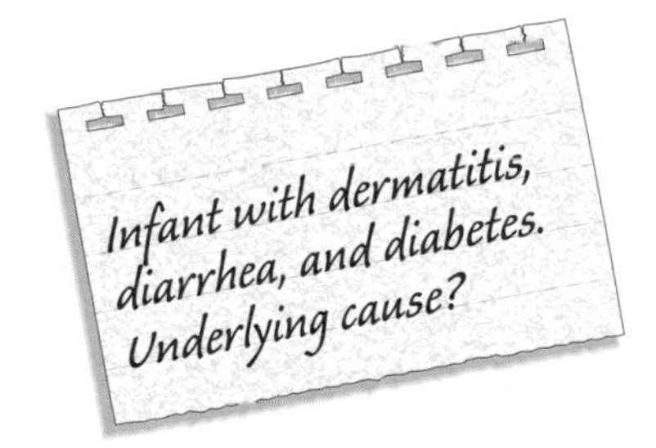

Infant with dermatitis,
diarrhea, and diabetes.
Underlying cause?

When first seen at the clinic, Billy weighed 5 kg (the third centile for age is 6.3 kg). He had diffuse eczema and sparse hair (Fig. 18.2). His cervical and axillary lymph nodes and spleen were enlarged. Laboratory tests revealed a normal white blood cell count of 7300 μl^{-1}, a normal hemoglobin of 11.3 g dl^{-1}, and a normal platelet count of 435,000 μl^{-1}. The percentage of eosinophils in the blood was high at 15% (normal <5%), and IgE was also elevated, at 1345 IU ml^{-1} (normal <50 IU ml^{-1}). Autoantibodies were found against glutamic acid decarboxylase (the GAD65 antigen) and against pancreatic islet cells. Billy was started on insulin therapy, which controlled his blood glucose level.

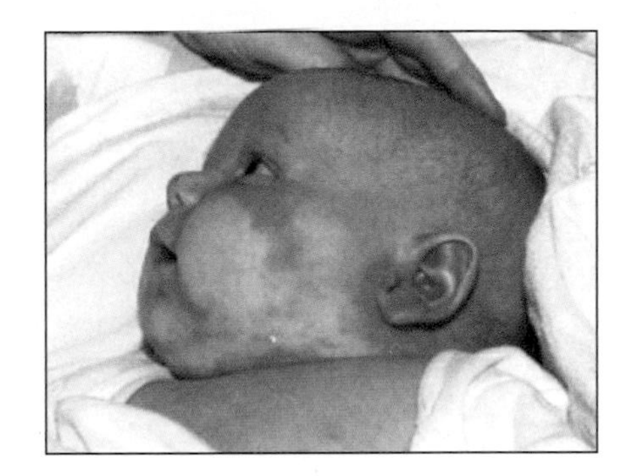

Fig. 18.2 Eczematous rash on the face of a baby boy with IPEX. Photograph courtesy of Talal Chatila, UCLA.

Because of the persistent diarrhea and failure to thrive, Billy required parenteral (intravenous) nutrition to maintain his weight. An endoscopy was ordered, to ascertain the cause of his persistent diarrhea, and a duodenal biopsy revealed almost total villous atrophy—an absence of villi in the lining of the duodenum—with a dense infiltrate of plasma cells and T cells (Fig. 18.3).

When Billy's mother was questioned, she revealed that there had been another son, who had died in infancy with severe diarrhea and a low platelet count. On the basis of Billy's symptoms and the family history, IPEX (immune dysregulation, polyendocrinopathy, enteropathy X-linked) was suspected. A FACS analysis of Billy's peripheral blood mononuclear cells revealed a lack of both CD4 CD25 cells and CD4 Foxp3-positive cells. Sequencing of Billy's *FOXP3* gene revealed a missense mutation, confirming the diagnosis.

With the diagnosis established, immunosuppressive therapy, including cyclosporin and tacrolimus, was started. Billy's diarrhea, glucose control, and eczema all improved markedly. After several months, however, his symptoms began to return and he stopped gaining weight. Shortly afterwards, he developed thrombocytopenia (a deficiency of blood platelets) and anti-platelet antibodies were detected.

The decision was made for Billy to be given a bone marrow transplant from his 5-year-old HLA-identical sister. In the weeks of conditioning leading up to the transplant, the diarrhea and eczema resolved. After transplantation, full engraftment of his sister's

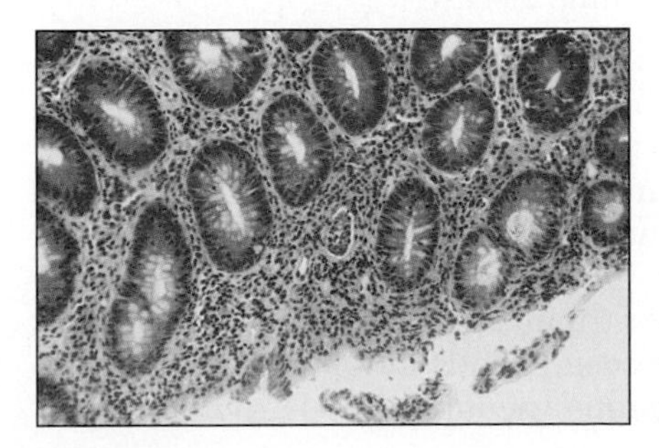

Fig. 18.3 Photomicrograph of duodenal biopsy from a child with IPEX. The section was stained with hematoxylin and eosin. Note the dense mononuclear cellular infiltrate. Photograph courtesy of Talal Chatila, UCLA.

stem cells was established. Two weeks after transplantation, anti-platelet, anti-GAD65, and anti-islet cell antibodies could not be detected. A year after the transplant, Billy continues to be symptom-free, although analysis for chimerism reveals that only 30% of his T cells are derived from his sister's cells.

Immune dysregulation, polyendocrinopathy, enteropathy X-linked disease (IPEX).

IPEX is a very rare disease caused by mutations in the gene for the forkhead transcription factor Foxp3, which is essential for the function of CD4 CD25 T_{reg} cells. Foxp3 expression is restricted to a small subset of TCRα:β T cells and defines two pools of regulatory T cells: $CD4^+$ $CD25^{high}$ T cells and a minor population of $CD4^+$ $CD25^{lo/neg}$ T cells. Ectopic expression of *Foxp3 in vitro* and *in vivo* is sufficient to convert naive murine CD4 T cells to T_{reg} cells. In contrast, overexpression of *FOXP3* in naive human $CD4^+$ $CD25^-$ T cells *in vitro* will not generate potent suppressor activity, suggesting that additional factors are required. Foxp3 expression and suppressor function can, however, be induced in human $CD4^+$ $CD25^-$ $Foxp3^-$ cells by cross-linking of the T-cell receptor and stimulation via the co-stimulatory receptor CD28, or after antigen-specific stimulation. This suggests that *de novo* generation of T_{reg} cells in the periphery may be a natural consequence of the human immune response.

T_{reg} cells are anergic *in vitro*. They fail to secrete IL-2 or proliferate in response to ligation of their T-cell receptors, and depend on the IL-2 generated by activated CD4 T cells to survive and exert their function. An *in vitro* assay that measures the ability of CD4 CD25 T cells to suppress CD4 T-cell proliferation is commonly used to test for T_{reg} function (Fig. 18.4). How T_{reg} cells suppress immune responses *in vivo* is still unclear. There is some evidence for contact-dependent inhibition, whereas other studies suggest that regulatory T cells exert their function by secreting immunosuppressive cytokines such as IL-10 or transforming growth factor-β (TGF-β), or by directly killing their target cells in a perforin-dependent manner.

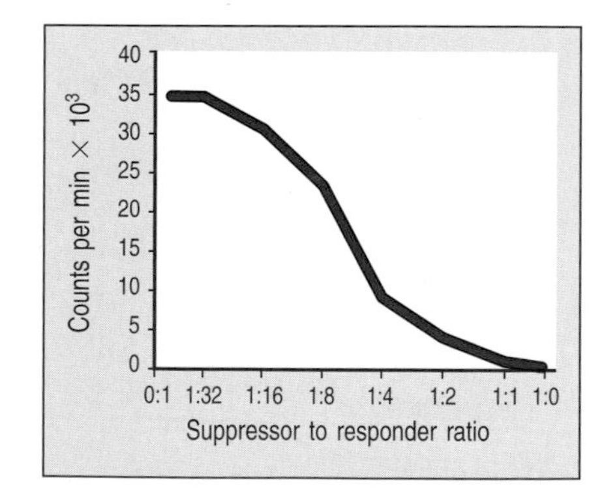

Fig. 18.4 Results of a functional assay for regulatory T cells from a normal individual. CD4 responder cells and CD4 CD25 T regulatory (suppressor) cells were mixed, together with antigen-presenting cells, at the ratios shown on the horizontal axis. The cells were stimulated with immobilized plate-bound anti-CD3 and soluble anti-CD28 for 3 days, then assessed for proliferation as measured by the incorporation of ^{3}H-labeled thymidine into DNA.

Several lines of evidence show that Foxp3 is crucial for the development and function of CD4 CD25 T_{reg} cells in mice. A mutation in *Foxp3* is responsible for an X-linked recessive inflammatory disease in the *Scurfy* mutant mouse. Male mice hemizygous for the mutation succumb to a CD4 T-cell-mediated lymphoproliferative disease characterized by wasting and multi-organ lymphocytic infiltration. T_{reg} cells are absent in *Scurfy* mice and in mice that have another spontaneous mutation in the *Foxp3* gene. In addition, specific ablation of *Foxp3* in T cells only is sufficient to induce the full lymphoproliferative autoimmune syndrome observed in the *Foxp3*-knockout mice. The Scurfy phenotype can be rescued by the introduction of a *Foxp3* transgene or by bone-marrow reconstitution, demonstrating the causative role of *Foxp3* in pathogenesis. Thus, the lack of *Foxp3*-expressing T_{reg} cells alone is sufficient to break self-tolerance and induce autoimmune disease.

In humans, missense or frameshift mutations in *FOXP3* result in loss of function of T_{reg} cells and uninhibited T-cell activation. As seen in Billy's case, the most common symptoms are an intractable watery diarrhea, leading to failure to thrive, dermatitis, and autoimmune diabetes developing in infancy. The diarrhea is due to widespread inflammation of the gut, including the colon (colitis), that results in villous atrophy, which reduces the absorptive capacity of the intestinal lining and thus contributes to wasting. Other diseases of immune dysregulation that are also seen include autoimmune thrombocytopenia, neutropenia, anemia, hepatitis, nephritis, hyperthyroidism or hypothyroidism, and food allergies. Autoantibodies also accompany these autoimmune diseases. Affected patients may also suffer more frequent infections, including sepsis, meningitis, or pneumonia, although the reason for the increased susceptibility to infection is unclear. Patients generally have normal immunoglobulin levels (except for the elevated IgE), and their ability to make specific antibody is intact.

Questions.

1 *Billy continued to remain asymptomatic even though he eventually only had around 30% engraftment of his sister's bone marrow stem cells (as judged from the proportions of T cells). Why is full engraftment not necessary in patients with IPEX?*

2 *Why did Billy's diarrhea improve while he was being prepared for transplantation?*

3 *Intravenous immunoglobulin (IVIG) has been used to treat IPEX. How might this be an effective therapy?*

4 *The occurrence of colitis in IPEX suggests that T_{reg} cells may be implicated in its pathogenesis and that they might be used therapeutically in the more common forms of colitis. Is there experimental data to support this claim?*

5 *What other gene mutations can give rise to a clinical picture similar to IPEX?*

CASE 19 Autoimmune Lymphoproliferative Syndrome (ALPS)

Increased survival of lymphocytes as a result of a mutation in Fas.

When antigen-specific lymphocytes are activated through their antigen receptors, they undergo blast transformation and then begin to increase their numbers exponentially by cell division. This clonal expansion can continue for up to 7 or 8 days, so that lymphocytes specific for the infecting antigen increase vastly in number and can come to predominate in the population. In the response to certain viruses, 50% or more of the CD8 T cells at the peak of the response are specific for a single virus-derived peptide:MHC class I complex. After clonal expansion, the activated lymphocytes undergo their final differentiation into effector cells; these remove the pathogen from the body and so terminate the antigenic stimulus.

When an infection has been overcome, activated effector T cells are no longer needed, and cessation of the antigenic stimulus prompts them to undergo programmed cell death, or apoptosis (Fig. 19.1). Apoptosis is widespread in the immune system and can be induced by several mechanisms; for example, the granule proteins released by cytotoxic T cells kill their target cells by inducing apoptosis. Another well-defined apoptotic pathway is that triggered by the interaction of the receptor molecule Fas with its ligand, called Fas ligand (FasL) (Fig. 19.2), which induces apoptosis in the Fas-bearing cell. FasL is a member of the tumor necrosis factor (TNF) family of membrane-associated cytokines, and Fas is a member of the TNF receptor (TNFR) family. Both Fas and FasL are normally induced on lymphocytes and other cell types during the course of an adaptive immune response. Apoptosis induced by cytotoxic T cells bearing FasL is a minor mechanism of cytotoxicity, whereas apoptosis in lymphocytes themselves, induced through Fas, seems to be an important mechanism of lymphocyte homeostasis, as this case shows. Finally, apoptosis can also be induced through a mitochondria-dependent mechanism (the so-called 'intrinsic pathway' of apoptosis), in which cell damage, cytokine deprivation, and other mechanisms result in an increased release of cytochome *c* contained in mitochondria, and the activation of caspase 9.

Topics bearing on this case:

- Lymphocyte survival
- Fas–Fas ligand interactions
- Apoptosis
- Lymphocyte activation
- Autoimmune disease
- TUNEL staining

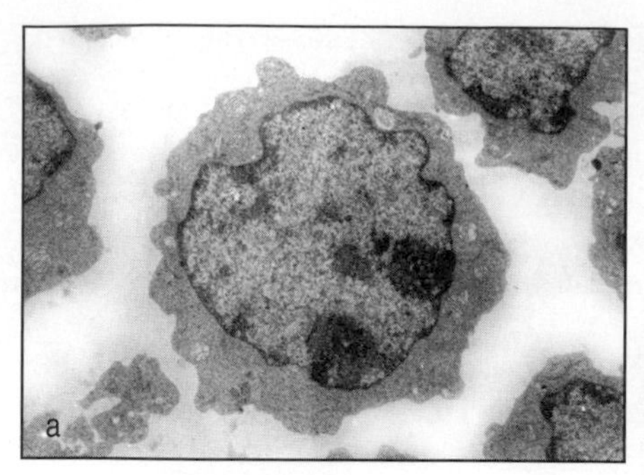

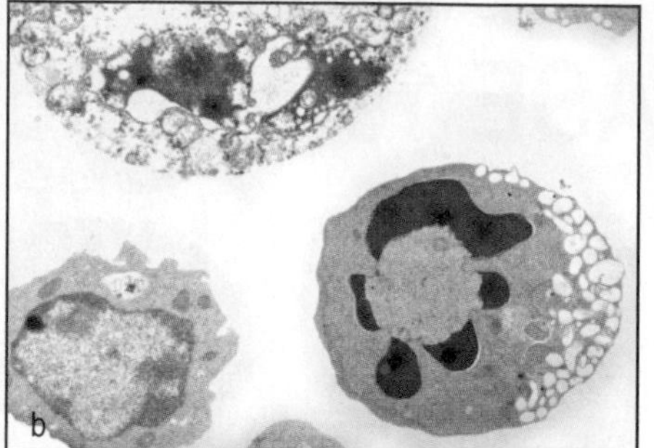

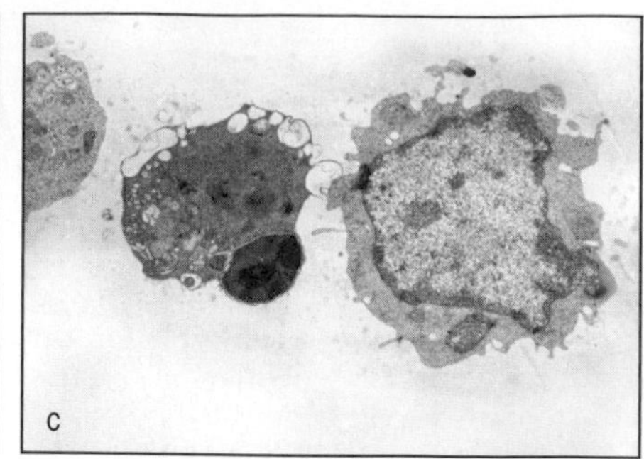

Fig. 19.1 Apoptosis. Apoptosis is a form of induced 'cell suicide' in which the cell undergoes chromatin compaction and DNA fragmentation, followed by cell shrinkage and internal degradation. Panel (a) shows a healthy cell with a normal nucleus. Early in apoptosis (panel b), the chromatin in the nucleus becomes condensed (red) and, although the cell sheds membrane vesicles, the integrity of the cell membrane is retained, in contrast to the necrotic cell in the upper part of the same field. In late stages of apoptosis (panel c), the cell nucleus (middle cell) is very condensed, no mitochondria are visible, and the cell has lost much of its cytoplasm and membrane through the shedding of vesicles. Photographs (× 3500) courtesy of R. Windsor and E. Hirst.

The case of Ellen O'Hara: uncontrolled lymphocyte proliferation in the absence of infection or malignancy.

18-month-old girl, enlarged spleen. Order blood tests.

Ellen O'Hara was born after a normal and uncomplicated pregnancy, was breast fed, and received her routine immunizations without any adverse reactions. At 18 months old, during a routine check-up by her pediatrician, she was found to have an enlarged spleen (splenomegaly) and extensive enlargement of her lymph nodes (lymphadenopathy) (Fig. 19.3). According to her parents, she had had no unusual infections and seemed to be growing and developing normally.

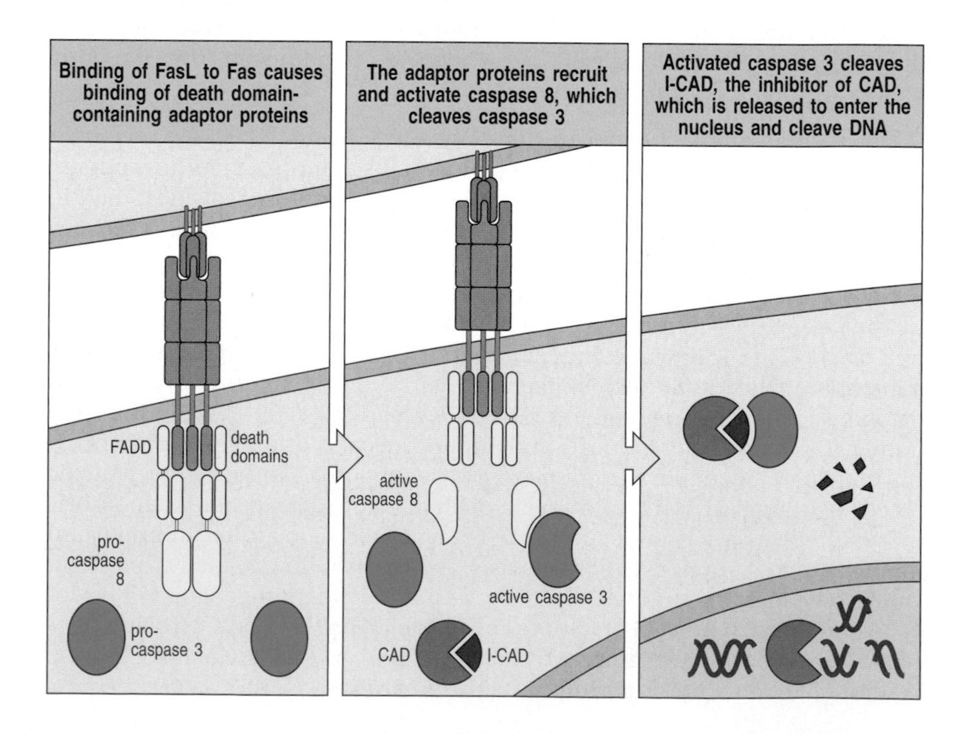

The adaptor proteins recruit and activate caspase 8, which cleaves caspase 3
Activated caspase 3 cleaves I-CAD, the inhibitor of CAD, which is released to enter the nucleus and cleave DNA

Fig. 19.2 Binding of FasL to Fas initiates the process of apoptosis in the Fas-bearing cell. Binding of trimeric FasL to trimeric Fas brings the death domains in the Fas cytoplasmic tails together. A number of adaptor proteins containing death domains bind to the death domains of Fas, in particular the protein FADD, which in turn interacts through a second death domain with the protease caspase 8. Clustered caspase 8 can transactivate, cleaving caspase 8 itself to release an active caspase domain that in turn can activate other caspases. The ensuing caspase cascade culminates in the activation of the caspase-activatable DNase (CAD), which is present in all cells in an inactive cytoplasmic form bound to an inhibitory protein called I-CAD. When I-CAD is broken down by caspases, CAD can enter the nucleus, where it cleaves DNA into the 200 bp fragments that are characteristic of apoptosis.

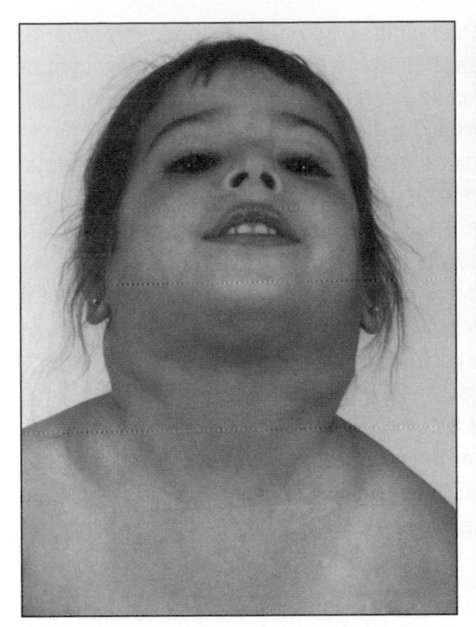
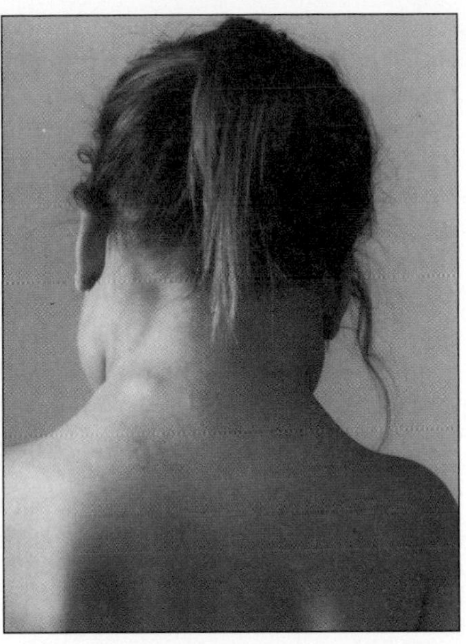

Fig. 19.3 Lymphadenopathy in ALPS. Young girl with ALPS with very enlarged lymph nodes in her neck. Photograph courtesy of Jennifer Puck.

Laboratory tests revealed that Ellen's white blood cell count was 12,500 μl^{-1}, of which 9175 were lymphocytes (normal 3000–7500). Her serum immunoglobulins were all elevated: IgG, 4000 mg dl^{-1} (normal 520–1500); IgM, 400 mg dl^{-1} (normal 40–200); and IgA, 1660 mg dl^{-1}. Flow cytometry analysis of her lymphocytes revealed that 29% were CD19-positive B cells (normal 5–15%; CD19 is a component of the B-cell co-receptor complex) and 65% were CD3-positive T cells (normal 61–84%; CD3 is a component of the T-cell receptor complex). Of the CD3-positive T cells, 14% carried the co-receptor protein CD4 and 18% the co-receptor CD8. She thus had many $CD3^{+}4^{-}8^{-}$ T cells. Of these, the vast majority expressed TCR$\alpha\beta$ (the $\alpha\beta$ form of the T-cell-receptors) and hence were TCR$\alpha\beta^{+}$ double-negative (DN) T cells (normally, TCR$\alpha\beta^{+}$ DN T cells are either absent or constitute less than 2% of circulating T cells). A biopsy of a lymph node from Ellen's neck showed extensive enlargement of the follicles (hyperplasia) and a marked increase in the numbers of immunoblasts and plasma cells in the paracortical area. No infectious agents were cultured from the lymph node, despite the fact that the observed changes resembled those caused by a viral infection. Although more than 50% of the T cells in the lymph node were double negatives, no chromosomal abnormality was found on karyotyping, and there was no evidence of oligoclonality of the T-cell receptor, thus ruling out a malignancy.

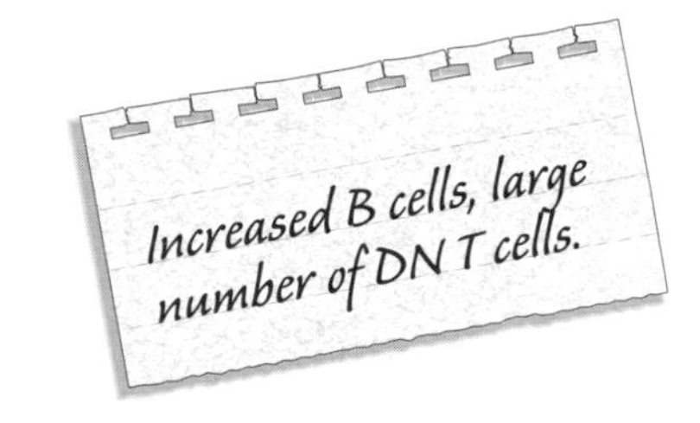

Increased B cells, large number of DN T cells.

In the absence of evidence of infection or malignant disease, autoimmune lymphoproliferative disease was diagnosed and Ellen received the anti-inflammatory steroid prednisone and the immunosuppressant drug cyclosporin A. Her lymph nodes rapidly reduced in size after this therapy, but enlarged again when therapy was discontinued.

Ellen continued to grow and develop normally, and when she reached adolescence the size of her lymph nodes decreased spontaneously. At 18 years of age, repeat blood counts revealed that her platelet count was 75,000 μl^{-1} (normal 150,000–250,000). An autoantibody against platelets was found in her serum. A diagnosis of idiopathic thrombocytopenic purpura (low platelet numbers accompanied by red or purplish-red spotty skin discoloration due to local hemorrhages) was made. She was treated with the steroid dexamethasone, and the condition resolved. At age 32, Ellen's blood neutrophil count fell to <1000 μl^{-1} (normal 2500–5000). She was found to have developed an autoantibody against granulocytes.

Ellen's family history was informative in that her paternal grandfather had splenomegaly and generalized lymphadenopathy as a child, and his spleen was removed at age 25. At age 60, he developed a B-cell lymphoma. Ellen's father also had

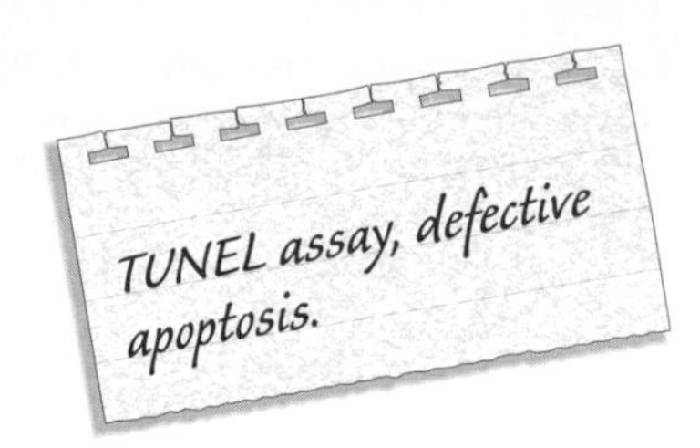

splenomegaly and lymphadenopathy but no clinical symptoms. When blood lymphocytes from Ellen's father and paternal grandfather were examined by flow cytometry, a large number of double-negative T cells were found. In contrast, her brother, mother, and maternal grandparents had normal T cells. The TUNEL assay for apoptotic cells (Fig. 19.4) was performed on blood mononuclear cells from Ellen, her parents, and her paternal grandfather. The cells were first stimulated *in vitro* with phytohemagglutinin for 3 days, and growth of the resulting T-cell blasts was continued for 3 weeks by the addition of IL-2 to the cultures. The cultures were then divided; half were exposed to an antibody to Fas, which mimics the function of FasL. The percentage of cells undergoing apoptosis was then counted. Sixty percent of her mother's T cells underwent apoptosis, whereas only 2% of Ellen's cells, <1% of her father's cells, and 1.4% of her paternal grandfather's cells demonstrated programmed cell death (normal controls 35–70%). The *FAS* and *FASL* genes were examined in DNA samples from Ellen, her father, and her paternal grandfather. An identical single-base transversion, causing a premature termination codon, was found in one of the alleles of the *FAS* gene in these DNA samples. The *FAS* genes in Ellen's mother and brother were normal.

Autoimmune lymphoproliferative syndrome (ALPS).

ALPS is characterized by splenomegaly and lymphadenopathy from early childhood, and, frequently, autoimmunity. Affected individuals can develop autoimmune hemolytic anemia, neutropenia, thrombocytopenia, and hepatitis (inflammation of the liver) and are at increased risk of developing lymphoma. Most patients with ALPS are heterozygous for a dominant mutation in the *FAS* gene, and their activated T cells do not undergo Fas-mediated apoptosis *in vitro*, as is the case for Ellen and her father and grandfather. Patients with ALPS due to *FAS* mutations also have elevated serum levels of FasL, IL-10, and vitamin B_{12}; these can be used as reliable biomarkers, along with the increase in DN T lymphocytes. In some cases, ALPS is due to somatic mutations of *FAS* that occur in an early lymphoid progenitor. Because of the impairment of apoptosis, the proportion of lymphocytes carrying the somatic mutations may increase over time, and is particularly high among DN T cells.

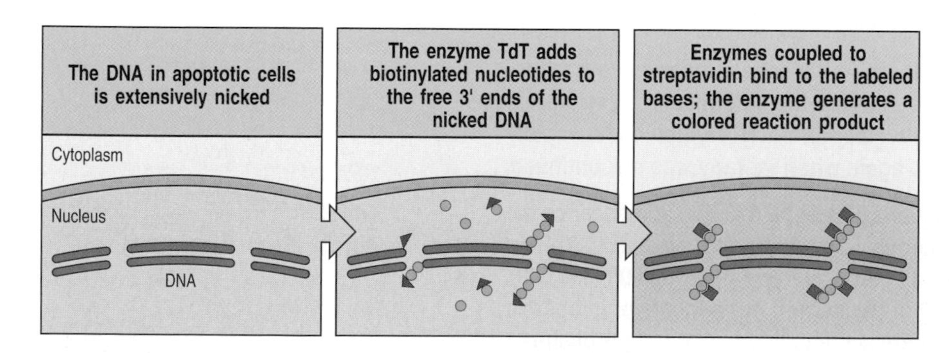

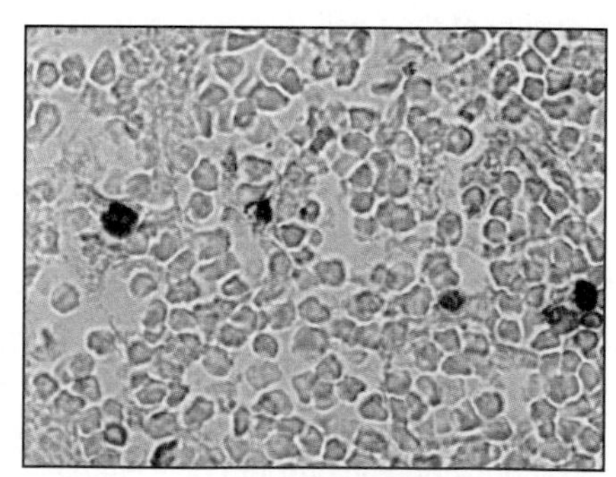

Fig. 19.4 The TUNEL assay. When cells undergo apoptosis, their DNA becomes fragmented and they can be revealed by labeling the fragmented DNA by using the enzyme terminal deoxynucleotidyltransferase (TdT). TdT adds nucleotides to the ends of DNA fragments; biotin-labeled nucleotides (usually dUTP) are most commonly added in this assay (second panel). The biotinylated DNA can be detected by using streptavidin, which binds to biotin, coupled to enzymes that convert a colorless substrate into a colored insoluble product (third panel). Cells stained in this way can be detected by light microscopy, as shown in the photograph of apoptotic cells (stained red) in the thymic cortex. Photograph courtesy of R. Budd and J. Russell.

Other patients with ALPS have been found to have mutations in the genes encoding FasL or caspase 10, an enzyme involved in triggering apoptosis via the Fas pathway. In one case, a gain-of-function mutation of the *NRAS* gene was identified that resulted in impaired induction of apoptosis in response to IL-2 deprivation. The *NRAS* mutation in this patient resulted in impaired induction of the pro-apoptotic molecule Bim, which controls mitochondrial stability upon cytokine deprivation.

Treatment of ALPS is mostly based on immune suppression and the surveillance of tumors. Splenectomy should be reserved for severe cases, because of the risk of infections by encapsulated bacteria (see Case 16).

The clinical and immunologic features of ALPS bear a striking resemblance to a lymphoproliferative disease observed in mice with *lpr* or *gld* mutations. The *lpr* phenotype results from the absence of Fas, whereas the *gld* phenotype is caused by a mutation in FasL. A progressive accumulation of DN T cells is observed in both these strains of mice (note that these circulating $CD3^+$ DN T cells should not be confused with the immature $CD3^-$ $CD4^-$ $CD8^-$ 'double-negative' thymocytes that are a normal stage of T-cell development in the thymus). The mice make antibodies against double-stranded DNA, similar to the situation in human systemic lupus erythematosus (see Case 37). Consistent with these findings in mice, patients with ALPS have defective T-cell apoptosis and abnormal accumulations of DN T cells. When B cells are activated, they also express Fas and become susceptible to Fas-mediated apoptosis. Thus, activated B cells in ALPS are not properly eliminated. The serum concentrations of immunoglobulins increase (hypergammaglobulinemia), the number of B cells is increased (B-cell lymphocytosis), and pathological autoantibody production ensues. Because T cells and B cells are not eliminated normally, patients with ALPS are predisposed to develop lymphomas. Autoimmunity may result because Fas-mediated killing is a mechanism for removing autoreactive B cells.

Questions.

1 *Patients with ALPS are heterozygous for the mutation in* FAS *or* FASL*; they have one normal allele and one mutant allele. How do you explain the dominant inheritance?*

2 *Ellen's great-aunt (her paternal grandfather's sister) was found to have the same FAS mutation as Ellen, yet she had no symptoms. How can this be explained?*

3 *It is advantageous for viruses to inhibit apoptosis so that the host cells in which they thrive do not get eliminated by apoptosis induced by recognition by cytotoxic T cells. How might a virus accomplish this?*

4 *When Fas is activated by FasL it associates with and activates caspase 8 (see Fig. 19.2). When the gene encoding caspase 8 is knocked out in mice, this proves to be lethal at the fetal stage. Would it be worthwhile to search for caspase 8 mutations in patients with ALPS when there is no mutation in* FAS *or* FASL*?*

CASE 20 Hyper IgE Syndrome

A defect in T_H17 lineage differentiation.

After antigen recognition and concomitant activation by specific sets of cytokines, naive CD4 T cells differentiate into distinct lineages of helper or regulatory effector T cells, characterized by distinct biologic functions and cytokine expression profiles. T_H17 cells are a recently described lineage of differentiated CD4 T cells, characterized by expression of the cytokines IL-17A (IL-17), IL-17F, IL-21, and, in some cases, IL-22 (Fig. 20.1). IL-17 and IL-17F have similar activities that include the induction in epithelial cells of chemokines that attract neutrophils and of β-defensins (antimicrobial peptides). T_H17 cells have a role in the immune response to extracellular bacteria and fungi, and in autoimmune diseases.

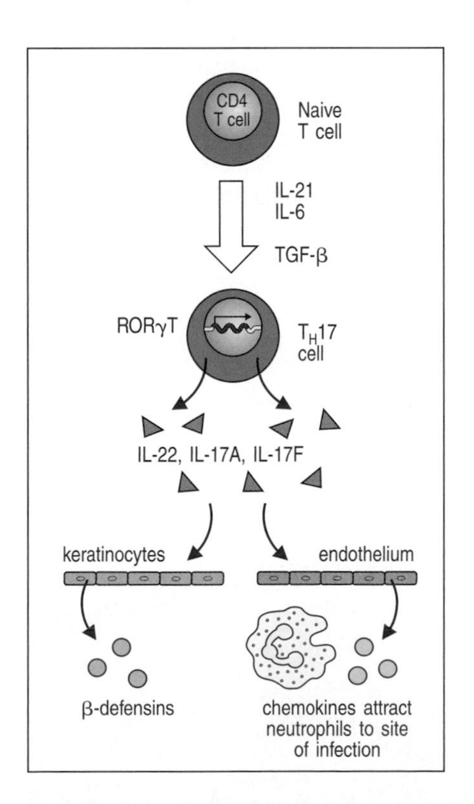

Fig. 20.1 Differentiation and actions of T_H17 cells. Effector T_H17 cells develop from antigen-activated naive CD4 T cells in the presence of the cytokines IL-6, IL-21, and TGF-β. They are characterized by expression of the transcription factor RORγT and produce the cytokines IL-17A (IL-17), IL-17F, and IL-22. The cytokines produced by T_H17 cells act on skin to induce the production of antimicrobial defensins, and on endothelial cells to induce the production of chemokines that recruit neutrophils to sites of infection.

Topics bearing on this case:

- Neutrophil function
- T-cell subsets and their differentiation
- T_H17 cell function
- IgE class antibody

This case was prepared by Raif Geha, MD, in collaboration with Lisa Bartnikas, MD.

In humans, IL-6 has a key role in the differentiation of T_H17 cells, in synergy with IL-1β, IL-21, and possibly TGF-β. IL-23 is important in amplifying the generation of T_H17 cells and sustaining their survival. IL-6, IL-21, and IL-23 all act on the same class of heterodimeric receptors, for which binding of their ligand activates Janus kinases (JAKs) associated with the receptor that phosphorylate the intracellular portions of the receptor. The transcription factor STAT3, which in inactive form is located in the cytoplasm, is recruited to the phosphorylated receptor and is itself phosphorylated (Fig. 20.2). Phosphorylated STAT3 dissociates from the receptor and forms homodimers that translocate to the nucleus and act as transcription factors to induce expression of the retinoic acid-related orphan receptor γT (RORγT). Differentiation of activated CD4 T cells into T_H17 cells is determined by RORγT expression.

The case of Stephen Dedalus: impaired T_H17 differentiation associated with elevated IgE and recurrent infections.

Stephen Dedalus was first taken to the Children's Hospital at 3 years of age for evaluation of his eczema. Stephen was born after an uncomplicated pregnancy, but within the first few days of life he developed significant seborrheic dermatitis (a flaky, red skin rash commonly affecting the scalp) and neonatal acne. By the first month of life, he developed eczema that was at times severe. Aggressive topical skin care was recommended. His parents were both healthy.

Over the next few years, Stephen experienced recurrent infections, but typically was afebrile (did not develop a fever) during these episodes. He had recurrent skin infections, with cultures growing *Staphylcoccus aureus*. Although he developed pustules, they were not particularly red or tender. He also had an episode of *S. aureus* bursitis (inflammation of the sac that contains joint fluid) of the knee. He had two episodes of pneumonia, one being complicated by a small pneumothorax (collapsed lung due to air leakage from the lung into the chest cavity). He had 10 episodes of rhinosinusitis (nose and sinus infection) and a chronic rhinorrhea (nasal discharge) that was typically yellow or green. He had recurrent episodes of otitis media (middle-ear infection)

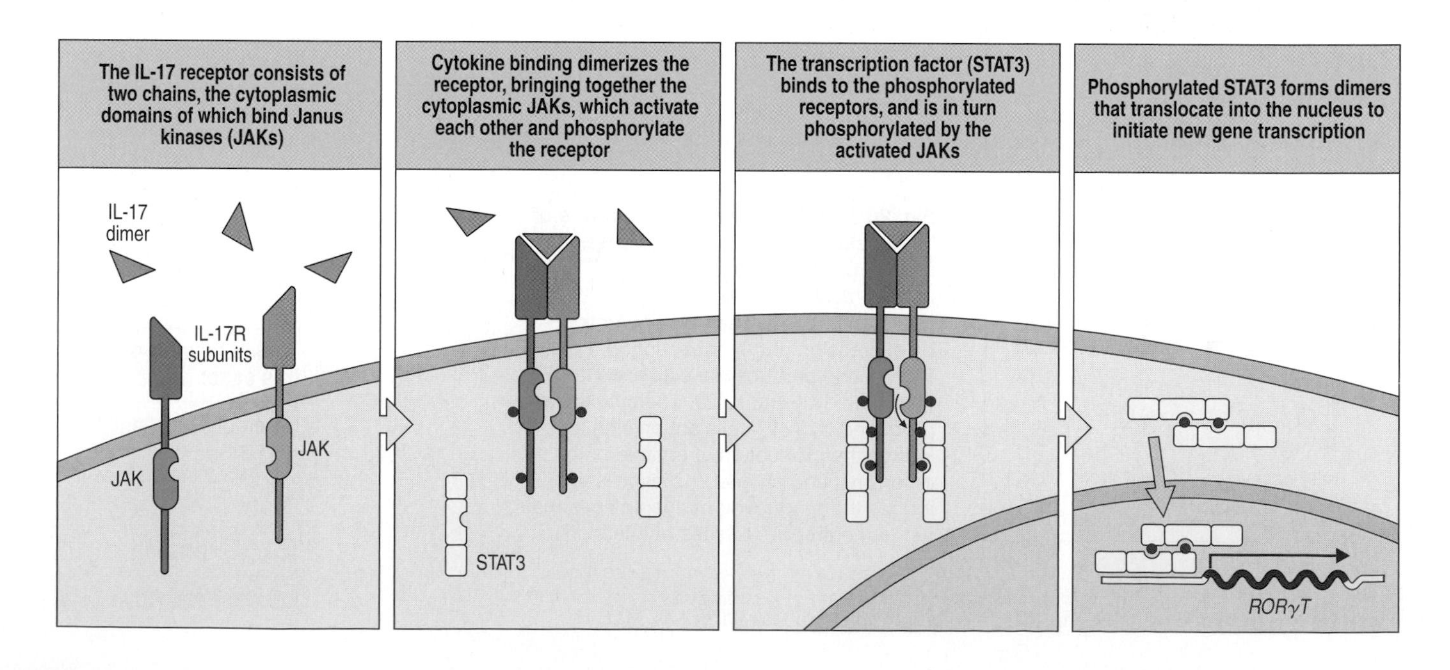
The IL-17 receptor consists of two chains, the cytoplasmic domains of which bind Janus kinases (JAKs)
IL-17 dimer
IL-17R subunits
JAK
JAK
Cytokine binding dimerizes the receptor, bringing together the cytoplasmic JAKs, which activate each other and phosphorylate the receptor
STAT3
The transcription factor (STAT3) binds to the phosphorylated receptors, and is in turn phosphorylated by the activated JAKs
Phosphorylated STAT3 forms dimers that translocate into the nucleus to initiate new gene transcription
RORγT

Fig. 20.2 STAT3 is involved in signaling pathways from some cytokine receptors. Many cytokines act via receptors that are associated with cytoplasmic Janus kinases (JAKs). The receptor consists of at least two chains, each associated with a specific JAK (first panel). Binding of dimeric ligand results in dimerization of the receptor chains, bringing together the JAKs, which can phosphorylate and activate each other. The activated JAKs then phosphorylate tyrosines in the receptor tails (second panel). Members of the STAT (signal transducers and activators of transcription) family of proteins (which contain SH2 domains) bind to the tyrosine-phosphorylated receptors and are themselves phosphorylated by the JAKs (third panel). After phosphorylation, STAT proteins form a dimer by the SH2 domains binding phosphotyrosine residues on the other STAT and translocate to the nucleus (last panel), where they bind to and activate the transcription of a variety of genes important for adaptive immunity. Different STATs are associated with different receptors. STAT3, for example, is activated by the receptors for the cytokines IL-6 and IL-10, among others.

and needed tympanostomy tubes inserted on multiple occasions. These are ventilation tubes inserted into the eardrum to relieve obstruction of drainage from the middle ear to the nose resulting from scarring due to frequent middle-ear infections. Stephen fractured his right arm on two occasions in the setting of relatively minor trauma.

Stephen was admitted to Children's Hospital at 5 years old for repeat tympanostomy tube placement. At the time he was experiencing chronic fluid drainage from his ears that grew *S. aureus*. His physical examination revealed a prominent forehead and wide nasal bridge. Thick white rhinorrhea was present, and healing patches of eczema were noted on his skin. Given Stephen's history of multiple infections, an immunologist was consulted. A complete blood count revealed 1040 cells μl^{-1} white blood cells (normal 5970–1049 cells μl^{-1}), of which 42% were neutrophils (normal 32–75%), 46% lymphocytes (normal 11–54%), and 7% eosinophils (normal 1–4%). Serum IgG was 1417 mg dl^{-1} (normal 600–1500 mg dl^{-1}), IgA was 70 mg dl^{-1} (normal 50–150 mg dl^{-1}), IgM was elevated at 145 mg dl^{-1} (normal 22–100 mg dl^{-1}), and IgE was markedly elevated at 36,698 mg dl^{-1} (normal 0–200 mg dl^{-1}).

Sequencing of Stephen's *STAT3* gene revealed a heterozygous missense mutation in the highly conserved DNA-binding domain. CD4 T cells were isolated from Stephen's peripheral blood, and intracellular staining for IL-17A, a marker of T_H17 lineage differentiation, was assessed by flow cytometry, revealing significantly decreased IL-17A expression compared with a healthy control (Fig. 20.3). Stephen was started on intravenous antibiotics after repeat tympanostomy tube placement. He was discharged to complete a course of antibiotics for his otitis media. He was then started on bactrim prophylaxis, which led to a marked decrease in the frequency of infections.

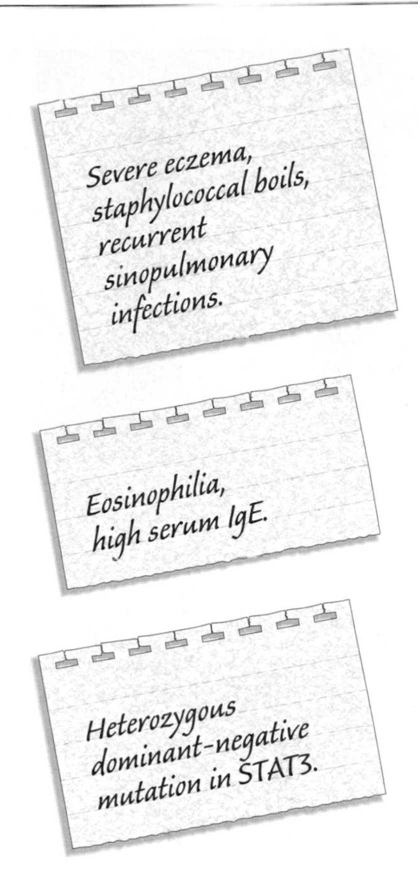

Severe eczema, staphylococcal boils, recurrent sinopulmonary infections.
Eosinophilia, high serum IgE.
Heterozygous dominant-negative mutation in STAT3.

Hyper IgE syndrome.

Hyper IgE syndrome (HIES) is a rare primary immunodeficiency characterized by a triad of symptoms—recurrent staphylococcal skin abscesses, pneumonia, and very high serum IgE levels. It is also called Job's syndrome, in reference to the Biblical figure Job, who was "smote with sore boils." It was first described in 1966 by S. D. Davis and Ralph Wedgwood in two red-headed sisters with recurrent staphylococcal skin abscesses that lacked typical features of inflammation such as redness and warmth, thus coining the term "cold abscesses." The disease was further characterized a few years later by R. Buckley and colleagues, who noted the association between these cold

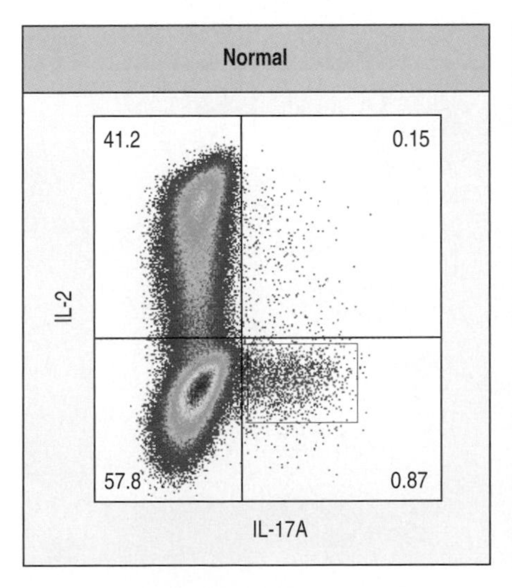

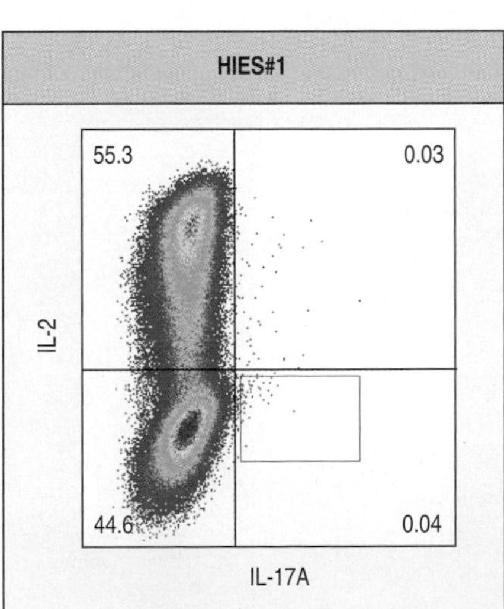

Fig. 20.3 T cells from a patient with hyper IgE syndrome due to a mutation in STAT3 fail to differentiate into T_H17 cells *in vitro*. T_H17 cells (boxed) are evident in the normal sample but absent from HIES sample. Photograph courtesy of Cindy Ma.

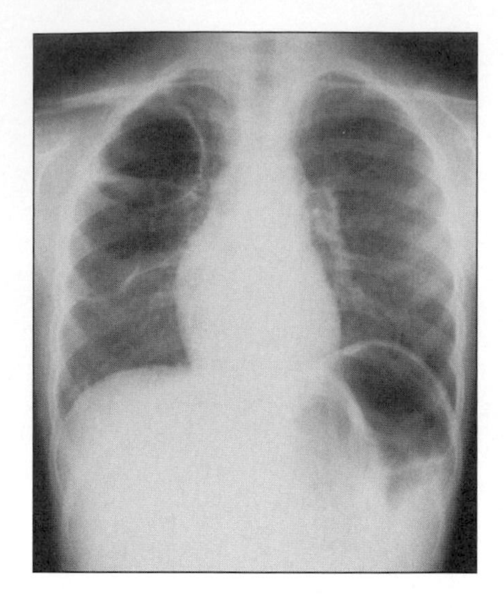

Fig. 20.4 Chest radiograph of a patient with hyper IgE syndrome and STAT3 deficiency, showing a pneumatocele in the right upper lobe of the lung.

abscesses, severe dermatitis, and high serum IgE levels and proposed the name "hyper IgE syndrome." Patients with HIES are extremely susceptible to infections with *S. aureus* and the yeast *Candida albicans* and typically do not mount strong inflammatory responses such as fever to infections.

Several immunologic and non-immunologic phenomena have been described in HIES. An eczematous rash in the neonatal period is typically the first clinical manifestation. Staphylococcal skin abscesses occur in the majority of patients and begin early in life. Recurrent pneumonias occur and are often complicated by the formation of pneumatoceles (thin-walled air-filled cysts in the lungs that can result from scarring) (Fig. 20.4) and dilation of the large airways. The bacteria causing pneumonia include *S. aureus, Streptococcus pneumoniae*, and *Haemophilus influenzae*. Systemic signs of illness such as fever are often absent. Mucocutaneous candidiasis is common. A typical facial appearance develops during childhood, characterized by facial asymmetry, a broad nose, deep-set eyes, and a prominent forehead. Musculoskeletal abnormalities are common and include scoliosis (curvature of the spine), fractures occurring in the setting of minimal trauma, and hyperextensibility of the joints. Most patients have delayed shedding of their primary teeth, requiring surgical extraction (Fig. 20.5). There is an increased incidence of malignancies, including lymphoma and leukemia.

HIES is a clinical diagnosis based on a constellation of clinical and laboratory findings. A scoring system was developed by the National Institutes of Health, and patients with scores greater than 40 are more likely to have HIES. Both autosomal dominant and autosomal recessive forms of HIES have been described. In all cases the final common pathway is impaired T_H17 differentiation. T_H17 cells are important for maintaining mucosal and epithelial surface immunity against extracellular fungi and bacteria. They recruit neutrophils and macrophages to sites of infection by inducing epithelial cells expressing the IL-17 receptor to produce chemotactic factors for neutrophils (such as the chemokine CXCL8) and macrophages. In addition, T_H17 cells produce

Fig. 20.5 Abnormal dentition in patients with hyper IgE syndrome and STAT3 deficiency.

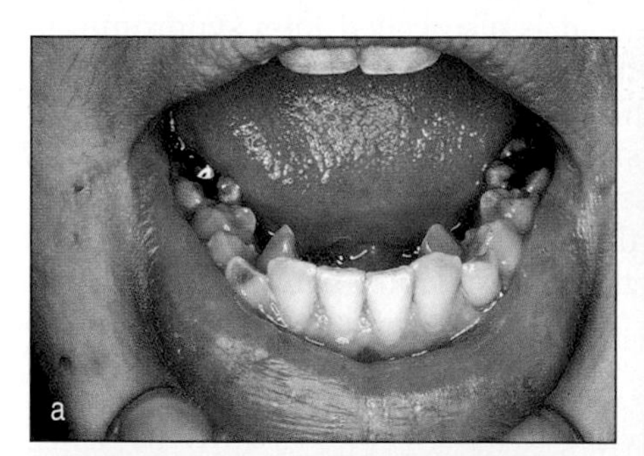

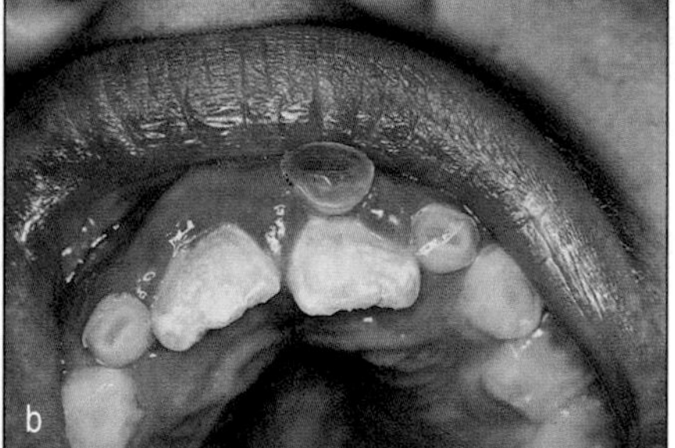

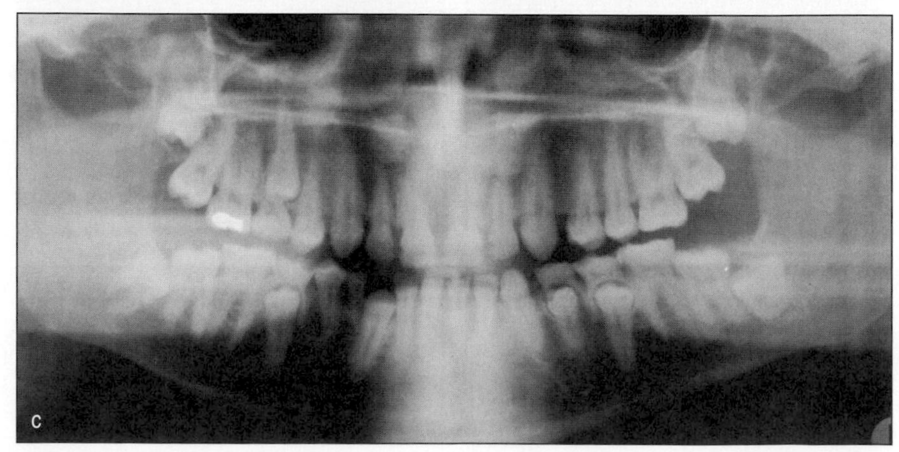

IL-22, which promotes the synthesis by keratinocytes of β-defensins, which are important in skin immunity. The mechanism by which the impaired T_H17 differentiation leads to elevated serum IgE is not yet understood.

The autosomal dominant form of HIES is the most common and has both sporadic and familial inheritance. It is a multisystem disease, and its features include skeletal and dental anomalies as well as recurrent bacterial infections. Heterozygous mutations in the gene for the transcription factor STAT3, located on chromosome 17, have been identified in the majority of patients with autosomal dominant HIES, resulting in a dominant-negative effect, as in Stephen. Most mutations occur within the DNA-binding, Src-homology (SH2), and transactivation domains of STAT3.

Autosomal recessive HIES is not typically associated with skeletal anomalies. In addition to bacterial infections, patients develop recurrent and severe infections of skin and mucous membranes with herpes simplex virus and the molluscum contagiosum virus, and are prone to develop squamous cell carcinomas. Mutations in the genes encoding tyrosine kinase 2 (*TYK2*) and dedicator of cytokinesis 8 (*DOCK8*) have been implicated in autosomal recessive HIES.

Mouse models of STAT3 deficiency have been developed. The homozygous mouse *STAT3* knockout was embryonic lethal, possibly explaining why no humans who are homozygous for STAT3 deficiency have been identified. Mice with tissue-specific *STAT3* deletions have been developed and show poor wound healing, eosinophilia, impaired β-defensin production, and pulmonary cyst formation, corresponding to some of the non-immunologic phenotypes seen in humans with HIES.

Questions.

1 *Would a bone marrow transplant be useful in correcting the immunologic defects seen in HIES?*

2 *Some authors have found that patients with HIES due to STAT3 mutation had a fourfold decrease in expression of RORγT. Explain how this could be possible.*

3 *Administration of interferon-gamma (IFN-γ) has been used in HIES in an attempt to decrease serum IgE. The rationale is that IFN-γ counteracts IgE class switching induced by cytokines produced by T_H2 cells. Only a mild decrease in serum IgE was observed in some trials, and no reduction in others. What could be the reason for this?*

4 *Can you think of possible ways in which STAT3 deficiency could cause high levels of IgE?*

CASE 21 Ataxia Telangiectasia

A defect in double-stranded DNA repair.

Throughout an individual's life their DNA is subject to damage by ionizing radiation, ultraviolet radiation and reactive oxygen species. These agents cause double-strand breaks in DNA, which are repaired by one of two main mechanisms: nonhomologous end joining (NHEJ) or homologous recombination repair (Fig. 21.1). NHEJ can join DNA ends with little sequence homology to each other and can occur during the G_0, G_1, and M phases of the cell cycle. In contrast, homologous recombination repair uses a sister chromatid as a template for repairing the broken DNA. Because of the need for a sister chromatid, homologous recombination repair occurs only during the S and G_2 phases of the cell cycle, after a chromosome has been replicated. Some of the same DNA repair proteins are used in both NHEJ and homologous recombination repair.

In many, but not all, instances, repair of DNA double-strand breaks begins with the recruitment of a protein complex, the MRE11:RAD50:NBS1 (MRN) complex, to the site of the break. The MRN complex facilitates the accumulation of the additional proteins necessary for DNA repair. One of the first proteins recruited by the MRN complex is the ATM (ataxia telangiectasia mutated) protein, which is a serine/threonine kinase. ATM exists as an inactive dimer constitutively bound to a phosphatase that negatively regulates ATM kinase activity. When ATM binds to double-strand DNA breaks, the phosphatase is released. ATM then undergoes autophosphorylation and dissociation into active monomers (Fig. 21.2).

On activation, ATM phosphorylates a large number of downstream proteins involved in DNA repair. One known substrate of ATM that is involved in NHEJ is Artemis, a nuclease that processes free DNA ends in preparation for ligation. Other substrates of ATM, such as the MRN complex itself, are important in homologous recombination repair. Activation of these ATM substrates results in cell-cycle arrest and DNA repair, or in apoptosis if the DNA cannot be repaired. Redundancy in cellular DNA repair mechanisms permits some DNA repair to occur even in the complete absence of ATM. However, cells lacking

Topics bearing on this case:
V(D)J recombination
Class-switch recombination
Lymphocyte development

This case was prepared by Raif Geha, MD, in collaboration with Janet Chou, MD.

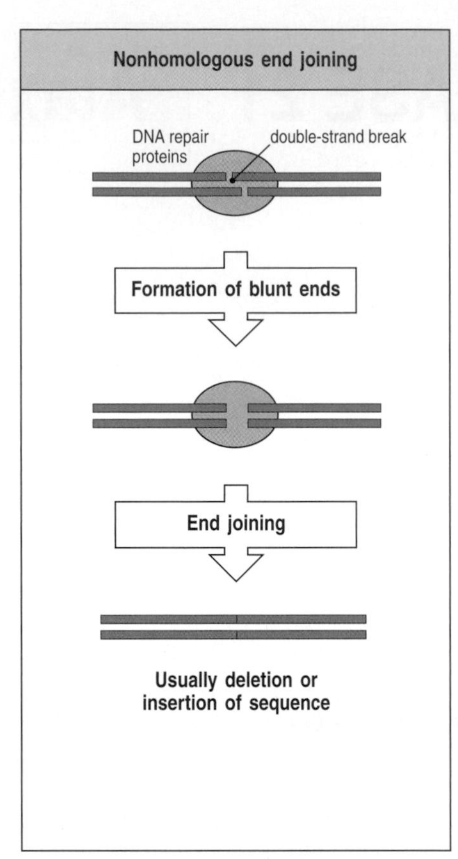

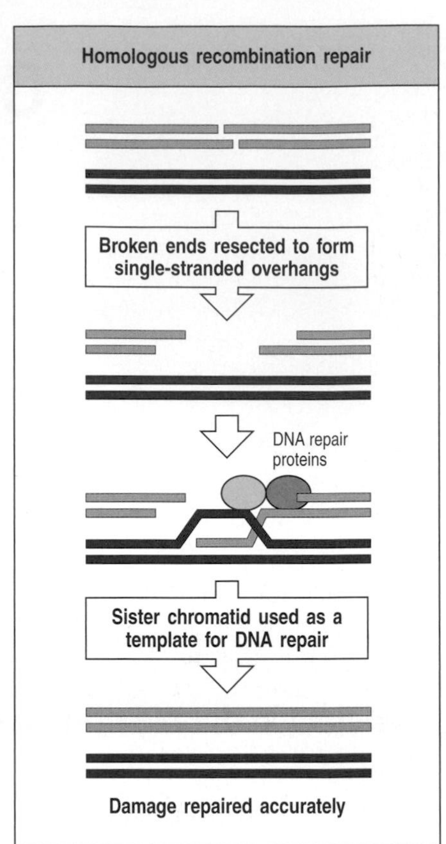

Fig. 21.1 Two methods of repairing double-strand DNA breaks. In nonhomologous end joining (NHEJ) (left), a complex of DNA repair proteins binds to the broken DNA ends. The broken ends are first processed by nucleases and DNA polymerases to remove or fill in single-stranded overhangs, and the blunt ends are then ligated. NHEJ usually leads to the deletion or insertion of nucleotides at the site of the break. An NHEJ-type repair process is used in the ligation of gene segments in V(D)J recombination and in class-switch recombination. In DNA repair by homologous recombination (right), nucleases recruited by a complex of DNA repair proteins transform the broken ends into longer single-stranded overhangs. One strand invades the DNA of the sister chromatid, which is used as a template to repair the DNA break. Unlike NHEJ, homologous recombination uses sequence homology to repair the DNA.

ATM do not repair every break, and so double-strand breaks accumulate over time, which is manifested as an abnormal sensitivity of cells to ionizing radiation and the accumulation of chromosomal abnormalities.

In addition to its use as a general means of DNA repair, some parts of the NHEJ pathway are involved in the rejoining of gene segments to create the signal and coding joints during V(D)J recombination in the immunoglobulin and T-cell receptor genes. NHEJ is also thought to be the means by which the double-strand breaks produced in the immunoglobulin C-region genes during class-switch recombination are resolved. Both ATM-dependent and ATM-independent pathways operate in these repair processes. As we will see in this case, a genetic deficiency of ATM can result in a constellation of symptoms resulting from the inability to repair DNA, including immunodeficiency.

Fig. 21.2 Activation of ATM. The MRN complex binds to double-strand DNA breaks and recruits the protein kinase ATM, which exists in the cytoplasm as an inactive dimer complexed with an inhibitory phosphatase PP2A. ATM activation involves the release of PP2A, which enables the autophosphorylation of ATM and its dissociation into enzymatically active monomers.

The case of Basil Ransom: recurrent infections with difficulty in walking.

Basil Ransom was 6 years old when he was referred to the Children's Hospital because of a history of recurrent ear infections and two pneumonias. Even after placement of ear tubes to ventilate the middle ear and prevent the accumulation of fluid and infections, he continued to have monthly infections. On physical examination he was noted to have a wobbly gait (ataxia) and two dilated superficial blood vessels (telangiectasias) on his conjunctivae and the outer part of his ear (the pinna) (Fig. 21.3). The pediatrician tested Basil's serum immunoglobulins and found total IgG to be low at 219 mg dl^{-1} (normal 600–1500 mg dl^{-1}) with no detectable IgG2 and IgG4, no detectable IgA, and an elevated level of IgM of 205 mg dl^{-1} (normal 22–100 mg dl^{-1}). Basil

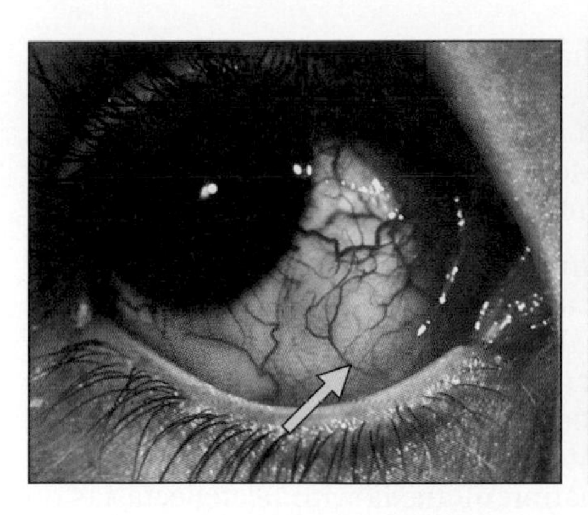
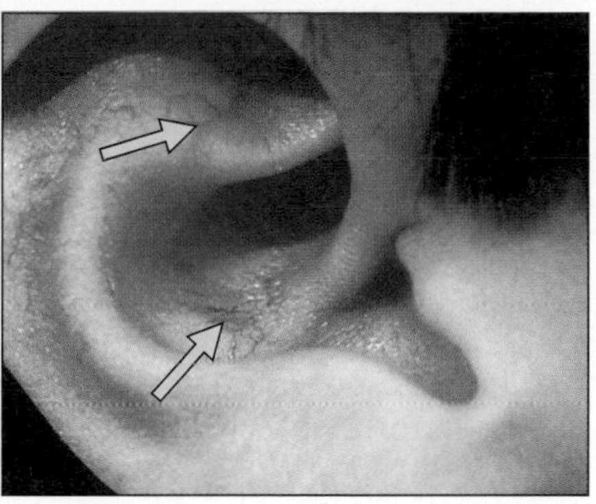

Fig. 21.3 Telangiectasias on the conjunctiva and pinna of the ear in a patient with ataxia telangiectasia.

had a positive antibody titer to only 1 of 14 pneumococcal (*Streptococcus pneumoniae*) subtypes tested, and no detectable antibodies against *Haemophilus influenzae,* although he had received both the pneumococcal and *H. influenzae* vaccines. His antibody titer to tetanus was normal.

Recurrent infections and telangiectasias on conjunctivae and ears.

At the Children's Hospital, a fluorescence-activated cell sorting (FACS) analysis of Basil's peripheral blood mononuclear cells (PBMCs) was notable for low numbers of CD4 T cells (459 cells μl^{-1}; normal 700–2200), CD8 T cells (221 cells μl^{-1}; normal 490–1300), and B cells (300 cells μl^{-1}; normal 390–1400). Proliferation of Basil's PBMCs in response to phytohemagglutinin and pokeweed mitogens and to tetanus and diphtheria antigens was normal. Because of the history of recurrent infections, ataxia, telangiectasias, hypogammaglobulinemia, and lymphopenia, the diagnosis of ataxia telangiectasia was considered and blood alpha-fetoprotein (AFP) was measured. AFP is a plasma protein produced by the liver that is elevated in ataxia telangiectasia, although not exclusively so. Basil's AFP level was elevated, at 64 ng ml^{-1} (normal 0–15 ng ml^{-1}).

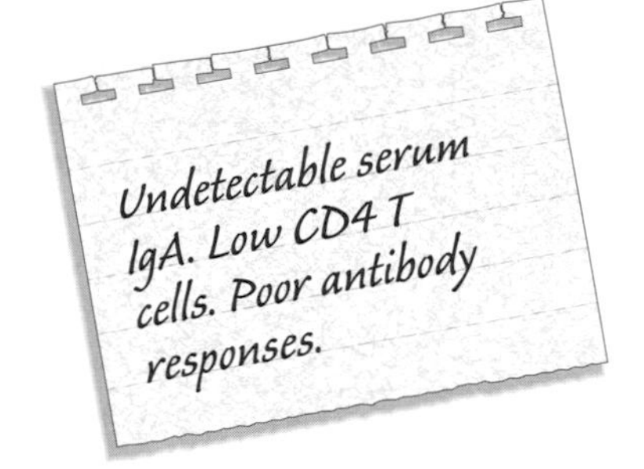

Undetectable serum IgA. Low CD4 T cells. Poor antibody responses.

To assess radiosensitivity, lymphoblastoid B-cell lines were established from Basil's blood and from a normal individual as a control. Peripheral blood lymphocytes were isolated and transformed with Epstein–Barr virus. For each sample, the transformed lymphoblastoid cells were split into two plates; one plate was exposed to 1 gray of radiation and the other was not irradiated. After 10 days in culture, the number of surviving colonies was counted. The survival fraction was calculated by comparing the number of surviving colonies in the irradiated plate with that of the nonirradiated plate. The normal control had a survival fraction of 45%, whereas Basil's survival fraction was only 12% (Fig. 21.4). Sequencing of Basil's *ATM* gene revealed that he was a compound heterozygote for two different mutations; that is, he had one mutation in one *ATM* allele and a different mutation in the other allele.

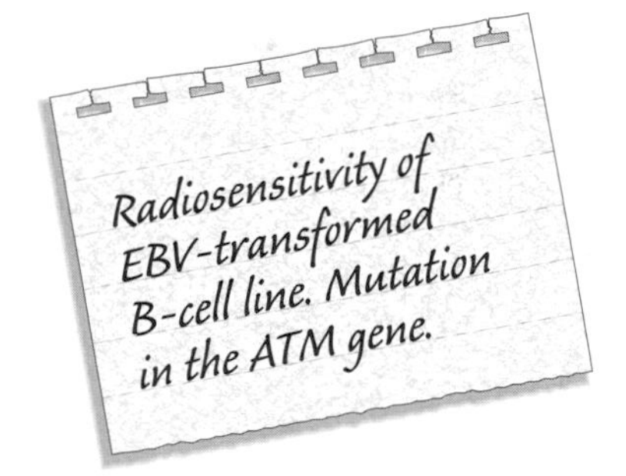

Radiosensitivity of EBV-transformed B-cell line. Mutation in the ATM gene.

Ataxia telangiectasia.

Ataxia telangiectasia is an autosomal recessive disorder characterized by progressive cerebellar ataxia and neurodegeneration, oculocutaneous telangiectasias, primary immunodeficiency, and sensitivity to ionizing radiation. It is caused by mutations in the *ATM* gene. These mutations are typically

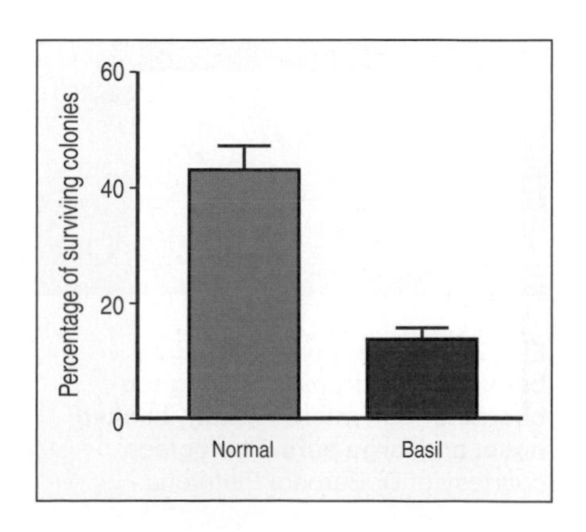

Fig. 21.4 Radiosensitivity assay of lymphoblastoid cell lines from Basil and a normal control. The histogram shows the percentage of lymphocytes surviving after irradiation (the survival fraction). The survival fraction for each sample is calculated from the number of surviving colonies on the irradiated plate divided by the number of colonies on a nonirradiated plate.

splice-site or truncating mutations that result in a nonproduction of ATM protein. Less commonly, leaky splicing mutations or defects in the promoter region result in low-level expression of functional ATM protein, resulting in a milder form of ataxia telangiectasia. Most patients, like Basil, are compound heterozygotes for two different *ATM* mutations.

Ataxia is the earliest sign of the condition, usually beginning at 2–3 years of age. Although the cerebellum is affected first and most severely, the accumulation of damaged DNA in neuronal cells becomes more widespread and results in a loss of gross and fine motor skills, dysphagia (difficulty in swallowing), dysarthria (speech difficulty), abnormal eye movements, and peripheral neuropathy and nerve dysfunction. Although ataxia is associated with several neurologic disorders, the combination of ataxia with telangiectasias on the conjunctivae or pinnae is characteristic of ataxia telangiectasia. Another identifying feature of the disease is the elevated AFP level, which is usually more than two standard deviations above normal.

Patients with ataxia telangiectasia can have a variety of defects in humoral and cellular immunity. *ATM* mutations may affect B-cell and T-cell development and function, leading to low numbers of B cells and T cells, most probably owing to deleterious effects on V(D)J recombination. As a result of *ATM* mutations that impair class-switch recombination, patients may have elevated IgM levels but decreased IgA, IgG, or IgE levels and a poor response to polysaccharide antigens. Consequently, these patients experience frequent infections of sinuses and lungs (sinopulmonary infections) by bacteria such as *S. pneumoniae* and *H. influenzae*. Lymphocyte proliferation in response to mitogens and antigens can vary from normal to impaired. However, unlike patients with irradiation-sensitive severe combined immunodeficiency (IR-SCID) resulting from a deficiency of NHEJ components that are essential for V(D)J recombination, patients with ataxia telangiectasia are not typically at increased risk of opportunistic infections, indicating the preservation of some immune function, especially T-cell function.

Patients with ataxia telangiectasia have a significantly increased risk of leukemias and lymphomas. As T cells and B cells routinely incur DNA double-strand breaks during antigen-receptor gene formation and class-switch recombination, the translocations most commonly found in these leukemias and lymphomas involve T-cell receptor genes and immunoglobulin genes on the one hand and oncogenes on the other. Translocations juxtaposing an active T-cell enhancer in the *TCR* locus with an oncogene offer a growth advantage to these abnormal cells, resulting in clonal expansion. Although lymphoma and leukemia constitute the majority of cancers in patients with ataxia telangiectasia, there is also an increased risk of solid organ cancers, such as breast cancer. The defective repair of double-strand breaks caused by ionizing radiation is the primary cause of tumors in those patients.

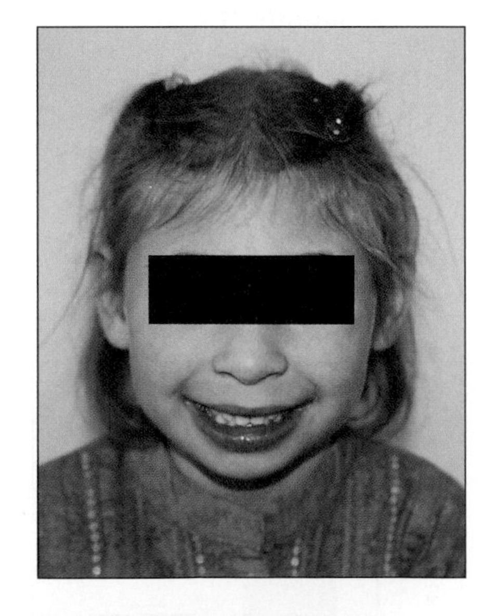

Fig. 21.5 Patient with Nijmegen breakage syndrome showing the characteristic microcephaly, beaked nose, and large ears. Photograph courtesy of Dr Barbara Pietrucha.

Mutations in proteins that cooperate with ATM cause clinical phenotypes that overlap with ataxia telangiectasia. Hypomorphic mutations (mutations causing decreased, but not absent, production and/or function of the protein) in the *MRE11* gene, which encodes a component of the MRN complex, result in a disease resembling a mild form of ataxia telangiectasia with delayed onset and slower progression of symptoms. Another component of the MRN complex, nibrin, binds to double-strand DNA breaks. Mutations in the gene encoding nibrin, *NBS1*, cause Nijmegen breakage syndrome, which is characterized by radiosensitivity, an increased risk of leukemia and lymphoma, and immunodeficiency. Patients with NBS also have microcephaly, mental retardation, and short stature, but do not have ataxia (Fig. 21.5).

Medical interventions for ataxia telangiectasia are supportive, because no targeted treatment exists for this disease. The median age of death is 25 years, caused most often by cancer or progressive pulmonary disease resulting from recurrent infections. Therefore, antibiotic prophylaxis should be considered in patients with repeated sinopulmonary infections; pulmonary function should be monitored. Infusions of gamma globulin should be administered to patients with hypogammaglobulinemia or impaired specific antibody production. Most importantly, exposure to ionizing radiation in diagnostic studies should be minimized.

Questions.

1 *Although patients with ataxia telangiectasia do not suffer from opportunistic infections, there are patients with a syndrome of opportunistic infections, radiosensitivity, and absent or very low numbers of T cells and B cells. What gene defects could cause this clinical picture?*

2 *Most patients with complete ATM deficiency have the classical clinical presentation of ataxia telangiectasia. However, there are rare patients who lack the ATM protein but have a much milder phenotype with delayed onset of symptoms, mild ataxia, and no history of sinopulmonary infections. What could be a possible explanation for this variability in clinical presentation?*

3 *Why do patients with complete ATM deficiency have a gradual onset of symptoms?*

4 *Radiosensitivity testing can be slow because of the need to establish a lymphoblastoid cell line and to culture the cells. Western blotting for ATM protein is a faster diagnostic tool, because most mutations causing ataxia telangiectasia are associated with complete absence of the ATM protein. If a Western blot of peripheral blood lymphocytes shows no ATM protein, does this indicate that the patient has ataxia telangiectasia, even in the absence of clinical symptoms?*

CASE 22 Warts, Hypogammaglobulinemia, Infections, and Myelokathexis Syndrome (WHIM Syndrome)

Chemokines direct the migration of leukocytes to where they are needed.

Leukocytes generated in the bone marrow are released into the bloodstream, from which they populate peripheral lymphoid organs and patrol the periphery to detect invading pathogens. Effector leukocytes circulate in the bloodstream and can be specifically recruited to sites of infection when required. Naive lymphocytes continuously recirculate from the bloodstream into peripheral lymphoid organs and back to the bloodstream (Fig. 22.1). Leukocyte migration, or 'homing,' to lymphoid organs and infected tissues is controlled mainly through interactions between chemokines and chemokine receptors.

Chemokines produced at sites of inflammation and/or infection attract leukocytes that express specific receptors for that chemokine, which promotes their extravasation. In contrast, chemokines constitutively expressed by stromal cells within the bone marrow and lymphoid organs contribute to immune homeostasis and maintain the architecture of lymphoid tissues.

The chemokine CXCL12 [stromal derived factor 1 (SDF-1)] is a key chemokine that directs the homing of both precursor and mature leukocytes to the bone marrow. It binds to the receptor CXCR4 expressed by leukocytes. Levels of CXCR4 on the surfaces of leukocytes vary depending on the maturation and activation status of the cells, thereby modulating their responsiveness to CXCL12 and their tendency to home to the bone marrow.

In particular, CXCR4 is critical in guiding hematopoietic stem cells to the bone marrow during embryogenesis. In postnatal life, it helps maintain hematopoietic stem cells in the bone marrow stem-cell niche. In leukocytes of the myeloid lineage, CXCR4 has a bimodal pattern of expression; higher levels are present on the surfaces of immature myeloid progenitors and senescent neutrophils than on the surfaces of mature leukocytes. The CXCR4 ligand CXCL12 is produced at high levels by bone marrow stromal cells, such that the interaction of CXCL12 with CXCR4 promotes the retention of immature myeloid progenitors in the bone marrow (permitting their maturation), and also favors the elimination of senescent neutrophils in the periphery by recruiting them to the bone marrow, where they undergo apoptosis. Concurrently with this dual pattern of expression of CXCR4 during their maturation, myeloid cells also show variable levels of expression of the chemokine receptor CXCR2, which

This case was prepared by Luigi Notarangelo, MD, in collaboration with Anna Virginia Gulino, MD.

Topics bearing on this case:
Neutrophil function
Chemokines and their receptors
Leukocyte homing
Leukocyte development

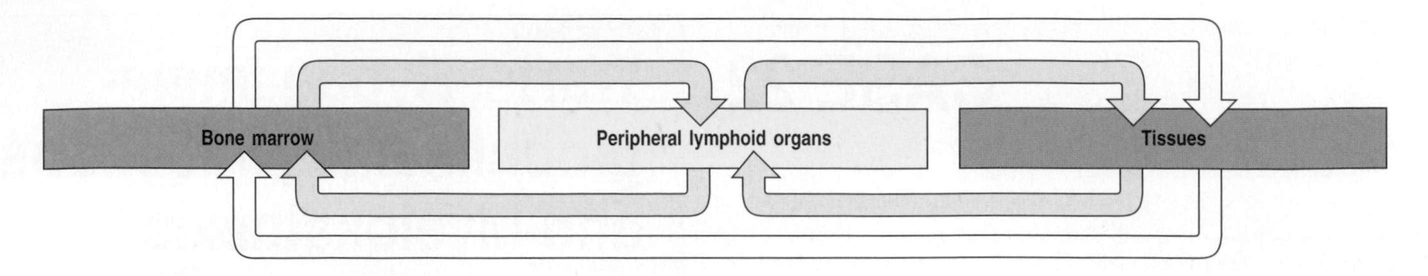

Fig. 22.1 Recirculation of leukocytes between the bone marrow, peripheral lymphoid organs, and peripheral tissues.

also has an important role in modulating the circulation of myeloid cells by its interaction with several ligands, including CXCL1 and CXCL2.

CXCR4 is also important for the recirculation and homing of lymphoid lineage cells; in particular, along with CXCR5 (the receptor for the chemokines CXCL19 and CXCL21), it modulates the migration of B lymphocytes within the follicles and germinal centers of the lymph nodes. CXCR4 also has an important role in T lymphocytes. In particular, it is a co-receptor for HIV-1, which facilitates the infection of CD4 T cells by the virus.

As we shall see in the case described below, mutations in the *CXCR4* gene in humans are associated with an immunodeficiency that results from deficiencies in leukocyte trafficking as a result of defective interaction between chemokine and receptor.

The case of Sue Bridehead: recurrent pyogenic infections with severe chronic neutropenia but hypercellular marrow.

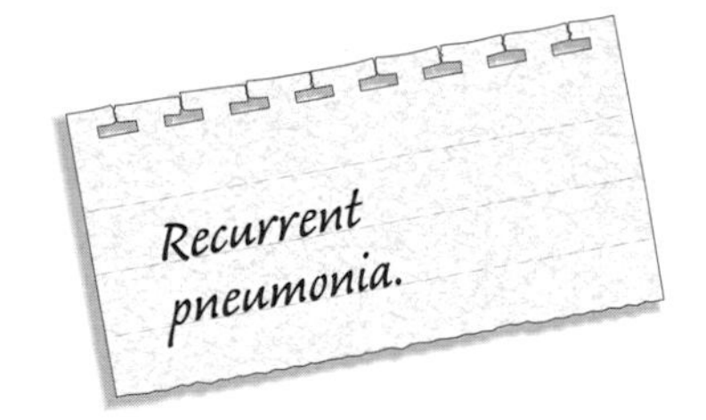

Sue was born with the congenital heart defect Tetralogy of Fallot. Severe neutropenia (absolute neutrophil count (ANC) 350 μl^{-1}; normal value 3000–5800 μl^{-1}) was detected at 2 years of age during work-up for secondary surgical correction of her heart defect, but she was not investigated further. Between 2 and 4 years of age, Sue had two episodes of pneumonia, which were treated with amoxicillin and clavulanic acid by her primary care physician, but were not investigated further. At the age of 4 years, Sue was admitted to Children's Hospital Boston because of a third episode of bacterial pneumonia. She had had a fever (temperature 38.7°C) and a cough for the past 3 days. Rales (crackles) could be heard at the right lower lobe, and a murmur at the left sternal border. She had very mild dyspnea (shortness of breath) with oxygen saturation of 96%. Sue was at the 10th centile for both height and weight and had no dysmorphic features. Her tonsils were small, there was no hepatosplenomegaly, and the rest of her physical examination was unremarkable. She was the only child of healthy non-consanguineous parents. Other than the bacterial pneumonias, she had not suffered significantly from infectious diseases and had received immunizations according to the normal schedule.

A chest X-ray confirmed right inferior lobar pneumonia. Laboratory investigations showed a normal ANC (4700 μl^{-1}) but elevated C-reactive protein (CRP; 17 mg ml^{-1}). Sue was started on antibiotic treatment with amoxicillin and clavulanate, and fever subsided after 48 hours. She was discharged, but a week later she was found to be leukopenic (white blood cell count 1500 μl^{-1}) with significant neutropenia (ANC 300 μl^{-1}) and lymphopenia (absolute lymphocyte count (ALC) 1000 μl^{-1}; normal more than 2000 μl^{-1}). Hemoglobin levels and platelet count were normal. CRP had also normalized. She was found to be hypogammaglobulinemic, with IgG levels of 225 mg dl^{-1} (normal 345–1236 mg dl^{-1}), IgA 12 mg dl^{-1} (normal 14–159 mg dl^{-1}), and IgM 33 mg dl^{-1} (normal 43–207 mg dl^{-1}). These tests were repeated after a month, and confirmed the

presence of leukopenia with severe neutropenia and moderate lymphopenia, along with hypogammaglobulinemia.

In spite of the fact that Sue had received a full course of immunizations, her antibody titers to tetanus toxoid, *Haemophilus influenzae*, and *Streptococcus pneumoniae* were nonprotective. The severe neutropenia prompted her physicians to take a bone marrow aspirate, which showed very active myelopoiesis with myelokathexis—that is, the accumulation of senescent neutrophils characterized by pyknotic and hypersegmented nuclei (Fig. 22.2). The possibility of WHIM syndrome was considered. Sequence analysis of the *CXCR4* gene revealed heterozygosity for the R334X nonsense mutation. Sue was treated with daily human recombinant granulocyte colony-stimulating factor (hrG-CSF, 5 μg per kg body weight) and regular administration of intravenous immunoglobulin. She is now 12 years old and has not had any other severe bacterial infection; however, she has recently begun to develop multiple warts, which are refractory to topical treatment and tend to recur after curettage.

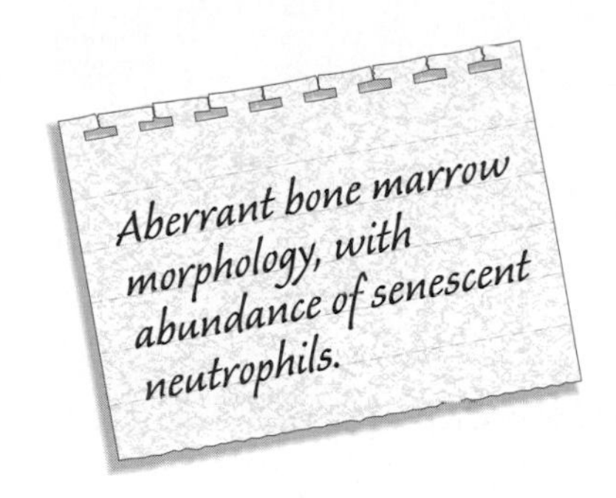

Aberrant bone marrow morphology, with abundance of senescent neutrophils.

WHIM syndrome.

WHIM syndrome is very rare, with a frequency less than 1 case per million individuals. The gene responsible for WHIM syndrome has been mapped to chromosome 2. It encodes the CXCR4 protein, a G-protein-coupled chemokine receptor containing seven transmembrane domains (Fig. 22.3), which specifically binds the chemokine CXCL12. Most patients with WHIM syndrome carry a heterozygous mutation that leads to premature truncation or disruption of the sequence of the carboxy-terminal tail of the CXCR4 protein. The mutant protein is normally synthesized and expressed at the cell surface. It may form homodimers or heterodimer complexes with the normal CXCR4 protein (encoded by the wild-type allele). These dimers maintain the ability to deliver intracellular activating signals but are refractory to β-arrestin-dependent endocytosis. Hence, WHIM-causing CXCR4 mutations result in increased chemokine receptor signaling. As a result, neutrophils and lymphocytes from WHIM patients show an increased chemotactic response to CXCL12.

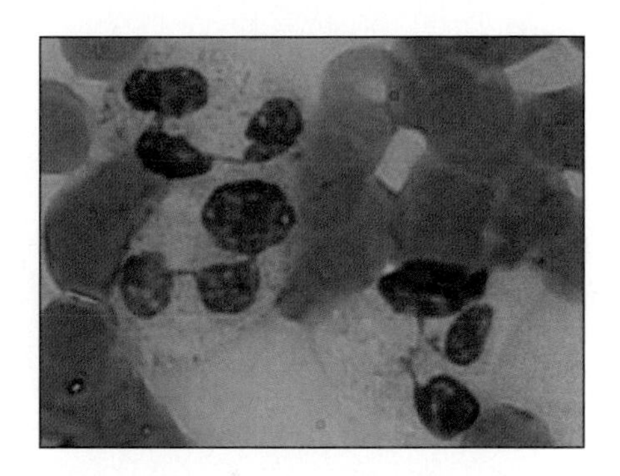

Fig. 22.2 Bone marrow examination in patients with WHIM syndrome reveals accumulation of senescent, hypersegmented neutrophils. Three neutrophils with hypersegmented nuclei (stained red with dark-staining chromatin) can be seen. Staining with May–Grünwald–Giemsa. Photograph courtesy of Antonio Regazzoli and Lucia Notarangelo, Spedali Civili Brescia, Italy. Magnification ×40.

A small minority of patients are phenocopies. Some of these patients have low levels of the kinase G-protein-coupled receptor kinase 3 (GRK3), which mediates CXCL12-dependent internalization and desensitization of CXCR4, and hence acts as a negative regulator of CXCR4 signaling. However, no mutations of the *GRK3* gene have been identified so far in these patients.

Although the acronym WHIM was coined after the description of a series of patients who shared all the typical clinical and laboratory features of the disease (warts, hypogammaglobulinemia, infections, and myelokathexis), after the discovery of the gene defect it has become obvious that patients can also present with an incomplete phenotype. In particular, warts may not become evident for several years after birth.

Peripheral leukopenia, and especially a marked reduction of the absolute neutrophil count, is almost invariably present. These peripheral data contrast with a hypercellularity in the bone marrow due to an expansion of mature and apoptotic neutrophils. The Greek-derived term 'myelokathexis' indicates white blood cell retention. The neutropenia may be somewhat ameliorated during acute infections, when pro-inflammatory signals outweigh the CXCL12-mediated retention of neutrophils in the bone marrow. Furthermore, G-CSF and granulocyte–macrophage colony-stimulating factor (GM-CSF), epinephrine, and glucocorticoids also rapidly stimulate leukocyte release. Daily administration of hrG-CSF is often used in the treatment of patients with WHIM and a history of multiple severe infections.

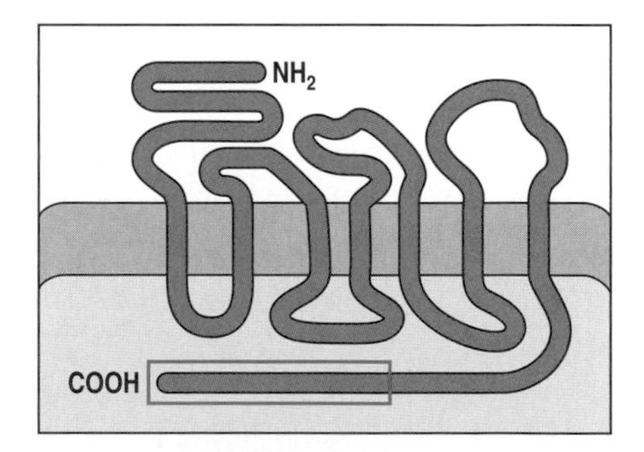

Fig. 22.3 Structure of the chemokine receptor CXCR4, a G-protein-coupled receptor. The structure is shown schematically. Note the seven membrane-spanning domains. The boxed area indicates the carboxy-terminal domain of the protein where heterozygous and frameshift mutations that cause WHIM syndrome are clustered.

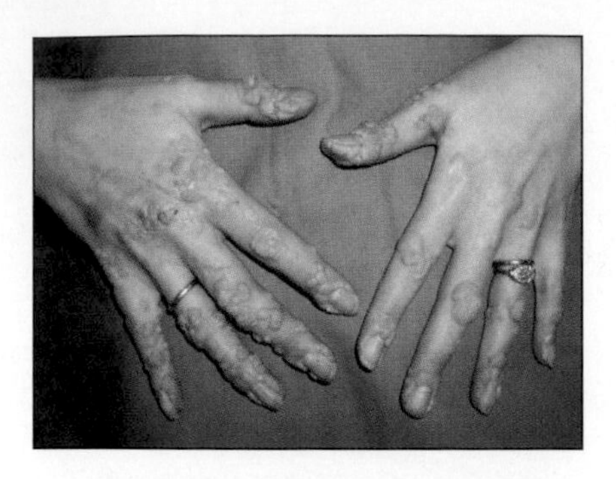

Fig. 22.4 Typical warts in a patient with WHIM syndrome.

Recurrent bacterial infections are common in WHIM syndrome, and are mostly due to common pathogens such as *H. influenzae*, *Staphylococcus aureus*, and *S. pneumoniae*. Infections often involve the respiratory tract, but deep-seated abscesses can also occur. During the acute phase of infection, the absolute neutrophil count can be normal or even elevated, indicating that WHIM neutrophils may leave the bone marrow in response to strong inflammatory signals.

Hypogammaglobulinemia is frequently present, and the count of circulating B lymphocytes is often decreased. In particular, there is a reduction in the number of switched memory ($CD27^+$ IgD^-) B cells. Patients with WHIM syndrome can mount specific antibody responses after immunization, but the titer of specific antibodies often wanes with time. Immunoglobulin replacement therapy should be considered in patients with hypogammaglobulinemia and is effective in reducing the risk of bacterial infections. Patients with WHIM syndrome also have a reduction in the number of naive T cells, and a relative expansion of effector memory T cells, which often show a restricted repertoire.

Most, but not all, patients with WHIM syndrome develop persistent and treatment-refractory warts (Fig. 22.4), which result from infection by the human papillomavirus (HPV). Typically, the hands are involved, but warts may appear on any part of the body. Treatment with intravenous immunoglobulin or hrG-CSF does not cause regression of the warts. Genital warts (condylomata acuminata) affecting the anogenital tract may predispose to epithelial cancer. For a long time, it was thought that patients with WHIM syndrome are uniquely susceptible to HPV but not to other viruses. However, an increased frequency of infections by herpes simplex virus (HSV), human herpesvirus 8 (HHV8, associated with Kaposi disease), the Epstein–Barr virus (EBV), and the viral skin infection molluscum contagiosum has recently been shown. EBV infection can lead to lymphoma, especially involving the skin (cutaneous lymphoma). A defect in the trafficking of effector cells (T cells and NK cells) and antigen-presenting cells (such as dendritic cells) in WHIM could explain the abnormal vulnerability to viruses affecting the skin.

Finally, the incidence of congenital cardiac malformations in patients with WHIM syndrome (7%) is substantially higher than in the general population (0.8%), which is consistent with an established role for CXCR4 in cardiac patterning.

Questions.

1 *Sue's laboratory features included neutropenia and hypogammaglobulinemia. Is WHIM syndrome the only immunodeficiency that presents with this association? What features might you look for to distinguish WHIM syndrome from other conditions that present with hypogammaglobulinemia and neutropenia?*

2 *What are the most likely reasons that neither of Sue's parents showed any clinical features of WHIM?*

3 *What mechanisms might account for the increased susceptibility to viral cutaneous infections of patients with WHIM syndrome?*

4 *What kind of pharmacological intervention could you suggest that would interfere with the CXCL12–CXCR4 interaction? Can you think of a useful clinical application for such an intervention?*

CASE 23 X-linked Hypohidrotic Ectodermal Dysplasia and Immunodeficiency

Immunodeficiency due to a defective component in an intracellular signaling pathway required for both innate and adaptive immunity.

Intracellular signaling molecules are vital to the normal maturation of B and T lymphocytes. In Case 1 we saw that X-linked agammaglobulinemia results from a failure of normal B-cell maturation because of mutations in a tyrosine kinase (Bruton's tyrosine kinase, Btk) that is vital to normal calcium movements in pre-B cells. In its absence, pre-B cells fail to progress to mature B cells and thus never become immunoglobulin-secreting plasma cells. Other signaling molecules are involved in the transmission of co-stimulatory signals to lymphocytes. In Case 2 we saw that X-linked hyper IgM syndrome results from the failure of B cells to undergo class-switch recombination and somatic hypermutation because they do not receive a signal from helper T cells, owing to a failure of these cells to express CD40 ligand.

Although some signaling molecules, such as CD40 ligand, are peculiar to lymphocytes, others are found in many cell types. A crucial step in the activation of many different cells is the induction of the transcription of specific genes by transcription factors of the NFκB family. When not required, NFκB and other family members reside in the cytoplasm, where they are maintained in an inactive state in a complex with their inhibitor protein, IκB. When the cell is activated by the appropriate external signal, a cascade of protein kinase activity is induced in the cytoplasm. This causes the assembly and activation of the IκB kinase (IKK) complex (Fig. 23.1), which consists of three components—the kinases IKKα and IKKβ along with another protein, the NFκB essential

Topics bearing on this case:
Macrophage activation
Isotype switching
Activation of NFκB
T-cell help in antibody response
CD40 ligand and isotype switching

This case was prepared by Raif Geha, MD, in collaboration with Douglas McDonald, MD, PhD.

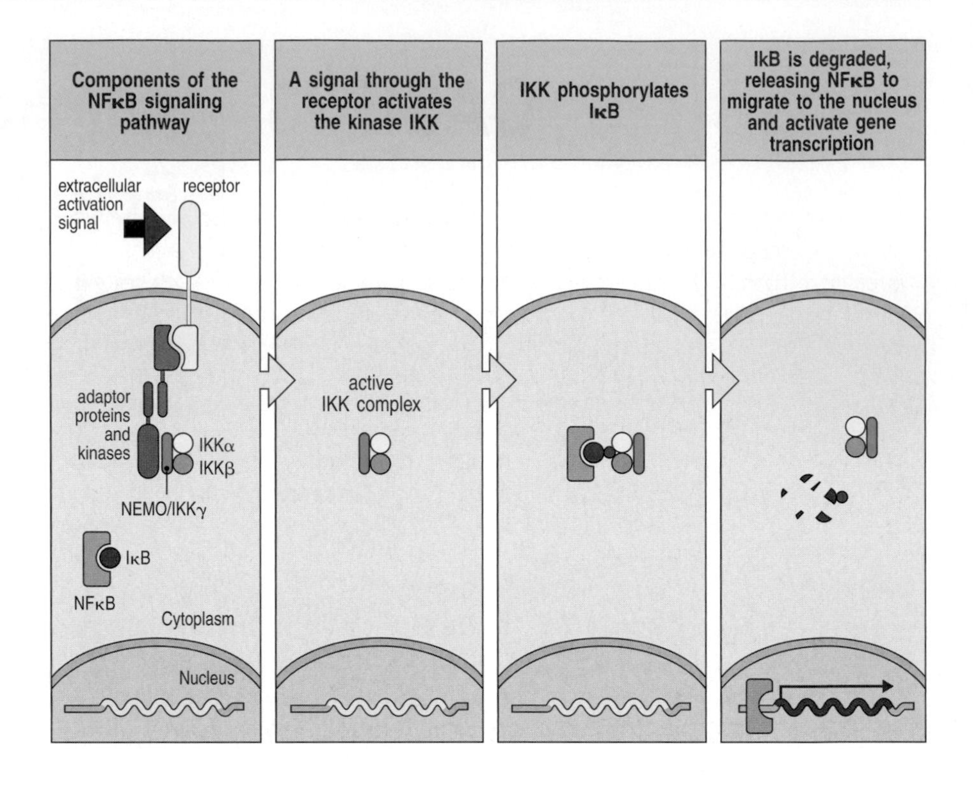

Fig. 23.1 The NFκB activation pathway. Various receptors can activate a series of intracellular adaptor molecules and kinases that ultimately induce the formation of the IKK complex (consisting of IKKα, IKKβ, and NEMO/IKKγ). The IKK complex then phosphorylates the inhibitory protein IκB, resulting in its dissociation from cytoplasmic NFκB and its ultimate ubiquitination and degradation. Once free from the inhibitory influence of IκB, NFκB translocates to the nucleus, where it induces the transcription of specific genes.

modifier (NEMO), also known as IKKγ. The active IKK complex phosphorylates IκB, which targets IκB for degradation by the ubiquitin–proteasome system, thus freeing NFκB from its inhibitor. A nuclear localization signal is uncovered on NFκB that enables the transcription factor to move into the nucleus, where it binds to particular nucleotide sequences in the promoters of selected genes and initiates their transcription.

NFκB is activated by the ligation of a variety of receptors, including many with fundamental roles in development, in inflammation, and in the generation of immune responses. For example, the binding of the pro-inflammatory cytokines tumor necrosis factor-α (TNF-α) or interleukin 1 (IL-1) to their receptors results in NFκB activity. The Toll-like receptors (TLRs) on macrophages, which are key components of the innate immune system and recognize microbial components and other danger signals, also work through NFκB activation. And NFκB is activated in B cells after stimulation via CD40.

Functional NFκB is measured in the laboratory by an electrophoretic mobility-shift assay (EMSA) that detects its presence in the nucleus. Cell nuclei are isolated by ultracentrifugation, and their NFκB content is measured by assessing the capacity of the nuclear extract to bind to a synthetic radiolabeled oligonucleotide probe and thus to retard its migration in gel electrophoresis (Fig. 23.2).

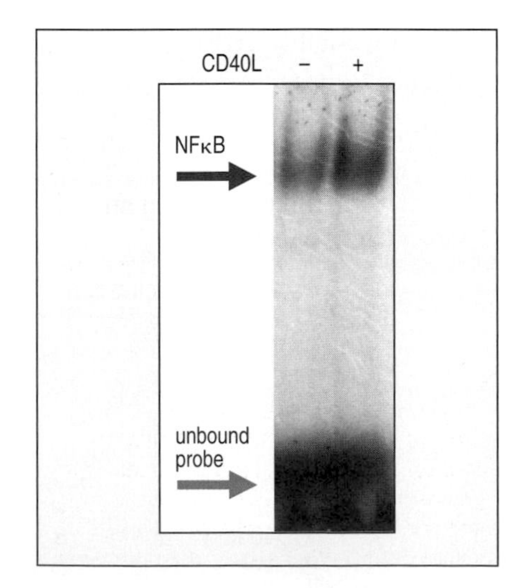

Fig. 23.2 Electrophoretic mobility-shift assay (EMSA). Nuclei were isolated from normal B cells by ultracentrifugation, and nuclear extracts were mixed with a ^{32}P-radiolabeled DNA probe that binds to NFκB. Free probe, not associated with NFκB, can be seen at the bottom of the gel, whereas probe bound to NFκB is retarded in the gel as a result of the higher molecular weight of the protein–probe complex. Although unstimulated B cells contain a baseline level of NFκB in their nuclei (as seen in the left lane), ligation of CD40 by CD40 ligand (CD40L) results in a higher level of NFκB in the nucleus (as evidenced by the darker band in the right lane).

The case of Robert Teixiera: severe bacterial infections and pointed teeth.

Nine-month-old boy
with strep meningitis.
Had been immunized.

Robert was born after an uneventful pregnancy. He received his normal immunizations at 2, 4, and 6 months, including a pneumococcal vaccine that contained seven different serotypes of *Streptococcus pneumoniae*, without any ill effects. At 9 months old he developed recurrent fever associated with nasal discharge, which was diagnosed as a viral infection of the upper respiratory tract. He began to show less interest in his bottle, cried frequently, and became increasingly difficult to console.

Three weeks after the fevers began, he had a seizure and was taken to the local emergency department. The seizure was controlled with an intravenous anticonvulsive medicine and the physician noted that his skin felt warm and had a reticular (lacy) pattern of blood vessels. Robert's neck was rigid and he screamed when his head was moved. A complete blood count revealed an abnormally high white blood cell count of 48,200 μl^{-1}, 90% of which were neutrophils (normal 5000 μl^{-1}). Spinal fluid was sampled by lumbar puncture and was free-flowing and cloudy. Microscopic and chemical analysis revealed 12 red blood cells per high-power field (normal 0); 11,382 white blood cells per high-power field (normal <6), of which 85% were neutrophils; a protein concentration of 410 mg dl^{-1} (elevated compared with normal); and a glucose concentration of 39 mg dl^{-1} (low compared with normal).

A Gram stain of the spinal fluid showed paired Gram-positive cocci. Robert was treated with ceftriaxone (an extended-spectrum cephalosporin antibiotic) and admitted to intensive care. He responded rapidly to the antibiotic and improved daily. A culture of his spinal fluid grew *S. pneumoniae*, which expressed the p14 cell-wall polysaccharide. An immunology consultation was requested because children receiving the seven-valent pneumococcal vaccine should be protected against strains of *S. pneumoniae* carrying this polysaccharide.

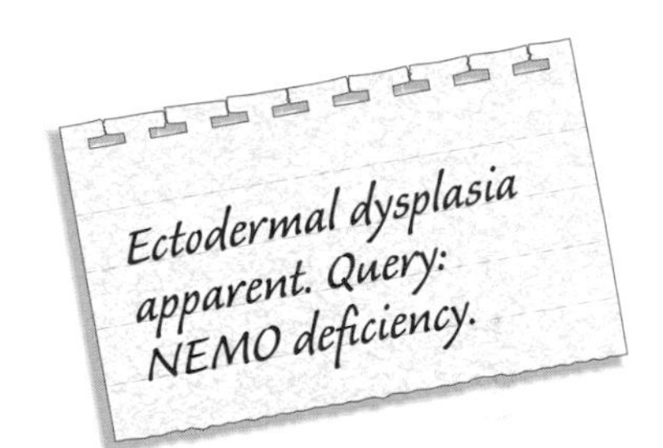

Laboratory analysis of Robert's serum immunoglobulin showed 170 mg dl^{-1} IgG (normal 400–1300 mg dl^{-1}), 34 mg dl^{-1} IgM (normal 30–120 mg dl^{-1}) and 184 mg dl^{-1} IgA (normal 20–230 mg dl^{-1}). He had a normal absolute lymphocyte count and normal percentages of T cells and B cells. Specific antibodies against 15 different pneumococcal polysaccharides or against tetanus toxoid (with which he had also been immunized) were not present in his serum. Because of the severity of his infection, his hypogammaglobulinemia and lack of specific antibody, intravenous immunoglobulin replacement therapy was begun at a dose of 400 mg kg^{-1} body weight every 3 weeks.

As Robert grew, it became clear that he had dysplastic ectoderm (that is, an abnormality in the growth of the structures produced from the ectoderm). His hair never grew properly; his first teeth erupted at 21 months and were pointed and conical (Fig. 23.3). A skin biopsy revealed a lack of eccrine sweat glands (hypohidrosis). On the basis of these findings, his hypogammaglobulinemia and his clinical history, the diagnosis of ectodermal dysplasia with immunodeficiency was made. A mutation in the *NEMO* gene was suspected. DNA sequence analysis revealed a point mutation that had caused an amino acid change at position 417 at the carboxy-terminal end of the NEMO protein.

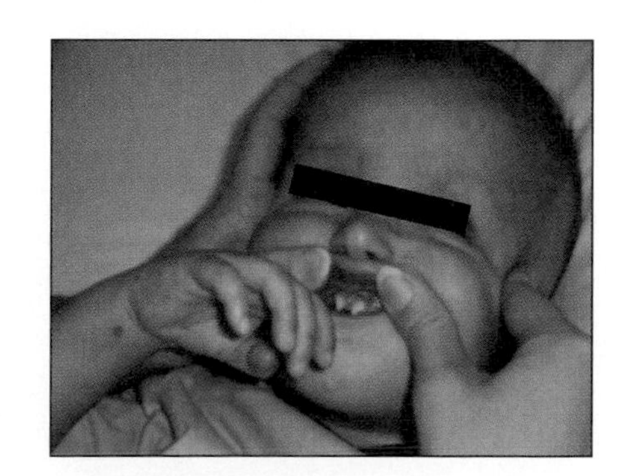

Fig. 23.3 Infant with X-linked ectodermal dysplasia and immunodeficiency due to NEMO deficiency. The features of ectodermal dysplasia include frontal bossing, deep-set eyes, fine or sparse hair, and conical or missing teeth.

Robert remained well until he was 2 years old, when he developed a hyperpigmented lacy rash on his back. A biopsy revealed diffuse granulomatous inflammation, and an acid-fast stain was positive for bacilli. A culture of the skin sample grew *Mycobacterium avium*. The cutaneous atypical mycobacterial infection was treated with a combination of antibiotics chosen on the basis of the *in vitro* sensitivity of the isolated mycobacteria. After several months of antibiotic treatment the rash had

disappeared, but repeated attempts to discontinue the antibiotics failed, because the rash always returned. Robert is currently well on immunoglobulin replacement therapy and continuous treatment with anti-mycobacterial antibiotics.

NEMO deficiency and X-linked hypohidrotic ectodermal dysplasia with immunodeficiency.

X-linked ectodermal dysplasia with immunodeficiency is caused by hypomorphic mutations in the *NEMO* gene that lead to the production of a protein with impaired function. This results in immunodeficiency because the end point of the intracellular signaling pathways leading from a variety of receptors important in activating immune responses is activation of NFκB and the induction of gene transcription. In the absence of any component of IKK (such as NEMO), IκB is not properly phosphorylated and it remains bound to NFκB, thus preventing the translocation of NFκB to the nucleus. A complete absence of NEMO function, amorphic mutation, is lethal in the embryo. The receptors leading to NFκB activation include the Toll-like receptors on macrophages, which recognize bacterial lipopolysaccharide (LPS), peptidoglycans, and bacterial DNA (CpG), and whose engagement stimulates innate immunity. Other receptors include CD40, which is important in adaptive immunity. In patients with a hypomorphic *NEMO* mutation, the interaction between CD40 on B cells and CD40 ligand on T cells occurs but NFκB activation is impaired, leading to diminished activation of mature naive B cells. This results in impaired antibody synthesis and impaired switch class recombination (as assessed by a lack of upregulation of the surface markers CD23 and CD54 (ICAM-1) and decreased IgE synthesis; Fig. 23.4).

Patients with mutations in *NEMO* have ectodermal dysplasia because a receptor that is required for ectodermal development also depends on IKK function and NFκB activation to transmit a signal to the nucleus. The receptor is the ectodysplasin A receptor (a member of the TNF receptor superfamily), which binds ectodysplasin A (a member of the TNF superfamily). Although defects in the genes for these two proteins result in ectodermal dysplasia, there is no concomitant immunodeficiency because the function of the IKK complex is intact. Interestingly, several patients with immune deficiency due to hypomorphic mutations in *NEMO*, but without ectodermal dysplasia, have been identified. Furthermore, distinct hypomorphic *NEMO* mutations have been characterized that differentially impair innate and adaptive immune functions.

The *NEMO* gene is located on the X chromosome, and thus the disease is transmitted by mothers to their sons. The gene contains 10 exons and is 419 amino acids long. Most of the mutations in the *NEMO* gene that cause ectodermal dysplasia are located in the tenth exon, which encodes a zinc-finger domain that is a site of protein–protein interaction.

Women who carry a single copy of a *NEMO* gene with a point mutation generally do not have overt symptoms, although they express both normal NEMO and mutant NEMO. Some may have missing teeth, mildly abnormal patterns of hair or unusual birthmarks. However, a major frameshift or deletion mutation in *NEMO* results in a condition known as incontinentia pigmenti, which is characterized by dermal scarring and hyperpigmentation in heterozygotes. This disorder does not affect the immune system, because the normal *NEMO* gene allows appropriate immune-system development and function. In fact,

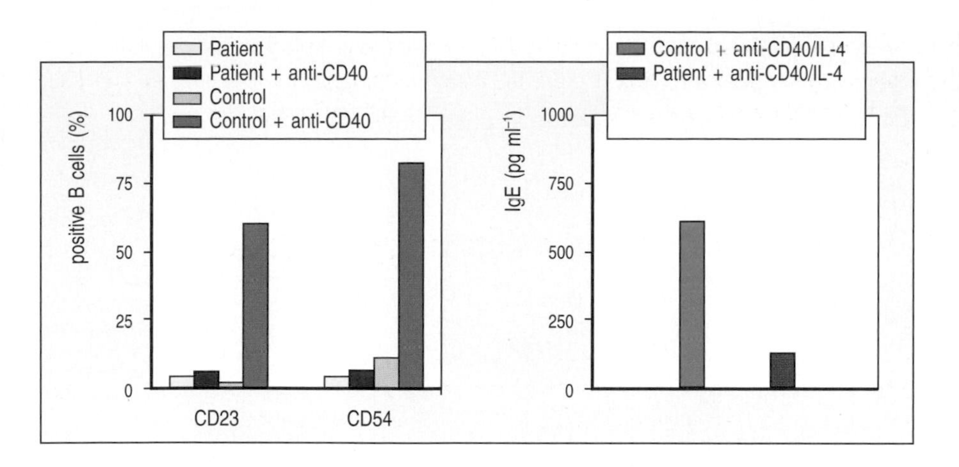

Fig. 23.4 Impairment of NFκB activation and B-cell activation in a patient with ectodermal dysplasia and immunodeficiency due to a mutation in NEMO. Left panel: B-cell activation can be assessed by increases in surface expression of CD23 (the low-affinity IgE receptor) or CD54 (ICAM-1). The presence of these molecules on the surface of B cells is measured by flow cytometry. In a healthy individual, stimulation through CD40 results in the majority of B cells expressing CD23 and CD54 on the cell surface. The gray bars show the unstimulated B cells, and the blue bars show the stimulated B cells. No increased expression of CD23 and CD54 after CD40 ligation is seen in the patient's B cells: compare the unstimulated and stimulated patient's B cells (yellow and red, respectively). Right: CD40 ligation causes IgE isotype switching and IgE synthesis in normal B cells (green) but not in B cells from a NEMO patient (purple), further demonstrating their inability to transmit a signal from CD40.

the immune cells of these women display nonrandom X-chromosome inactivation, and all their immune-system cells contain the normal *NEMO* gene. Any male offspring of women with incontinentia pigmenti who receive the mutant *NEMO* gene die *in utero*.

Questions.

1 *Why are boys with NEMO mutations susceptible to infection with opportunistic mycobacteria such as* M. avium*?*

2 *Patients with NEMO mutations also suffer from chronic viral infections such as cytomegalovirus (CMV). Why is this?*

3 *Although B cells from some patients with NEMO mutations fail to undergo immunoglobulin class switching in response to CD40 ligation, these patients have residual serum levels of IgG and IgA. Why is this?*

4 *Recently a patient has been discovered to have a gain-of-function mutation in IκBα. This patient, who also has ectodermal dysplasia, has a more severe immunological defect than patients with the NEMO defect. The gain-of-function mutation results from the change of a serine residue to an isoleucine residue, so that the IκBα chain cannot be phosphorylated. Why do you think the immunodeficiency is more severe in this defect?*

5 *You follow a family in your immunodeficiency clinic. There are two brothers and a maternal male cousin who suffer from recurrent pneumonias. One patient has had an infection with an atypical mycobacterium. None of the boys has ectodermal dysplasia. Could this family still have a hypomorphic NEMO mutation? You have sequenced NEMO in cDNA generated from one of the boys to rule out a mutation. No mutation was found. Does the lack of a mutation in the patient's NEMO message rule out NEMO as a cause of the immune deficiency?*

CASE 24 Interferon-γ Receptor Deficiency

The destruction of intracellular microorganisms in macrophages.

Certain pathogens such as mycobacteria, *Listeria*, *Leishmania*, and *Salmonella* take up residence in macrophages and are thereby protected from elimination by antibody or cytotoxic T cells. These microorganisms can be eliminated only when their host macrophages are activated and produce increased amounts of nitric oxide, oxygen radicals, and other microbicidal molecules (Fig. 24.1). The activation of macrophages in an adaptive immune response is masterminded by T_H1 cells (Fig. 24.2); the most important cytokine involved in macrophage activation is interferon (IFN)-γ.

When macrophages take up microorganisms by phagocytosis, they secrete interleukin (IL)-12. This cytokine is necessary for the induction of IFN-γ synthesis by T cells and natural killer (NK) cells. IL-12 furthermore favors the maturation of T_H1 cells, which activate macrophages, and suppresses the maturation of T_H2 cells, which secrete a cytokine, IL-10, involved in the deactivation of macrophages.

IFN-γ acts at a receptor on the macrophage surface. This receptor is composed of two different types of polypeptide chain—IFN-γ receptor 1 (IFN-γR1) and IFN-γR2, each of which is associated with a tyrosine kinase—JAK1 and JAK2 respectively. When a dimer of IFN-γ binds to two molecules of IFN-γR1, it causes them to associate with two IFN-γR2 chains (Fig. 24.3) and this initiates a signaling pathway that eventually results in changes in gene transcription.

As we shall see in this case, a mutation in the gene encoding the IFN-γR1 chain has drastic effects on the capacity to fight certain pathogens.

Topics bearing on this case:
Microbicidal action of phagocytes
Macrophage activation
Cytokine receptor signaling pathway

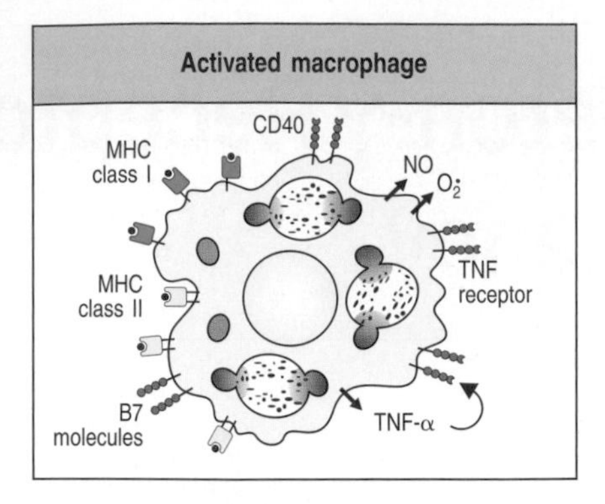

Fig. 24.1 Activated macrophages undergo changes that greatly increase their antimicrobial effectiveness and amplify the immune response. Once the macrophage is activated, lysosomes fuse with the intracellular vesicles within which the pathogenic bacteria (red) reside, which exposes the microorganisms to degradative enzymes and other microbicidal agents. Activated macrophages also increase their expression of receptors for tumor necrosis factor (TNF), and secrete TNF-α. This autocrine stimulus synergizes with interferon (IFN)-γ secreted by T_H1 cells to increase the antimicrobial action of the macrophage, in particular by inducing the production of nitric oxide (NO) and oxygen radicals ($O_2^{\bullet}$). The macrophage also increases the expression of CD40, which by interaction with the CD40 ligand on T cells upregulates the expression of B7 proteins and increases the expression of class II MHC molecules on the macrophage, thus allowing the further activation of resting CD4 T cells.

The case of Clarissa Dalloway: a relentless infection due to mycobacteria.

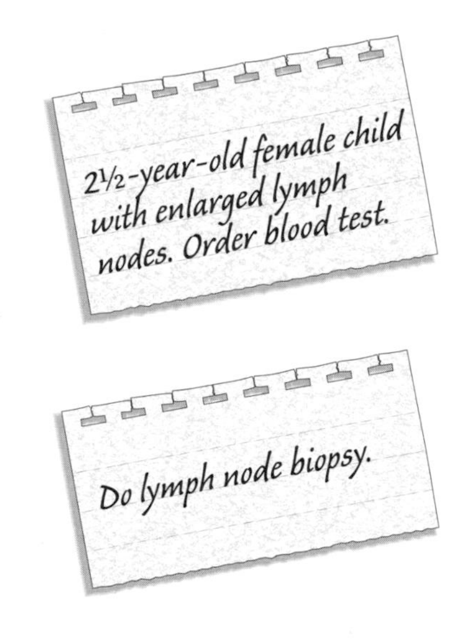

Clarissa Dalloway was the first child born to a couple who lived in an isolated fishing village on the coast of Maine. Her parents thought that they were distantly related to each other. The fishermen of this village were all descended from English settlers who came there in the late 17th century, and there was much intermarriage in the community.

Clarissa was well at birth and developed normally until she was around 2½ years old. Her mother then noticed that she was not eating well, had diarrhea, and was losing weight. She took Clarissa to a pediatrician in the nearest town of Bath, Maine. The pediatrician, Dr Woolf, noted enlarged lymph nodes and ordered an ultrasound and CT scan of the chest and abdomen. These showed enlarged lymph nodes in the mesentery and para-aortic region, and Dr Woolf referred Clarissa to the Children's Hospital in Boston.

Blood tests revealed a white blood cell count of 9400 μl^{-1}, of which 55% were neutrophils, 30% lymphocytes, and 15% monocytes (slightly elevated). Her serum IgG was 1750 mg dl^{-1}, IgA 450 mg dl^{-1}, and IgM 175 mg dl^{-1} (these immunoglobulin values are all elevated).

It was decided to biopsy the enlarged lymph nodes. On histological examination they showed marked proliferation of histiocytes, and many neutrophils were seen in the lymph node. There was no granuloma formation and no giant cells were seen. An acid-fast stain for mycobacteria (see, for example, Fig. 48.4) revealed numerous microorganisms, and *Mycobacterium avium intracellulare* was cultured from the lymph nodes as well as from the blood.

Despite appropriate antibiotic treatment for the mycobacterial infection, Clarissa eventually developed infiltrates in the lungs and progressive enlargement of the spleen. At 6 years old she developed sepsis and *Salmonella paratyphi* was cultured from her blood. She was successfully treated with antibiotics for this infection, but soon afterward she developed meningitis and died. *M. avium intracellulare* was cultured from the cerebrospinal fluid.

A detailed family history revealed that Clarissa had three male cousins, two of whom were brothers, who had died of mycobacterial infections. In one case *M. chelonei* had

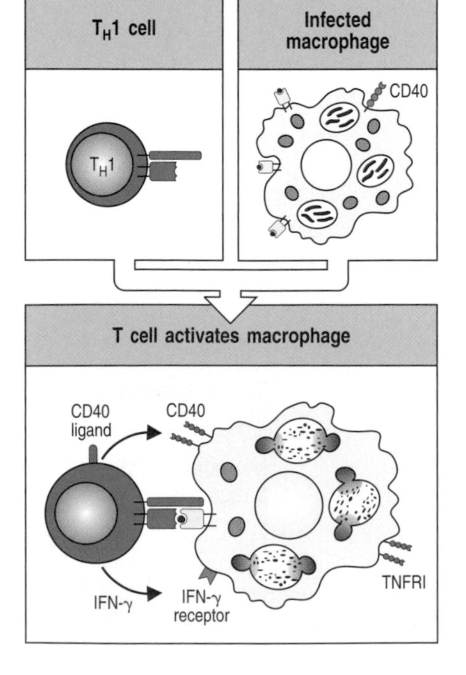

Fig. 24.2 T_H1 cells activate macrophages to become highly microbicidal. When a T_H1 cell specific for a bacterial peptide contacts an infected macrophage, the T cell is induced by IL-12 to secrete the macrophage-activating factor IFN-γ and to express CD40 ligand. Together, these newly synthesized T_H1 proteins activate the macrophage.

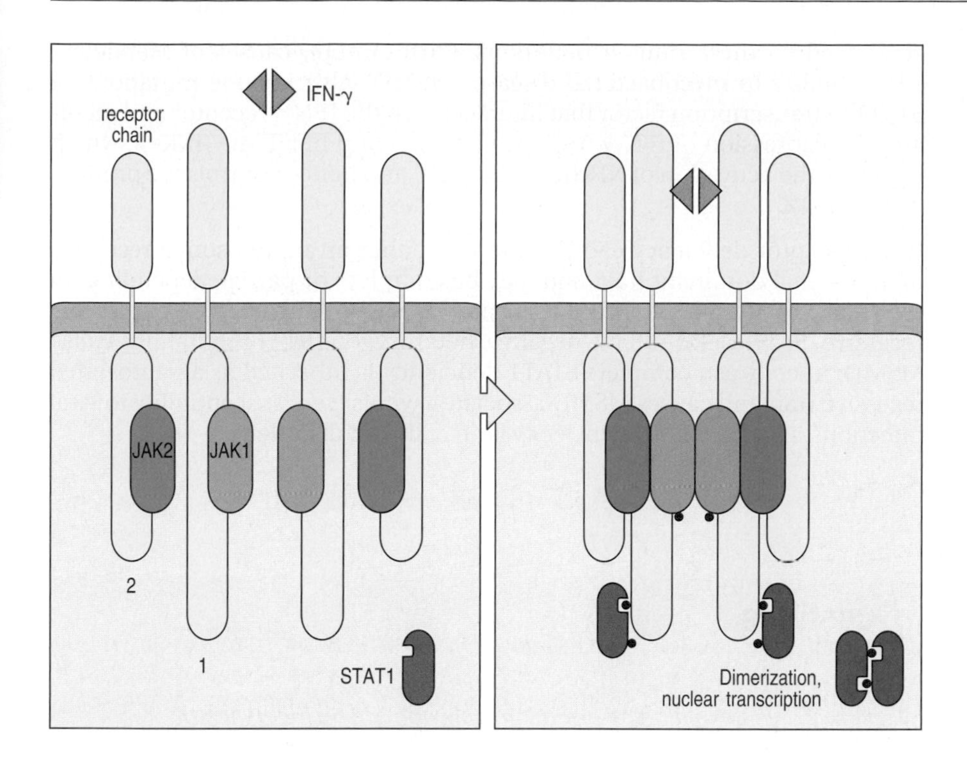

Fig. 24.3 Diagram of the IFN-γ receptor and its activation of the JAK–STAT signaling pathway. A dimer of IFN-γ causes clustering of the four chains of the IFN-γ receptor. The receptor 1 chains become phosphorylated by JAK1, forming a site to which the transcription factor STAT1α can bind. This in turn becomes phosphorylated, dimerizes, and is then transported to the nucleus, where it upregulates the various genes that produce the activated macrophage phenotype. The IFN-γR1 chain is encoded on chromosome 6, whereas the receptor 2 chain is encoded on chromosome 21.

been cultured from lymph nodes; *M. fortuitum* and *M. avium intracellulare* had been cultured from blood and lymph nodes of the two brothers.

Interferon-γ receptor deficiency.

It is estimated by the World Health Organization that 1.7 billion human beings currently alive have been infected with mycobacteria. Yet only a very small fraction of these individuals develop disease as a result of this infection. The AIDS epidemic has markedly increased the incidence of mycobacterial disease due to *M. tuberculosis,* and to other mycobacterial species such as *M. avium intracellulare,* which are collectively called 'atypical mycobacteria.' In general, atypical mycobacteria do not cause disease in immunologically normal human beings, except for swollen lymph nodes in which these mycobacteria survive. Infection with atypical mycobacteria causes a positive tuberculin reaction to develop.

Although there are many ethnic differences in susceptibility to *M. tuberculosis,* no single genetic factor had ever been found in humans to explain this susceptibility until Clarissa Dalloway and her family were found to have a genetic defect in the IFN-γR1 gene (*IFNGR1*). The genetic defect was ascertained in this family by mapping the susceptibility gene to chromosome 6q22, the map location of the receptor gene.

This finding prompted the examination of children who had developed progressive mycobacterial infection after immunization with the BCG (Bacille Calmette–Guérin) vaccine, and who did not have severe combined immunodeficiency (see Case 5). More kindreds with the IFN-γR1 deficiency were discovered as a result of this lead. Subsequently, a similar susceptibility to mycobacterial infection has been found in patients with defects in synthesis of the p40 subunit shared by IL-12 and IL-23, or of the IL-12 receptor (IL-12R) β1 chain. Thus the dependence of IFN-γ synthesis on IL-12 has been neatly confirmed by these human mutations. Furthermore, causes of Mendelian susceptibility to mycobacterial disease (MSMD) also include mutations of STAT1 (a transcription factor that interacts with the IFN-γ receptor and mediates the expression of IFN-γ-responsive genes) and of NEMO (IKK-γ), which regulates the activation of NFκB (see Case 23) and hence controls the production of IL-12.

IFN-γ receptor deficiency can be inherited either as an autosomal recessive or autosomal dominant trait, and may be complete or partial, depending on the nature of the genetic defect. IL-12 and IL-12Rβ1 deficiency are autosomal recessive diseases. Dominant-negative, heterozygous STAT1 mutations cause MSMD; in contrast, complete STAT1 deficiency is inherited as an autosomal recessive trait and causes MSMD associated with severe susceptibility to viral infections. Finally, NEMO deficiency is an X-linked disorder.

Questions.

1 *How do you explain the alarming increase in the incidence of mycobacterial infections in patients with AIDS?*

2 *Clarissa Dalloway and her cousins had positive tuberculin skin tests. Would you have predicted this?*

3 *Clarissa and her cousins also did not develop granulomas as a result of infection with atypical mycobacteria. On one occasion mycobacteria were surprisingly cultured from Clarissa's blood. What do these observations tell us?*

4 *Clarissa also had a problem with salmonella, but she had no problem with pneumococcal infection or with any viruses, such as chickenpox. How do you explain this?*

CASE 25 Severe Congenital Neutropenia

A lack of neutrophils, leading to difficulty in fighting infections.

Neutrophils are phagocytic cells that have a crucial role in the response to many types of infections, especially infections with extracellular bacteria. They are classified as innate immune cells and are the first cells to be recruited to sites of infection and inflammation, into which they migrate in large numbers from the bloodstream. Once activated, neutrophils are short-lived; after ingesting microbes and killing them, neutrophils die and are broken down by the action of macrophages, producing the pus that is characteristic of many extracellular bacterial infections (Fig. 25.1). Neutrophils recognize their microbial targets mostly through invariant receptors, such as Fcγ receptors and Toll-like receptors. The binding of antibody-coated ('opsonized') microorganisms to Fcγ receptors on the neutrophil cell surface promotes internalization of these targets by phagocytosis, followed by intracellular killing, a process that involves the production of reactive oxygen species and the activation of proteolytic enzymes and antimicrobial peptides contained in neutrophil granules (see Case 26).

Neutrophils are generated in the bone marrow, where hematopoietic stem cells differentiate into myeloblasts, relatively large cells with a large, oval nucleus and a small amount of cytoplasm. Promyelocytes represent the next step in myeloid differentiation. Promyelocytes also have a round nucleus with diffuse chromatin, but their nucleoli tend to become less prominent as differentiation proceeds. Mature neutrophils contain several types of membrane-enclosed granules, which store enzymes and antimicrobial peptides. The primary or azurophilic granules appear at the promyelocyte stage, whereas the secondary granules become apparent only at the next step of differentiation, the myelocyte stage. Secondary granules are smaller than the primary granules and arise from the convex surface of the Golgi complex. Primary and secondary granules differ in their content. Primary granules contain myeloperoxidase, defensins, cathepsin G, and neutrophil elastase. In contrast, secondary granules contain lactoferrin and cathelicidin.

The morphological features typical of neutrophils start to become apparent when myelocytes differentiate into metamyelocytes, which are characterized by an indented nucleus containing dense and clumped chromatin. Primary, secondary, and tertiary granules fill the cytoplasm, and the cell loses its ability to divide. The nucleus becomes horseshoe-shaped, and the chromatin is further condensed, a feature of the so-called 'band neutrophil' stage. Finally, as the neutrophil matures, nuclear constrictions appear and the nucleus takes on the characteristic lobed shape.

Topic bearing on this case:

Neutrophil function

This case was prepared by Luigi Notarangelo, MD, in collaboration with Itai Pessach, MD.

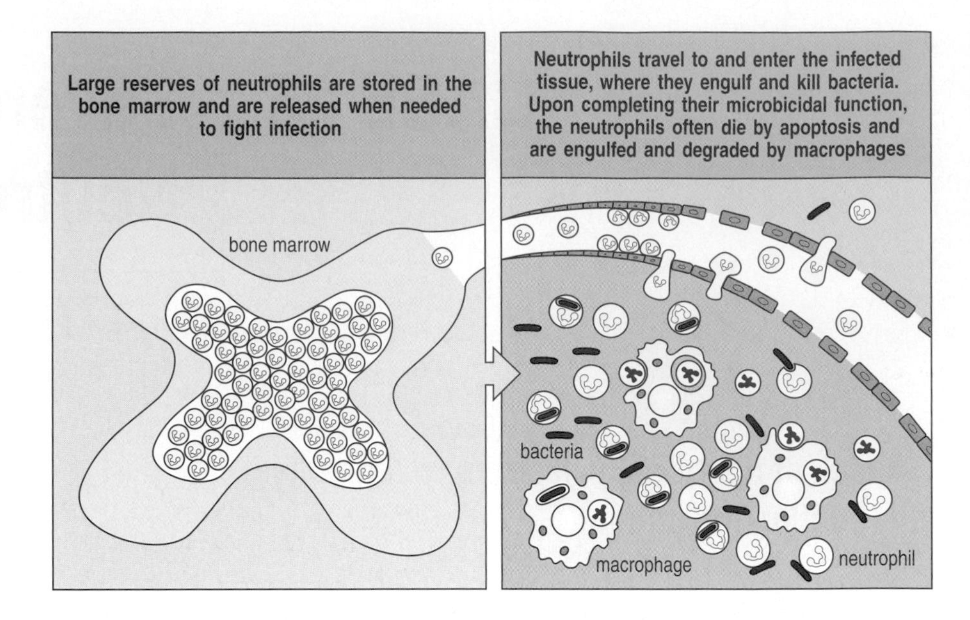

Large reserves of neutrophils are stored in the bone marrow and are released when needed to fight infection
Neutrophils travel to and enter the infected tissue, where they engulf and kill bacteria. Upon completing their microbicidal function, the neutrophils often die by apoptosis and are engulfed and degraded by macrophages
bone marrow
bacteria
macrophage
neutrophil

Fig. 25.1 Neutrophils are stored in the bone marrow and move in large numbers to sites of infection, where they act and then die. After one round of ingestion and killing of bacteria, a neutrophil dies. The dead neutrophils are eventually mopped up by long-lived tissue macrophages, which break them down. The creamy material known as pus is composed of dead neutrophils.

In healthy adults more than 50×10^9 neutrophils are produced by the bone marrow and released into the circulation every day. This high rate of production is required to balance the rapid rate of neutrophil destruction and to compensate for the relatively short half-life of the neutrophil—around 8 hours. The normal range of neutrophil count varies with age and ethnic group and is highly influenced by various conditions such as stress, infections, and medication.

An increase in the number of neutrophils in peripheral blood (neutrophilia) is often seen during infections of various kinds, as well as during stress or inflammation. A decrease in the neutrophil count (neutropenia, from *penia*, the Latin word for 'deficiency' or 'poverty') can reflect either decreased production of neutrophils as a result of bone marrow suppression or malfunction, or increased consumption or destruction of neutrophils. This case shows the dramatic effects of a severe lack of neutrophils, and illustrates the essential role of neutrophils in controlling infections due to extracellular bacteria.

The case of Michael Henchard: near death in infancy as a result of invasive bacterial infection.

Michael Henchard, the first son of a 25-year-old healthy mother, was born after 38 weeks' gestation. He remained in the nursery for 3 days, with no complications, and was then discharged and sent home. After 2 weeks, during which he fed and grew well, his mother noticed a slight redness around the stump of his umbilical cord. She did not think much of it at first, but after 24 hours the redness had spread and the umbilical area had become hard and swollen. Michael became very irritable, refused to nurse, and developed a fever (38.7°C).

His mother took him to the pediatrician, who diagnosed Michael with omphalitis (an infection of the umbilical stump) and immediately referred him for treatment at the nearest emergency room. On arrival there, Michael was lethargic and hypotonic. His temperature was 40.2°C and his heart rate above 180 beats per minute. His breathing was fast and shallow and his blood pressure was low (60/35 mmHg). His abdomen was distended and the abdominal wall was very warm to the touch, hard and erythematous (reddened).

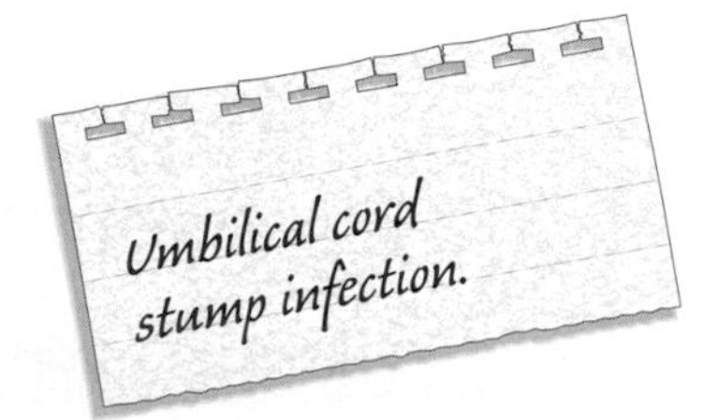

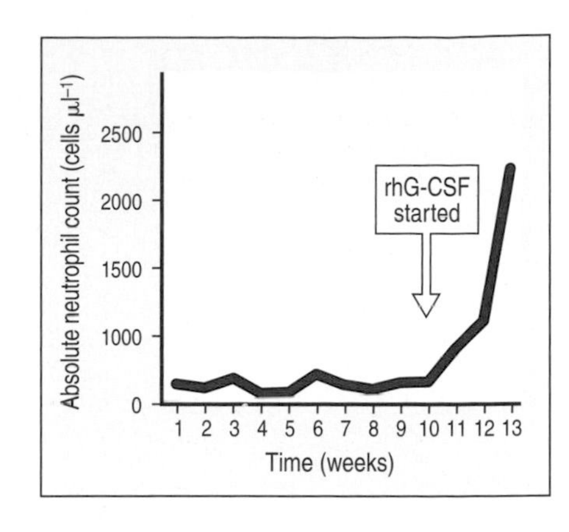

Fig. 25.2 Michael's absolute neutrophil count during hospitalization and treatment. On admission, Michael's absolute neutrophil count (ANC) was 174 cells μl^{-1}, well below the 500 cells μl^{-1} that qualifies as severe neutropenia. Although his infection was controlled by antibiotics over several weeks, his ANC remained low until he was treated with recombinant human G-CSF (rhG-CSF), a growth factor that promotes the development of neutrophils, after the diagnosis of severe congenital neutropenia.

The emergency room team immediately placed a venous catheter and treated Michael with fluids and broad-spectrum antibiotics. Blood cultures and a peri-umbilical swab were performed to identify pathogens. A complete blood count revealed severe neutropenia with an absolute neutrophil count (ANC) of 174 neutrophils μl^{-1}. Abundant Gram-positive cocci, identified as *Staphylococcus aureus*, were cultured from the umbilical stump swab as well as from the blood.

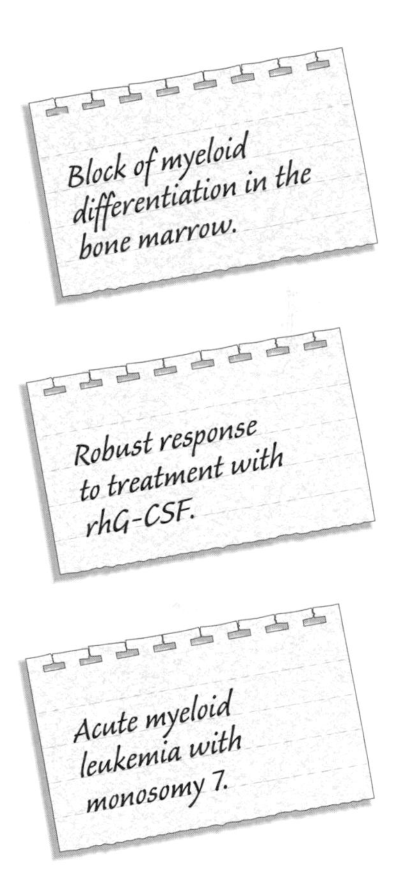

Block of myeloid differentiation in the bone marrow.
Robust response to treatment with rhG-CSF.
Acute myeloid leukemia with monosomy 7.

Michael was transferred to intensive care, where he remained for almost 2 weeks. He was treated with the antibiotic vancomycin, and with fluids and other medication to support his cardiovascular system. His condition gradually improved and the symptoms of infection in the umbilical area subsided. Nevertheless, his neutrophil count remained very low (Fig. 25.2). His physicians therefore wondered whether his neutropenia was secondary to the infection (that is, due to bone marrow suppression) or whether he suffered from congenital neutropenia. To address this question, a bone marrow aspirate was performed. This showed normal numbers of granulocyte progenitors but a severe block in neutrophil differentiation at the promyelocyte stage and an almost complete lack of more mature forms (Fig. 25.3). This suggested a diagnosis of severe congenital neutropenia (SCN).

Treatment with recombinant human granulocyte colony-stimulating factor (rhG-CSF), at a dose of 5 $\mu g\ kg^{-1}$ per day, was initiated. Within 2 weeks, the neutrophil count had increased to 700 cells μl^{-1} (see Fig. 25.2). Michael was discharged from the hospital and maintained on daily injections of rhG-CSF. He remained mostly free of infection. A few months later, genetic testing showed that Michael carried a heterozygous mutation (a substitution of arginine for glycine at position 185) in the *ELA2* gene, which encodes the enzyme neutrophil elastase. Neither of his parents carried this mutation.

At the age of 6 years, Michael was diagnosed with acute myeloid leukemia. Leukemic blast cells showed monosomy of chromosome 7 and were found to carry a somatic mutation in the gene encoding the G-CSF receptor. Michael received chemotherapy, followed by allogeneic hematopoietic stem-cell transplantation from a matched unrelated donor. He is now 13 years old and remains in good health, with full remission. His neutrophil count has normalized.

Severe congenital neutropenia (SCN).

Michael presents the classical phenotype of severe congenital neutropenia (SCN). This is a genetically and phenotypically heterogeneous group of hereditary disorders, with an estimated frequency of around two cases per million live births. Typically, patients with SCN present very early in life, usually within the first few months, mostly with recurrent, often severe, bacterial infections that involve the skin, the umbilical stump, soft tissues, lungs, and

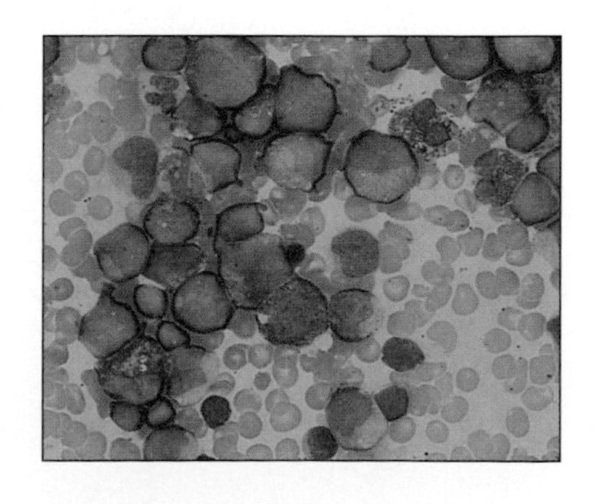

Fig. 25.3 Bone marrow aspirate from a patient with severe congenital neutropenia, showing the presence of myeloid progenitors but a lack of myelocytes and metamyelocytes. Photograph courtesy of Antonio Regazzoli.

deep organs, or with sepsis. Patients with SCN are also at high risk of invasive fungal infections, including those caused by *Aspergillus* and *Candida* species.

SCN is characterized by marked neutropenia. To be called 'severe,' the absolute neutrophil count (ANC) must be below 500 cells μl^{-1}, but in SCN it is usually even lower—less than 200 cells μl^{-1}. In contrast, monocyte and eosinophil counts are normal or elevated. Serum levels of immunoglobulins are often increased.

The differential diagnosis of SCN includes various forms of neutropenia due to an impaired production or accelerated destruction of neutrophils. Viral infections, but also some bacterial infections, in particular *Salmonella* and bacterial infection of the blood (sepsis), can cause a suppression of myelopoiesis that can last for several weeks. Impaired generation of neutrophils can also be part of a broader spectrum of genetically determined disorders of bone marrow function, known as bone marrow failure syndromes. These include Fanconi anemia, dyskeratosis congenita, Diamond–Blackfan anemia, Shwachman–Diamond syndrome, cartilage hair hypoplasia, and others. These disorders typically also show other hematological abnormalities, and may present with distinctive extra-hematological features (for example short stature in Fanconi anemia, and diarrhea in Shwachman–Diamond syndrome). Finally, neutropenia can also reflect myelodysplasia, a condition of ineffective production of myeloid cells in the bone marrow. Myelodysplasia may progress to leukemia, with replacement of the bone marrow hematopoietic matrix by clonal proliferation of leukemic cells. Monosomy 7 is a chromosomal abnormality that is frequently observed in patients with myelodysplasia and is associated with a higher risk of leukemic transformation.

Accelerated destruction of neutrophils can reflect an immune mechanism, as observed in the phenomenon of autoimmune neutropenia, which is often seen in systemic autoimmune diseases such as systemic lupus erythematosus (see Case 37). Transplacental passage of anti-neutrophil antibodies from an autoimmune mother may cause alloimmune neutropenia in the infant for up to several months after birth. Neutropenia may also be secondary to hypersplenism, a condition of spleen enlargement, associated with retention and destruction of neutrophils in the spleen. Neutropenia is also associated with warts, hypogammaglobulinemia, infections, and myelokathexis (retention of mature neutrophils in the bone marrow) syndrome (WHIM syndrome; see Case 22).

A diagnosis of SCN is supported by characteristic findings of 'maturation arrest' at the promyelocyte or myelocyte stage in the bone marrow. Several genetic defects have recently been described in patients with SCN, including mutations in the genes encoding elastase 2 (*ELA2*), the anti-apoptotic protein HCLS1-associated X1 (*HAX1*), the p14 endosomal protein (*MAPBPIP*), the transcription factor growth factor independent-1 (*GFI1*), and the β3A subunit of the adaptor protein complex 3 (AP-3) (*AP3BP1*). Activating mutations of the Wiskott–Aldrich syndrome gene (*WAS*) have also been shown to cause SCN. For some of these mutations, the neutropenia is associated with other manifestations. For instance, neurological problems are frequently seen in patients with *HAX1* mutations.

Severe neutropenia can also be present in other genetic disorders as part of a broader phenotype, as observed in patients with Chediak–Higashi syndrome (see Case 15), Griscelli syndrome type 2, Hermansky–Pudlak syndrome type 2, Barth syndrome, Cohen syndrome, Pearson's syndrome, and others.

Conditions of chronic neutropenia must be distinguished from other situations in which the neutropenia is intermittent and follows a cyclic pattern. In the latter, neutrophil counts fluctuate between normal and very low, with the lowest ANC usually occurring every 18–21 days (Fig. 25.4). These conditions are termed 'cyclic neutropenia.' Determination of the ANC once a week for several weeks is important to document the cycling and establish the diagnosis. Cyclic neutropenia has been associated with *ELA2* mutations; however, mutations in the same gene may also cause chronic neutropenia.

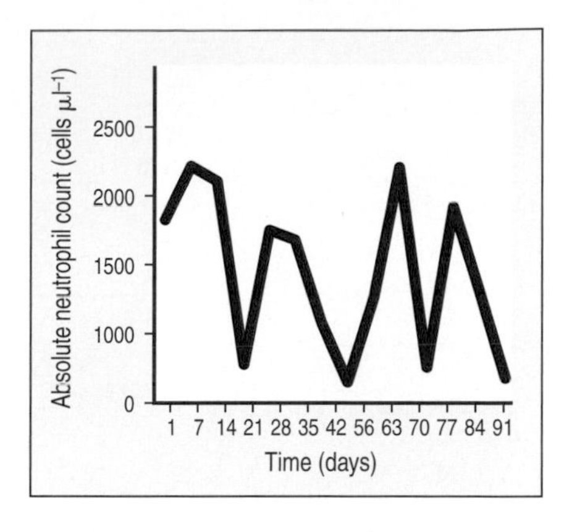

Fig. 25.4 Representative pattern of ANC in a patient with cyclic neutropenia.

In the past, patients with SCN tended to die during the first years of life as a result of overwhelming infections. Since the introduction of treatment with rhG-CSF, the vast majority of the patients can attain relatively normal neutrophil counts, with reduced risk of developing severe infections. The mechanism by which rhG-CSF might increase the neutrophil count is not well defined, but may reflect increased stimulation of early myeloid progenitors. However, some patients fail to respond to rhG-CSF, and others (especially those with *ELA2* mutations) are at increased risk of developing myelodysplasia and acute myeloid leukemia. Secondary mutations in the G-CSF receptor that cause premature truncation of the intracellular tail of the receptor increase the risk of acute myeloid leukemia. In such cases, and in those who fail to respond to rhG-CSF, allogeneic hematopoietic stem-cell transplantation remains the only curative approach.

Questions.

1 *How might one differentiate between increased peripheral destruction of neutrophils and decreased production in the bone marrow as a cause of neutropenia?*

2 *Why are neutrophil transfusions only rarely used in the treatment of SCN?*

3 *Why do somatic mutations that truncate the intracytoplasmic tail of the G-CSF receptor cause an increased risk of leukemia?*

CASE 26 Chronic Granulomatous Disease

A specific failure of phagocytes to produce H_2O_2 and superoxide.

One of the most important mechanisms for destroying invading microbes in both the innate and adaptive immune responses is their phagocytosis by macrophages and neutrophils, after which they are killed rapidly within the phagocytic vacuoles, or phagosomes. Uptake of microorganisms by phagocytes is enhanced by the opsonization of the particle—that is, coating it with complement, or, in the case of adaptive immune response, with antibody and complement (Fig. 26.1). The microbicidal actions of phagocytes occur through a variety of mechanisms. One of the most important of these involves the production of hydrogen peroxide (H_2O_2) and superoxide radicals (O_2^-), by a complex process that results in changes to the phagosomal pH and membrane potential, as well as the production of active bactericidal factors. These changes have a key role in the creation of the highly bactericidal environment within the phagosome that facilitates the activation and function of the

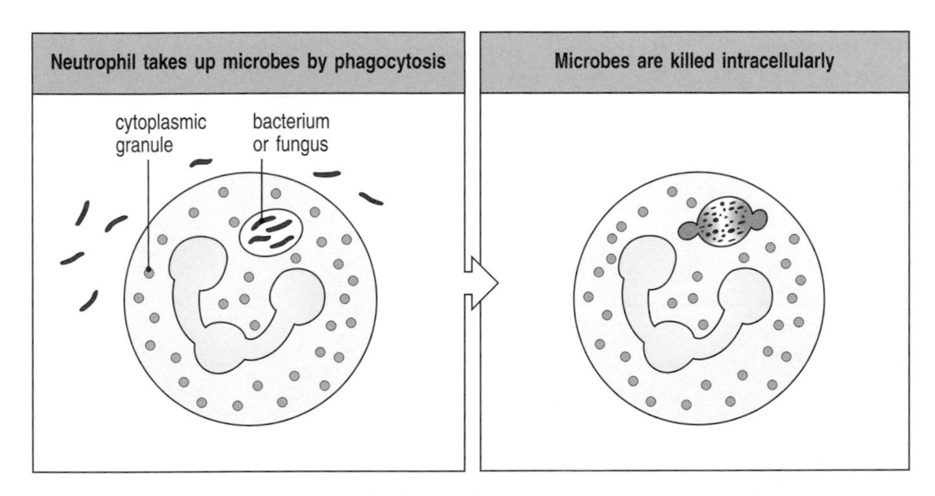
Neutrophil takes up microbes by phagocytosis
Microbes are killed intracellularly
cytoplasmic granule
bacterium or fungus

Fig. 26.1 Activation of a neutrophil. Microbes (red) are ingested by a phagocyte and enter the cytoplasm in a phagocytic vacuole, or phagosome. The phagosome fuses with a cytoplasmic granule containing microbicidal enzymes. The enzymes are released into the phagosome, where they kill and degrade the microbe.

Topics bearing on this case:
Microbicidal action of phagocytes
Leukocyte migration
Chronic inflammation

This case was prepared by Raif Geha, MD, in collaboration with Itai Pessach, MD.

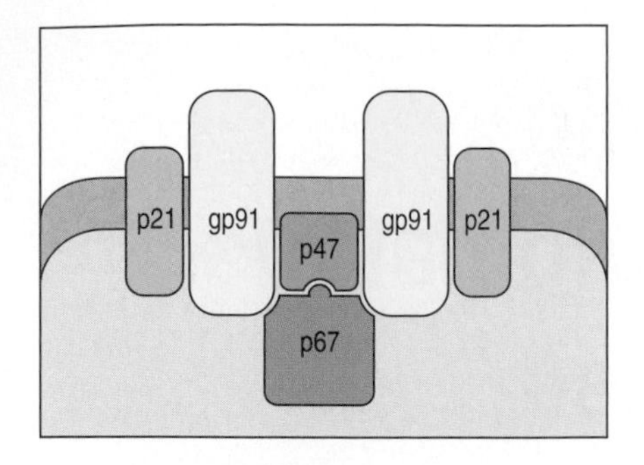

Fig. 26.2 The NADPH oxidase complex. NADPH is a large, multisubunit enzyme complex whose assembly in the phagosome membrane is triggered by a phagocytic stimulus. For simplicity, only four major components are shown here: gp91^{phox} and p21^{phox}, which together form the complex known as cytochrome b_{558} in the granule membranes, and p67^{phox} and p47^{phox}, which normally reside in the cytoplasm. On delivery of a phagocytic stimulus to the cell, p47^{phox} becomes hyperphosphorylated and binds to p67^{phox} and other cytosolic components, and these components migrate to the phagosome membrane, where they assemble with cytochrome b_{558} to form the complete NADPH oxidase.

enzymes that are released into the phagosome. A complex enzyme, NADPH oxidase, catalyzes the initial reaction, which results in the generation of superoxide radicals:

$$NADPH + 2O_2 = NADP^+ + 2O_2^- + H^+$$

The superoxide then undergoes dismutation to produce H_2O_2:

$$2H^+ + 2O_2^- = H_2O_2 + O_2$$

NADPH oxidase is a large enzyme complex that is assembled in the phagosome membrane from component parts originally residing in the granule membranes and the cytosol. The membrane complex cytochrome b_{558} contains the catalytic site of the NADPH oxidase and comprises a heavy chain, gp91^{phox}, and a light chain, p21^{phox} (Fig. 26.2). Several other subunits, including p47^{phox}, p67^{phox}, p40^{phox}, and Rac2, reside in the cytoplasm in unstimulated phagocytes. On stimulation of the phagocyte, the cytosolic subunits translocate to the membrane to interact with cytochrome b_{558} to form the active NADPH oxidase that shuttles electrons from cytosolic NADPH to molecular oxygen (O_2). The genes encoding p47^{phox}, p67^{phox}, and p21^{phox} map to autosomal chromosomes, whereas gp91^{phox} is encoded on the short arm of the X chromosome. Different genetic defects affect various components of the NADPH oxidase enzyme, but all result in a disease with a common phenotype, called chronic granulomatous disease (CGD). The commonest form of the disease is X-linked CGD, which is caused by mutations in the gene encoding gp91^{phox}.

The diagnosis of CGD can easily be ascertained by taking advantage of the metabolic defect in the phagocytic cells. A dye called nitro blue tetrazolium (NBT) is pale yellow and transparent. When it is reduced, it becomes insoluble and turns a deep purple. To test for CGD, a drop of blood is suspended on a slide and is treated with a phagocytic stimulus (for example phorbol 12-myristate 13-acetate (PMA), an activator of the oxidase); a drop of NBT is then added to the blood. Neutrophils, which are the main phagocytes in blood, take up the NBT and the PMA at the same time. In normal blood the NBT is reduced to a dark purple, insoluble formazan, easily seen in the phagocytic cells; in CGD blood no dye reduction is seen (Fig. 26.3). The dye dihydrorhodamine (DHR) can be used in the same way. In this method, whole blood is stained with DHR, and the cells are then stimulated with PMA. Normal phagocytes will produce superoxide radicals that reduce DHR to the highly

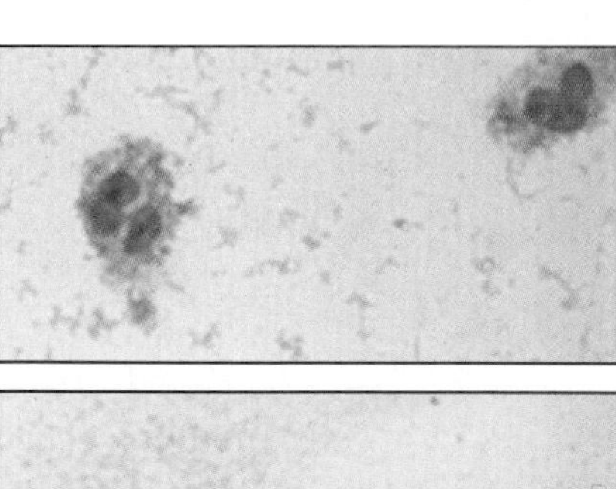

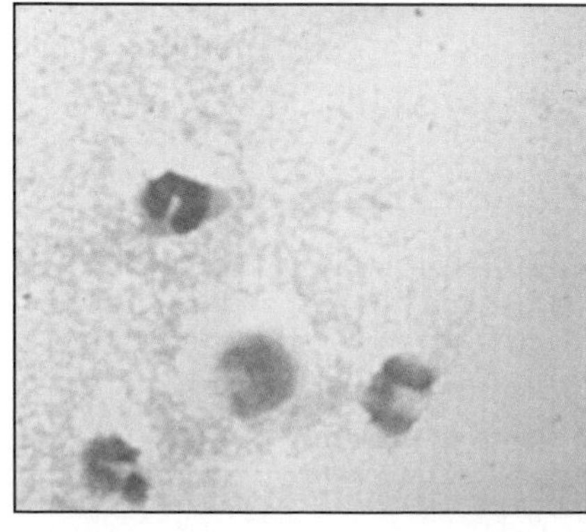

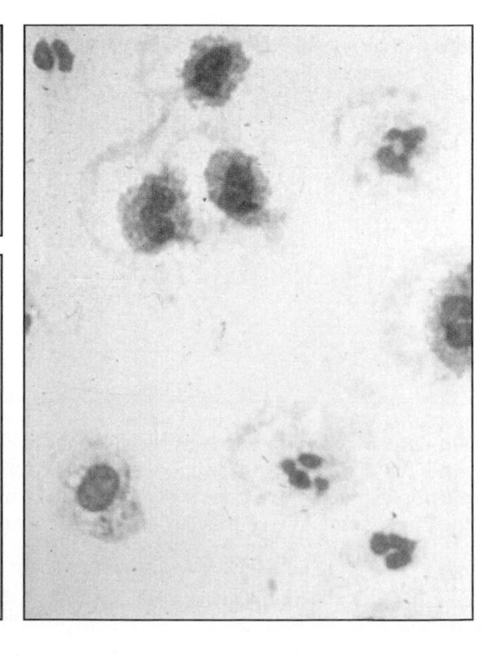

Fig. 26.3 Neutrophils stimulated with phorbol myristate acetate in the presence of nitro blue tetrazolium (NBT) dye. The upper left panel shows two normal neutrophils stained purple. The lower left panel shows neutrophils from a patient with CGD that failed to reduce NBT. The neutrophils in the right panel are from a heterozygous carrier of chronic granulomatous disease. Half the neutrophils have reduced NBT (purple), and half do not stain. Photographs courtesy of P. Newberger.

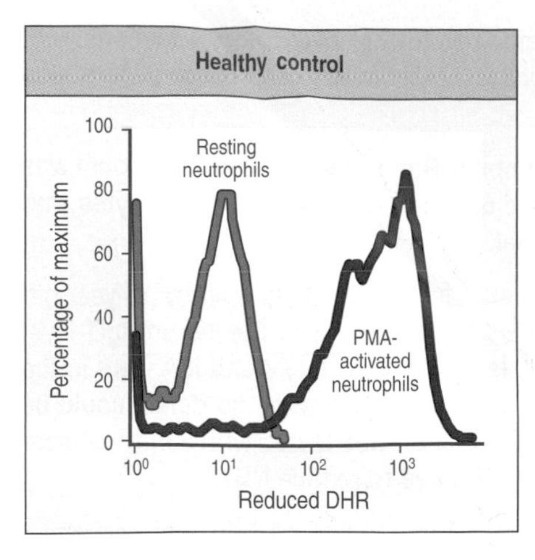

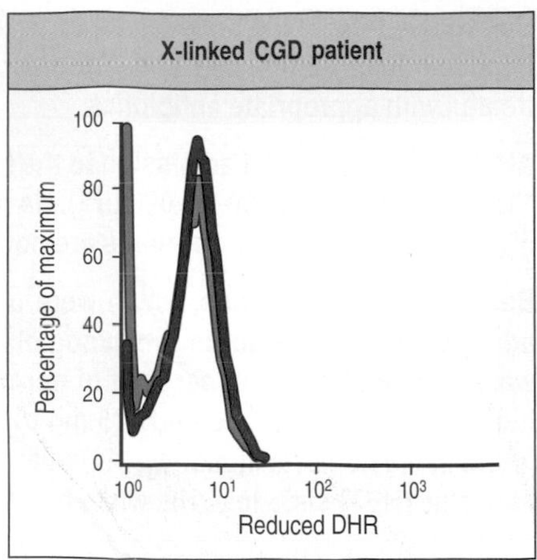

Fig. 26.4 DHR assay for analysis of NADPH oxidase activity in a healthy individual and a patient with X-linked CGD. Whole blood was stained with dihydrorhodamine and activated by addition of phorbol 12-myristate 13-acetate (PMA). The samples were then analyzed by FACS, selecting neutrophils by their surface characteristics. Normal phagocytes produce superoxide radicals that reduce DHR to rhodamine, a highly fluorescent molecule. In normal individuals, almost 100% of the neutrophils become activated, causing reduction of the DHR, whereas in CGD patients the cells fail to reduce the DHR and hence remain non-fluorescent.

fluorescent molecule rhodamine. The fraction of fluorescent cells represents those cells that were able to produce superoxides, and it can be measured with a fluorescence-activated cell sorting (FACS) machine (Fig. 26.4).

The case of Randy Johnson: a near death from exotic bacteria and fungi.

Randy Johnson, a 15-year-old high-school student, had a summer job working with a gardening crew. His job entailed spreading bark mulch in garden beds. At the end of August of that summer he rapidly developed severe shortness of breath, a persistent cough, and chest pain. Because of impending respiratory failure he was admitted to the hospital. A radiological examination of Randy's chest was performed and this revealed the presence of large 'cotton ball' densities in both lungs (Fig. 26.5). One of these lesions was aspirated with a fine needle; when the aspirate was stained, numerous fungal hyphae were seen. A culture of the aspirate grew *Aspergillus fumigatus*.

Fifteen-year-old boy with impending respiratory failure.

Randy was started on intravenous amphotericin B and assisted mechanical ventilation through a tracheotomy. He slowly improved over the course of two months.

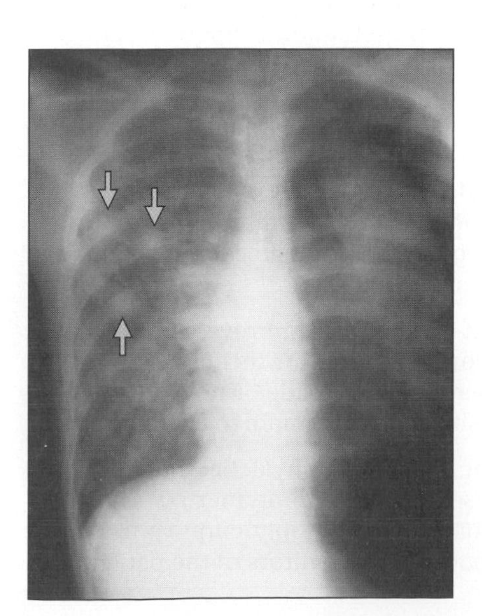

Fig. 26.5 A radiograph of aspergillus pneumonia. Arrows point to 'cotton ball' densities characteristic of fungi.

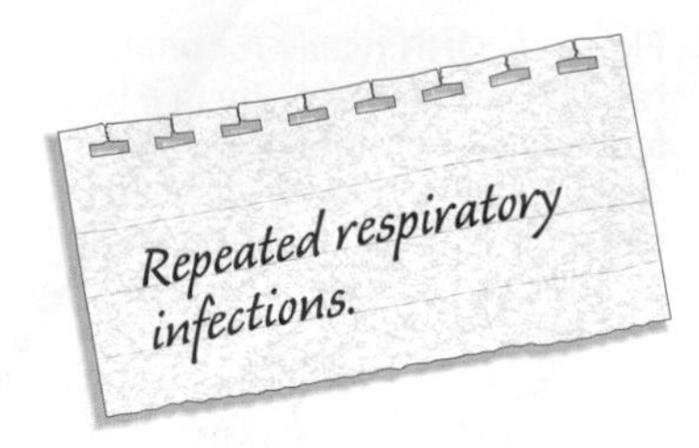
Repeated respiratory
infections.

During this time in the hospital, he contracted two further respiratory infections, with *Pseudomonas aeruginosa* and *Streptococcus faecalis*. These infections were also treated with appropriate antibiotics.

At the time of his initial admission to the hospital, Randy's white blood cell count was 11,500 μl^{-1} (normal 5000–10,000 μl^{-1}). He had 65% neutrophils, 30% lymphocytes, and 5% monocytes: these proportions are normal.

Because of his infections, which were unusual for a seemingly healthy 15-year-old adolescent, Randy's serum immunoglobulins were measured. The serum IgG level was 1650 mg dl^{-1} (the upper limit of normal is 1500 mg dl^{-1}). IgM and IgA were in the high normal range at 250 and 175 mg dl^{-1}, respectively. Because no defect could be found in his humoral immunity, his white cell function was tested with a nitro tetrazolium blue (NBT) slide test. His white blood cells failed to reduce NBT.

Randy had four older sisters and two brothers. They all had an NBT test performed on their blood. One, his brother Ralph, aged 9 years, also failed to reduce NBT. His mother reported that Ralph had had a perirectal abscess in infancy but had otherwise been well.

Randy's mother and one sister had a mixed population of neutrophils in the NBT test; about half of the phagocytes reduced NBT and the other half did not (see Fig. 26.3). His other three sisters and other brother gave normal NBT reduction.

Further studies of Randy's and Ralph's granulocytes revealed that the rate of H_2O_2 production was 0.16 and 0.028 nmol min^{-1} per 10^6 cells (normal 6.35 nmol min^{-1} per 10^6 cells). The cytochrome *b* content in both brothers' granulocytes was less than 1.0 pmol mg^{-1} protein (normal control 101 pmol mg^{-1} protein). Both Ralph and Randy were started on treatment with injections of interferon-γ.

Chronic granulomatous disease (CGD).

Randy and Ralph Johnson present a phenotype of X-linked chronic granulomatous disease. Males affected with this defect commonly show undue susceptibility to infection in the first year of life. The most common infections are pneumonia, infection of the lymph nodes (lymphadenitis), and abscesses of the skin and of the viscera such as the liver. Granuloma formation occurs when microorganisms are opsonized and ingested by phagocytic cells but the infection cannot be cleared. The persistent presentation of microbial antigens by phagocytes induces a sustained cell-mediated response by CD4 T cells, which recruit other inflammatory cells and set up a chronic local inflammation called a granuloma. This can occur in normal individuals in response to intracellular bacteria that colonize macrophages (Fig. 26.6); in CGD the intracellular microbicidal mechanisms are defective, so that many infections can persist and set up chronic inflammatory reactions resulting in granulomas. The fungi and bacteria that are most frequently responsible for infections and granuloma formation in CGD are listed in Fig. 26.7.

It was discovered quite by chance that interferon-γ improves the resistance of males with X-linked CGD to infection. The basis for this effect is still not understood. This cytokine does not increase superoxide or peroxide production in the deficient phagocytes; its positive effects remain to be determined.

Other conservative forms of treatment include prophylactic antibiotics and aggressive treatment of infections. In severe cases, bone marrow transplantation is indicated and is currently considered the only curative treatment for CGD by replacing the defective blood-cell progenitors of the patient with

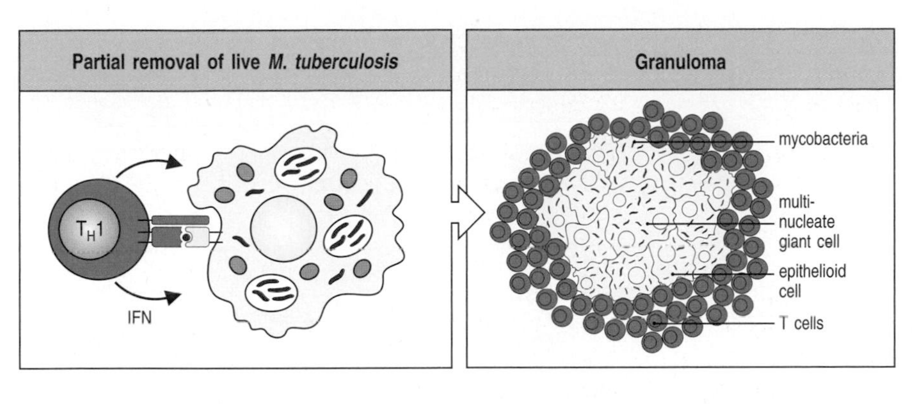

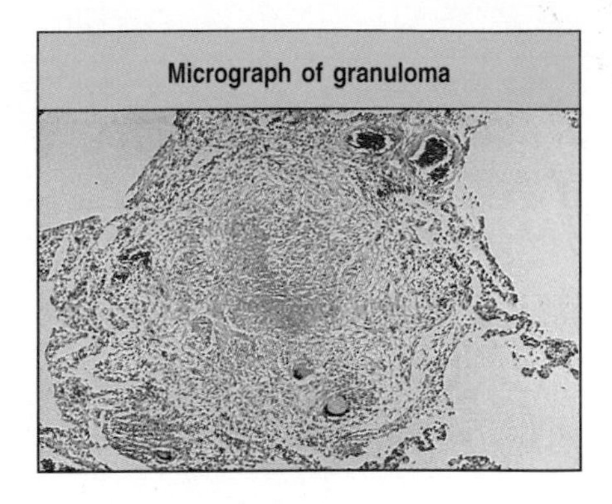

Fig. 26.6 Granulomas form when an intracellular pathogen or its constituents cannot be eliminated completely. The pathogen illustrated here is the intracellular bacterium *Mycobacterium tuberculosis*. In the left panel, mycobacteria (red) are shown resisting the effects of macrophage activation by a T_H1 CD4 T cell. The middle panel is a schematic diagram of a granuloma, which forms as the result of a localized inflammatory response that develops because of the sustained activation of chronically infected phagocytes. The granuloma consists of a central core of infected macrophages, and may include multinucleate giant cells, which are fused macrophages, surrounded by large macrophages often called epithelioid cells. These may, in turn, be surrounded by T cells, many of which are CD4-positive. Granulomas frequently form in CGD because macrophages are unable to destroy the bacteria they ingest, and infected macrophages accumulate. The right panel shows a micrograph of a granuloma. Photograph courtesy of J. Orrell.

normal progenitors from the donor. Nevertheless, bone marrow transplantation carries significant risks and complications, and its success depends on the availability of an HLA-matched donor.

In recent years, gene therapy for CGD has been attempted. Progenitor blood cells were collected from the patients and were infected with a virus containing a normal copy of the particular gene that was mutated in that patient. The corrected cells were then injected back into the patients. This resulted in the development of a population of normal phagocytes that were able to produce superoxides. Although these studies are only preliminary, this approach holds great promise and the prospect for a possible future cure for CGD, as well as for other hematologic and immunologic disorders.

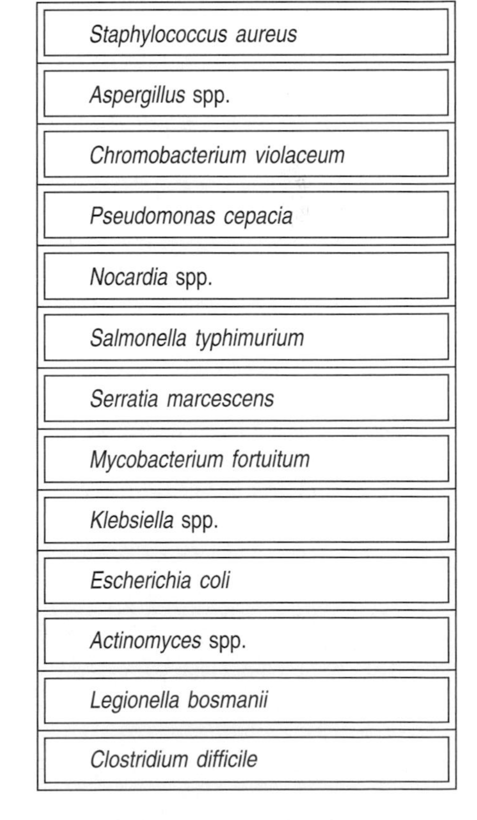

Staphylococcus aureus
Aspergillus spp.
Chromobacterium violaceum
Pseudomonas cepacia
Nocardia spp.
Salmonella typhimurium
Serratia marcescens
Mycobacterium fortuitum
Klebsiella spp.
Escherichia coli
Actinomyces spp.
Legionella bosmanii
Clostridium difficile

Fig. 26.7 Table of fungal and bacterial infectious agents most commonly responsible for infections in CGD.

Questions.

1 *In Fig. 26.3, 50% of Randy's mother's neutrophils have reduced the NBT and 50% have not. How do you explain this?*

2 *Children with CGD do not have problems with pneumococcal infections. Could you hypothesize why this is so?*

3 *The immunoglobulin levels in Randy's blood were somewhat elevated. How do you explain this?*

4 *A child presents with a clinical picture of CGD. In addition to defiant neutrophil superoxide production, her neutrophils exhibit defective chemotaxis. What gene in the membrane cytochrome b_{558} complex would you suspect may be deficient?*

CASE 27 Leukocyte Adhesion Deficiency

The traffic of white blood cells.

Newly differentiated blood cells continually enter the bloodstream from their sites of production: red blood cells, monocytes, granulocytes, and B lymphocytes from the bone marrow, and T lymphocytes from the thymus. Under ordinary circumstances, red blood cells spend their entire lifespan of 120 days in the bloodstream. However, white blood cells (leukocytes) are destined to emigrate from the blood to perform their effector functions. Lymphocytes recirculate through secondary lymphoid tissues, where they are detained if they encounter an antigen to which they can respond; macrophages migrate into the tissues as they mature from circulating monocytes; effector T lymphocytes and large numbers of granulocytes are recruited to extravascular sites in response to infection or injury. For example, it is estimated that each day 3 billion neutrophils enter the oral cavity, the most contaminated site in our body.

The process by which leukocytes migrate from the bloodstream to sites of infection is fairly well understood and involves interactions between adhesion molecules on the leukocyte surface and those on endothelial cells (Fig. 27.1). First, leukocyte flow is retarded by interactions between selectins, whose expression is induced on activated vascular endothelium, and fucosylated glycoproteins (for example sialyl-Lewis^x) on the leukocyte surface. Tight binding of leukocytes to the endothelial surface is then triggered by the actions of chemokines, such as CXCL8 (formerly known as IL-8), which activate the leukocyte integrins (for example LFA-1 and Mac-1) to adhere more tightly to their receptors. Crossing the endothelial cell wall also involves interactions between the leukocyte integrins and their receptors, while the subsequent direction of migration follows a concentration gradient of chemokines (for

Topics bearing on this case:
Complement receptors
Migration and homing of leukocytes
Phagocytic cell defects
Adhesion molecules in T-cell interactions
Bone marrow transplantation

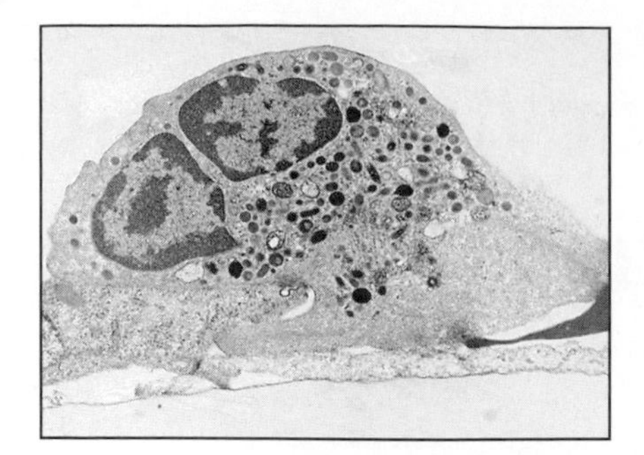

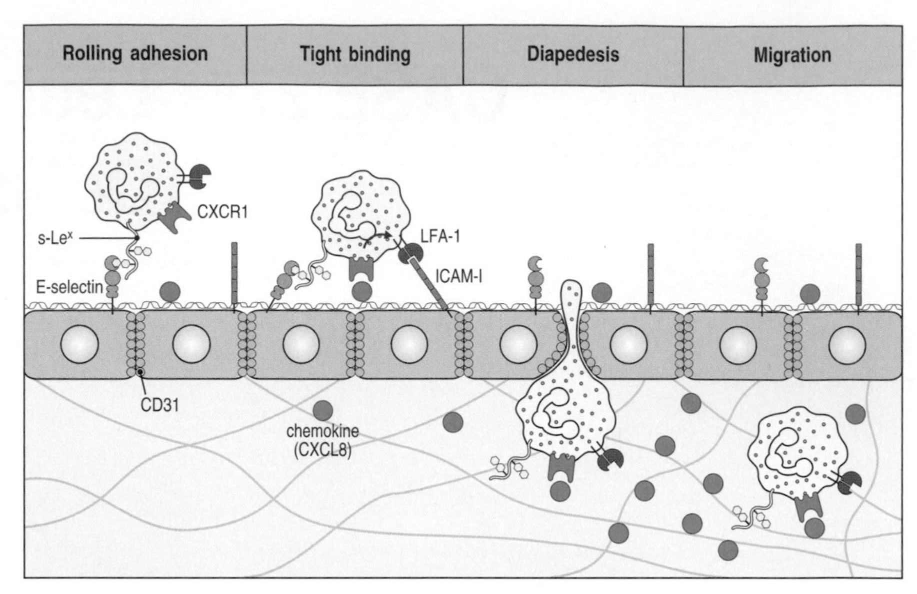

Fig. 27.1 Phagocytic leukocytes and effector T leukocytes are directed to sites of infection through interactions between adhesion molecules induced by chemokines. The first step (right panel) involves the reversible binding of leukocytes to vascular endothelium through interactions between selectins induced on the endothelium and their carbohydrate ligands on the leukocyte, shown here for E-selectin and its ligand the sialyl-Lewis^x moiety (s-Le^x). This interaction cannot anchor the cells against the shearing force of the flow of blood; instead, they roll along the endothelium, continually making and breaking contact. The binding does, however, allow stronger interactions, which occur as a result of the induction of the adhesion molecule ICAM-1 on the endothelium and the activation of its receptors, the integrins LFA-1 and Mac-1 (not shown) on the leukocyte. This binding is mediated by chemokines (such as CXCL8, which is recognized by the receptor CXCR1 on the leukocyte) retained on heparan sulfate proteoglycans on the endothelial cell surface. Tight binding between these molecules arrests the rolling and allows the leukocyte to extravasate (leave the bloodstream) by squeezing between the endothelial cells forming the wall of the blood vessel (a process known as diapedesis). The leukocyte integrins LFA-1 and Mac-1 are required for extravasation and for migration toward chemoattractants. Adhesion between molecules of CD31, expressed on both the leukocyte and the junction of the endothelial cells, is also thought to contribute to diapedesis. Finally, the leukocyte migrates along a concentration gradient of chemokines (here shown as CXCL8) secreted by cells at the site of infection. The electron micrograph shows a neutrophil that has just started to migrate between two endothelial cells (bottom of photo). Note the pseudopod that the neutrophil has inserted between adjacent endothelial cells. The dark mass at the bottom right is an erythrocyte that has become trapped underneath the neutrophil. Photograph (×5500) courtesy of I. Bird and J. Spragg.

example CXCL8) produced by cells already at the site of infection or injury. The process by which lymphocytes home to secondary lymphoid tissue is very similar, except that it is initiated by mucin-like addressins on lymphoid venules binding to L-selectin on the surface of naive lymphocytes (Fig. 27.2). The various adhesion molecules involved in leukocyte trafficking and other important leukocyte cell–cell interactions are described in Fig. 27.3.

An excellent opportunity to study the role of integrins is provided by a genetic defect in CD18, the common β chain of the three β_2 integrins: LFA-1 (CD11a:CD18), Mac-1 (CD11b:CD18, which is the complement receptor CR3), and gp150,95 (CD11c:CD18, which is the complement receptor CR4). Children with this genetic defect suffer from leukocyte adhesion deficiency (LAD). They have recurrent pyogenic infections and problems with wound healing, and if they survive long enough they develop severe inflammation of the gums (gingivitis). Surprisingly, children with LAD are not unduly susceptible to opportunistic infections. This implies normal T-cell function despite the absence of LFA-1, which was thought to be important for T-cell adhesion to antigen-presenting cells (Fig. 27.4). The capacity to form antibodies is also unimpaired, showing that adequate collaboration between T cells and B cells can also occur without LFA-1. Finally, because neutrophils are released as normal from the bone marrow into the bloodstream but their migration into tissues is impaired, patients with LAD have a remarkable leukocytosis with neutrophilia.

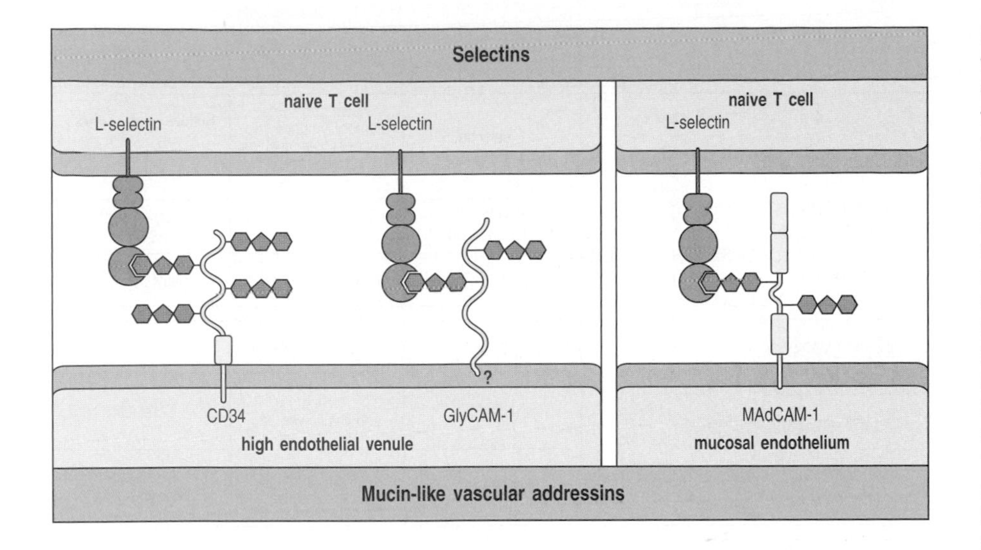

Fig. 27.2 L-selectin and the mucin-like vascular addressins direct the homing of naive lymphocytes to lymphoid tissues. L-selectin is expressed on naive T cells, which bind to the vascular addressins CD34 and GlyCAM-1 on high endothelial venules to enter lymph nodes. The relative importance of CD34 and GlyCAM-1 in this interaction is not clear. GlyCAM-1 is expressed exclusively on high endothelial venules but has no transmembrane region and it is unclear how it is attached to the membrane; CD34 has a transmembrane anchor and is expressed on endothelial cells but not exclusively on high endothelial venules. The addressin MAdCAM-1 is expressed on mucosal endothelium and guides entry into mucosal lymphoid tissue. L-selectin recognizes carbohydrate moieties on the vascular addressins.

The case of Luisa Ortega: a problem of immobile white blood cells.

Luisa Ortega was born at full term and weighed 3.7 kg. She was the second child born to the Ortegas. At 4 weeks of age, Luisa was taken by her parents to her pediatrician because she had swelling and redness around the umbilical cord stump (omphalitis), and a fever of 39°C. Her white blood cell count was 71,000 μl^{-1} (normal 5000–10,000 μl^{-1}). She was treated in the hospital with intravenous antibiotics for 12 days and then discharged home with oral antibiotics. At the time of discharge her white blood count was 20,000 μl^{-1}. Cultures obtained from the inflamed skin about the umbilical stump before antibiotic treatment grew *Escherichia coli* and *Staphylococcus aureus.*

Luisa, 4 weeks old, presents with acute omphalitis.
Family history of overwhelming infection.

The Ortegas had had a baby boy 3 years before Luisa's birth. At 2 weeks of age he developed a very severe infection of the large intestine (necrotizing enterocolitis). Separation of his umbilical cord was delayed. He subsequently suffered from multiple skin infections and he died of staphylococcal pneumonia at 1 year of age. Just before his death his white blood cell count was recorded at 75,000 μl^{-1}.

Because of the previous family history, Luisa was referred to the Children's Hospital. At the time of her admission to the Children's Hospital she seemed normal on physical examination, and radiographs of the chest and abdomen were normal.

Cultures of urine, blood, and cerebrospinal fluid were negative. Her white blood count was 68,000 μl^{-1} (very high). Of her white cells, 73% were neutrophils, 22% lymphocytes, and 5% eosinophils (this distribution of cell types is in the normal range but the absolute count for each is abnormally high). Her serum IgG concentration was 613 mg dl^{-1} (normal), her IgM was 89 mg dl^{-1} (normal), and her IgA 7 mg dl^{-1} (normal). The concentration of complement component C3 in her serum was 185 mg dl^{-1} and that of C4 was 28 mg dl^{-1} (both normal).

A Rebuck skin window was performed. In this procedure, the skin of the forearm is gently abraded with a scalpel blade and a coverslip is placed on the abrasion. After 2 hours the coverslip is removed and replaced by another coverslip every subsequent 2 hours for a total of 8 hours. In this way, the migration of immune cells into the damaged skin can be monitored. In Luisa's case, no white cells accumulated on the coverslips. All types of leukocytes, however, were present in abnormally high numbers in her blood. Of her blood lymphocytes, 53% (7930 μl^{-1}) were T cells (as measured by

Fig. 27.3 Adhesion molecules in leukocyte interactions. Several structural families of adhesion molecules play a part in leukocyte migration, homing, and cell–cell interactions: the selectins, mucin-like vascular addressins, the integrins, proteins of the immunoglobulin superfamily, and the protein CD44. The figure shows schematic representations of an example from each family, a list of other family members that participate in leukocyte interactions, their cellular distribution, and their partners in adhesive interactions. The nomenclature of the different molecules in these families is confusing because it often reflects the way in which the molecules were first identified rather than their related structural characteristics. Thus, although all the ICAMs are immunoglobulin-related, and all the VLA molecules are β_1 integrins, the CD nomenclature reflects the characterization of leukocyte cell-surface molecules by raising monoclonal antibodies against them, and embraces adhesion molecules in all the structural families. The LFA molecules were defined through experiments in which cytotoxic T-cell killing could be blocked by monoclonal antibodies against cell-surface molecules on the interacting cells, and there are LFA molecules in both the integrin and the immunoglobulin families. Alternative names for each of the molecules shown are given in parentheses. Sialyl-Lewis^x, which is recognized by P- and E-selectin, is an oligosaccharide present on cell-surface glycoproteins of circulating leukocytes. The ligand DC-SIGN, which binds to ICAM-3, is expressed on dendritic cells. It is suspected to have a major role in interactions between dendritic cells and T cells, and serves as a co-receptor for HIV.

Family	Example	Name	Tissue distribution	Ligand
Selectins Bind carbohydrates. Initiate leukocyte–endothelial interaction	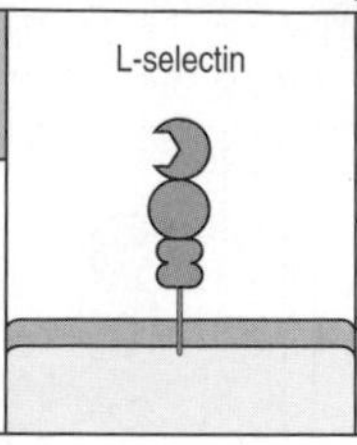 L-selectin	L-selectin (MEL-14, CD62L)	Naive and some memory lymphocytes, neutrophils, monocytes, macrophages, eosinophils	Sulfated sialyl-Lewis^x, GlyCAM-1, CD34, MAdCAM-1
		P-selectin (PADGEM, CD62P)	Activated endothelium and platelets	Sialyl-Lewis^x, PSGL-1
		E-selectin (ELAM-1, CD62E)	Activated endothelium	Sialyl-Lewis^x
Mucin-like vascular addressins Bind to L-selectin. Initiate leukocyte–endothelial interaction	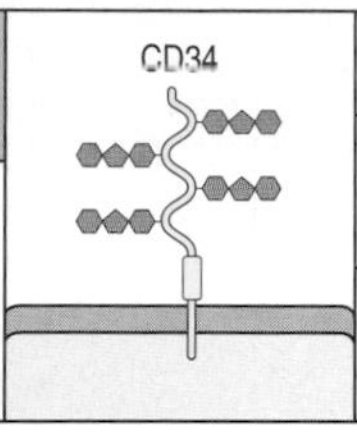 CD34	CD34	Endothelium	L-selectin
		GlyCAM-1	High endothelial venules	L-selectin
		MAdCAM-1	Mucosal lymphoid tissue venules	L-selectin, integrin α_4:β_7
Integrins Bind to cell-adhesion molecules and extracellular matrix. Strong adhesion	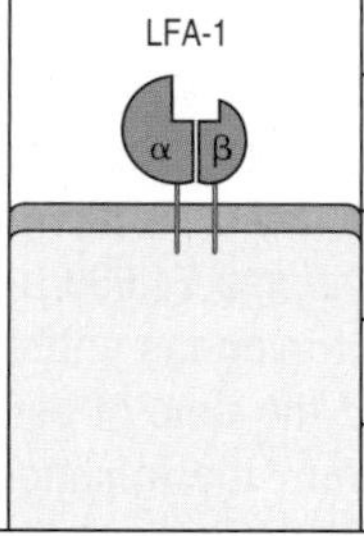 LFA-1	α_L:β_2 (LFA-1, CD11a/CD18)	Monocytes, T cells, macrophages, neutrophils, dendritic cells	ICAMs
		α_M:β_2 (Mac-1, CR3, CD11b/CD18)	Neutrophils, monocytes, macrophages	ICAM-1, iC3b, fibrinogen
		α_X:β_2 (CR4, p150-95, CD11c/CD18)	Dendritic cells, macrophages, neutrophils	iC3b
		α_4:β_1 (VLA-4, LPAM-2, CD49d/CD29)	Lymphocytes, monocytes, macrophages	VCAM-1 Fibronectin
		α_5:β_1 (VLA-5, CD49d/CD29)	Monocytes, macrophages	Fibronectin
		α_4:β_7 (LPAM-1)	Lymphocytes	MAdCAM-1
		α_E:β_7	Intraepithelial lymphocytes	E-cadherin
Immunoglobulin superfamily Various roles in cell adhesion. Ligand for integrins	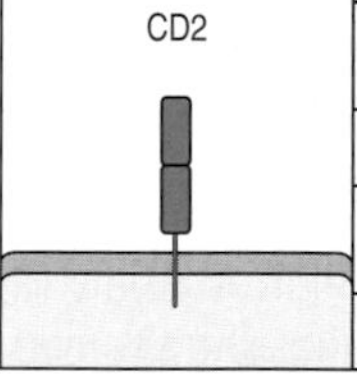 CD2	CD2 (LFA-2)	T cells	LFA-3
		ICAM-1 (CD54)	Activated vessels, lymphocytes, dendritic cells	LFA-1, Mac-1
		ICAM-2 (CD102)	Resting vessels	LFA-1
		ICAM-3 (CD50)	Naive T cells	DC-SIGN
		LFA-3 (CD58)	Lymphocytes, antigen-presenting cells	CD2
		VCAM-1 (CD106)	Activated endothelium	VLA-4

CD3 expression); of these, 36% were CD4 and 16% CD8 (normal proportions), 25% (3740 μl^{-1}) were B cells (as measured by antibody against CD19), and 14% were NK cells (as measured by CD16 expression). B-cell and NK-cell counts were both elevated.

Proliferation of Luisa's T cells in response to phytohemagglutinin (PHA) was slightly depressed. Further flow-cytometric analysis revealed that Luisa's CD3$^+$ T cells failed to express CD18, whereas CD3$^+$ cells normally expressed this molecule (Fig. 27.5). Luisa's blood mononuclear cells were stimulated with PHA and examined after 3 days of incubation with a monoclonal antibody against CD11a (the α chain of LFA-1). No LFA-1 could be detected on the cell surface.

Luisa was treated with bone marrow transplantation. Initially, she received chemotherapy with busulfan, cyclophosphamide, and anti-thymocyte globulin (ATG). After this therapy, she was given purified CD34$^+$ bone marrow stem cells from her mother at the dose of 5×10^6 CD34$^+$ cells per kg body weight. In addition, Luisa received

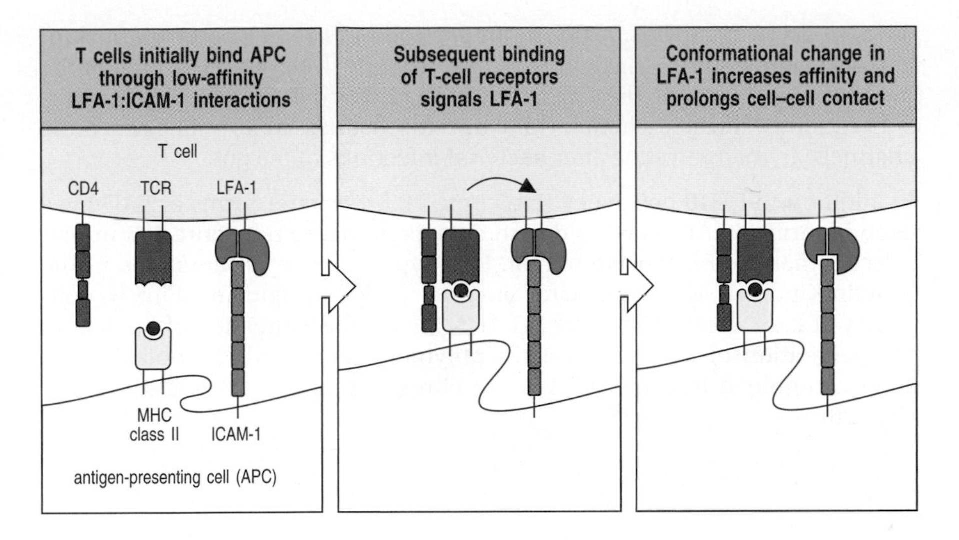

Fig. 27.4 Transient adhesive interactions between T cells and antigen-presenting cells are stabilized when specific antigen recognition induces high-affinity binding by LFA-1. When a T cell binds to its specific ligand on an antigen-presenting cell (APC), intracellular signaling through the T-cell receptor (TCR) induces a conformational change in LFA-1 that causes it to bind with higher affinity to ICAM-3 on the antigen-presenting cell. The cell shown here is a CD4 T cell.

immunosuppressive treatment with cyclosporin A. Twenty-eight days after the transplant, Luisa's myeloid cells were shown to be of donor origin (full chimerism) and all of them expressed CD18 at normal density. Within six months, lymphocyte counts had significantly increased and were also found to be of donor origin. She subsequently did well clinically and her white blood count remained stable at about 5000–7800 μl^{-1}.

Leukocyte adhesion deficiency (LAD).

Children such as Luisa, who lack the leukocyte integrin subunit CD18, are subject to recurrent, severe bacterial infections that are eventually fatal if untreated. Bone marrow transplantation has been very successful in rescuing infants with LAD from certain death. In such patients, encapsulated bacteria are coated as usual with antibody and complement. However, the neutrophils and monocytes that would normally be recruited to the site of infection are trapped in the bloodstream and cannot emigrate into the tissues because they lack the CD18-containing integrins LFA-1 and Mac-1/CR3. In a normal individual, the first coverslip in a Rebuck skin window would contain many neutrophils. Monocytes begin to appear at 4 hours and by 8 hours the coverslip contains predominantly monocytes and very few neutrophils. In Luisa's case the coverslips had no cells because her leukocytes were unable to emigrate from the bloodstream and onto the coverslip. For this reason, the levels of white cells in the bloodstream were very high—a characteristic of LAD. The ability to deal with pyogenic bacteria is further compromised because of the vital role of CR3-mediated uptake of these opsonized bacteria by neutrophils. The role of CR4, the third member of the β_2-integrin family, is less well understood but, like CR3, it binds fragments of complement component C3, and it is thought to have a role in the uptake of bacteria by macrophages.

The importance of the daily, massive neutrophil emigration into the oral cavity is well illustrated by individuals with LAD, who, when they survive, invariably

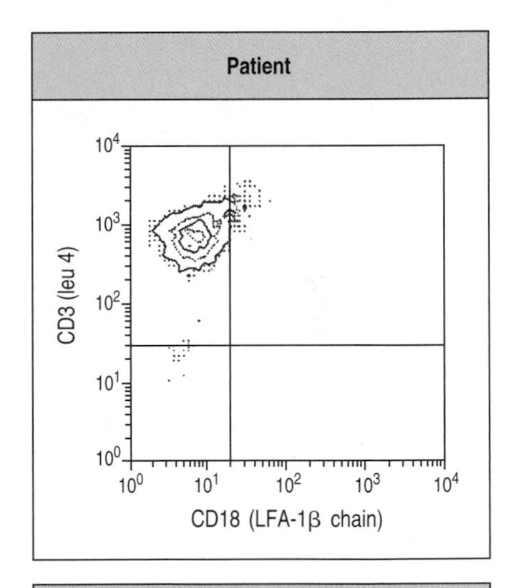

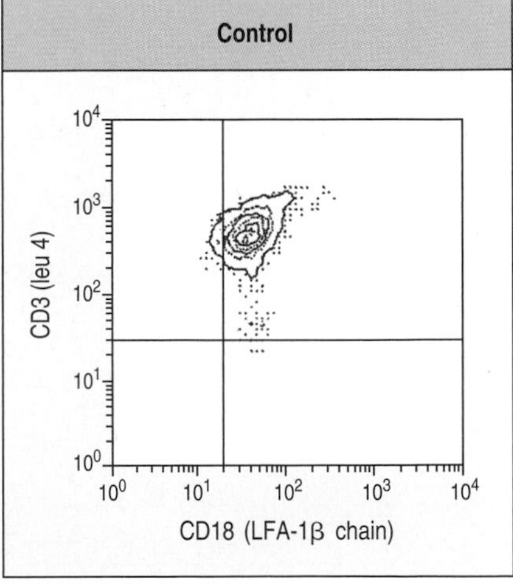

Fig. 27.5 Flow cytometric analysis of CD18 expression on Luisa's lymphocytes (upper panel) compared with that of a control subject (lower panel). The vertical axis measures fluorescence from labeled antibody against CD3; the horizontal axis measures fluorescence from labeled antibody against CD18. In the control subject, $CD3^+$ cells also express CD18; in contrast, Luisa's $CD3^+$ cells fail to express CD18 molecules.

develop severe gingivitis. A poorly understood consequence of the lack of leukocyte emigration is the failure to heal wounds. Delayed separation of the umbilical cord is the earliest manifestation of this defect in wound healing. Subsequently, affected children may develop fistulas (abnormal connecting channels) in their intestine after bacterial infections of the gut.

In addition to CD18 deficiency (LAD type 1), other rarer forms of LAD have been identified. LAD type 2 is due to defects in fucose transport that impair selectin-mediated leukocyte rolling. LAD type 3 is due to mutations in the protein Kindlin 3, which regulates integrin-mediated signaling. Finally, deficiency of Rac2, a small Rho-type GTPase, affects the expression of L-selectin, the generation of superoxide, actin polymerization, and chemotaxis, and shows combined features of LAD and chronic granulomatous disease (see Case 26).

Questions.

1 *Can you deduce the inheritance pattern of LAD type 1 from this case?*

2 *In some families with LAD, a mild phenotype is observed and leukocytes of affected individuals express less than 10% of the normal amount of CD18. The mild phenotype is usually due to a splice-site defect in the gene encoding CD18, whereas, in the severe form of the disease, nonsense or frameshift mutations, or deletions of the CD18 gene, are identified. Do these observations help you to predict the likely effects of treatment with monoclonal antibodies against CD18? Can you suggest situations in which such treatment might be therapeutically useful?*

3 *Can you suggest why the homing of T cells to lymphoid tissue is normal in patients with LAD and why these individuals do not suffer from susceptibility to opportunistic infections?*

4 *What manifestation of defective wound healing occurred in this case?*

5 *LAD type 2 is caused by mutations in the protein GDP-fucose transporter 1, which is important in the transport of fucose into the Golgi apparatus. Can you explain why such a defect should cause LAD?*

6 *Shortly after Luisa received a bone marrow transplant from her mother it was ascertained that she had complete chimerism of myeloid and lymphoid cells. How was this ascertained?*

7 *Why was Luisa treated with busulfan, anti-thymocyte serum, and cyclophosphamide before the bone marrow transplant?*

8 *Why was the response of Luisa's T cells to PHA depressed?*

9 *How does LAD differ from chronic granulomatous disease (Case 26)?*

CASE 28 Recurrent Herpes Simplex Encephalitis

The control of viral infections by type I interferons.

In addition to lymphocyte-mediated defenses against viral infection (see Cases 14 and 45), the type I interferons—IFN-α and IFN-β—provide another ubiquitous cell-mediated antiviral defense mechanism. The induction of interferons is transcriptionally regulated, and is elicited by a variety of upstream receptors. Among these, the Toll-like receptors TLR-3, TLR-7, TLR-8, and TLR-9, which all recognize viral components, have a prominent role. TLR-3 binds double-stranded RNA; TLR-7 and TLR-8 recognize single-stranded RNA; and TLR-9 binds double-stranded DNA. These TLRs are present in the membranes of endosomes, and are therefore prompted to bind nucleic-acid intermediates generated during intracellular viral replication. All these TLRs require association with the endosomal membrane protein UNC93B for signaling, but differ in the pathway components downstream of the receptor. TLR-7, TLR-8, and TLR-9 signal through the adaptor protein MyD88 and the serine/threonine kinases IRAK4 (IL-1-receptor-associated kinase 4; see Case 29) and IRAK1, activating the IKKα:β:γ protein kinase complex, which in turn leads to activation of the transcription factor NFκB (the mechanism of NFκB activation is shown in more detail in Fig. 23.1). Activation of TLR-7, TLR-8, and TLR-9 can also induce the interferon regulatory factor IRF-7 by an alternative signaling pathway that involves MyD88 and IRAK4 but is independent of IKK (Fig. 28.1).

In contrast, TLR-3 signals through the adaptor protein TRIF, activating a signaling pathway that uses the protein kinases TBK1 and IKKε and activates IRF3 and IRF7 (see Fig. 28.1). IRF3, IRF7, and NFκB can all switch on the type I interferon genes, thus linking the activation of TLRs recognizing different viral replication intermediates to a common outcome—the production of type I interferons.

After IFN-α and IFN-β are produced and released by the virus-infected cell, they bind to their common receptor, a heterodimer of IFNαR1 and IFNαR2, on the cell surface. This results in activation of the JAK1 and TYK2 kinases and

Topics bearing on this case:
Toll-like receptor signaling
Type I interferons

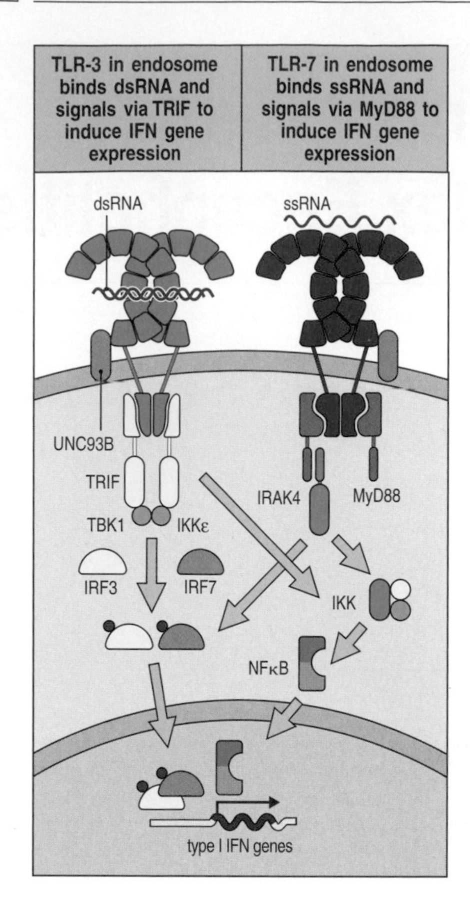

Fig. 28.1 Toll-like receptor-mediated induction of type 1 interferons in response to viral infection. Viruses induce production of type 1 interferons (IFN-α and IFN-β) by triggering activation of endosome-associated TLR-3, TLR-7, TLR-8, or TLR-9. The endoplasmic reticulum membrane protein UNC93B is essential for signaling by all these receptors. TLR-3 signaling occurs through a pathway dependent on the adaptor TRIF and primarily activates interferon-regulatory factors IRF3 and IRF7 (left panel; see text for details), whereas TLR-7 (shown here), TLR-8 and TLR-9 signal through a pathway dependent on the adaptor MyD88 that involves IRAKs and primarily activates the classical NFκB pathway. Cross-talk between the pathways means that IRF3, IRF7, and NFκB can in principle all be activated via either set of receptors.

of the transcription factor ISGF3, a heterotrimeric complex that comprises STAT1, STAT2, and IRF9. ISGF3 binds to the interferon-stimulated response element (ISRE) within the promoter of various type I-IFN-dependent genes, thereby inducing their transcription and triggering antiviral activity and destruction of the virus (Fig. 28.2).

The case of Mercédès Mondego: relapsing fever and lateralized seizures.

Mercédès was born at term after an uneventful pregnancy. Her parents are immigrants from a small village in French Guiana. At birth, Mercédès was of normal weight and length, and she grew normally, reaching her milestones according to the calendar. At the age of 6 months, she spiked a high fever (39.7°C) accompanied by vomiting and right hemiclonic seizures. She was immediately brought to the emergency room, where the seizures were treated with diazepam.

Initial analysis of the cerebrospinal fluid (CSF) was normal. However, electroencephalographic (EEG) tracing carried out at day 3 revealed spike-waves in the left temporal lobe. Cerebral magnetic resonance imaging (MRI) showed hyperintensity of the signal in the left temporal lobe. Results of a repeat CSF analysis at day 5 suggested viral meningoencephalitis, with 54 cells per μl (normal less than 3 cells per μl), 94% of which were lymphocytes, and only a slight increase in protein concentration (55 mg dl^{-1}; normal 20–50 mg dl^{-1}). The CSF tested positive for herpes simplex virus type 1 (HSV-1) antigen, and HSV-1 infection was confirmed by a positive polymerase chain reaction (PCR) for HSV-1 DNA in the CSF. Other laboratory tests showed an increased white blood cell count (15,000 μl^{-1}) with relative lymphocytosis (72%), and normal serum levels of IgG (650 mg dl^{-1}), IgA (32 mg dl^{-1}), and IgM (51 mg dl^{-1}).

Mercédès was treated with intravenous acyclovir for 3 weeks and gradually recovered. No significant infections were recorded in the following months. One month after the episode, she had detectable anti-HSV-1 antibodies. At 14 months of age, she was immunized against measles, mumps, and rubella (MMR) without adverse consequences, and was able to mount a protective antibody response. Mercédès began daycare at 2 years old and continued to do well, both in her growth and her development. She suffered from the common viral infections of early childhood, especially during the winter, but the frequency of infections was no different from that of her peers. At the age of 3 years 4 months, she again developed high fever (39.4°C),

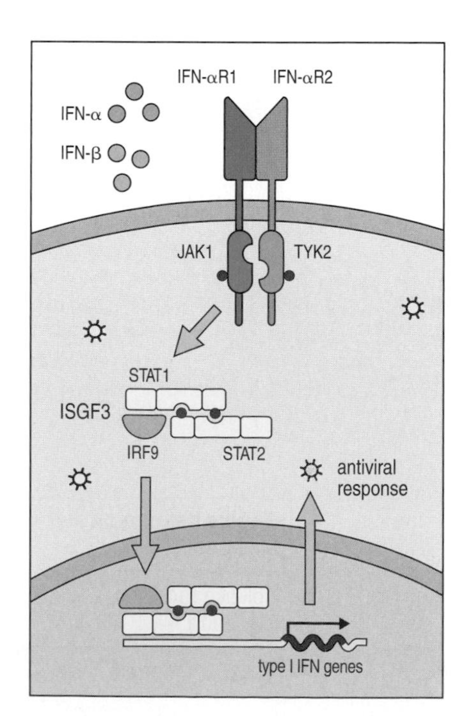

Fig. 28.2 Cellular responses to type I interferons. Type I interferons bind to the IFN-α receptor (IFNαR), which is composed of two different chains. This receptor triggers activation of Janus kinases JAK1 and TYK2 and formation of the transcription factor ISGF3 (a heterotrimer of STAT1, STAT2 and IRF9), which drives the expression of type I IFN-dependent genes that encode proteins that mediate the antiviral response.

right-sided tonico-clonic seizures, paralysis of the right face and arm (brachio-facial paralysis) and photophobia (an aversion to light). CSF analysis revealed a recurrence of meningoencephalitis (180 cells μl^{-1}, with 97% lymphocytes; 62 mg dl^{-1} protein), and PCR for HSV-1 in the CSF was positive. A brain MRI revealed a new lesion in the left parietal lobe and in the left thalamus. Mercédès was treated with intravenous acyclovir for 3 weeks, with progressive clinical improvement. Genetic testing for possible causes of relapsing herpes simplex encephalitis (HSE) revealed a homozygous single-nucleotide deletion in exon 1 of the *UNC93B* gene. Mercédès is now 5 years old and the paralysis of the right face and arm remain. She has not suffered other episodes of severe viral infection.

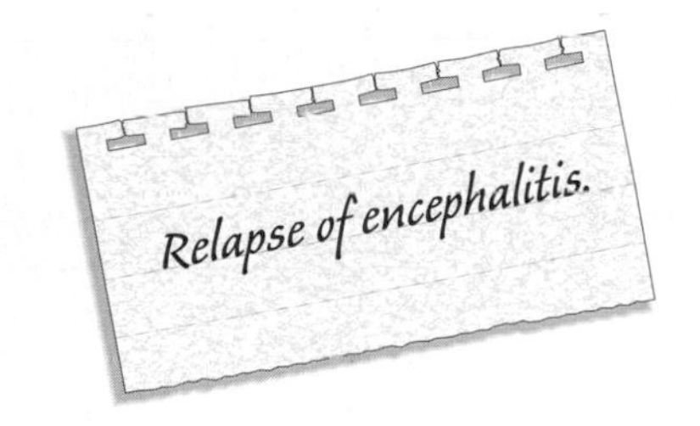
Relapse of encephalitis.

Recurrent herpes simplex encephalitis.

HSV-1 is a double-stranded DNA virus, and is typically associated with infection of the oral mucosa (causing oral ulcers) or of the eye (causing conjunctivitis and keratitis). After replication at the initial site of infection, the virus is transported through sensory neurons to the trigeminal nerves and ganglia, where it establishes a latent infection. Reactivation of the virus manifests as herpes labialis (cold sores) in about 30% of the infected population. HSV-1 infection is very common in the general population; about 85% of adults have detectable antibodies against HSV-1.

Although HSV-1 infection is usually benign, on rare occasions the virus invades the brain, causing herpes simplex encephalitis (HSE), which affects between two and four individuals in 1 million each year in the United States. The virus invades the brain through the olfactory tract and trigeminal nerves, and infects both neuronal and glial cells, causing a necrotizing encephalitis. The temporal and parietal lobes are typical targets.

For a long time HSE was thought to be simply an acquired disease, but it has recently been shown to follow a pattern of Mendelian inheritance in some families. Various genetic defects have been shown to cause HSE, all of which result from defects in the production of, or the response to, type I interferons.

Although *MyD88*-deficient mice are prone to HSE, patients with MyD88 or IRAK4 deficiency suffer from recurrent infections with pyogenic bacteria but do not show increased susceptibility to HSE or to other viral infections (see Case 29), even if their fibroblasts fail to produce type I interferons in response to stimulation of TLR-7, TLR-8, or TLR-9. This indicates that these IRAK4-dependent TLR-mediated interferon responses are redundant for protective immunity to HSV-1 (and other viruses) *in vivo* in humans.

In contrast, defects along the TLR-3 signaling pathway are associated with increased susceptibility to recurrent HSE. Although HSV-1 is a DNA virus, double-stranded RNA is generated during its replication, and it is this RNA intermediate that is recognized by endosomal TLR-3. Genetic defects in *TLR3*, *UNC93B*, and *TRAF3* have been identified among patients with HSE. Furthermore, patients with NEMO (IKKγ) deficiency show increased susceptibility to HSE, along with a more complex clinical and immunological phenotype that also includes susceptibility to mycobacterial disease (see Case 23). Interestingly, these patients do not show disseminated HSV-1 disease, nor do they show increased susceptibility to other viral infections, suggesting that HSE results from the inability of cells of the central nervous system to respond to HSV-1 through TLR3-dependent mechanisms.

HSE can also result from genetic defects in the transcription factor STAT1. In addition to activation by the type I interferon receptor, STAT1 can be activated via a different receptor that responds to the cytokine IFN-γ (which does not have antiviral activity but which promotes T_H1 activation of macrophages, among other functions). Activation of this receptor leads to formation of the gamma-activated factor (GAF), a heterodimer of STAT-1, that promotes transcription of IFN-γ-dependent genes. Three genetic forms of STAT1 deficiency are known. Autosomal recessive, null mutations of the *STAT1* gene result in increased susceptibility to severe viral disease (including HSE) and to mycobacterial infection (resulting from the failure of macrophage activation), because the formation of both ISGF3 and GAF is impaired. In contrast, dominant-negative, heterozygous mutations of *STAT1* increase susceptibility to mycobacterial disease but not to severe viral infections (and hence do not cause HSE). In fact, in patients with heterozygous, dominant-negative mutations of *STAT1*, the formation of sufficient amounts of ISGF3 is still possible and, consequently, type I interferon-dependent anti-viral responses are not or are only marginally impaired. Finally, heterozygous gain-of-function mutations of *STAT1* cause chronic mucocutaneous candidiasis.

Questions.

1 *Why do patients with TLR-3 signaling defects present increased susceptibility to HSE and not to other viral infections that generate double-stranded RNA intermediates during viral replication?*

2 *How might we take advantage of the identification of genetic defects leading to HSE to help in the treatment of the patients?*

CASE 29 Interleukin 1 Receptor-Associated Kinase 4 Deficiency

An inability of the innate immune system to detect the presence of pathogens.

The presence of an invading microorganism is first recognized by tissue macrophages at the site of infection. Pathogens carry characteristic chemical structures on their surface and in their nucleic acids that are not present in human cells. These pathogen-associated molecular patterns (PAMPs) are recognized by invariant pattern-recognition receptors (PRRs) on macrophages, and also on neutrophils, which take up the pathogen and destroy it. PRRs are also present on dendritic cells, whose function is to present antigen to and activate naive T cells; dendritic cells form the link between the innate and the adaptive immune response. Among the PRRs are the Toll-like receptors (TLRs), which are present on the cell surface and in the membranes of endosomes, and which enable the cells of the innate immune system to detect and respond to a wide variety of pathogens.

There are 10 different TLRs in humans, which between them recognize microbial nucleic acids, such as unmethylated bacterial DNA and long double-stranded RNA, as well as molecules specific to particular classes of microorganisms, such as the bacterial lipopolysaccharide (LPS or endotoxin) characteristic of Gram-negative bacteria and the protein flagellin of bacterial flagella (Fig. 29.1). All these molecules are essential for the microorganism's survival and are therefore relatively invariant. The limited number of TLRs can thus detect the presence of many different microorganisms by recognizing features that are typical of the different groups of pathogens—bacteria, viruses, fungi, and parasites.

The TLRs are signaling receptors, and their activation leads to enhancement of the antimicrobial activity of macrophages and neutrophils, and the secretion of cytokines by macrophages, which help attract more macrophages and neutrophils out of the blood and into the site of infection. Upon engagement of their TLRs, dendritic cells undergo a brief period of macropinocytosis that is important for taking up and processing microbial antigens for subsequent presentation to naive T cells. Signaling via TLRs also induces the maturation of tissue dendritic cells and their migration to peripheral lymphoid tissues such as lymph nodes, where they encounter and activate antigen-specific T cells. During maturation, conventional dendritic cells upregulate the production of co-stimulatory molecules such as CD40, B7.1 (CD80), and B7.2 (CD86), which are essential for T-cell activation, and they produce chemokines and cytokines that help induce adaptive immune responses, such as the pro-inflammatory cytokines tumor necrosis factor-α (TNF-α), interleukin-1 (IL-1), IL-6, and

Topics bearing on this case:
Innate immunity
Toll-like receptors
Toll-like receptor/IL-1R signaling pathways
Dendritic cells and the initiation of adaptive immune responses

This case was prepared by Raif Geha, MD, in collaboration with Douglas McDonald, MD, PhD.

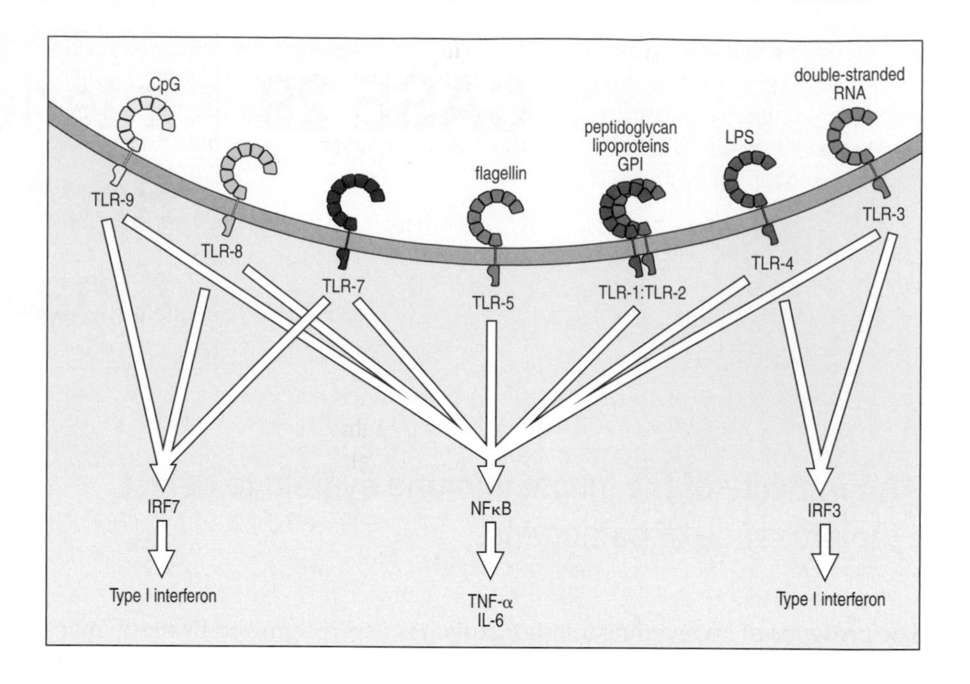

Fig. 29.1 Toll-like receptors and cytokine production. Toll-like receptors (TLRs) and their representative ligands are shown. Simplified signaling pathways are shown below each receptor, leading to the production of cytokines characteristic of each TLR. TLR-7, TLR-8, and TLR-9 are present in the membranes of endosomes, where they recognize ligands generated in the endosomal pathway from microbes engulfed by phagocytosis or macropinocytosis. The other TLRs are expressed on the cell surface. GPI, glycosylphosphatidylinositol (*Trypanosoma cruzi*); CpG, unmethylated DNA; LPS, bacterial lipopolysaccharide.

IL-12 (see Case 23). Another type of dendritic cell, the plasmacytoid dendritic cell, produces large amounts of the antiviral type I interferons (IFN-α and IFN-β) in response to viral infections.

The stimulation of dendritic cells by PAMPs such as LPS and unmethylated CpG in bacterial DNA is particularly critical for their ability to induce the development of T_H1 effector cells. The exposure of dendritic cells to pro-inflammatory cytokines is sufficient to enable them to induce the proliferation (clonal expansion) of antigen-specific naive T cells, but the upregulation of co-stimulatory molecules and of IL-12 production as a result of TLR stimulation is essential for dendritic cells to be able to induce the differentiation of T_H1 effector functions, such as the secretion of IFN-γ, in CD4 T cells. Thus, TLRs have a vital role in the development of both innate and adaptive immune responses.

The TLRs and the IL-1 receptor signal through a common intracellular signaling pathway, and this case shows how a defect in this pathway has serious effects on a person's ability to respond to some classes of pathogens.

The case of Douglas Mooster: a 6-year-old boy with recurrent pneumococcal meningitis.

Douglas is an only child and was referred to the immunology clinic at the Boston Children's Hospital at the age of 6 years after two episodes of pneumococcal meningitis (the first a year previously) caused by serotypes 3 and 14, respectively. As a result of the meningitis, Douglas had suffered a stroke and seizures, and had been left with mild hearing loss and mild learning impairment. When he was first seen at the immunology clinic, his overall neurological function was good.

Frequent middle-ear infections (otitis media) were a significant part of Douglas's medical history, and myringotomy tubes had been inserted in both ears when he was 2 years old to aerate the middle ear. At 11 months old he had developed intestinal intussusceptions (telescoping of a segment of the intestine into an adjacent segment), complicated by perforation of the intestine and formation of an abscess in the

peritoneum. He had also suffered from boils (furuncles) of the scalp and several skin infections, all of which had responded well to treatment with antibiotics. His medical history also noted an unusually weak febrile response, as Douglas usually developed only low-grade fevers late in the course of these illnesses. He seemed to have no particular susceptibility to viral infections.

On physical examination, he appeared thin but otherwise well. His tonsils were slightly enlarged, as were the lymph nodes of the anterior cervical (neck) chains (lymphadenopathy). His abdomen was soft and not tender, although the liver could be felt about 3 cm below the lower rib on the right (the liver is normally non-palpable). The spleen was not enlarged and his skin showed no rash or other unusual features.

Laboratory tests showed a normal complete blood count, complement function, and immunoglobulin titer. Specific antibody titers showed that Douglas had made protective immune responses to protein antigens, such as tetanus toxoid, after his routine immunizations. Douglas had been given the vaccine Pneumovax as part of the evaluation at the immunology clinic; this vaccine contains polysaccharide antigens from 23 different serotypes of *Streptococcus pneumoniae* (pneumococcus), and he had not made responses to any of the 14 serotypes that were tested for at the clinic. He had also failed to mount a protective antibody titer after immunization with polysaccharide antigens from *Neisseria meningitidis* (meningococcus), a common cause of meningitis. He had, however, normal absolute numbers of B cells and T cells.

Adaptive immunity mostly normal. Test for TLR function.

Given Douglas's susceptibility to pneumococcal infections, with no obvious underlying immunological cause, and his poor febrile response to infection, his Toll-like receptor (TLR) function was evaluated. A sample of peripheral blood cells (PBCs) produced little or no TNF-α in response to known TLR ligands when tested in the laboratory (Fig. 29.2). The integrity of the signal transduction pathways leading from the TLRs was evaluated *in vitro* by testing the cells' responses to IL-1, as the IL-1 receptor (IL-1R) stimulates the same signaling pathway. Stimulation of the cells with IL-1 produced no activation of MAP kinases and no phosphorylation of IκBα, important intermediate steps in these pathways (Fig. 29.3). These results were consistent with a defect very early in the pathway, and a defect in the IL-1R-associated kinase 4 (IRAK4) was suspected. Sequencing of Douglas's DNA for the *IRAK4* gene revealed a homozygous nonsense mutation within the kinase domain, resulting in the production of a truncated protein with no kinase activity.

Interleukin 1 receptor-associated kinase 4 deficiency (IRAK4 deficiency).

The IL-1 receptor (IL-1R) and the TLRs share a common signal transduction pathway that involves the adaptor protein MyD88, the signaling intermediate TRAF-6, and the receptor-associated protein kinases IRAK1 and IRAK4. Activation of these receptors leads to recruitment of MyD88 to the receptor, followed by the recruitment and activation of IRAK4, the initial protein kinase

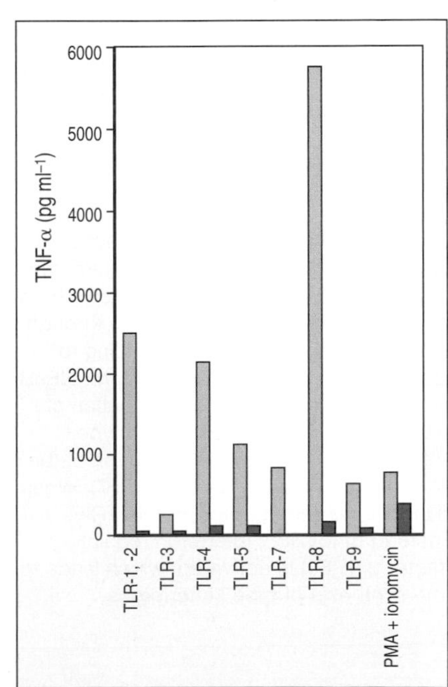

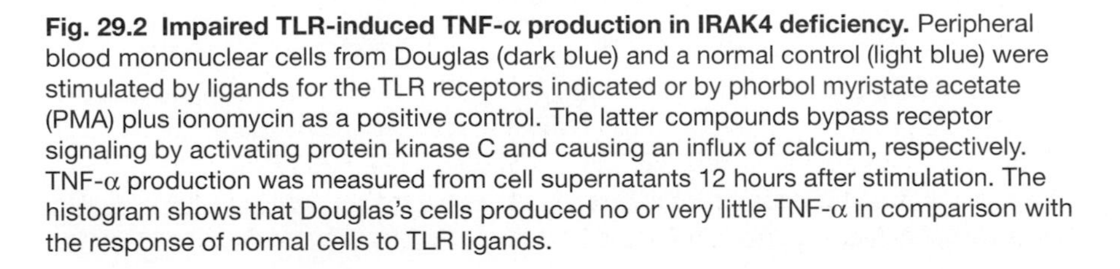
Fig. 29.2 Impaired TLR-induced TNF-α production in IRAK4 deficiency. Peripheral blood mononuclear cells from Douglas (dark blue) and a normal control (light blue) were stimulated by ligands for the TLR receptors indicated or by phorbol myristate acetate (PMA) plus ionomycin as a positive control. The latter compounds bypass receptor signaling by activating protein kinase C and causing an influx of calcium, respectively. TNF-α production was measured from cell supernatants 12 hours after stimulation. The histogram shows that Douglas's cells produced no or very little TNF-α in comparison with the response of normal cells to TLR ligands.

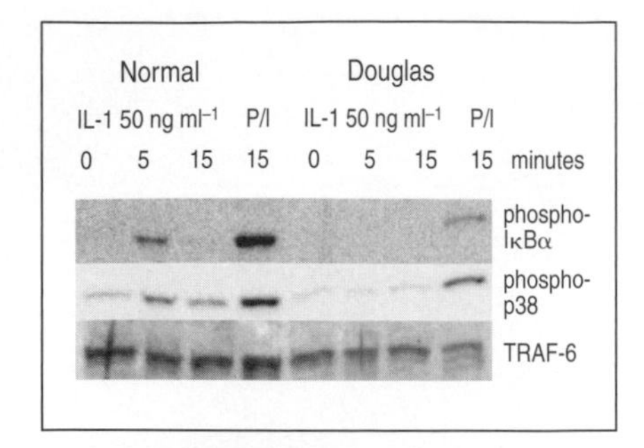

Fig. 29.3 Impaired IL-1-induced phosphorylation of p38 MAP kinase and IκBα in IRAK4 deficiency. Peripheral blood mononuclear cells from Douglas and a normal control were stimulated with IL-1 or PMA plus ionomycin (P/I) as a positive control for the times indicated. Cell lysates were subjected to Western blotting with antibodies against phosphorylated p38 MAP kinase or phosphorylated IκBα. No phospho-IκBα or phospho-p38 were present in Douglas's blood cells. Antibody against TRAF-6 is shown as a protein-loading control. Courtesy of D. McDonald.

in the TLR signal transduction pathway. This is followed by recruitment of IRAK1 and TRAF-6 to the pathway. This pathway can then diverge, one branch activating the MAP kinases, the other leading to activation of the transcription factor NFκB (Fig. 29.4). The steps leading to the activation of NFκB can be seen in more detail in Fig. 23.1. Activation of the classical NFκB pathway results in the production of pro-inflammatory cytokines such as IL-6, IL-12, and TNF-α (see Fig. 34.2). A functional IRAK4 is critical to the ability to make responses to TLR ligands, because signaling via the NFκB pathway is blocked in its absence.

There are exceptions, however, because TLR-3 and TLR-4 can also signal via a pathway that does not use IRAK4. They can use a second, MyD88-independent, pathway involving the adaptor proteins TRIF (for TLR-3) or TRIF/TRAM (for TLR-4). This pathway results in the activation of the interferon regulatory factor 3 (IRF3) and the production of type I interferons (see Fig. 29.4). Interferon production is an essential function of the innate response to certain viruses, including the herpesvirus family and the human immunodeficiency virus (HIV).

Type I interferon production is also stimulated in plasmacytoid dendritic cells and monocytes by the activation of TLR-7, TLR-8, and TLR-9. Like the other TLRs, these receptors can signal via the IRAK/TRAF-6 pathway, leading to the production of pro-inflammatory cytokines. IRAK1 can, however, also phosphorylate and activate the transcription factor IRF7, which induces the expression of the gene encoding IFN-α. Virus-induced activation of IRF7 and subsequent IFN-α production is essential for host defense against viruses. Indeed, herpes simplex virus infection in mice lacking IRF7 is lethal.

In addition to its role in innate immunity, IRAK4 may also be directly involved in the development of optimal adaptive immune responses. Antigen-stimulated T cells from mice with inactivated IRAK4 kinase secrete reduced amounts of IL-17, a cytokine that is important in antibacterial immunity. Recently, mice lacking IRAK4 were found to have reduced splenic and peripheral expansion of CD8 T cells in response to infection with lymphocytic choriomeningitis virus, suggesting that IRAK4 may be required for optimal antiviral CD8 T-cell

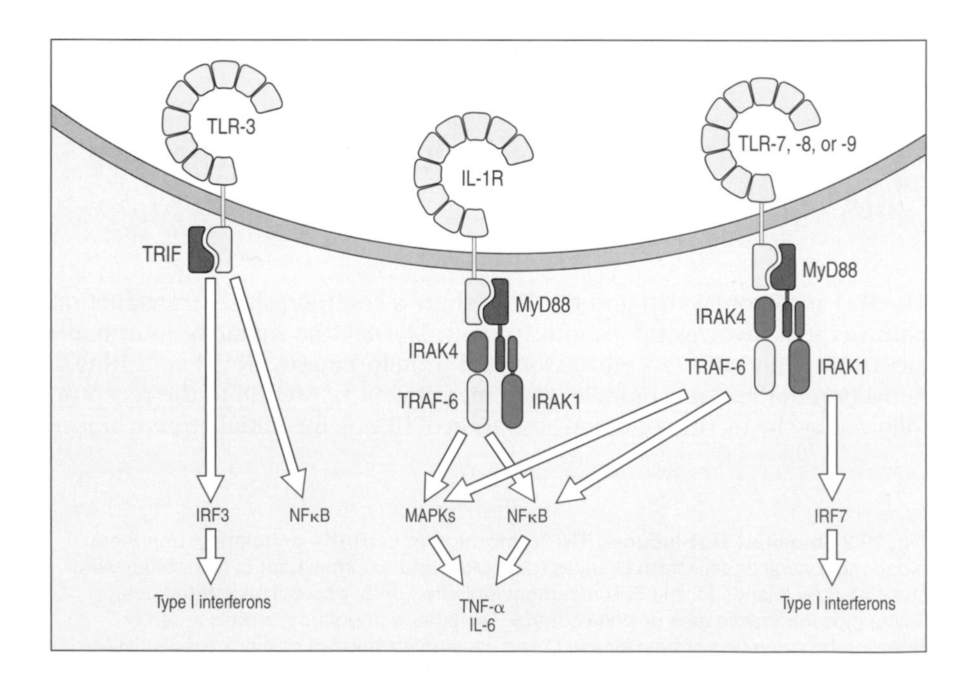

Fig. 29.4 TLR signaling pathways. Signaling through IL-1R and all TLRs can activate IRAK4 and IRAK1 (as shown here for IL-1R and the TLRs 7, 8, or 9) through the adaptor protein MyD88, leading to activation of the transcription factor NFκB and cytokine production. Stimulation of TLRs 7, 8, or 9 can also lead to type I interferon production via activation of the interferon regulatory factor 7 (IRF7). When TLR-3 uses the adaptor protein TRIF instead of MyD88, interferon regulatory factor 3 (IRF3) is activated, which leads to the synthesis of type I interferons.

responses *in vivo*. An important point to note is that IRAK4-deficient mice are susceptible to both bacterial and viral infections, whereas the IRAK4-deficient human patients identified so far seem to be primarily susceptible to pyogenic bacteria.

More than 28 patients with IRAK4 deficiency have now been identified as a result of their severe, recurrent infections with pyogenic bacteria. By routine immunological criteria they appear normal, with normal numbers of B and T cells, normal serum immunoglobulin levels, and protective antibody titers to protein antigens, but variably impaired antibody titers to polysaccharide antigens. A longitudinal study of IRAK4-deficient patients has observed that their susceptibility to invasive bacterial infections decreases with age, becoming similar to that of the normal population by 14 years old. The reason for this clinical improvement with age is unknown, but compensation by the adaptive immune system has been hypothesized. To completely understand the role of IRAK4 in human innate and adaptive immunity will require the identification and evaluation of more patients with this defect.

Questions.

1 *Douglas's immune deficiency is characterized by susceptibility to severe, invasive pneumococcal infection. What other immunodeficiencies are associated with similar susceptibility to pneumococci?*

2 *What clue in the history of a patient with recurrent pneumococcal or staphylococcal infections would favor a possible diagnosis of IRAK4 deficiency?*

3 *What is the clinical course of patients with known defects in IRAK4?*

4 *Why might IRAK-deficient patients fail to respond to immunization with polysaccharide antigens?*

5 *IRAK4 seems to be involved in signaling from all TLRs. Why has increased susceptibility to viral infections not yet been documented in human patients?*

6 *How should IRAK4-deficient patients be managed to prevent the recurrence of invasive infection?*

CASE 30 Congenital Asplenia

The role of the spleen in immunity.

The adaptive immune response occurs mainly in the peripheral lymphoid tissues—the lymph nodes, the gut-associated lymphoid tissue, and the spleen (Fig. 30.1). Pathogens and their secreted antigens are trapped in these tissues and presented to the naive lymphocytes that constantly pass through. Microorganisms that enter the body through the skin or the lungs drain to regional lymph nodes, where they stimulate an immune response. Microorganisms and food antigens that enter the gastrointestinal tract are collected in the gut-associated lymphoid tissue. Microbes that enter the bloodstream stimulate an immune response in the spleen.

The spleen is organized to accomplish two functions (Fig. 30.2). In addition to being a peripheral lymphoid organ, it acts as a filter of the blood to remove aged or abnormal red cells and other extraneous particles that may enter the bloodstream, including microorganisms. In the absence of a functioning spleen, these aged and abnormal red blood cells can be seen in a peripheral blood smear in the form of pitted red blood cells and Howell–Jolly bodies (nuclear remnants in red blood cells that are usually removed by the spleen) (Fig. 30.3).

The lymphoid function of the spleen is performed in the white pulp, and the filtration function by the red pulp. Many microorganisms are recognized directly and engulfed by the phagocytes of the red pulp. Others are not removed efficiently until they are coated by antibodies generated in the white pulp. In experimental animals, an immune response (as measured by

Topics bearing on this case:

- Circulation of lymphocytes through peripheral lymphoid tissues
- Toxoid vaccines
- Hemagglutination tests

This case was prepared by Raif Geha, MD, in collaboration with Itai Pessach, MD.

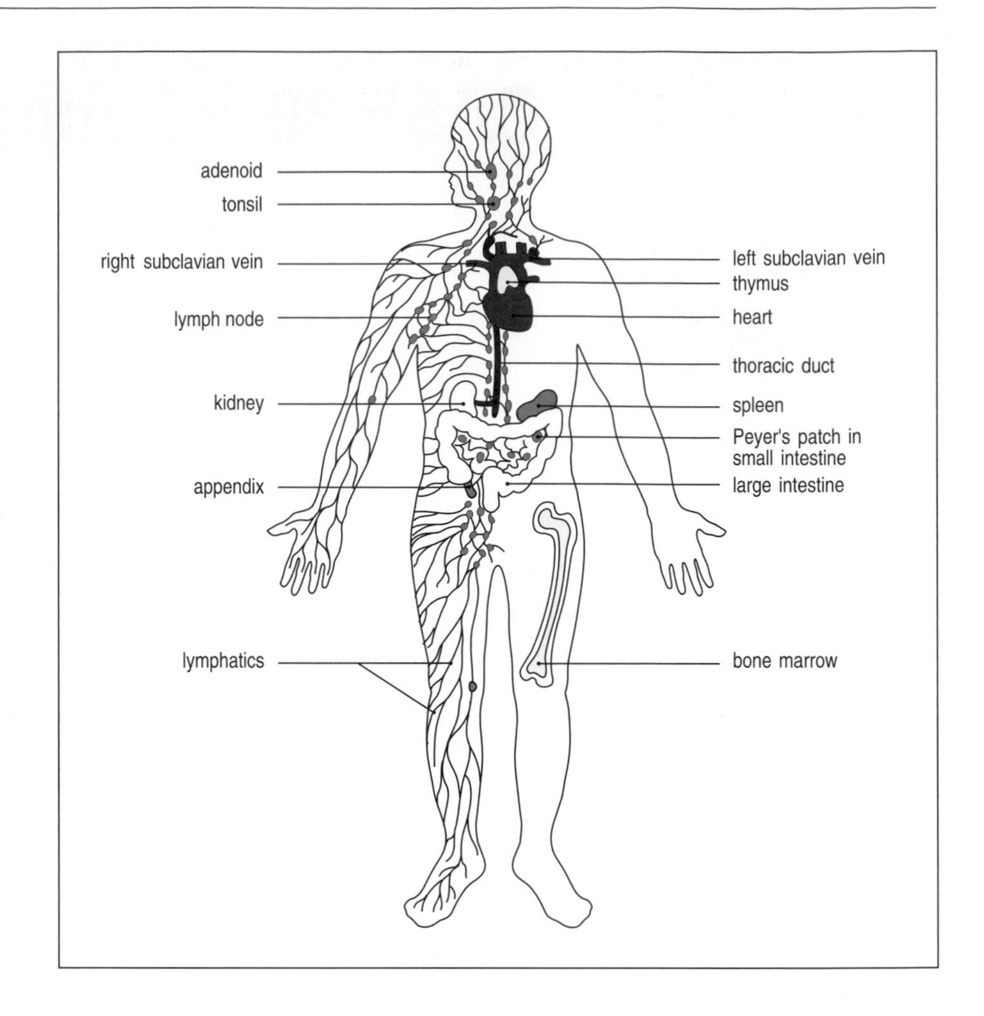

Fig. 30.1 The distribution of lymphoid tissues in the body. Lymphocytes arise from stem cells in bone marrow, and differentiate in the central lymphoid organs (yellow)—B cells in bone marrow and T cells in the thymus. They migrate from these tissues through the bloodstream to the peripheral lymphoid tissues (blue)—the lymph nodes, spleen, and mucosa-associated lymphoid tissues such as tonsils, Peyer's patches, and appendix. These are the sites of lymphocyte activation by antigen. Lymphatics drain extracellular fluid as lymph through the lymph nodes and into the thoracic duct, which returns the lymph to the bloodstream by emptying into the left subclavian vein. Lymphocytes that circulate in the bloodstream enter the peripheral lymphoid organs, and are eventually carried by lymph to the thoracic duct, where they reenter the bloodstream.

antibody formation) can be detected in the white pulp of the spleen about 4 days after the intravenous injection of a dose of microorganisms. The clearance of antibody- and complement-coated bacteria or viruses by the phagocytic cells of the red pulp of the spleen is very rapid. Rapid clearance from the blood is important because it prevents these bacteria from disseminating and causing infections of the meninges (meningitis), the kidney (pyelonephritis), the lung (pneumonia), or other distant anatomical sites.

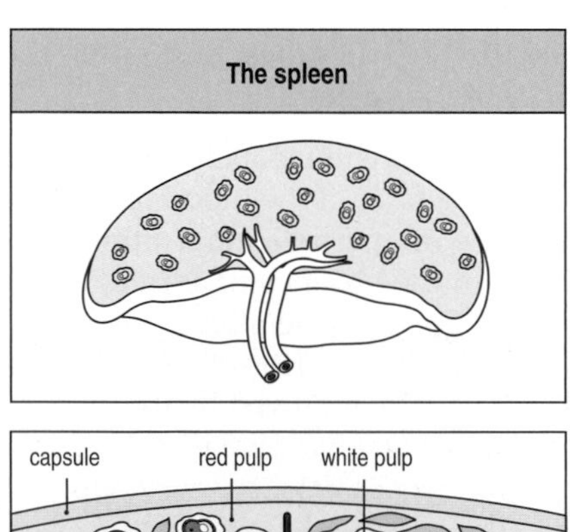

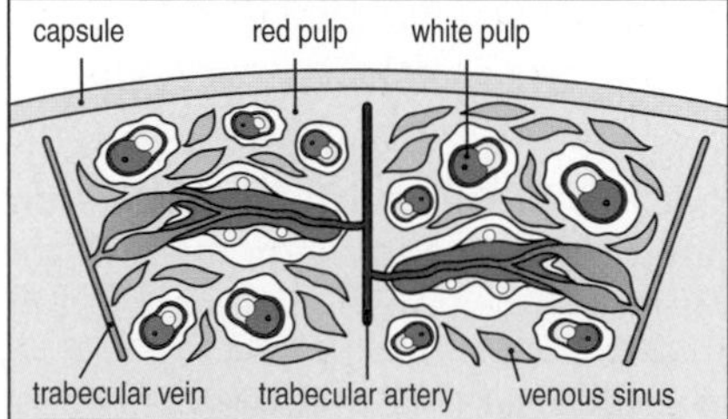

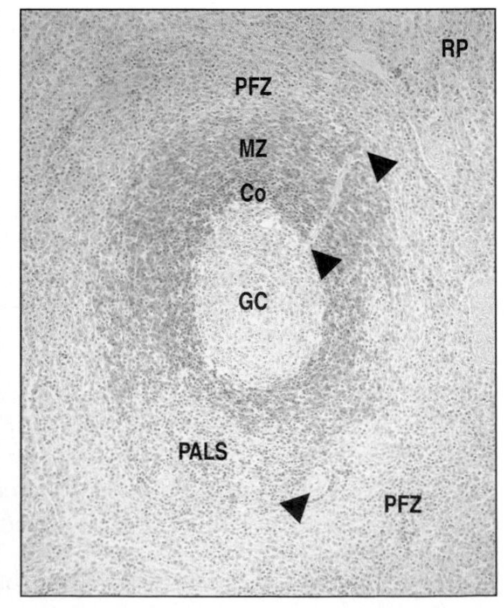

Fig. 30.2 Schematic views and light micrograph of a section of human spleen. The schematic at top left shows that the spleen consists of red pulp (pink areas), which is a site of red blood cell destruction, interspersed with the lymphoid white pulp. An enlargement of an area of white pulp is shown below. The follicle (yellow in the middle panel) and T-cell areas (blue) are surrounded by the perifollicular zone (PFZ) (palest yellow). The light micrograph on the right shows a transverse section of white pulp immunostained for mature B cells. The follicular arteriole emerges in the periarteriolar lymphoid sheath (PALS) of T cells (lower arrowhead in the bottom panel), traverses the follicle, goes through the marginal zone and opens into the perifollicular zone (upper arrowheads). Co, follicular B-cell corona; GC, germinal center (activated B cells); MZ, marginal zone (B cells); RP, red pulp; arrowheads, central arteriole. Photograph courtesy of N.M. Milicevic.

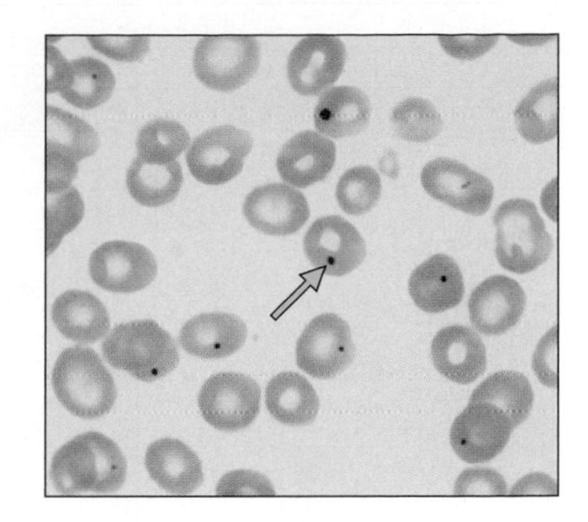

Fig. 30.3 The arrow marks a red blood cell with Howell–Jolly body. These inclusions are formed by the retention of nuclear remnants in red cells, which are usually removed by the spleen but can be found in patients with asplenia or significantly decreased splenic function.

Bacteria enter the bloodstream all the time, such as when we brush our teeth or when we have a local infection, for example of the skin or middle ear. Normally, these bacteria are disposed of efficiently by the spleen. When, for one reason or another, the spleen is not present, serious, even fatal, infections occur.

The case of Susan Vanderveer: a fatality because of an absent spleen.

Mr and Mrs Vanderveer owned a farm in the Hudson Valley in lower New York State. They were both descended from Dutch settlers who came to the Hudson Valley in the mid-17th century. There were multiple consanguineous marriages among their ancestors, and Mr and Mrs Vanderveer were distantly related to each other. At the time of this case, they had five children—three girls and two boys. Their youngest daughter, Susan, was 10 months old when she developed a cold, which lasted for 2 weeks. On the 14th day of her upper respiratory infection, she became sleepy and felt very warm. Her mother found that her temperature was 41.7°C. When Susan developed convulsive movements of her extremities, she was rushed to the emergency room but she died on the way to the hospital. Post-mortem cultures of blood were obtained, and also from her throat and cerebrospinal fluid. All the cultures grew *Haemophilus influenzae*, type b. At autopsy Susan was found to have no spleen.

Susan Vanderveer, age 10 months, dead on arrival in Emergency.

At the time of Susan's death her 3-year-old sister, Betsy, also had a fever, of 38.9°C. She complained of an earache, and her eardrums were found to be red. She had no other complaints and no other abnormalities were detected on physical examination. Her white blood count was 28,500 cells μl^{-1} (very elevated). Cultures from her nose, throat, and blood grew out *H. influenzae*, type b. She was given ampicillin intravenously for 10 days in the hospital and was then sent home in good health. Her cultures were negative at the time of discharge from the hospital. She was seen by a pediatrician on three occasions during the following year for otitis media (inflammation of the middle ear), pneumonia, and mastoiditis (inflammation of the mastoid bone behind the ear).

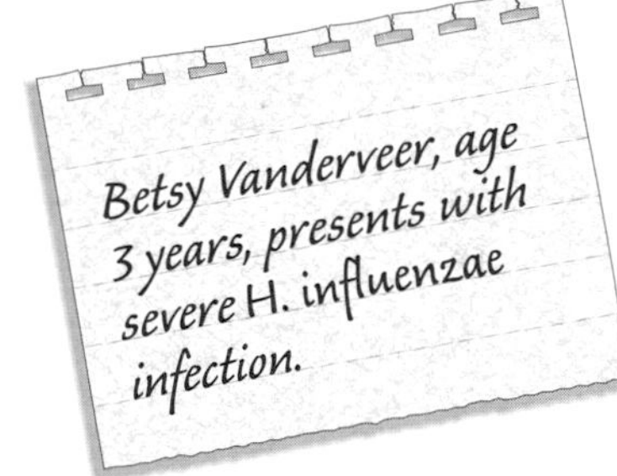

Betsy Vanderveer, age 3 years, presents with severe H. influenzae infection.

David, Susan's 5-year-old brother, had been admitted to the hospital at 21 months of age with meningitis caused by *Streptococcus pneumoniae*. He had responded well to antibiotic therapy and had been discharged. Another occurrence of pneumococcal meningitis at 27 months of age had also been followed by an uneventful recovery after antibiotics. He had had pneumonia at age 3½ years. At the time of Susan's death he was well.

The two other children of the Vanderveers, a girl aged 8 years and a newborn male, were in good health.

All the Vanderveer children had received routine immunization at ages 3, 4, and 5 months with tetanus and diphtheria toxoids and killed *Bordetella pertussis* to protect against tetanus, diphtheria, and whooping cough, which are three potentially fatal diseases caused by bacterial toxins (Fig. 30.4). Serum agglutination tests were used to test their antibody responses to these and other immunogens. Samples of serum from both Betsy and David caused hemagglutination (the clumping of red blood cells) when added to red blood cells (type O) coated with tetanus toxoid. Hemagglutinating antibodies against tetanus toxoid were seen at serum dilutions of 1:32 for both Betsy

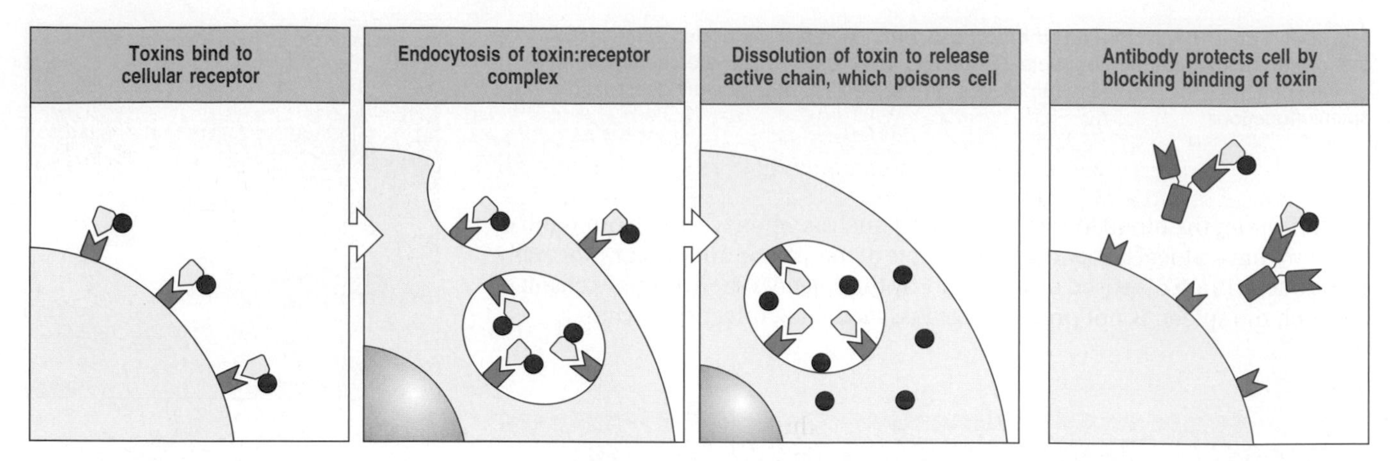

Fig. 30.4 Neutralization by antibodies protects cells from toxin action. Secreted bacterial toxins usually contain several distinct moieties. One piece of the toxin must bind a cellular receptor, which allows the molecule to be internalized. A second part of the toxin molecule then enters the cytoplasm and poisons the cell. In some cases, a single molecule of toxin can kill a cell. Antibodies that inhibit toxin binding can prevent, or neutralize, these effects. Protective antibodies can be generated by subcutaneous immunization with toxoids. Toxoids are toxins rendered harmless by treatment with denaturing agents, such as formalin, which destroy their toxicity but not their ability to generate neutralizing antibodies. In the case of the DPT vaccine, the killed *Bordetella pertussis* cells act as an adjuvant, which enhances the immune response to all components of the vaccine by delivering activating signals to antigen-presenting cells.

and David, and were found at a similar titer in their 8-year-old sister. All three children were given typhoid vaccine subcutaneously, and 4 weeks later samples of their sera were tested for the ability to agglutinate killed *Salmonella typhosa*. The results indicated a normal immune response. David had an agglutination titer of 1:16, Betsy 1:32, and their normal 8-year-old sister 1:32. All three children were given 1 ml of a 25% suspension of sheep red cells intravenously. David had a titer of 1:4 for hemagglutinating antibodies against sheep red blood cells before the injection. He was tested again 2 and 4 weeks later and there was no increase in titer. Betsy had an initial titer of 1:32 and her titer did not increase either. The 8-year-old normal sister had a preimmunization titer of 1:32. She was tested 2 and 4 weeks after the immunization, when she was found to have a hemagglutinating titer of 1:256 against sheep red blood cells.

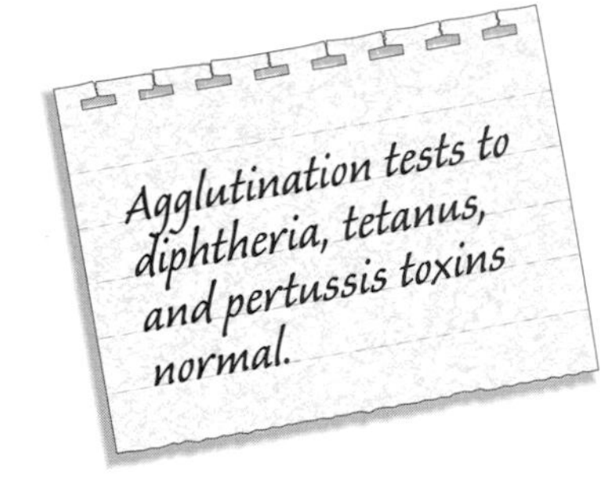

Agglutination tests to diphtheria, tetanus, and pertussis toxins normal.

All the children and their parents were injected intravenously with radioactive colloidal gold (^{198}Au), which is taken up by the reticuloendothelial cells of the liver and spleen within 15 minutes after the injection. A scintillation counter then scans the abdomen for radioactive gold. The pattern of scintillation revealed that Betsy and David had no spleens (Fig. 30.5).

Asplenia and splenectomy.

Significant impairment of splenic function can be either congenital, where the spleen is absent or dysfunctional at birth, or acquired as a result of conditions that damage the spleen such as trauma or sickle-cell anemia, and which often lead to its surgical removal (splenectomy). Congenital asplenia is further divided into two main categories. The more prevalent is syndromic asplenia, in which the lack of splenic tissue is part of a more complex genetic syndrome affecting other systems as well. In these syndromes splenic defects are usually associated with significant heart defects and heterotaxia, a condition in which malformations arise as a result of lateralization defects of organs in the thorax and the abdomen. Several human genes, including *ZIC3, LEFTYA, CRYPTIC,*

ACVR2B, and *CFC1*, all of which have important roles in directing lateralization, have been shown to be associated with these syndromes.

The second category is isolated congenital asplenia. This is a group of conditions in which the only abnormality is the lack of splenic tissue. Only a relatively small number of cases of true isolated congenital asplenia have so far been described. Most of these cases were diagnosed after episodes of overwhelming pneumococcal infections, either *post mortem* or while screening family members of affected individuals, as in the Vanderveer family. The genetic defect causing asplenia has not yet been identified. Most familial cases described so far follow an autosomal dominant pattern of inheritance, although families with an autosomal recessive or X-linked pattern of inheritance have also been reported.

The Vanderveer family is unusual in that three of their first four children were born without a spleen. After the events described in this case, the Vanderveers had three more children. One of the boys and the girl were also born without a spleen; the other boy had a normal spleen. This family provides us with an uncomplicated circumstance in which to examine the role of the spleen. The major consequence of its absence is a susceptibility to bacteremia, usually caused by the encapsulated bacteria *Streptococcus pneumoniae* or *Haemophilus influenzae*. This susceptibility is caused by a failure of the immune response to these common extracellular bacteria when they enter the bloodstream.

Surgical removal of the spleen is quite common. The capsule of the spleen may rupture from trauma, for example in an automobile accident. In such cases, the spleen has to be removed very quickly because of blood loss into the abdominal cavity. The spleen may also be removed surgically for therapeutic reasons in certain autoimmune diseases, or because of a malignancy in the spleen. After splenectomy, patients, particularly children, are susceptible to bloodstream infections by microorganisms against which they have no antibodies. Microorganisms against which the host has antibodies are removed quickly from the bloodstream by the liver, where the Kupffer cells complement the role of the red pulp of the spleen. Antibodies against the encapsulated bacteria that commonly cause bloodstream infections persist for a very long time in the bloodstream of exposed individuals, even in the absence of a spleen (for reasons that are not fully understood). Adults who already have antibodies against these microorganisms are therefore much less vulnerable to bacteremia than children who have not yet developed such antibodies. Fortunately, effective vaccines against both *S. pneumoniae* and *H. influenzae*, type b, have been developed, and are now part of the routine vaccinations given to many children worldwide, thus protecting asplenic children from some of the severe infections to which they are prone. Nevertheless, specific precautions, including prophylactic antibiotic treatment, are recommended to most individuals with an absent or non-functional spleen.

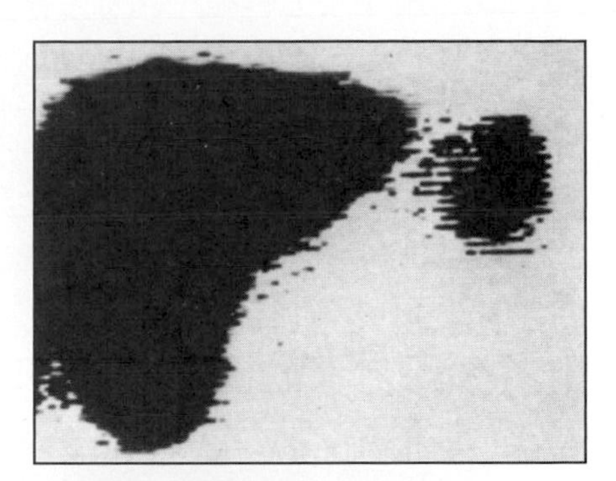

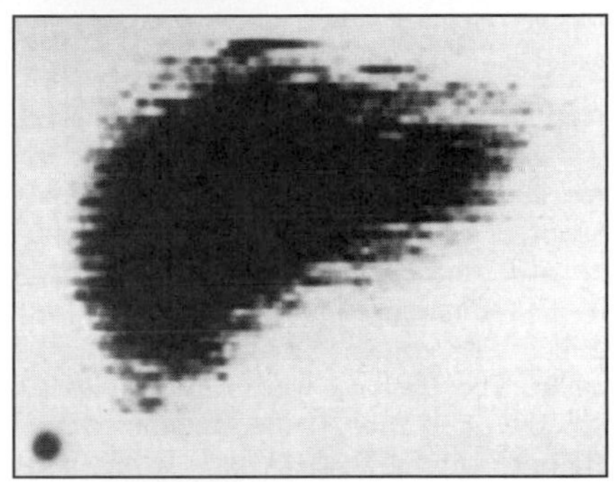

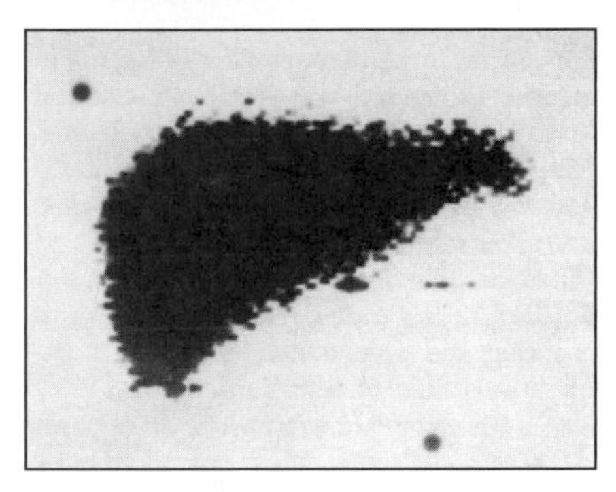

Fig. 30.5 A scintillation scan of the abdomen after intravenous injection with radioactive colloidal gold (^{198}Au) reveals that Betsy and David Vanderveer have no spleens. The top panel shows an abdominal scan of Betsy's mother. The large mass on the left is the liver and the small mass on the right is the spleen. The reticuloendothelial cells of both liver and spleen take up the labeled gold within 15 minutes after the injection. No spleen is seen in either Betsy (middle panel) or David (bottom panel).

Questions.

1 *Nicholas Biddleboy, a 5-year-old boy, has had his spleen removed after a sledding accident, during which both he and his sled struck a tree trunk. In the emergency room of a nearby hospital, it was determined that his spleen had ruptured. The surgeon, after removal of a spleen that had indeed ruptured, calls you for an immunology consultation. What do you advise?*

2 *Why did David and Betsy have normal responses to the typhoid vaccine but not to the sheep red blood cells?*

3 *The Vanderveer family is unique in the medical literature. The parents, who were distantly related, were normal and had normal spleens. Five of their eight children were born without spleens. Of these, only Betsy subsequently had children—four boys and one girl. They are all normal and have spleens. What is the inheritance pattern of congenital asplenia in this family? According to Mendelian laws how many of the eight Vanderveer children would be expected to have no spleen?*

CASE 31 Hereditary Angioedema

Regulation of complement activation.

Complement is a system of plasma proteins that participates in a cascade of reactions, generating active components that allow pathogens and immune complexes to be destroyed and eliminated from the body. Complement is part of the innate immune defenses of the body and is also activated via the antibodies produced in an adaptive immune response. Complement activation is generally confined to the surface of pathogens or circulating complexes of antibody bound to antigen.

Complement is normally activated by one of three routes: the classical pathway, which is triggered by antigen:antibody complexes or antibody bound to the surface of a pathogen; the lectin pathway, which is activated by mannose-binding lectin (MBL) and the ficolins; and the alternative pathway, in which complement is activated spontaneously on the surface of some bacteria. The early part of each pathway is a series of proteolytic cleavage events leading to the generation of a convertase, a serine protease that cleaves complement component C3 and thereby initiates the effector actions of complement. The C3 convertases generated by the three pathways are different, but evolutionarily homologous, enzymes. Complement components and activation pathways, and the main effector actions of complement, are summarized in Fig. 31.1.

The principal effector molecule, and a focal point of activation for the system, is C3b, the large cleavage fragment of C3. If active C3b, or the homologous but less potent C4b, accidentally becomes bound to a host cell surface instead of a pathogen, the cell can be destroyed. This is usually prevented by the rapid hydrolysis of active C3b and C4b if they do not bind immediately to the surface where they were generated. Protection against inappropriate activation of complement is also provided by regulatory proteins.

One of these, and the most potent inhibitor of the classical pathway, is the C1 inhibitor (C1INH). This belongs to a family of serine protease inhibitors (called serpins) that together constitute 20% of all plasma proteins. In addition to being the sole known inhibitor of C1, C1INH contributes to the

This case was prepared by Raif Geha, MD, in collaboration with Arturo Borzutzky, MD.

Topics bearing on this case:
Classical pathway of complement activation
Inhibition of C1 activation
Alternative pathway of complement activation
Inflammatory effects of complement activation
Regulation of C4b

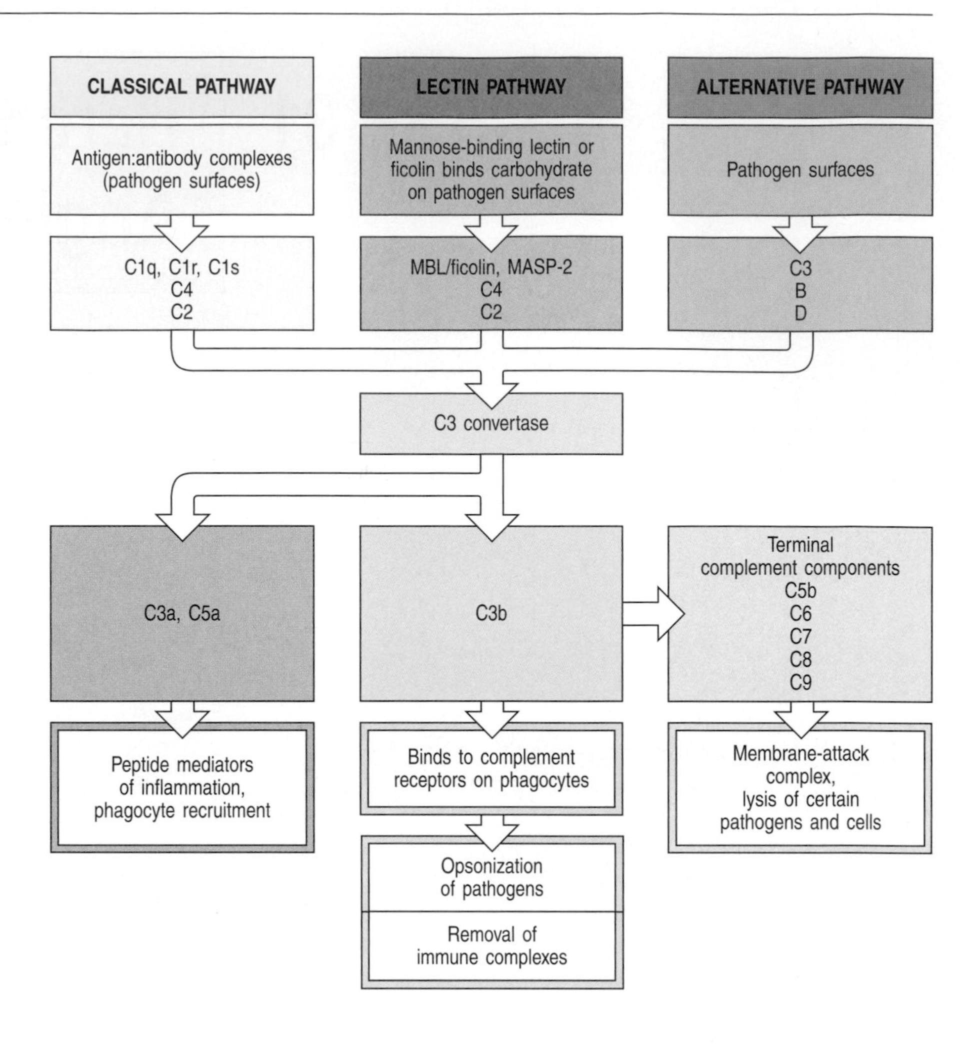

Fig. 31.1 Overview of the main components and effector actions of complement. The early events of all three pathways of complement activation involve a series of cleavage reactions that culminate in the formation of an enzymatic activity called a C3 convertase, which cleaves complement component C3 into C3b and C3a. The production of the C3 convertase is the point at which the three pathways converge and the main effector functions of complement are generated. C3b binds covalently to the bacterial cell membrane and opsonizes the bacteria, enabling phagocytes to internalize them. C3a is a peptide mediator of local inflammation. C5a and C5b are generated by the cleavage of C5b by a C5 convertase formed by C3b bound to the C3 convertase (not shown in this simplified diagram). C5a is also a powerful peptide mediator of inflammation. C5b triggers the late events in which the terminal components of complement assemble into a membrane-attack complex that can damage the membrane of certain pathogens. Although the classical complement activation pathway was first discovered as an antibody-triggered pathway, it is now known that C1q can activate this pathway by binding directly to pathogen surfaces, as well as paralleling the lectin activation pathway by binding to antibody that is itself bound to the pathogen surface. In the lectin pathway, MASP stands for mannose-binding lectin-associated serine protease.

regulation of serine proteases of the clotting system and of the kinin system, which is activated by injury to blood vessels and by some bacterial toxins. The main product of the kinin system is bradykinin, which causes vasodilation and increased capillary permeability.

C1INH intervenes in the first step of the complement pathway, when C1 binds to immunoglobulin molecules on the surface of a pathogen or antigen:antibody complex (Fig. 31.2). Binding of two or more of the six tulip-like heads of the C1q component of C1 is required to trigger the sequential activation of the two associated serine proteases, C1r and C1s. C1INH inhibits both of these proteases, by presenting them with a so-called bait-site, in the form of an arginine bond that they cleave. When C1r and C1s attack the bait-site they covalently bind C1INH and dissociate from C1q. By this mechanism, the C1 inhibitor limits the time during which antibody-bound C1 can cleave C4 and C2 to generate C4b2a, the classical pathway C3 convertase.

Activation of C1 also occurs spontaneously at low levels without binding to an antigen:antibody complex, and can be triggered further by plasmin, a protease of the clotting system, which is also normally inhibited by C1INH. In the absence of C1INH, active components of complement and bradykinin are produced. This is seen in hereditary angioedema (HAE), a disease caused by a genetic deficiency of C1INH.

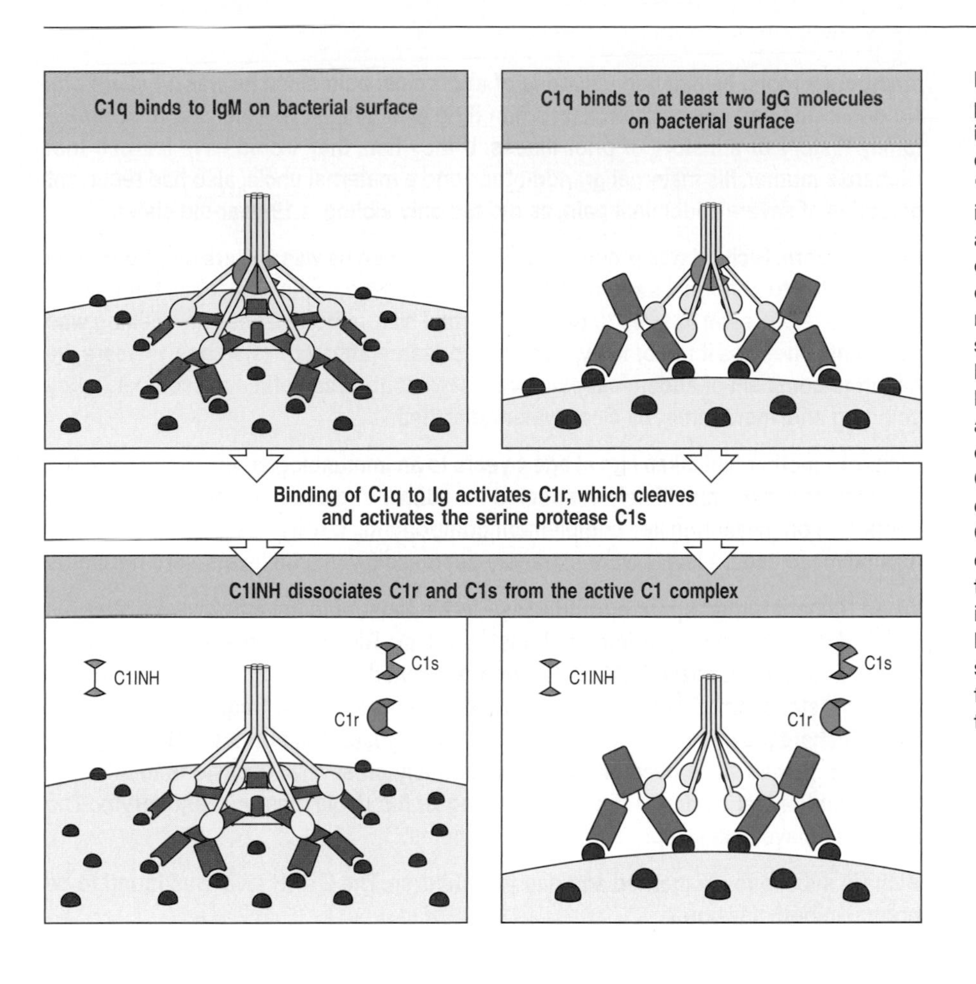

C1q binds to IgM on bacterial surface
C1q binds to at least two IgG molecules on bacterial surface

C1INH dissociates C1r and C1s from the active C1 complex

Fig. 31.2 Activation of the classical pathway of complement and intervention by C1INH. In the left panel, one molecule of IgM, bent into the 'staple' conformation by binding several identical epitopes on a pathogen surface, allows binding by the globular heads of C1q to its Fc pieces on the surface of the pathogen. In the right panel, multiple molecules of IgG bound to the surface of the pathogen allow binding by C1q to two or more Fc pieces. In both cases, binding of C1q activates the associated C1r, which becomes an active enzyme that cleaves the proenzyme C1s, a serine protease that initiates the classical complement cascade. Active C1 is inactivated by C1INH, which binds covalently to C1r and C1s, causing them to dissociate from the complex. There are in fact two C1r and two C1s molecules bound to each C1q molecule, although for simplicity this is not shown here. It takes four molecules of C1INH to inactivate all the C1r and C1s.

The case of Richard Crafton: a failure of communication as well as of complement regulation.

Richard Crafton was a 17-year-old high-school senior when he had an attack of severe abdominal pain at the end of a school day. The pain came as frequent sharp spasms and he began to vomit. After 3 hours, the pain became unbearable and he went to the emergency room at the local hospital.

At the hospital, the intern who examined him found no abnormalities other than dry mucous membranes of the mouth, and a tender abdomen. There was no point tenderness to indicate appendicitis. Richard continued to vomit every 5 minutes and said the pain was getting worse.

A surgeon was summoned. He agreed with the intern that Richard had an acute abdominal condition but was uncertain of the diagnosis. Blood tests showed an elevated red blood cell count, indicating dehydration. The surgeon decided to proceed with exploratory abdominal surgery. A large midline incision revealed a moderately swollen and pale jejunum but no other abnormalities were noted. The surgeon removed Richard's appendix, which was normal, and Richard recovered and returned to school 5 days later.

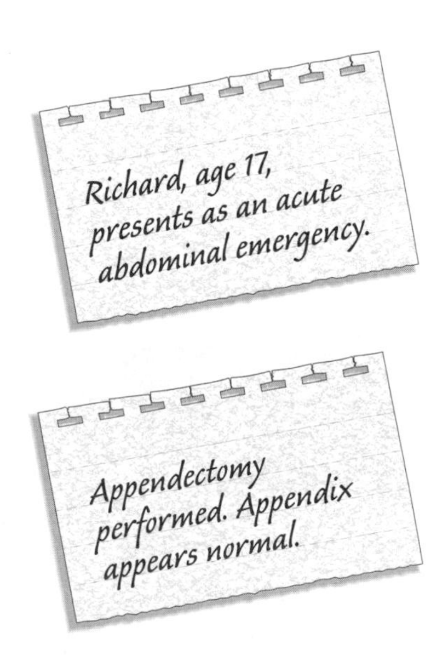

Richard, age 17, presents as an acute abdominal emergency.
Appendectomy performed. Appendix appears normal.

What Richard had not mentioned to the intern or to the surgeon was that, although he had never had such severe pains as those he was experiencing when he went to the emergency room, he had had episodes of abdominal pain since he was 14 years old. No one in the emergency room asked him if he was taking any medication, or took a family history or a history of prior illness. If they had, they would have learned that Richard's mother, his maternal grandmother, and a maternal uncle, also had recurrent episodes of severe abdominal pain, as did his only sibling, a 19-year-old sister.

Family history of colic.

As a newborn, Richard was prone to severe colic. When he was 4 years old, a bump on his head led to abnormal swelling. When he was 7, a blow with a baseball bat caused his entire left forearm to swell to twice its normal size. In both cases, the swelling was not painful, nor was it red or itchy, and it disappeared after 2 days. At age 14 years, he began to complain of abdominal pain every few months, sometimes accompanied by vomiting and, more rarely, by clear, watery diarrhea.

Richard's mother had taken him at age 4 years to an immunologist, who listened to the family history and immediately suspected hereditary angioedema. The diagnosis was confirmed on measuring key complement components. C1INH levels were 16% of the normal mean and C4 levels were markedly decreased, while C3 levels were normal.

When Richard turned up for a routine visit to his immunologist a few weeks after his surgical misadventure, the immunologist, noticing Richard's large abdominal scar, asked what had happened. When Richard explained, he prescribed daily doses of Winstrol (stanozolol). This caused a marked diminution in the frequency and severity of Richard's symptoms. When Richard was 20 years old, purified C1INH became available; he has since been infused intravenously on several occasions to alleviate severe abdominal pain, and once for swelling of his uvula, pharynx, and larynx. The infusion relieved his symptoms within 25 minutes.

Richard subsequently married and had two children. The C1INH level was found to be normal in both newborns.

Hereditary angioedema.

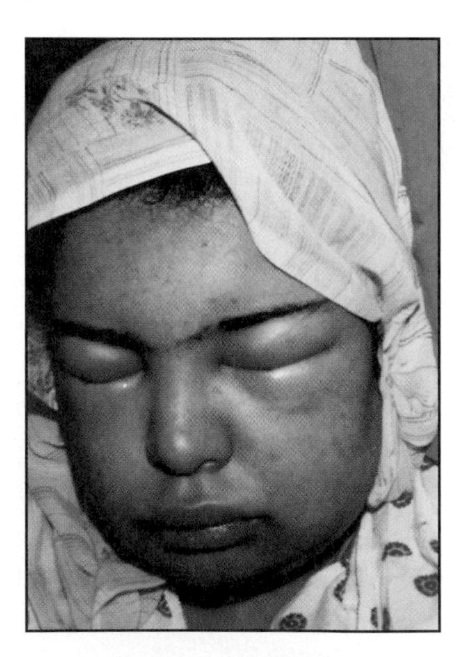

Fig. 31.3 Hereditary angioedema. Transient localized swelling that occurs in this condition often affects the face.

Individuals like Richard with a hereditary deficiency of C1INH are subject to recurrent episodes of circumscribed swelling of the skin (Fig. 31.3), intestine, and airway. Attacks of subcutaneous or mucosal swelling most commonly affect the extremities, but can also involve the face, trunk, genitals, lips, tongue, or larynx. Cutaneous attacks cause temporary disfigurement but are not dangerous. When the swelling occurs in the intestine it causes severe abdominal pain, and obstructs the intestine so that the patient vomits. When the colon is affected, watery diarrhea may occur. Swelling in the larynx is the most dangerous symptom, because the patient can rapidly choke to death. HAE attacks do not usually involve itching or hives, which is useful to differentiate this disease from allergic angioedema. However, a serpiginous, or linear and wavy, rash is sometimes seen before the onset of swelling symptoms. Such episodes may be triggered by trauma, menstrual periods, excessive exercise, exposure to extremes of temperature, mental stress, and some medications such as angiotensin-converting enzyme inhibitors and oral contraceptives.

HAE is not an allergic disease, and attacks are not mediated by histamine. HAE attacks are associated with activation of four serine proteases, which are normally inhibited by C1INH. At the top of this cascade is Factor XII, which directly or indirectly activates the other three (Fig. 31.4). Factor XII is normally activated by injury to blood vessels, and initiates the kinin cascade, activating

Fig. 31.4 Pathogenesis of hereditary angioedema. Activation of Factor XII leads to the activation of kallikrein, which cleaves kininogen to produce the vasoactive peptide bradykinin; it also leads to the activation of plasmin, which in turn activates C1. C1 cleaves C2, whose smaller fragment C2b is further cleaved by plasmin to generate the vasoactive peptide C2 kinin. The red bars represent inhibition by C1INH.

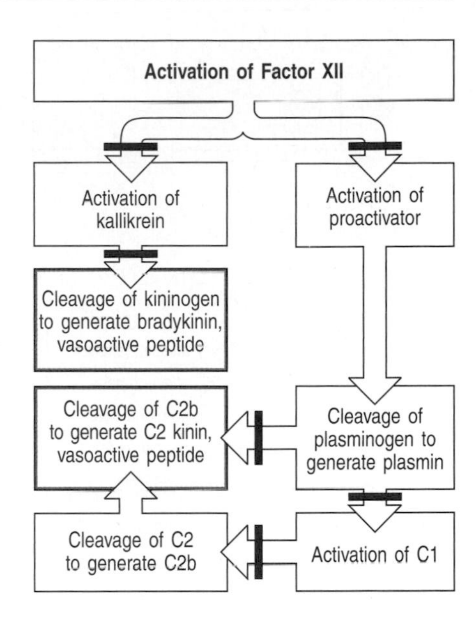

kallikrein, which generates the vasoactive peptide bradykinin. Factor XII also indirectly activates plasmin, which, as mentioned earlier, activates C1 itself. Plasmin also cleaves C2b to generate a vasoactive fragment called C2 kinin. In patients deficient in C1INH, the uninhibited activation of Factor XII leads to the activation of kallikrein and plasmin; kallikrein catalyzes the formation of bradykinin, and plasmin produces C2 kinin. Bradykinin is the main mediator responsible for HAE attacks by causing vasodilation and increasing the permeability of the postcapillary venules by causing contraction of endothelial cells so as to create gaps in the blood vessel wall (Fig. 31.5). This is responsible for the edema; movement of fluid from the vascular space into another body compartment, such as the gut, causes the symptoms of dehydration as the vascular volume contracts.

Treatment of HAE can focus on preventing attacks or on resolving acute episodes. Purified or recombinant C1INH is an effective therapy in both these settings. A kallikrein inhibitor and a bradykinin receptor antagonist have also been developed to target the kinin cascade and bradykinin activity.

Questions.

1 *Activation of the complement system results in the release of histamine and chemokines, which normally produce pain, heat, and itching. Why is the edema fluid in HAE free of cellular components, and why does the swelling not itch?*

2 *Richard has a markedly decreased amount of C4 in his blood. This is because it is being rapidly cleaved by activated C1. What other complement component would you expect to find decreased? Would you expect the alternative pathway components to be low, normal, or elevated? What about the terminal components?*

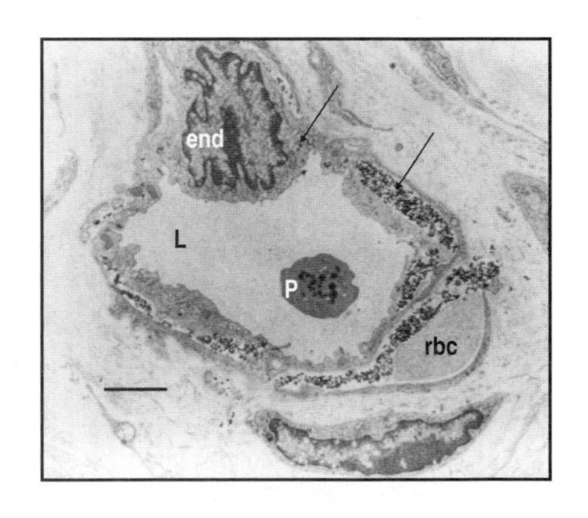

Fig. 31.5 Contraction of endothelial cells creates gaps in the blood vessel wall. A guinea pig was injected intravenously with India ink (a suspension of carbon particles). Immediately thereafter the guinea pig was injected intradermally with a small amount of activated C1s. An area of angioedema formed about the injected site, which was biopsied 10 minutes later. An electron micrograph reveals that the endothelial cells in post-capillary venules have contracted and formed gaps through which the India ink particles have leaked from the blood vessel. L is the lumen of the blood vessel; P is a polymorphonuclear leukocyte in the lumen; rbc is a red blood cell that has leaked out of the blood vessel. Micrograph courtesy of Kaethe Willms.

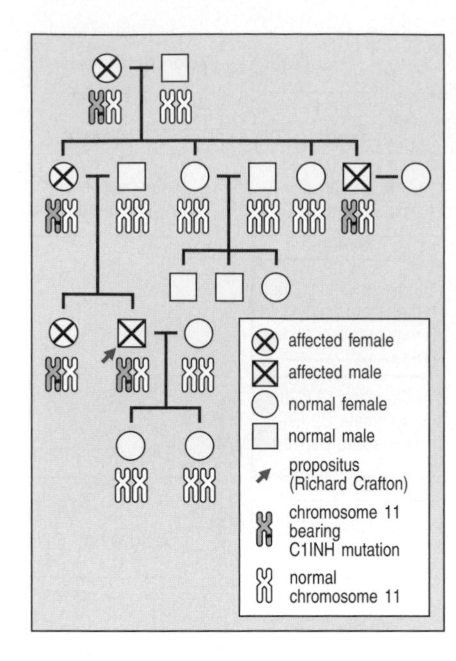

Fig. 31.6 The inheritance of hereditary angioedema in Richard's extended family.

3 *Despite the complement deficiency in patients with HAE, they are not unduly susceptible to infection. Why not?*

4 *What is stanozolol, and why was it prescribed?*

5 *Emergency treatment for HAE cases is sometimes necessary because of airway obstruction. In most cases, however, a patient with obstruction of the upper airways is likely to be suffering from an anaphylactic reaction. The treatment in this case would be epinephrine. How might you decide whether to administer epinephrine or intravenous C1INH?*

6 *Figure 31.6 shows Richard's family tree. What is the mode of inheritance (dominant or recessive, sex-linked or not) of HAE? Can Richard's two children pass the disease onto their offspring?*

CASE 32 Factor I Deficiency

The alternative pathway of complement activation is important in innate immunity.

The complement system plays a crucial part in the destruction and removal of microorganisms from the body. Pathogens coated with complement proteins are more efficiently phagocytosed by macrophages, and bacteria coated with complement can also be directly destroyed by complement-mediated lysis. The system of plasma proteins known collectively as complement can be activated in various ways (see Fig. 31.1), of which the so-called alternative pathway is important in innate or nonadaptive immunity. This pathway can be activated in the absence of antibody, although even low titers of IgM antibodies against an infecting microorganism will greatly amplify complement activation.

The complement protein C3 is the starting point of the alternative pathway. It is one of the more abundant globulins in blood and contains a highly reactive thioester bond. This is continuously being cleaved by spontaneous hydrolysis at a fairly low rate to form C3(H_2O) in the plasma (a process known as 'tick-over'). C3(H_2O) associates with fragment Bb of another alternative pathway component, factor B, to form a short-lived proteinase C3(H_2O)Bb, which is called the 'fluid-phase' C3 convertase; this cleaves many C3 molecules into C3a and the larger C3b fragment, which contains the thioester bond. C3b then bonds covalently, via the thioester bond, with the hydroxyl group of serine or threonine in a protein or the hydroxyl group of a sugar on a microbial surface. If C3b fails to attach to a microbial surface, the thioester bond is spontaneously hydrolyzed and the C3b is inactivated (Fig. 32.1).

Topics bearing on this case:
Alternative pathway of complement activation
Factor I cleavage of C3b
Opsonizing activity of C3b
C3a activation of mast cells

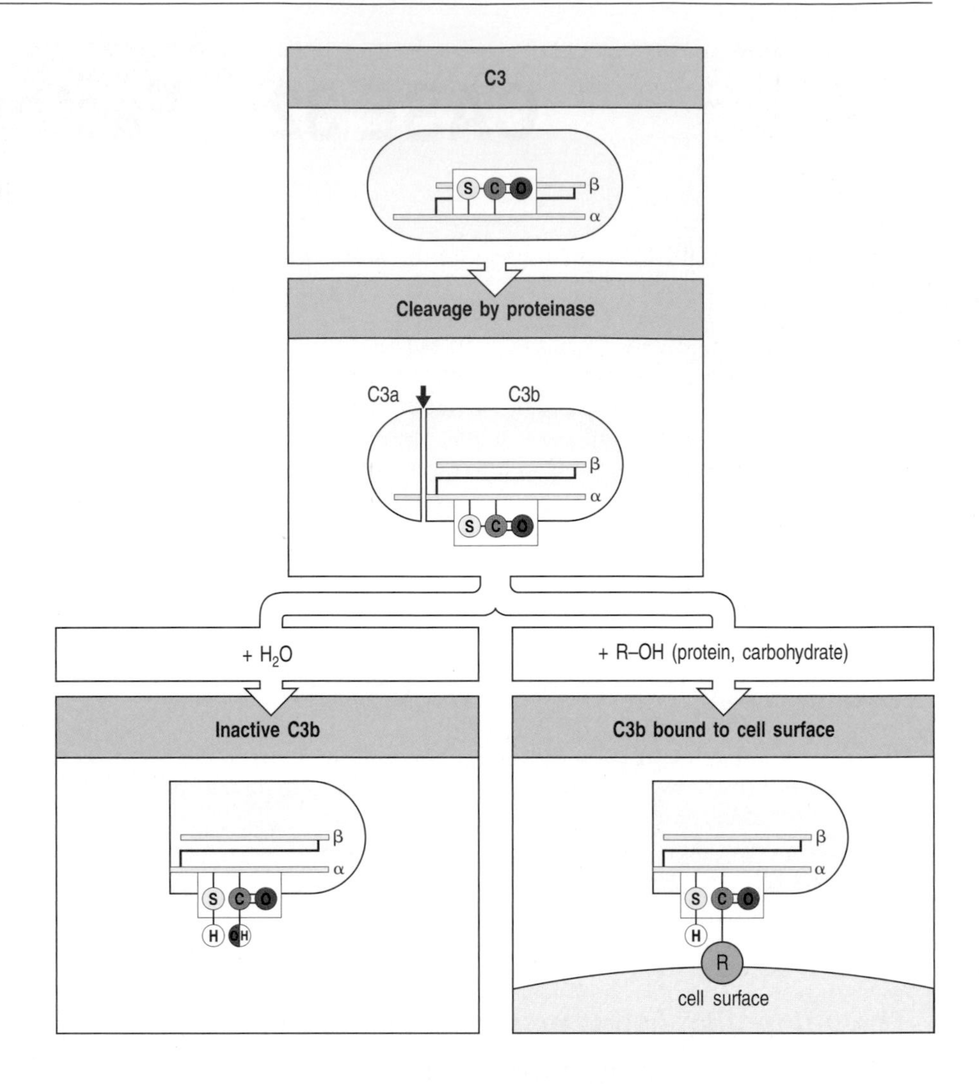

Fig. 32.1 Cleavage of C3 exposes a reactive thioester bond that enables the larger cleavage fragment, C3b, to bind covalently to the bacterial cell surface. Intact C3 has a shielded thioester bond that is exposed when C3 is cleaved by a proteinase to produce C3b. The highly reactive thioester bond in C3b can react with hydroxyl or amino groups to form a covalent linkage with molecules on the microbial surface. In the absence of such a reaction, the thioester bond is rapidly hydrolyzed, inactivating C3b.

The binding of C3b to a microbial surface stimulates the cleavage of yet more C3 molecules. Factor B binds to C3b on a bacterial surface, and in this bound state is cleaved by a preexisting blood proteinase, factor D, leaving the larger Bb fragment still bound to the C3b. The resulting C3bBb complex (like the C3(H_2O)Bb complex) is an active serine protease, known as the alternative pathway C3 convertase, which specifically cleaves native C3 to make more C3b and C3a (Fig. 32.2).

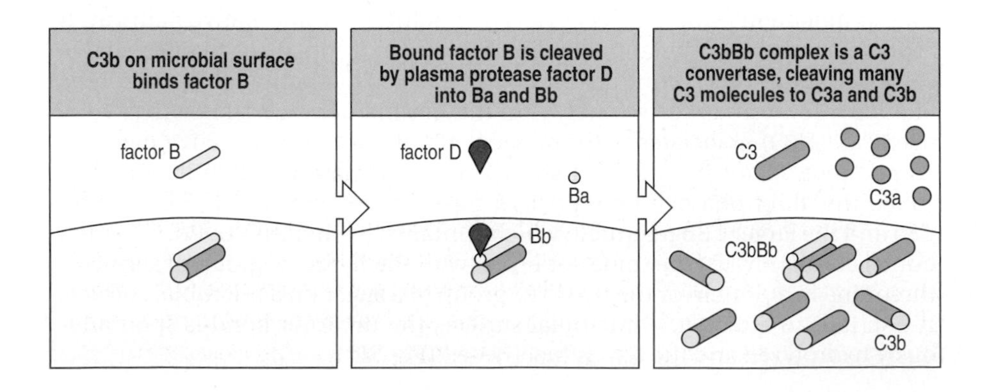

Fig. 32.2 The alternative pathway of complement activation leads to amplification of C3 cleavage. C3b deposited on a microbial surface can bind factor B, making it susceptible to cleavage by factor D. The C3bBb complex is the C3 convertase of the alternative pathway of complement activation, and its action results in the deposition of many more molecules of C3b on the pathogen surface.

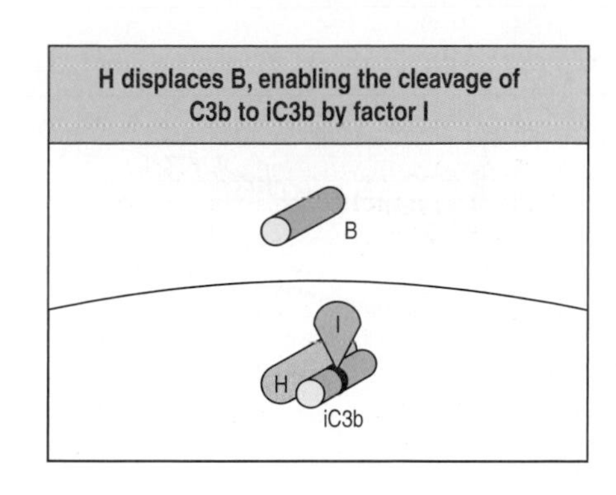

Fig. 32.3 The conversion of C3b to iC3b by factor I. On microbial surfaces, factor H displaces B from the C3bBb complex and factor I then cleaves C3b to produce iC3b. On host cell surfaces, the complement receptor CR1, which binds C3b, can substitute for factor H in this reaction.

C3b bound to a microbial surface acts as an opsonin by binding to a specific receptor, the complement receptor 3 (CR3), on phagocytes, facilitating the ingestion of C3b-coated particles. But before C3b can act as a ligand for CR3, and thus as an effective opsonin, it has to undergo a further cleavage to a fragment called iC3b, which is effected by a blood serine protease called factor I, acting in conjunction with the blood protein factor H, components of the alternative pathway (Fig. 32.3). iC3b acting at the receptor CR3 can activate neutrophils and macrophages in the absence of antibody (Fig. 32.4). Cleavage of C3b by factor I also has another critical effect. It inhibits the C3 convertase activity of the C3b complex, thus ensuring that supplies of C3 do not become depleted. On host cell surfaces, complement receptor 1 (CR1) can bind to C3b and serve as the cofactor instead of factor H.

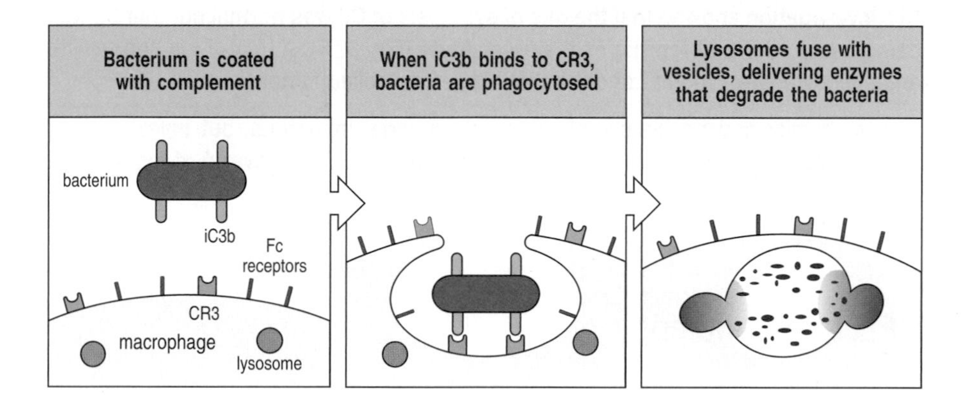

Fig. 32.4 iC3b binds to complement CR3 receptors on phagocytes and stimulates phagocytosis. iC3b on the surface of the pathogen binds to CR3 receptors on the phagocyte. Binding signals the phagocyte to engulf the particle and activates the internal destruction mechanisms.

The case of Morris Townsend: uncontrolled complement activation leads to susceptibility to infection and to hives.

Morris Townsend was admitted to the Brighton City Hospital at age 25 with pneumonia. This was his 28th admission to the hospital in his lifetime. From his first year onwards he had been repeatedly admitted for middle ear infections and mastoiditis. During these episodes, which were successfully treated with antibiotics, a variety of pyogenic (pus-forming) bacteria were cultured from his ears or mastoids, including *Staphylococcus aureus*, *Proteus vulgaris*, and *Pseudomonas aeruginosa*. At age 3 he had a tonsillectomy and adenoidectomy because of enlargement and chronic infection of his nasopharyngeal lymphoid tissue; at age 6 he had scarlet fever. He had also been admitted at other times with left lower lobe pneumonia (when *Haemophilus influenzae* had been cultured from his sputum), an abscess in the groin, acute sinusitis, a posterior ear abscess due to *Corynebacterium* species, skin abscesses with accompanying bloodstream infection (septicemia) due to β-hemolytic streptococci and, on one occasion, to septicemia due to *Neisseria meningitidis* (meningococcemia).

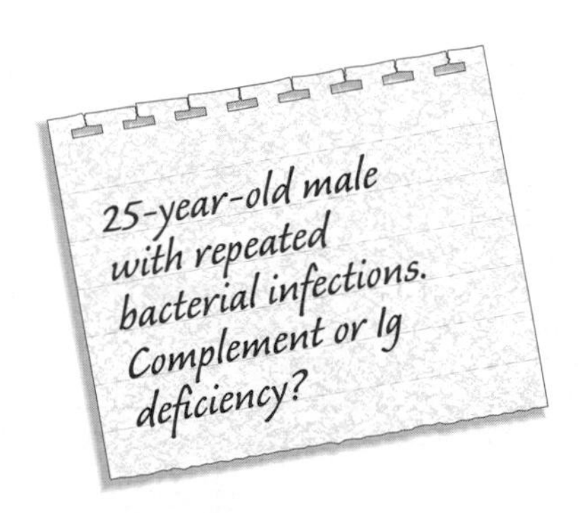

25-year-old male with repeated bacterial infections. Complement or Ig deficiency?

Ig normal; check C3.

On physical examination at his latest admission, Morris was found to be slightly obese but otherwise normally developed. His hearing was poor in both ears, and this was attributed to his recurrent ear infections and mastoiditis. He also told doctors that he developed hives all over his body after drinking alcohol or after taking a bath or shower.

A urine analysis yielded normal results. His hematocrit was 43% (normal) and his white cell count was 6000 μl^{-1}. His platelet count was 240,000 μl^{-1} (normal) and his blood clotted normally. His red blood cells gave a strong positive agglutination reaction with an antibody to C3 but no agglutination with an antibody to IgG or IgM. His serum IgG level was 915 mg dl^{-1}, IgA 475 mg dl^{-1}, and IgM 135 mg dl^{-1} (all normal). Morris responded normally to an injection of tetanus toxoid; his antibody titer rose from 0.25 to 8.0 hemagglutinating units ml^{-1}. He gave a positive delayed-type skin reaction to mumps and monilia antigens.

C3 synthesis normal; check factor I.

Genetic deficiency? Test family for factor I.

Serum levels of C3 were 27 mg dl^{-1} (normal values 97–204 mg dl^{-1}); of this, 8 mg dl^{-1} was C3 and 19 mg dl^{-1} was C3b. The serum levels of all other complement components were normal except for factor B, which was undetectable. His serum failed to kill a smooth strain of *Salmonella enterica* Newport, even after addition of C3 to the serum to render the C3 concentration normal. To investigate the turnover of C3, Morris was injected with a dose of C3 labeled with the radioactive tracer ^{125}I. The results of this investigation showed that the rate of synthesis of C3 was normal but that C3 was being broken down at four times the normal rate (Fig. 32.5). A test of his serum with an antibody against factor I showed that his serum lacked factor I.

Morris's family had no history of recurrent bacterial infections, but investigations showed reduced levels of factor I in both his parents and in several of his siblings (Fig. 32.6).

Factor I deficiency.

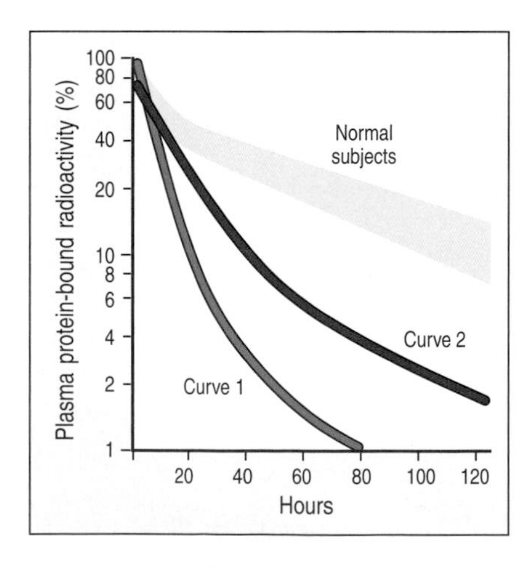

Fig. 32.5 The rate of disappearance of ^{125}I-labeled C3 from the plasma. The rate of disappearance of radioactively labeled C3 from the patient's plasma (curve 1) is much faster than that in normal subjects (shaded area; 11 normal subjects). Curve 2 shows the rate of disappearance of C3 in the patient after the infusion of 500 ml of normal plasma.

Patients such as Morris Townsend with a genetic deficiency in factor I were instrumental in deciphering the mechanism of activation of the alternative pathway of complement. Innate immunity is a first and highly effective means of defense against the common extracellular bacteria that cause pyogenic infections. The lack of factor I means that the alternative pathway C3 convertase is uninhibited and consumption of C3 is greatly accelerated, leading to C3 depletion. The lack of C3, and the nonproduction of iC3b, results in defective opsonization, which is the main means of removing and destroying these bacteria. Thus, factor I deficiency, like the genetic deficiency of C3 itself, results in a greatly increased susceptibility to infections with such bacteria. The clinical findings in factor I deficiency are not unlike those observed in X-linked agammaglobulinemia—a failure of opsonization results in frequent pyogenic infections (see Case 1).

The gene encoding factor I is on chromosome 4. The family of Morris Townsend provides a classic case of the inheritance of a recessive mutation in an autosomal gene (see Fig. 32.6). His parents, some of his siblings, and two of his nephews are heterozygous for the defect; they produce roughly half the normal amounts of factor I, which is sufficient to prevent any clinical symptoms. Morris seems to be the only family member who is homozygous for the defect and who thus exhibits symptoms.

Two interesting facts emerge from his clinical history. He sustained recurrent hives and had one bout of meningococcemia. It is easy to understand why he had hives. He was constantly cleaving C3 to C3a and C3b. C3a binds to mast cells and, among other things, causes the release of histamine, leading to hives. The interesting question is why he did not have hives all the time and why they became problematic only after the ingestion of alcohol or exposure to hot and cold water. We must suppose that he had tachyphylaxis, or end-organ unresponsiveness, to histamine. Only when histamine release was increased, as by alcohol consumption or sudden changes in ambient body temperature, did the symptoms appear.

Morris Townsend's meningococcemia in particular is symptomatic of a deficiency in components of the alternative pathway of complement action. There are two common human pathogens in the bacterial genus *Neisseria*: *Neisseria gonorrhoeae* and *Neisseria meningitidis*. The former causes the sexually transmitted disease gonorrhea; the latter causes septicemia and meningitis and can be rapidly fatal. Patients have died from septic shock within 20 minutes of the onset of the symptoms of meningococcemia. Patients with genetic defects in the alternative pathway of complement activation or in the terminal components of complement sustain overwhelming and repeated infection with *Neisseria*. Deficiencies in the alternative pathway components factor D and properdin were discovered because these patients developed recurrent meningococcemia. Similar observations have been made in patients with deficiencies of the later-acting C5, C6, C7, C8, and C9 components. These clinical observations highlight the importance of the bactericidal action of complement in controlling septicemia due to *Neisseria*.

Complement factor I and factor H are essential regulators of complement activation. While complete deficiency in either one of these factors results in increased occurrence of pyogenic infections (as in the case of Morris Townsend), heterozygous mutations of factor H or of factor I have been also identified in patients with atypical hemolytic uremic syndrome (aHUS), a glomerular thrombotic microangiopathy. Mutations of factor H may also cause membranoproliferative glomerulonephritis (with dense deposits in the glomeruli) and age-related macular degeneration. These diseases are characterized by an inability of mutated factor H to control C3 activation at the cell surface, whereas regulation of C3 activation in plasma proceeds normally.

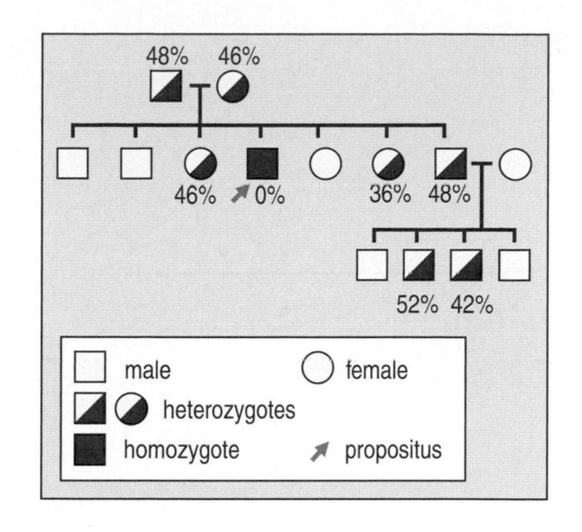

Fig. 32.6 Autosomal recessive inheritance of the factor I defect. The numbers indicate percentages of normal level of factor I in serum. The absence of a number indicates normal levels. The blue arrow indicates the patient, Morris Townsend.

Questions.

1 *Morris Townsend's clinical course has improved with age and he now has far fewer infections than he had as a child and adolescent. How do you explain this?*

2 *From the radiolabeled C3 experiment, we found that Morris Townsend catabolized the C3 very quickly but that his rate of synthesis of C3 was normal. What do you predict would happen if we repeated the experiment with radiolabeled factor B?*

3 *Morris Townsend was given a large dose of pure factor I intravenously. What changes would you predict to occur in his serum proteins?*

4 *What other genetic defect in the alternative pathway might lead to the same clinical and laboratory results as factor I deficiency?*

5 *Why did Morris's red blood cells agglutinate with antibody against C3?*

CASE 33 Deficiency of the C8 Complement Component

A loss of the lytic function of complement.

The complement system of plasma proteins is an effector mechanism of both innate and adaptive immunity that tags pathogens for destruction (see Cases 31 and 32). The assembly of the so-called terminal components of the complement system (C5 through C9; Fig. 33.1) on the surface of a bacterial cell or a human cell results in the formation of protein complexes that make pores in the cell membrane, leading to cell lysis and death.

Assembly of the terminal complex is initiated when the fifth component of complement, C5, binds to the cell surface and is cleaved by C5 convertase. This reaction releases C5a, a peptide with potent chemotactic activity, from the α, or heavy chain, of C5. The rest of the molecule, C5b, binds C6 to initiate the formation of the so-called membrane-attack complex of terminal complement components. The C5b6 complex binds one C7 molecule, resulting in the exposure of a hydrophobic site on C7 that enables the complex to sink partway into the lipid bilayer of a cell membrane. Complement component C8, unlike the single-chain C6 and C7, is composed of three chains, C8α, C8β, and C8γ. The C8γ chain binds to the C5a67 complex and enables the hydrophobic portion of C8 to embed itself in the cell membrane (Fig. 33.2). This last event induces the polymerization of 10–16 molecules of C9 to make a cylindrical structure, the membrane-attack complex, which forms a pore in the cell membrane. The diameter of the pore is approximately 100 Å, and through this channel sodium and water enter the cell, which swells until it bursts (lysis). The cell-surface protein CD59, which is found in most mammalian cell membranes, inhibits the action of C8 on C9, thereby preventing formation of the membrane-attack complex on the cells of the body but not on those of bacteria.

Curiously, humans with a genetic deficiency in C5, C6, C7, C8, or C9 have an increased susceptibility to systemically invasive infection with bacteria of the genus *Neisseria*. Two common pathogens, *N. meningitidis* and *N. gonorrhoeae*, belong to this genus; the former causes endemic and epidemic meningitis, and the latter the sexually transmitted disease gonorrhea.

Topics bearing on this case:

- Terminal complement components
- Classical pathway of complement activation

Fig. 33.1 The terminal complement components that assemble to form the membrane-attack complex.

The terminal complement components that form the membrane-attack complex		
Native protein	**Active component**	**Function**
C5	C5a	Small peptide mediator of inflammation (high activity)
	C5b	Initiates assembly of the membrane-attack system
C6	C6	Binds C5b; forms acceptor for C7
C7	C7	Binds C5b6; amphiphilic complex inserts in lipid bilayer
C8	C8	Binds C5b67; initiates C9 polymerization
C9	$C9_n$	Polymerizes to C5b678 to form a membrane-spanning channel, lysing cell

The case of Dolly Oblonsky: recurrent meningococcal infection leads to the discovery of a complement deficiency.

Female student, temperature, stiff neck, reddish-purple rash. Call physician urgently.

Dolly Oblonsky was doing well in her first year at university when she developed a cough and diarrhea. She felt very tired and achy and went to bed early in her dormitory room. The next morning she woke early with a severe headache. She felt sick and her neck seemed to be stiff. Her roommate took her to the university infirmary emergency room. She seemed ill and somewhat confused to the nurse on duty, who found that Dolly had a blood pressure of 70/40 (low), a pulse of 124 (fast), a respiratory rate of 24 (increased), and a temperature of 39.2°C (elevated). The nurse noticed a petechial rash (small areas of reddish-purple discoloration) on Dolly's chest and urgently summoned the physician on call, Dr Tolstoy. He found that Dolly had a red throat with moderate enlargement of the tonsils. No other physical symptoms were apparent except for the neck stiffness and the petechial rash on her palate, trunk, and extremities. Dr Tolstoy immediately placed an intravenous needle into a vein on the back of Dolly's left hand, obtained blood for blood counts and bacterial cultures, and started intravenous administration of the antibiotic Cephtriaxone because he suspected meningococcal meningitis. He then performed a lumbar puncture to obtain cerebrospinal fluid (CSF).

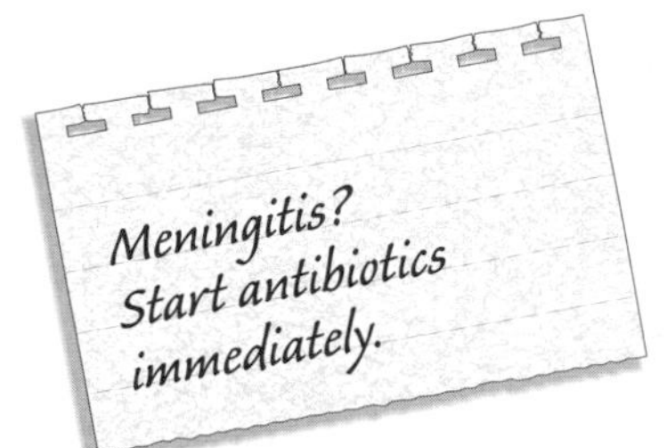
Meningitis?
Start antibiotics
immediately.

The blood cultures grew *Neisseria meningitidis* (meningococcus) serogroup C. The CSF proved sterile but contained 20 white blood cells μl^{-1} (an abnormally high number; CSF usually contains five or fewer white blood cells μl^{-1}). Dolly's hematocrit was 36.5% (slightly low) and her blood white-cell count was 8700 cells μl^{-1}, of which 90% were neutrophils and 10% were lymphocytes.

Dolly quickly improved on continued intravenous antibiotic therapy. The fever disappeared, she became alert, and her neck stiffness resolved over the next 72 hours. Blood taken after 24 hours of antibiotic treatment proved sterile on culture, and she was discharged from the infirmary.

Dolly told Dr Tolstoy that she had had meningococcal meningitis in her third year of high school. She was told at the time that she had positive CSF and blood cultures for *N. meningitidis* serogroup Y. Dr Tolstoy suspected that Dolly might have a

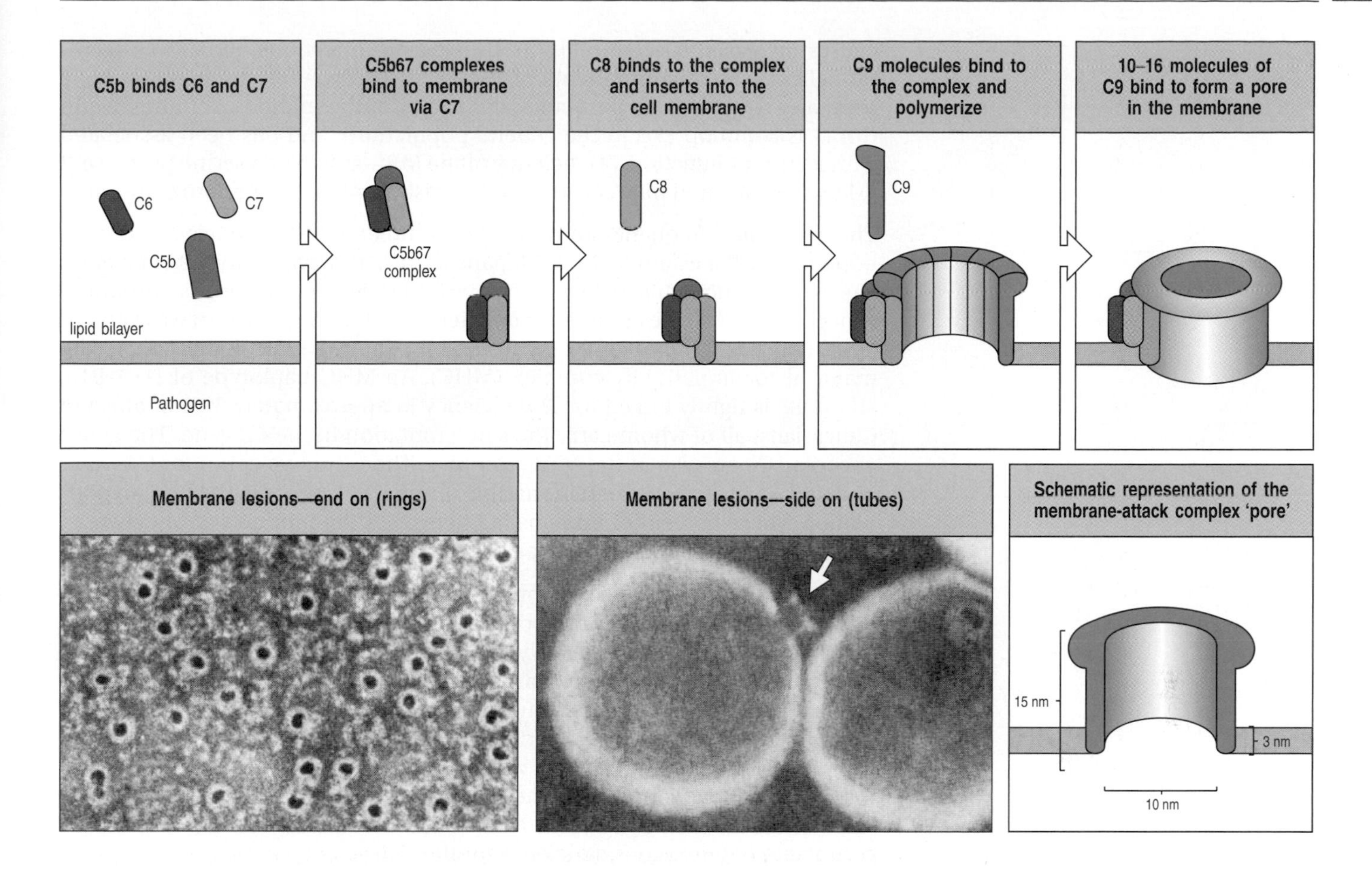

Fig. 33.2 Assembly of the membrane-attack complex generates a pore in the lipid bilayer membrane. The sequence of steps and their approximate appearance are shown here in schematic form. C5b triggers the assembly of a complex of one molecule each of C6, C7, and C8, in that order. C7 and C8 undergo conformational changes that expose hydrophobic domains that insert into the membrane. This complex causes moderate membrane damage in its own right, and also serves to induce the polymerization of C9, again with the exposure of a hydrophobic site. Up to 16 molecules of C9 are then added to the assembly to generate a channel 100 Å in diameter in the membrane. This channel disrupts the bacterial cell membrane, killing the bacterium. The electron micrographs show erythrocyte membranes with membrane-attack complexes in two orientations, end on and side on. Photographs courtesy of S. Bhakdi and J. Tranum-Jensen.

complement deficiency and sent a blood sample for a CH_{50} assay, which tests the ability of the blood to perform complement-mediated hemolysis. A week later he received the report: Dolly's CH_{50} was zero, which means that her serum contained no hemolytic complement activity.

Dolly was one of five sisters. Blood samples were obtained from all of them to assay their CH_{50}, which turned out also to be zero in her three younger sisters, aged 9, 14, and 17. The two youngest sisters were well; the 17-year-old sister had had meningococcal meningitis the year before. An older sister, aged 20, had a normal CH_{50}. The father was unavailable for testing, but the mother proved to have a half-normal CH_{50}. The sera of all the sisters and the mother were tested in a double-diffusion assay in agar to ascertain whether C5, C6, C7, C8, and C9 were present in their blood. Dolly's serum lacked C8, as did that of her three sisters with CH_{50} values of zero. Her mother had a half-normal level of C8. Her older sister had no evident deficiency of these complement components. All the family were given tetravalent meningococcal vaccine.

Complement component deficiencies.

Inherited deficiencies of all the components of the classical and alternative pathways have been found in humans. Deficiencies of C1q, C1r, C1s, C4, or C2 of the classical pathway result in immune-complex disease, which frequently proves fatal. Inherited defects in factor D, or properdin, of the alternative pathway, as well as defects in components of the membrane-attack complex, C5, C6, C7, C8, or C9, on which both pathways converge, result in increased susceptibility to neisserial infections, as illustrated by this case. C3 deficiency causes increased susceptibility to all pyogenic infections. Finally, congenital deficiencies of components of the lectin pathway of complement have been also described. In particular, deficiency of the mannose-binding lectin (MBL) is common (5% in the general population), and has been associated with recurrent infections. Deficiency of the MBL-associated serine protease-2 (MASP2) is rare and may cause increased risk of infections or autoimmunity.

There is a high frequency of particular complement deficiencies in certain populations. For example, 1 in 40 Japanese are heterozygous for C9 deficiency. One in 100 Caucasians are heterozygous for C2 deficiency, and, as one would expect, 1 in 10,000 are completely deficient in C2. This prediction was verified by testing blood donors in Manchester, UK. The gene encoding C2 is in the major histocompatibility complex (MHC). An MHC haplotype of HLA-B18, HLA-DR2 is tightly linked to C2 deficiency in approximately 90% of affected Caucasians, all of whom carry the same mutation in the *C2* gene. The genes encoding C4 are also in the MHC complex. There are two genes for C4—*C4A* and *C4B*—which are present in variable numbers of copies in the human population. Thirty percent of humans fail to express the products of one, two, or three of these genes, and thus may have a low C4 titer in their blood.

Deficiency of C8 in Caucasian populations is caused by a few well-known mutations in the gene encoding the β chain. C8 deficiency in African Americans in the USA, in contrast, is commonly due to mutations in the gene encoding the α chain. More than 90% of the mutations in C8β are due to C→T transitions in the *C8B* gene encoding C8β. Such a mutation was carried by Dolly and her sisters and was presumably inherited from her Caucasian father, who was not available for testing. The mother was of Filipino origin and was heterozygous for another type of mutation in the *C8B* gene not found in Caucasians—hence her half-normal levels of C8. Dolly and her three affected sisters thus have a compound heterozygous deficiency for the *C8B* gene.

Questions.

1 *Hemolytic complement levels were determined by measuring the CH_{50}. What is that, and how is the test performed?*

2 *Bacteria of the genus* Neisseria *are encapsulated in a thick carbohydrate capsule. Clinical evidence shows that the membrane-attack complex is vital in host defense against these bacteria. It is obvious that the hydrophobic components of the membrane-attack complex cannot penetrate the polysaccharides of the bacterial capsule. How, then, is complement involved in bacterial killing?*

3 *Patients who are completely deficient in C1q, C1r, C1s, or C4 sustain persistent and severe immune-complex disease such as glomerulonephritis which can ultimately be fatal. In contrast, patients who are C2-deficient usually sustain only minor forms of immune-complex disease. How do you explain this difference?*

CASE 34 Hereditary Periodic Fever Syndromes

Defects in intracellular receptors of innate immunity leading to an uncontrolled inflammatory response.

One of the first responses to infection is the generation of an inflammatory response by the innate immune system, which causes pain, redness, heat, and swelling (edema) in the infected tissues. Inflammation at the site of infection augments the killing of invading microorganisms by the front-line macrophages and causes changes in the walls of local blood vessels that enable additional effector molecules and cells, such as neutrophils, to enter the infected tissue from the blood (Fig. 34.1). Inflammation also induces local blood clotting, making a physical barrier to the spread of infection through the bloodstream. These changes are induced by a variety of inflammatory mediators that are released by macrophages and other cells of the innate immune system as a consequence of their recognition of pathogens.

Among these inflammatory mediators are pro-inflammatory cytokines that help to induce and maintain an inflammatory response. The main pro-inflammatory cytokines are interleukin-1β (IL-1β), tumor necrosis factor-α (TNF-α), and IL-6. In addition to inducing tissue inflammation, these cytokines also induce fever (Fig. 34.2). Fever is generally beneficial to host defense; most pathogens grow better at lower temperatures, and adaptive immune responses are more intense at elevated temperatures. IL-1β and IL-6 also enhance the production of innate immunity proteins called the acute-phase reactants, which are synthesized by the liver and released into the blood.

Because unregulated inflammation is harmful to the host, the production of pro-inflammatory cytokines is tightly controlled, so that they are only produced in significant amounts in response to infection. Cells involved in innate immunity, such as monocytes/macrophages, dendritic cells and neutrophils, detect the presence of pathogens by means of both cell-surface and intracellular receptors for microbial components. Intracellular bacterial infections of macrophages are detected by a family of intracellular proteins called the NOD-like receptors (NLRs). These include two subfamilies, distinguished by their domain composition and structure—the NODs themselves and the NLRPs. The NOD proteins detect bacterial cell-wall muramyl dipeptides and respond by triggering a signaling pathway that terminates in the activation of the transcription factor NFκB, which in turn switches on the genes for several of the pro-inflammatory cytokines.

The NLRP subfamily, with which this case is concerned, are found in monocytes/macrophages and neutrophils, where they act as sensors of cellular damage or stress, such as occurs on infection. Normally held inactive by accessory proteins in the cytoplasm, the NLRPs are thought to be activated by a variety of stimuli produced by stressed cells. For example, the toxins of bacteria such as *Staphylococcus aureus* and the intracellular pathogen *Listeria*

Topics bearing on this case:
Inflammatory responses
Acute-phase responses
NOD-like receptors in innate immunity
Cytokine processing
The autoinflammatory syndromes

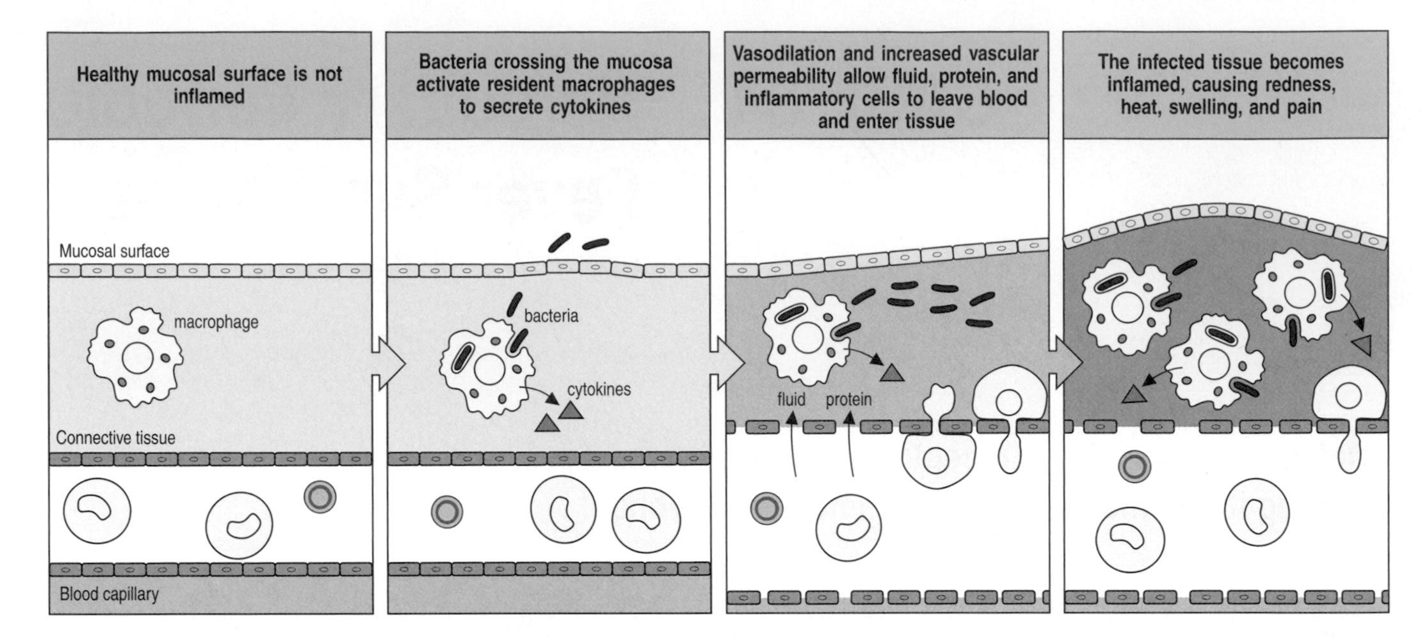

Fig. 34.1 Innate immune mechanisms establish a state of inflammation at sites of infection. Pathogens penetrating the skin or mucosal epithelium invade the underlying connective tissue. Here they activate resident effector cells such as macrophages to secrete chemokines, cytokines, and other pro-inflammatory mediators, which trigger the events leading to inflammation. The cells primarily attracted into the inflamed tissues are monocytes/macrophages and neutrophils, which contribute to enhancing and sustaining the state of inflammation.

monocytogenes cause an efflux of K^+ ions from affected cells. The resulting low levels of intracellular K^+ dissociate the inhibitory proteins from the NRLP. One of the best characterized of the NLRPs is NLRP3, also known as cryopyrin because it was initially recognized as being involved in inducing fever. In stressed cells, active NLRP3 forms part of a large protein complex called the inflammasome for its role in inducing inflammation (Fig. 34.3). In the inflammasome, NLRP3 is associated with the protease caspase 1, which is activated within the inflammasome and is responsible for processing IL-1β and another pro-inflammatory cytokine, IL-18, to give their active forms (Fig. 34.4). This explains the initial recognition of NLRP3 as a fever-inducer.

A group of clinical syndromes called the autoinflammatory syndromes are manifested as inflammation in the absence of any evidence of microbial infection. They are characterized by episodes of fever accompanied by systemic and localized inflammation, affecting abdominal structures, joints, and the skin in particular. They do not involve self-reactive T cells or self-reactive antibodies (autoantibodies) and so are not due to autoimmune reactions, another cause of unexplained inflammation. As the name 'autoinflammatory'

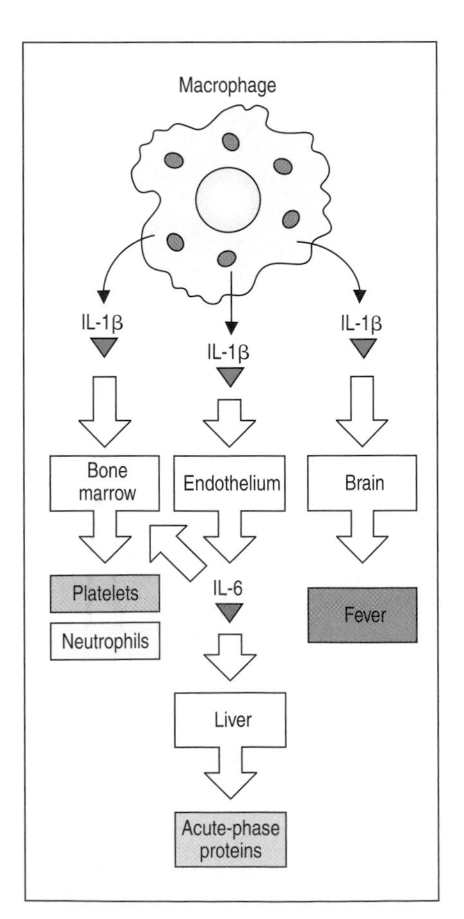

Fig. 34.2 Systemic effects of IL-1β. Active IL-1β is secreted by many cell types, including monocytes and macrophages. It enters the circulation and triggers IL-1 receptors on the hypothalamic vascular network, resulting in the synthesis of cyclooxygenase-2, which causes levels of prostaglandin E_2 in the brain to rise, thus activating the thermoregulatory center for fever production. In the periphery, IL-1β activates IL-1 receptors on endothelium, resulting in rashes and the production of IL-6. Circulating IL-6 stimulates liver cells to synthesize several acute-phase proteins, which cause an increased erythrocyte sedimentation rate. IL-1β also acts on the bone marrow to increase the mobilization of granulocyte progenitors and mature neutrophils, resulting in peripheral neutrophilia. It also indirectly increases IL-6-induced platelet production, resulting in thrombocytosis, and causes a decreased response to erythropoietin, which results in anemia.

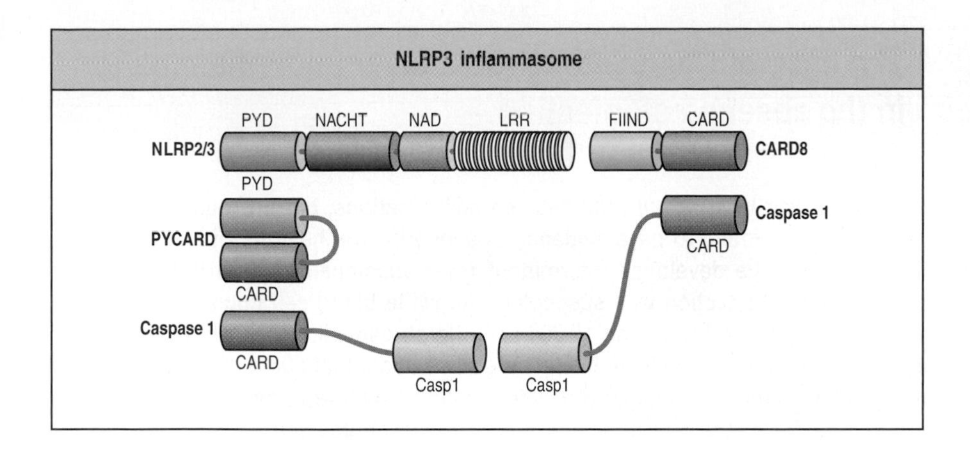

Fig. 34.3 Schematic structure of the NLRP3 inflammasome. NLRP3 has an amino-terminal pyrin domain (PYD), followed by a nucleotide-binding domain (NACHT), a NACHT-associated domain (NAD), and several leucine-rich repeats (LRRs). In the absence of stress or infection, the LRR domain is bound by inhibitory proteins that hold the NLRP inactive (not shown). Activated NRLP3 associates with the adaptor proteins PYCARD and CARD8, which have CARD domains that can recruit caspase 1 to the inflammasome.

suggests, these syndromes are driven by unchecked inflammatory responses. In normal individuals, the inflammatory cascade of reactions is tightly controlled, in great part by regulating the production of pro-inflammatory cytokines such as IL-1β. In contrast, individuals affected by autoinflammatory syndromes develop adaptive immune responses to stimuli such as infection or vaccination, but they are unable to control inflammation once it is under way, and inflammation may also flare up in response to minor stimuli that would not elicit an inflammatory response in unaffected individuals.

The autoinflammatory syndromes include relatively common conditions such as systemic-onset juvenile rheumatoid arthritis (see Case 35), but one category comprises rarer diseases known as the hereditary periodic fever syndromes, which are due to mutations in single genes. The genetic defects underlying these syndromes have been identified and have proved to be defects in pathways that normally control the inflammatory response by regulating the production of the pro-inflammatory cytokines. This case describes a hereditary periodic fever syndrome called neonatal-onset multisystem inflammatory disease/chronic infantile neurologic cutaneous and articular syndrome (NOMID/CINCA), which is due to mutations in the *NLRP3* gene.

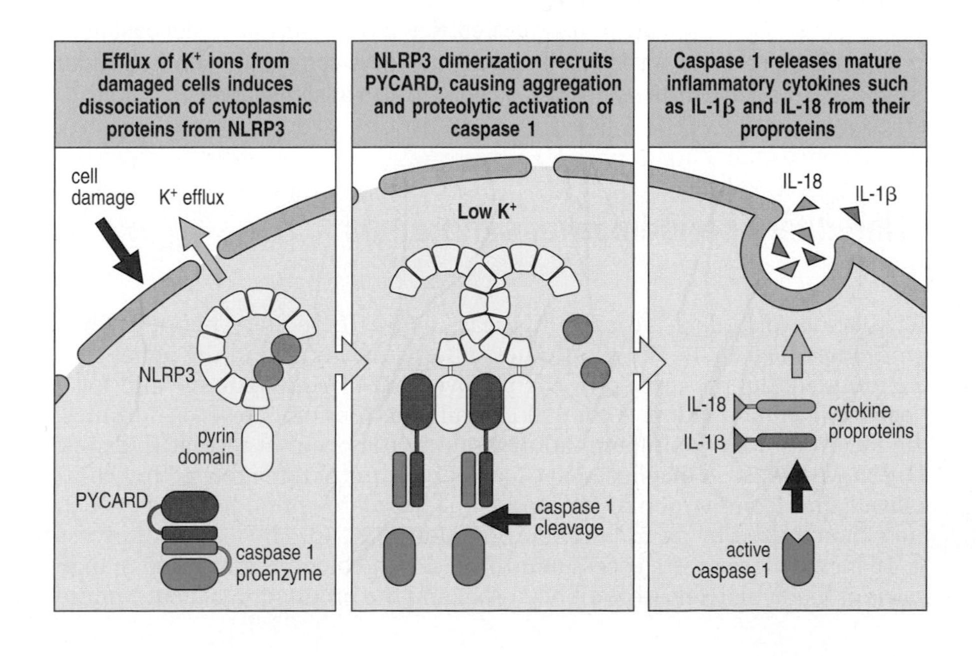

Fig. 34.4 NLRPs sense cellular damage and activate the processing of pro-inflammatory cytokines. Under normal physiologic conditions the LRR domain of NLRP3 associates with cytoplasmic proteins, which are presumed to prevent the dimerization of NLRP3 via the LRR region. When cells are injured or put under stress, the typical efflux of K^+ ions is thought to trigger the dissociation of these proteins from NLRP3, allowing its oligomerization (center panel). The pyrin domains of NLRP3 recruit complexes of the adaptor protein PYCARD, which associate with the caspase 1 proenzyme through their CARD (C) domains. Aggregation of the proenzymes causes them to undergo autoactivation via proteolytic cleavage to form active caspases. Active caspase 1 cleaves the proprotein forms of pro-inflammatory cytokines to release the mature cytokines, which can then be secreted.

The case of Chancellor Krook: fever and inflammation in the absence of infection.

Chancellor was born at full term with no complications, but on routine screening at birth he was found to have evidence of sensorineural hearing loss in both ears. At 1 week old, he developed intermittent fever accompanied by irritability, and a severe bacterial infection was suspected. His white blood cell count was elevated, at 21,000 μl^{-1} (normal less than 10,000 μl^{-1}). Cerebrospinal fluid (CSF) was sampled and also had an abnormally high white blood cell count, of 500 ml^{-1} (normal for age is less than 20 ml^{-1}). Although no bacteria could be cultured from samples of blood, urine, or CSF, Chancellor was kept in hospital and treated with antimicrobial drugs because bacterial or viral meningitis was suspected. Despite this treatment, the fevers continued and no cause could be found. His mother noticed a raised, red, itchy rash all over Chancellor's body, which his physicians diagnosed as urticaria (hives), attributing it to a reaction to the antibiotics he had been given. The rash persisted, however, and Chancellor continued to be intermittently feverish. He was eventually sent home without a diagnosis.

Infant with recurrent
fever and irritability.
No infection detectable.

The episodes of fever and irritability with no apparent source continued throughout Chancellor's first year, and his growth and development appeared unusually slow. Tests revealed progressive hearing loss, and he was given a hearing aid. In view of his continuing symptoms, Chancellor was referred to the rheumatology department at the Children's Hospital when he was 1 year old for testing for inflammatory disease.

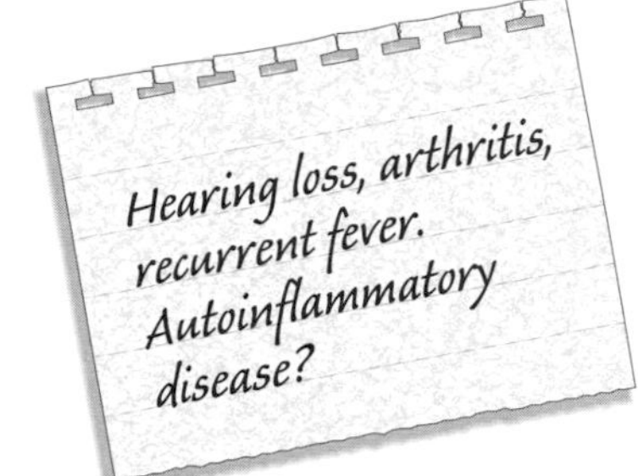
Hearing loss, arthritis,
recurrent fever.
Autoinflammatory
disease?

The examining physician noted Chancellor's prominent forehead and saddle-shaped nose (Fig. 34.5). His liver and spleen were enlarged, and examination of his joints revealed evidence of arthritis in the knees. A diffuse, red, urticaria-like rash covered his trunk, back, legs, and arms. Laboratory tests revealed an erythrocyte sedimentation rate of 35 mm h^{-1} (normal 0–20 mm h^{-1}). Sedimentation of red blood cells is hastened when the concentration of certain blood proteins is increased, among them the acute-phase reactants. One of these reactants, C-reactive protein, was measured and was found to be elevated, at 8 mg dl^{-1} (normal less than 0.50 mg dl^{-1}).

Mutation in NLRP3
gene.

On the basis of the constellation of symptoms shown by Chancellor—fever, rash, arthritis, hearing loss, and irritability—a diagnosis of NOMID/CINCA was made. Genetic testing revealed a mutation in the gene *NLRP3*, which confirmed the diagnosis. Chancellor was treated with anakinra (Kineret), a recombinant human IL-1 receptor antagonist, and his symptoms disappeared. Now 2 years old, Chancellor continues to require treatment with anakinra. His growth and development are now normal for his age, and his blood markers of inflammation are also at normal levels.

Hereditary periodic fever syndromes.

Ten distinct disorders are currently classified as hereditary periodic fever syndromes (Fig. 34.6). The syndromes differ clinically and in the genes that are mutated, but are unified by the involvement of both systemic and focal (localized) inflammation. Although classified as 'periodic' fever syndromes, the recurrent bouts of inflammation often do not occur at regular intervals. The NOMID/CINCA diagnosed in Chancellor is one of three related but clinically distinct 'cryopyrinopathies' among the hereditary periodic fevers, which are characterized by an excessive production of IL-1β. The effects of excess IL-1β include fever, rash, excess neutrophils in the blood (neutrophilia), high levels of blood platelets (thrombocytosis), and the production of acute-phase

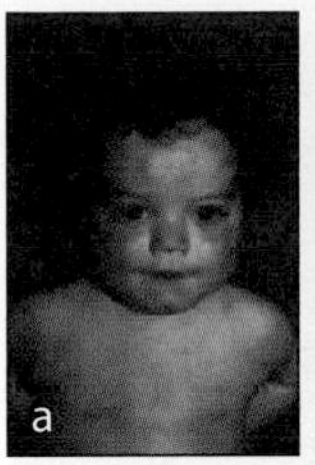

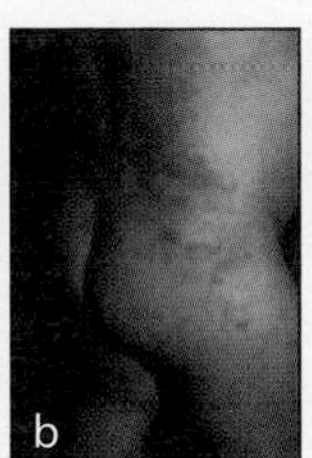

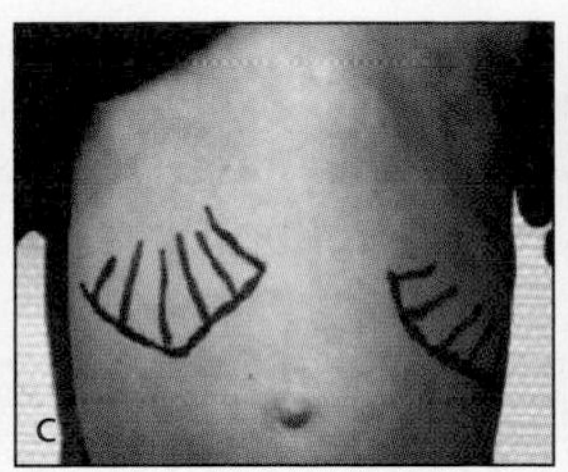

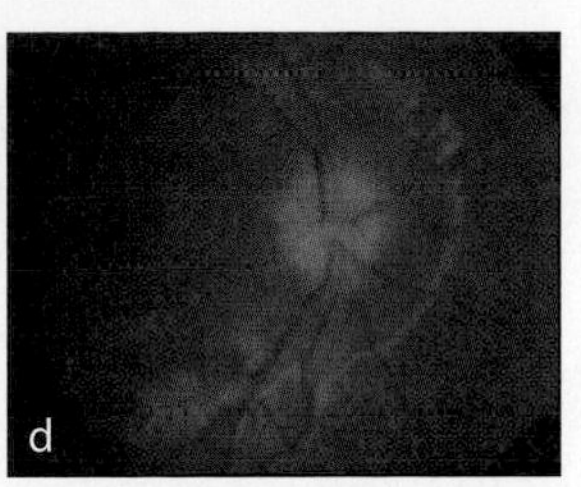

Fig. 34.5 Clinical features of NOMID/CINCA. Panel a, dysmorphic facial features: frontal bossing, saddle nose, and long philtrum of upper lip; panel b, urticarial rash; panel c, enlarged liver and spleen with distended abdomen (markings show the lower margins of the spleen (right) and the liver (left)); panel d, fundoscopic image of the optic disk showing papilledema.

reactants, all of which were seen in Chancellor's case. The other cryopyrinopathies comprise Muckle–Wells syndrome and familial cold autoinflammatory syndrome (see Fig. 34.6), and all three disorders result from mutations in the cryopyrin-encoding gene *NLPR3*. The exact mechanism by which the mutations found in patients with cryopyrinopathies promote inflammation is still under investigation. The mutant NLRP3 proteins that have been found are constitutively active and are hyperresponsive to microbial antigens or to physical stimuli such as cold. In either case this results in exaggerated secretion of IL-1β and IL-18.

NOMID/CINCA patients typically have intermittent fevers associated with persistent urticaria, arthritis-like symptoms and central nervous system involvement that ranges through papilledema (swelling of the optic disk, where the sensory fibers from the retina exit from the eye), increased numbers of inflammatory cells in the CSF, causing a chronic meningitis, to sensorineural hearing loss (deafness due to a defect in nerve conduction or perception by the brain of the auditory impulse).

Of the other hereditary periodic fevers, familial Mediterranean fever (FMF) is the most common and is associated with mutations in the gene *MEFV*, which encodes a protein called pyrin, named after the fever characteristic of the disease. TNF receptor-associated periodic syndrome (TRAPS, originally known as familial Hibernian fever) presents with longer episodes of fever, abdominal pain, and a rash. This disorder is due to mutations in the gene encoding the TNF receptor 55 kDa subunit (*TNFRSF1A*) that lead to excess TNF-α in the circulation. Hyperimmunoglobulinemia D with periodic fever syndrome (HIDS) was originally differentiated from FMF by the detection of elevated levels of IgD. Attacks of HIDS tend to last 7 days and recur every 4–8 weeks. Mutations in the gene *PSTPIP1*, which encodes CD2-binding protein 1 (CD2BP1), a pyrin-interacting protein, are associated with the syndrome of pyogenic arthritis, pyoderma gangrenosum, and acne (PAPA). These mutations accentuate the interaction between pyrin and CD2BP1, and it has been suggested that the sequestration of pyrin results in excess secretion of the pro-inflammatory cytokine IL-1β.

One of the most serious long-term complications of the periodic fever syndromes is the development of systemic amyloidosis, which occurs when misfolded fragments of serum amyloid, an acute-phase reactant, are deposited in tissues. The kidneys, gastrointestinal tract, adrenal glands, spleen, testes, and lung are most often affected, but the liver, heart, and thyroid gland can also be involved. Amyloidosis occurs most commonly in patients with FMF, but it can also affect patients with TRAPS, Muckle–Wells syndrome, or NOMID/CINCA. Treatment with the drug colchicine, and to a lesser extent with TNF-α inhibitors and the IL-1 receptor antagonist anakinra, has been shown to prevent the amyloidosis from developing.

Other autoinflammatory syndromes include Blau syndrome, which presents with uveitis, synovitis, skin rash, and cranial neuropathies, and is frequently associated with Crohn's disease. It is due to mutations in the gene *NOD2* that may interfere with the binding of NOD2 protein to bacterial cell-wall components and the subsequent signaling through NFκB. Deficiency of the IL-1 receptor antagonist IL-1RA leads to unopposed action of IL-1β and presents with neonatal-onset osteomyelitis, periostitis, and pustulosis. Finally, mutations of the *LPIN2* gene also lead to multifocal ostemyelitis, associated with anemia and skin inflammatory manifestations.

Disease	Clinical features	Functional defects	Inheritance	Defective gene (protein)
Familial Mediterranean fever (FMF)	Recurrent fever, serositis (inflammation of the pleural and/or peritoneal cavity) responsive to colchicine, amyloidosis. Predisposes to vasculitis and inflammatory bowel disease	Decreased production of pyrin leading to increased IL-1β processing and secretion; decreased macrophage apoptosis	Autosomal recessive	*MEFV* (pyrin)
TNF-receptor associated periodic syndrome (TRAPS)	Recurrent fever, serositis, rash, and ocular or joint inflammation	Reduced levels of soluble TNF receptor leading to excess TNF-α in the blood	Autosomal dominant	*TNFRSF1A* (TNF 55 kDa receptor/TNFR-I)
Hyper-IgD syndrome (HIDS)	Periodic fever and leukocytosis with high levels of IgD in serum	Mevalonate kinase deficiency affecting cholesterol synthesis; pathogenesis of disease unclear	Autosomal recessive	*MVK* (mevalonate kinase)
Pyogenic arthritis, pyoderma gangrenosum and acne (PAPA)	Periodic fever, myalgia (muscle pain), rash, acute-phase response	Sequestration of pyrin by CD2BP1 leading to increased IL-1β processing and secretion	Autosomal dominant	*PSTPIP* (CD2-binding protein 1/CD2BP1)
Muckle–Wells syndrome (MWS)*	Urticaria, hearing loss	Constitutive activation of cryopyrin leading to increased caspase 1 activation and IL-1β processing and secretion	Autosomal dominant	*NLRP3* (cryopyrin)
Familial cold autoinflammatory syndrome (FCAS)*	Non-pruritic urticaria, arthritis, chills, fever, and leukocytosis after exposure to cold			
Neonatal onset multisystem inflammatory disease/chronic infantile neurologic cutaneous and articular syndrome (NOMID/CINCA)*	Neonatal-onset rash, chronic meningitis, hearing loss, arthropathy with fever and inflammation responsive to the IL-1R antagonist anakinra			
Blau syndrome	Uveitis, synovitis, rash, cranial neuropathy. It is frequently associated with Crohn's disease	Mutations of *NOD*, possibly disrupting interaction with bacterial cell-wall muramyl peptides and NFκB signaling	Autosomal dominant	*NOD2* (NOD2)
Deficiency of IL-1 receptor antagonist (DIRA)	Neonatal-onset sterile multifocal osteomyelitis, periostitis, pustulosis	Mutations of IL-1R antagonist, leading to unopposed action of IL-1β	Autosomal recessive	*IL1RN* (IL-1 receptor antagonist)
Chronic recurrent multifocal osteomyelitis and congenital dyserythropoietic anemia	Multifocal chronic osteomyelitis, anemia, cutaneous inflammatory disorders	Undefined	Autosomal recessive	*LPIN2*

Fig. 34.6 The spectrum of hereditary autoinflammatory syndromes. The cells primarily affected in all these diseases are mainly granulocytes, especially neutrophils, and monocytes and macrophages. Chondrocytes are also affected in NOMID/CINCA. *The cryopyrinopathies. All three syndromes are associated with similar mutations in *NLRP3*, the gene for cryopyrin; the disease phenotype in any individual seems to depend on the modifying effects of other genes and environmental factors.

Questions.

1 *Why has the drug anakinra proved a successful treatment for some of the autoinflammatory syndromes?*

2 *What is the proposed mechanism of action for colchicine in the treatment of the autoinflammatory syndromes?*

3 *What cytokine, other than IL-1 and IL-18, would you expect to be elevated in the setting of caspase 1 activation?*

CASE 35 Systemic-Onset Juvenile Idiopathic Arthritis

Cytokine dysregulation leading to inflammatory disease.

Cytokines are small proteins that act over short distances to affect adjacent cells or that can be carried in the blood stream over large distances, affecting the entire body. The pro-inflammatory cytokines, such as interleukin 1β (IL-1β), tumor necrosis factor-α (TNF-α), and IL-6 elicit powerful inflammatory effects that lead to aberrant thermoregulation, metabolism, hematopoiesis, and tissue inflammation (Fig. 35.1).

IL-1β is a prototypic pro-inflammatory cytokine. Like many other cytokines, it is synthesized as an inactive precursor molecule, and cleavage of pro-IL-1β by caspase 1 is required for its activation and secretion (see Fig. 34.4). Its production and release are regulated at multiple levels, including this cleavage step. The activation of caspase 1 is dependent on the inflammasome, a multiprotein complex activated by diverse substances such as microbial constituents, vaccine adjuvants, pollutants, amyloid, and uric acid crystals. Several monogenic autoinflammatory disorders have been identified in which genes encoding inflammasome proteins are mutated. These disorders are characterized by recurrent fevers, elevation of inflammatory markers, and skin, joint, and organ inflammation (see, for example, Case 34).

Another potent inflammatory cytokine is IL-6. IL-6 is produced primarily by T cells and is important for T-cell differentiation and proliferation as well as the differentiation of B cells into plasma cells. A subset of helper T cells, T_H17 cells, produces IL-17 and TNF-α, as well as IL-6. T_H17 cells have been shown to have an important role in inflammation and in the induction of autoimmune diseases. Along with lymphocyte-specific effects, excess IL-6 induces symptoms of fever, anorexia, and fatigue in patients, as well as elevation of acute-phase reactants (CRP, serum amyloid A, and fibrinogen).

This case was prepared by Raif Geha, MD, in collaboration with Erin Janssen, MD.

Topics bearing on this case:
Cytokines and the inflammatory response
Cytokine processing
Acute-phase responses
T_H17 cells
Anti-cytokine therapeutic agents
The autoinflammatory syndromes

Fig. 35.1 Pro-inflammatory cytokines. IL-1β, IL-6, and TNF-α are powerful pro-inflammatory cytokines. They elicit responses not only from immune cells but also from various target organ systems. These cytokines promote a catabolic state with breakdown of muscle and fat, along with fever and an increased production of acute-phase proteins.

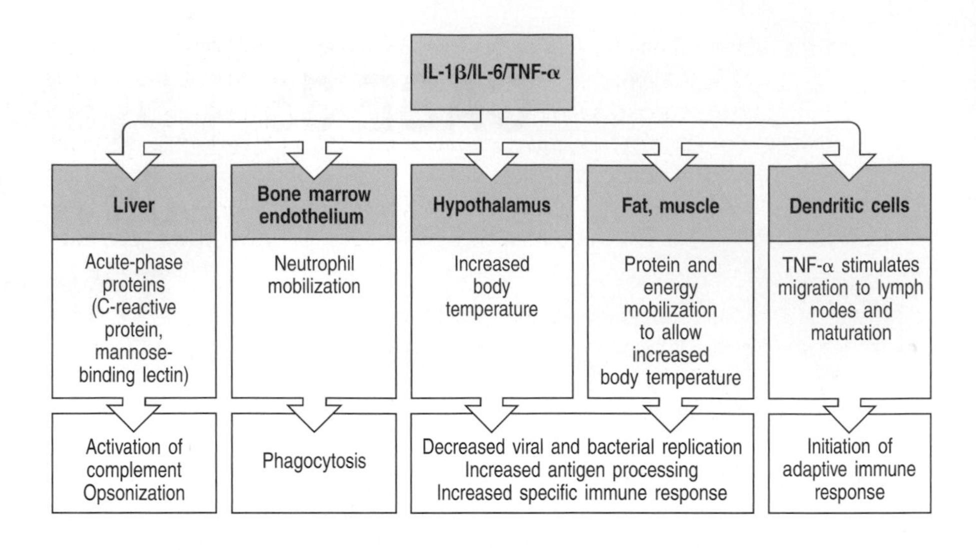

The case of Catherine Earnshaw: the fever that wouldn't stop.

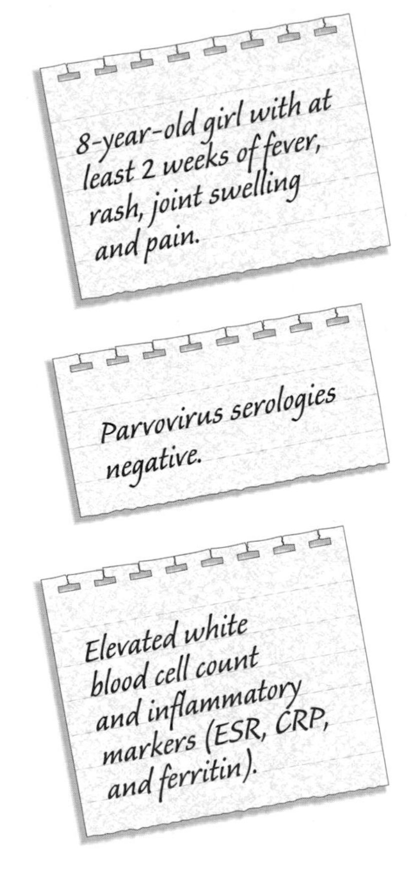

8-year-old girl with at least 2 weeks of fever, rash, joint swelling and pain.
Parvovirus serologies negative.
Elevated white blood cell count and inflammatory markers (ESR, CRP, and ferritin).

Catherine had enjoyed normal good health until she was 8 years old, when she developed recurrent high fevers and fatigue. She also found it difficult to get out of bed in the morning—she ached all over. Her parents noticed that a pink rash on her shoulders and arms appeared during the fevers. After more than a week of these symptoms, Catherine's parents were concerned that this was more than a typical childhood illness and took her to the pediatrician. Although Catherine had no fever or rash at the time, Dr Brontë noticed subtle swelling of several joints, along with pain on movement; she initially suspected a parvovirus infection. Parvoviral serologies were sent, and ibuprofen and acetaminophen were recommended for symptomatic relief. A few days later, Catherine's parvoviral serologies returned negative.

Catherine went back to the pediatrician after another week of twice-daily fevers, joint pain, and rash. Laboratory studies were sent at that visit. Catherine's white blood cell count was 12,500 μl^{-1} (high) and her hemoglobin was 10 g dl^{-1} (low). Erythrocyte sedimentation rate (ESR) was also elevated, at 82 mm h^{-1}. She was sent to the local children's hospital for further evaluation.

In the hospital, additional laboratory studies were carried out to look for infection, all of which were negative. The hematologists took a peripheral blood smear, which also appeared normal apart from a mild anemia and increased neutrophil count. Catherine continued to have intermittent high fevers (Fig. 35.2), and the hospital staff noticed a

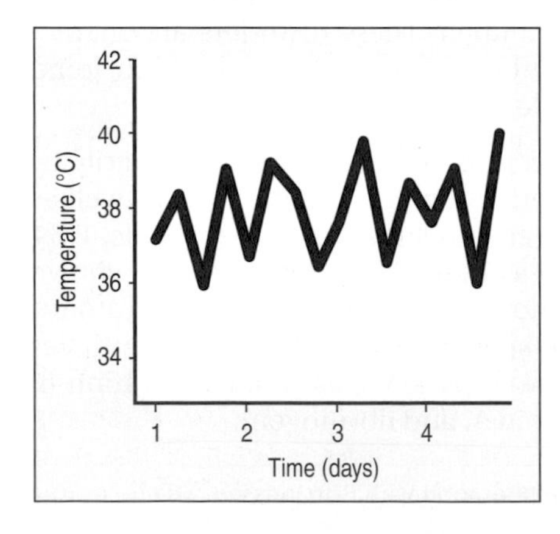

Fig. 35.2 A typical fever curve from a patient with systemic-onset juvenile idiopathic arthritis (sJIA). Patients with sJIA usually have spiking fevers once or twice each day. In between fevers, their body temperature is normal or even low. Patients often seem and feel quite well when they are not feverish.

light pink rash during these times (Fig. 35.3). Catherine continued to have difficulty getting out of bed in the morning, but was better able to move later in the day. The rheumatology service saw Catherine and noticed swelling and a decreased range of motion in her ankles, knees, and wrists. Her white blood cell count and ESR continued to be high. Levels of C-reactive protein (CRP) were also very high, at 28 mg dl^{-1}, and ferritin was 1949 ng ml^{-1} (high). Transaminases, electrolytes, creatinine, and immunoglobulins were normal. Tests for anti-nuclear antibodies (ANA) and rheumatoid factor were negative.

Catherine's symptoms, physical examination, and laboratory results were consistent with systemic-onset juvenile idiopathic arthritis (sJIA). Catherine was started on indomethacin (1 mg per kg body weight) twice daily to help control her joint pain and fevers. She was also started on 30 mg prednisone twice a day to treat her inflammation. Catherine improved significantly after 2 days on steroids and was discharged. She was followed in the rheumatology clinic over the next few months. Catherine initially did well, but when her steroid dose was reduced to 10 mg a day, the fevers, rash, and joint pain returned. The prednisone dose was increased again, and she was started on methotrexate at 20 mg weekly. After several weeks on methotrexate, the steroids were decreased again without any return in symptoms. She was then continued on methotrexate alone.

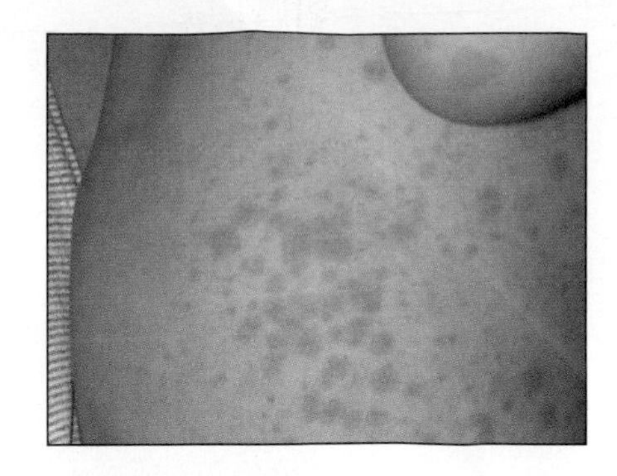

Fig. 35.3 The sJIA rash. The classic rash associated with sJIA is a flat, salmon-pink evanescent rash, which tends to appear at the same time as the fever. Photograph courtesy of E. Janssen.

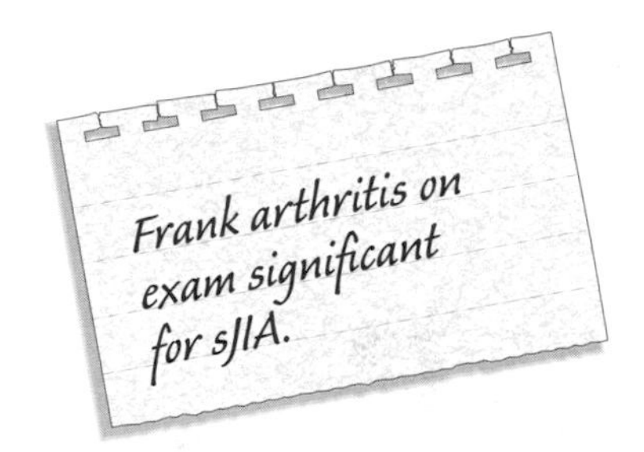
Frank arthritis on
exam significant
for sJIA.

Systemic-onset juvenile idiopathic arthritis.

Systemic-onset juvenile idiopathic arthritis (sJIA) is an autoimmune condition consisting of arthritis along with prominent systemic features such as high spiking fevers, fatigue, rash, and enlargement of lymph nodes and spleen. It was first described in the 1890s by George Frederick Still, a British pediatrician, and is often referred to as Still's disease. The annual incidence of chronic childhood arthritis (JIA) ranges from 1 to 20 per 100,000, and approximately 10% of children with JIA initially show these systemic symptoms.

sJIA is often difficult to diagnose. Laboratory tests give nonspecific results, and many of the manifestations, especially rash and fever, are commonly seen with childhood infections. The arthritis may be delayed in presentation, making correct diagnosis even more difficult. In addition to infection, malignancy also enters into the differential diagnosis, because in some cancers white blood cell counts are elevated and other cell types are decreased. Criteria have been established to help distinguish children with infections and other transient disorders, such as serum sickness, from those with true sJIA (Fig. 35.4). These criteria are quite stringent and can exclude children who only develop arthritis later in their presentation. They also exclude patients with a family history of psoriatic arthritis or other types of spondyloarthropathy, a group of arthritic diseases that tend to affect the spine and large joints of the legs often in conjunction with inflammation of tendon and ligament insertion sites, as well as patients positive for rheumatoid factor and/or HLA-B27. Clinical judgment must be used in evaluating children with these exclusion factors for possible sJIA.

Diagnostic criteria for systemic-onset juvenile idiopathic arthritis
Arthritis + daily fever for 2 weeks or more plus at least one of the following:
Evanescent red rash Generalized lymph-node enlargement Enlargement of the liver or spleen Serositis
Exclusions
Psoriasis in the patient or a first-degree relative Arthritis in a HLA B27-positive male older than 6 years A spondyloarthropathy or acute anterior uveitis in a first-degree relative The presence of IgM rheumatoid factor on two or more occasions at least 3 months apart

Fig. 35.4 Diagnostic criteria for sJIA from the International League of Associations for Rheumatology. sJIA is a clinical diagnosis and a diagnosis of exclusion. These criteria were published in 2001 to aid in diagnosing sJIA and in distinguishing it from other childhood febrile diseases.

In contrast to rheumatoid arthritis (Case 36) and other forms of JIA, there is no association with autoantibodies in sJIA. In addition, certain MHC class II associations and autoreactive T cells seen in other forms of arthritis are not present in sJIA. Overall, sJIA seems to have more in common with the inherited autoinflammatory conditions (see Case 34) than with other types of arthritis. In the active state of the disease, persistent activation of phagocytes and upregulation of inflammatory cytokines are seen. Indeed, polymorphisms in the promoters and coding sequences for the genes encoding TNF-α, IL-6, and macrophage migration inhibitory factor (MIF) have been associated with sJIA. The inciting cause of sJIA remains unknown.

Initial treatment of sJIA usually involves nonsteroidal anti-inflammatory medications along with corticosteroids. Long-term outcomes are highly variable. A subset of patients will improve after initial treatment and do not have further manifestations of disease, whereas up to one-third of patients have persistent disease. Long-term treatment often relies on traditional arthritis medications such as methotrexate or prolonged courses of corticosteroids.

Newer therapies for sJIA have exploited its relationship with inflammatory cytokines. TNF-α levels are elevated in sJIA, as they are in all forms of JIA, and levels of soluble TNF receptors are associated with sJIA disease severity. Inhibitors of TNF-α (see Case 36) have been used therapeutically in sJIA since the 1990s. Unfortunately, the response rate to these medications is significantly lower than in other types of JIA.

More recently, therapies relying on the inhibition of IL-1 and IL-6 have undergone clinical trials. Increased levels of IL-6 have been observed in the serum and joint fluid of patients with sJIA, and these levels correlate with disease activity. Spikes in IL-6 levels also parallel the fever curve seen in sJIA. Tocilizumab is a humanized monoclonal antibody directed against the IL-6 receptor that has recently been approved by the US Food and Drug Administration for the treatment of rheumatoid arthritis. Tocilizumab has shown promising results in sJIA refractory to TNF inhibitors and more conventional arthritis medications.

Aberrant IL-1 release has also been shown to have a role in many familial autoinflammatory disorders (see Case 34). Although levels of soluble IL-1 are probably not elevated in sJIA, this cytokine binds to various large serum proteins, and its total amount may be increased. Serum from patients with sJIA has been shown to induce IL-1 secretion as well as the transcription of various innate immunity genes from peripheral blood mononuclear cells from healthy individuals. According to individual reports as well as controlled trials, IL-1 receptor antagonists, such as anakinra, have been especially effective in controlling the systemic manifestations of sJIA.

Questions.

1 *Why is sJIA considered to be primarily an autoinflammatory syndrome, rather than simply an arthritis?*

2 *What might be some possible side effects of IL-1 or IL-6 inhibition?*

3 *What other conditions involve IL-1 dysregulation?*

4 *Why is Catherine's rash only present in association with fever?*

CASE 36 Rheumatoid Arthritis

A common, severe inflammatory disease of unknown cause.

The adaptive immune system can incite inflammation and cause extensive tissue damage and destruction. The utility of this response to the host is not always apparent, and the result may even be detrimental. Rheumatic disease—inflammation and damage to joints—is caused by a subset of autoimmune diseases (Fig. 36.1 and see Case 35). Of these, rheumatoid arthritis is one of the most common inflammatory diseases. This chronic, debilitating disease is chiefly characterized by inflammation of the synovium (the thin lining of a joint). As the disease progresses, the inflamed synovium causes invasion and damage of the cartilage, followed by erosion of the bone, and this ultimately leads to joint destruction (Fig. 36.2). Without proper treatment, patients with rheumatoid arthritis suffer chronic pain, loss of function, and disability.

Rheumatoid arthritis was originally considered to be a B-cell-driven autoimmune disease because of its association with IgM autoantibodies called rheumatoid factor, which are directed against the Fc portion of IgG. However, rheumatoid factor occurs in some healthy individuals and is absent from a subset of patients with rheumatoid arthritis, indicating that is not the primary or the sole mediator of disease pathology. The discovery that rheumatoid arthritis has an association with particular major histocompatibility complex (MHC) alleles suggested that T cells were also involved in the pathogenesis of this disease. The current thinking is that autoreactive CD4 T cells become activated by self antigen presented by dendritic cells or nonspecifically activated by pro-inflammatory cytokines produced by macrophages. Once activated, the autoreactive effector T cells recognize antigens within the joint. These activated T cells maintain and exacerbate the inflammatory reaction. Activated T cells may also provide help to potentially autoreactive B cells, leading to their differentiation into plasma cells producing autoantibodies.

Rheumatic diseases caused by autoimmunity
Systemic lupus erythematosus (SLE)
Rheumatoid arthritis
Juvenile arthritis
Sjögren's syndrome
Scleroderma (progressive systemic sclerosis)
Polymyositis–dermatomyositis
Behçet's disease
Ankylosing spondylitis
Reiter's syndrome
Psoriatic arthritis

Fig. 36.1 Rheumatic diseases are autoimmune in nature.

Topics bearing on this case:
Inflammatory responses
Cytokines and chemokines
Genetic associations with autoimmune disease
Anti-cytokine therapeutic agents

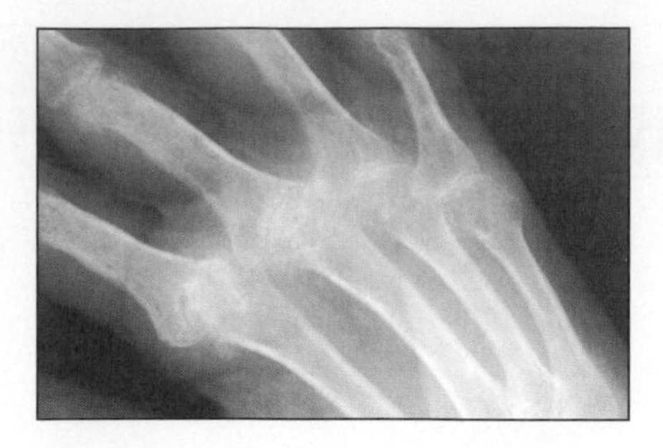

Fig. 36.2 X-ray of the right hand of a patient with rheumatoid arthritis. It shows extensive destruction and dislocation of the metacarpophalangeal joints.

The case of Ariadne Underwood: swollen joints that are stiff and painful in the morning portend a common diagnosis.

Ariadne Underwood teaches a class of sixth-graders. When she was 30 years old, Ariadne developed stiffness in her shoulders, hands, knees, and hips. The stiffness lasted for about an hour every morning. After her wrists and fingers began to swell, she saw her primary care physician. He suspected rheumatoid arthritis and advised her to take four aspirin tablets a day. He also obtained a blood sample, which was positive for rheumatoid factor.

As her joints grew stiffer and more painful, she was referred to a rheumatologist, Dr Shaw. He ascertained that she had no family history of rheumatoid disease. On physical examination, he felt spongy swelling indicating thickening of the synovium in both elbows, and decreased flexion and extension of her wrists. Her blood pressure, pulse, and respiratory rate were all normal, and she had no fever. Dr Shaw prescribed the nonsteroidal anti-inflammatory drug ibuprofen; she was instructed to take 600 mg four times a day. When Ariadne returned to see Dr Shaw 3 months later, she felt much better but her synovia had thickened further, so Dr Shaw added the antimalarial agent hydroxychloroquine sulfate (Plaquenil) to her daily medications.

On a return visit to Dr Shaw several years later, Ariadne had blood tests that revealed her rheumatoid factor titer to be 1:2560 (normal 0). Her hemoglobin was 11.4 g dl^{-1}, and her hematocrit was 33.4% (she was slightly anemic). Her white blood cell count was 4720 μl^{-1} (slightly low). Her sedimentation rate was 87 mm h^{-1} (very high). A urinalysis was normal. X-rays of her hands and feet revealed erosions in several joints. Ariadne was also having increased morning stiffness, and Dr Shaw decided to start her on methotrexate (25 mg) weekly with folic acid supplementation. She had only marginal improvement in her symptoms, and by the time she was 55 years old her feet were so deformed by the arthritis that she required surgery to enable her to walk normally.

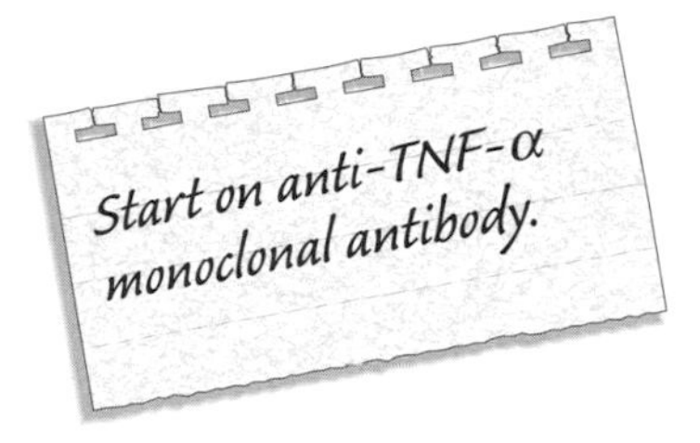
Start on anti-TNF-α
monoclonal antibody.

Because of her worsening symptoms, Dr Shaw decided start Ariadne on monthly intravenous infusions of a monoclonal antibody against tumor necrosis factor-α (TNF-α) called infliximab (Remicade). On this treatment, she felt remarkably better, her joint swelling resolved, and she had no further morning stiffness.

Rheumatoid arthritis.

There is no single diagnostic test for rheumatoid arthritis. As discussed above, rheumatoid factor is positive in many patients, although this is not a particularly sensitive or specific diagnostic character. Antibodies against citrullinated proteins (CCP) are also commonly found in patients with rheumatoid arthritis. Although these antibodies are highly specific for rheumatoid arthritis, a substantial fraction of patients lack anti-CCP. In lieu of a rigorous diagnostic test for rheumatoid arthritis, the American Rheumatism Association has set criteria for establishing a diagnosis (Fig. 36.3). Ariadne Underwood met all the criteria, with the exception of the presence of rheumatoid nodules.

There is a strong association between the MHC alleles HLA-DRB1*0404 and DRB1*0401 and rheumatoid arthritis in Caucasians. Genetic studies and, more recently, genome-wide association studies have found associations between rheumatoid arthritis and alleles of other genes (for example *PTPN22, STAT4, TRAF1, CTLA4, IL2, IL21,* and *CD40*), many of which encode proteins important for immune system function. Autoantigens such as type II collagen, glycoprotein 39, IgG, and citrullinated proteins have been proposed as

Criteria for the diagnosis of rheumatoid arthritis

- Morning stiffness lasting at least 1 hour before maximal improvement
- Arthritis of three or more joints simultaneously with swelling and/or fluid in the joints
- Arthritis in hand joints with swelling in the wrists, or metacarpophalangeal joints or proximal interphalangeal joints
- Symmetrical arthritis of the same joint areas
- Rheumatoid nodules
- Serum rheumatoid factor
- Typical radiographic changes

Fig. 36.3 The American Rheumatism Association criteria for the diagnosis of rheumatoid arthritis.

potential antigens, and injection of type II collagen has been shown to elicit arthritis in mice. The pathogenic role of these autoantigens in humans, however, remains to be ascertained.

Animal models have shed some light on the pathogenesis of rheumatoid arthritis. Mice bearing a transgenic T-cell receptor that recognizes a glucose-6-phosphate isomerase peptide on the K/BxN background spontaneously develop arthritis and produce antibodies against glucose-6-phosphate isomerase. The arthritis can be transferred to another animal by this antibody as long as the recipient has an intact alternative complement pathway. The arthritis also cannot be induced in mice lacking mast cells. Thus, in this model, a specific adaptive immune response to a widely distributed autoantigen amplifies the innate immune response to extreme hyperreactivity, and a disease that resembles human rheumatoid arthritis ensues.

In rheumatoid arthritis, inflammation of the synovial membrane, initiated by some unknown trigger, attracts autoreactive lymphocytes and macrophages to the inflamed tissue (Fig. 36.4). In the rheumatoid joints, the synovial membrane is thickened as a result of hyperplasia, and there is increased growth of blood vessels. The synovium is infiltrated by inflammatory cells, prominent among which are activated CD4 T cells. These T cells stimulate the activation of macrophages and synovial fibroblasts to produce pro-inflammatory cytokines such as TNF-α, interleukin-1 (IL-1), IL-6, IL-17, and interferon-γ (IFN-γ), the chemokines CXCL8 (IL-8) and CCL2 (MCP-1), and matrix metalloproteinases (MMPs), all of which are involved in the tissue destruction. Osteoclasts and chondrocytes also become activated and secrete enzymes that destroy cartilage and bone.

Several therapeutic agents used for rheumatoid arthritis are aimed at decreasing the amounts of pro-inflammatory cytokines. TNF-α is one of the dominant cytokines found in rheumatoid joints and is likely to be a prime agent in generating a proliferative synovium. Inhibition of TNF-α has proved beneficial in

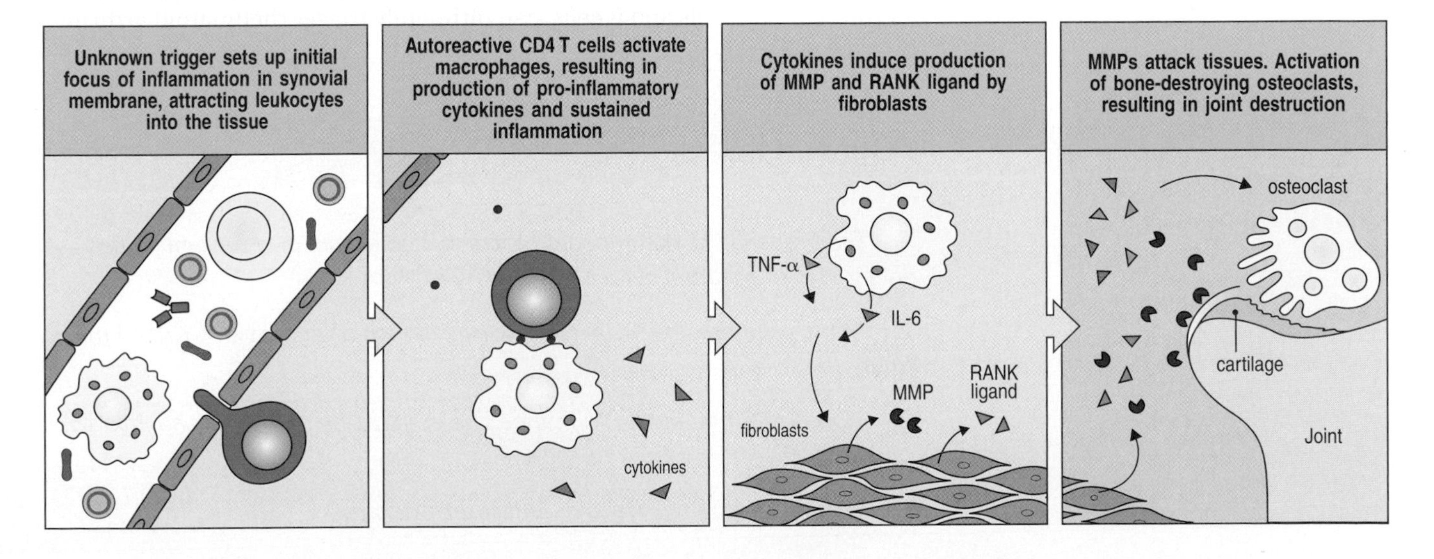
Unknown trigger sets up initial focus of inflammation in synovial membrane, attracting leukocytes into the tissue
Autoreactive CD4 T cells activate macrophages, resulting in production of pro-inflammatory cytokines and sustained inflammation
cytokines
Cytokines induce production of MMP and RANK ligand by fibroblasts
TNF-α
IL-6
MMP
RANK ligand
fibroblasts
MMPs attack tissues. Activation of bone-destroying osteoclasts, resulting in joint destruction
osteoclast
cartilage
Joint

Fig. 36.4 The pathogenesis of rheumatoid arthritis. Inflammation of the synovial membrane, initiated by some unknown trigger, attracts autoreactive lymphocytes and macrophages to the inflamed tissue. Autoreactive effector CD4 T cells activate macrophages, with the production of pro-inflammatory cytokines such as IL-1, IL-6, IL-17, and TNF-α. Fibroblasts activated by cytokines produce matrix metalloproteinases (MMPs), which contribute to tissue destruction. The TNF-family cytokine RANK ligand, expressed by T cells and fibroblasts in the inflamed joint, is the primary activator of bone-destroying osteoclasts. Antibodies against several joint proteins are also produced (not shown), but their role in pathogenesis is uncertain.

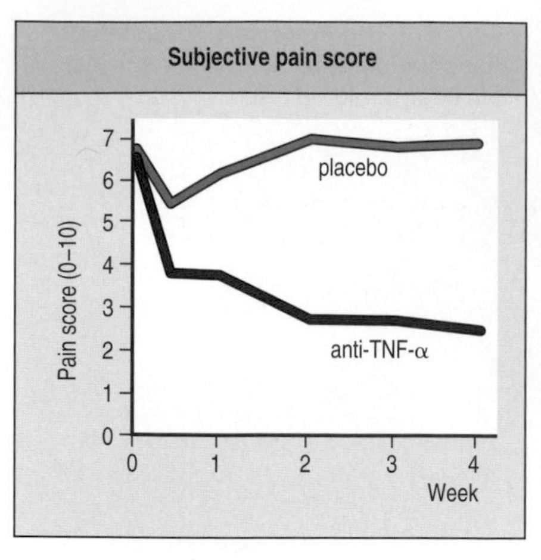

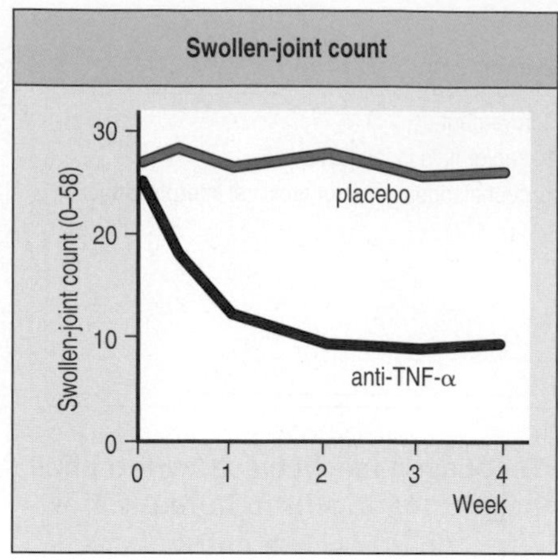

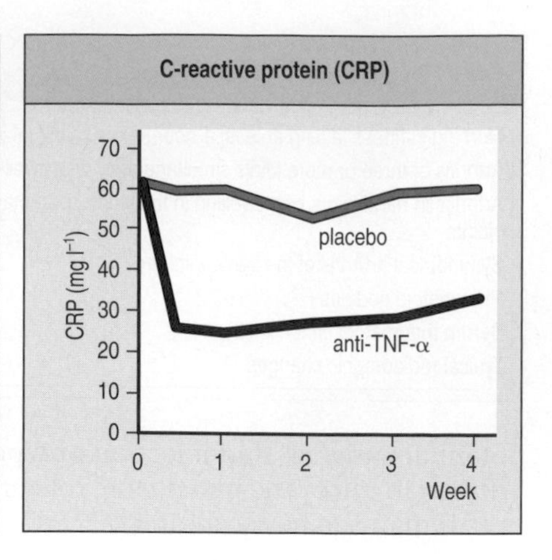

Fig. 36.5 The effects of treatment of rheumatoid arthritis with anti-TNF-α. For each parameter measured (pain, swollen joints, and level of C-reactive protein), the values for patients given a placebo treatment are depicted by the blue curve and the values for patients treated with anti-TNF-α antibody are depicted by the red curve.

subduing arthritic symptoms both in experimental animal models of arthritis and in human patients (Fig. 36.5). Several biologic agents that block TNF-α activity are in widespread clinical use. These chimeric antibodies and fusion proteins all act as a trap or sump for released TNF-α. The US Food and Drug Administration recently approved additional anti-cytokine therapies: anakinra (Kineret) is a recombinant protein antagonist of the IL-1 receptor, and tocilizumab is a humanized anti-IL-6 receptor antibody. Both are approved for treatment of moderate to severe rheumatoid arthritis in patients in whom more traditional arthritis medications have failed.

Although they account for the minority of lymphocytes in arthritic joints, B cells are likely to have a role in rheumatoid arthritis. B cells in the joints become activated by the binding of CD40 and B7.1 (CD80) by CD40 ligand and CD28, respectively, on T cells. The activated B cells secrete immunoglobulins, including rheumatoid factor (which activates complement), into the synovial fluid. The antibody rituximab (Rituxan), which is specific for the B-cell surface protein CD20, is used to deplete peripheral B cells. After rituximab infusion, peripheral B cells are not measurable in most patients for 6–12 months, and levels of rheumatoid factor are also decreased. Mean disease activity has also been shown to be significantly lower in rituximab-treated patients. B7.1 and B7.2 on antigen-presenting cells, including B cells, also bind to the T-cell surface protein CTLA-4, in this case inhibiting co-stimulation. Abatacept (Orencia), a fusion protein of CTLA-4 and the Fc portion of IgG, exploits this inhibitory signaling pathway. It has shown promising results in clinical trials and is now in use to treat arthritis refractory to TNF inhibitors.

Which cells or antigens trigger rheumatoid arthritis in humans is not yet known. It is likely that many different antigens, together with the nonspecific recruitment of T cells and B cells, can ultimately trigger rheumatoid arthritis.

Questions.

1 *The presence of rheumatoid factor in the serum of a patient is not diagnostic of rheumatoid arthritis. Why is this?*

2 *Mast cells are essential for the expression of arthritis in K/BxN mice. Why?*

3 *How do TNF-α and IL-1 enhance the infiltration of leukocytes into the joint space?*

4 *What might be the risks of infliximab therapy?*

CASE 37 Systemic Lupus Erythematosus

A disease caused by immune complexes.

Immune complexes are produced whenever there is an antibody response to a soluble antigen. As the immune response progresses, larger immune complexes form that trigger the activation of complement. These activated complement components then bind the triggering immune complexes. Large complexes are efficiently cleared by binding to complement receptor 1 (CR1) on erythrocytes, which convey the immune complexes to the liver and spleen. There, they are removed from the red-cell surface through interaction with a variety of complement and Fc receptors on Kupffer cells and other phagocytes (Fig. 37.1). When antigen is released repeatedly, there may be a sustained formation of small immune complexes; these complexes tend to be trapped in the small blood vessels of the renal glomeruli and synovial tissue of the joints.

The most common immune-complex diseases are listed in Fig. 37.2. In subacute bacterial endocarditis, bacteria reside for a protracted period on the heart valves. This infection and subsequent inflammation damage the valve. At the same time, the antibody response to the prolonged presence of the bacteria is intense, and immune complexes of IgG antibodies and bacterial antigens are formed. These complexes become trapped in the renal glomeruli and cause glomerulonephritis. The immunoglobulins in the immune complexes provoke the formation of anti-IgG IgM antibodies known as rheumatoid factor (see Case 36). In a similar fashion, viral hepatitis can become a chronic infection that provokes a marked IgG antibody response, with the consequent formation of virus-containing immune complexes and rheumatoid factor. The immune complexes can be entrapped in the renal glomeruli as well as in small blood vessels of the skin, nerves, and other tissues, where they cause inflammation of the blood vessels (vasculitis). The antibodies in the virus-containing immune complexes have the property of precipitating in the cold (less than 37°C) and are therefore termed cryoglobulins (see Case 38).

Topics bearing on this case:

- Clearance of immune complexes by complement
- Immune-complex disease
- Coombs' tests

This case was prepared by Raif Geha, MD, in collaboration with Erin Janssen, MD.

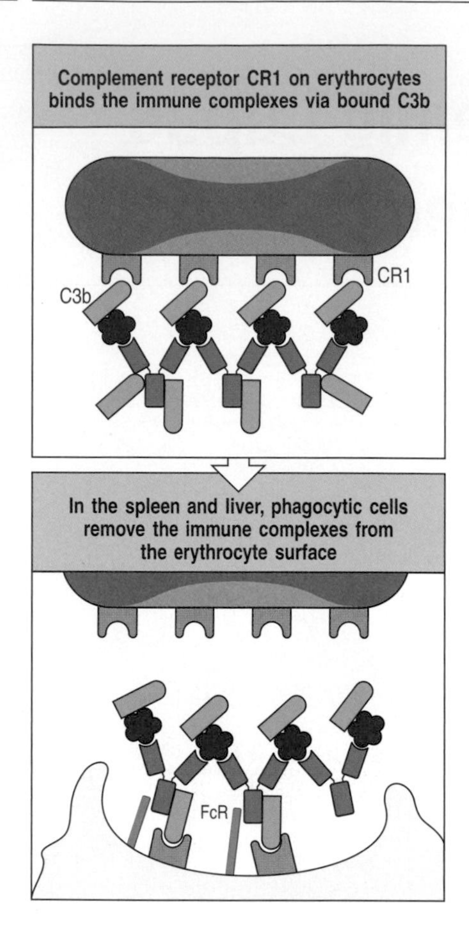

Fig. 37.1 Immune complexes are cleared from the circulation by binding to complement and Fc receptors. Immune complexes activate C3 in the serum, and bind activated complement components C3b, C4b, and C2a. C3b binds to complement receptors on erythrocytes, which transport the immune complexes to the spleen and liver, where complement receptors and Fc receptors on phagocytic cells bind to complement components and to the Fc portion of antibodies, and are thereby stimulated to engulf the complexes and degrade them.

The most prevalent immune-complex disease is systemic lupus erythematosus (SLE), which is characterized by the formation of antibodies against DNA. Every day, millions of nuclei are extruded from erythroblasts in the bone marrow as they mature into red blood cells (erythrocytes). This event, among others, provides a rich source of DNA in those individuals prone to making an immune response to DNA and developing SLE.

The case of Nicole Chawner: too much sun at the beach.

Sixteen-year-old girl, butterfly rash and symmetric morning stiffness.

Nicole Chawner was a healthy 16-year-old until this summer. A few days after excessive exposure to the sun on the beach, Nicole developed a red rash on her cheeks. She saw her family doctor, who recognized that the butterfly rash on her cheeks and bridge of her nose was typical of systemic lupus erythematosus (SLE) (Fig. 37.3).

He referred Nicole to the Children's Hospital, where she was asked about any other problems she might have noticed. Nicole said that when she woke up in the morning her fingers and knees were stiff, although they got better as the day wore on. Nicole had also noticed some symmetric swelling in her fingers.

A blood sample was taken from Nicole to ascertain whether she had anti-nuclear antibodies (ANA). These were positive, at a titer of 1:1280. Because of this result, further tests were performed for antibodies characteristically found in SLE. An elevated level of antibodies against double-stranded DNA was also found. Her serum C3 level was 73 mg dl^{-1} (normal 100–200 mg dl^{-1}). Her platelet count was normal at 225,000 μl^{-1}, and her direct and indirect Coombs tests were negative, as was a test for anti-phospholipid antibodies. A urine sample was also found to be normal.

Nicole was advised to take an antimalarial agent, hydroxychloroquine sulfate (Plaquenil), and to avoid direct sunlight. She did well for a while but, after a month, the

Fig. 37.2 Three autoimmune diseases that result in damage by immune complexes.

Immune-complex disease		
Syndrome	**Autoantigen**	**Consequence**
Subacute bacterial endocarditis	Bacterial antigen	Glomerulonephritis
Mixed essential cryoglobulinemia (see Case 38)	Rheumatoid factor IgG complexes (with or without hepatitis C antigens)	Systemic vasculitis
Systemic lupus erythematosus	DNA, histones, ribosomes, snRNP, scRNP	Glomerulonephritis, vasculitis, arthritis

morning stiffness in her fingers and knees worsened. She developed a fever of 39°C each evening accompanied by shaking chills. Enlarged lymph nodes were felt behind her ears and in the back of her neck. She also lost 4.6 kg over the course of the next 2 months.

When she returned to the hospital for a check-up, it was noted that her butterfly rash had disappeared. She had diffuse swelling of the proximal joints in her fingers and toes. Blood was drawn at this time, and the level of anti-DNA antibodies was found to have increased. The serum C3 level was 46 mg dl^{-1}. Nicole was advised to take 10 mg of prednisone twice a day, as well as 250 mg of the nonsteroidal anti-inflammatory drug naproxen twice a day. This quickly controlled her symptoms, and she remained well. At her next visit, her serum C3 level was 120 mg dl^{-1}.

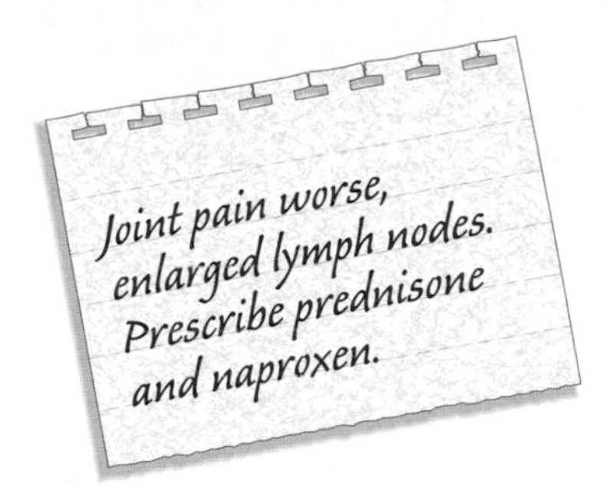
Joint pain worse,
enlarged lymph nodes.
Prescribe prednisone
and naproxen.

Systemic lupus erythematosus (SLE).

Systemic lupus erythematosus (SLE) is the most prevalent immune-complex disease in developed countries. For reasons that are not clear, it affects 10 times as many females as males. Patients with SLE usually have antibodies against multiple autoantigens. The most common autoantibody, which is found in the serum of 60% of all SLE patients, is against double-stranded DNA. Other commonly found antibodies are against small ribonucleoproteins. Autoantibodies against blood cells, such as platelets and red blood cells, as well as against the phospholipid complex that is formed by the activation of the proteins of the clotting system (antiphospholipid antibodies), are not infrequently seen. Most patients tend to have a range of these autoantibodies.

The immune complexes in SLE are small and tend to be trapped or formed inside tissues, primarily in the kidney and, to a lesser extent, in the synovial tissues of joints. For this reason, glomerulonephritis and arthritis are two of the most frequently encountered symptoms of SLE. These immune complexes fix complement efficiently, and tissue injury to the kidney or joints is mediated by activation of the complement system.

Cytokine signaling pathways have also been implicated in the pathogenesis of lupus. Type I interferons (IFN-α and IFN-β), which are important in suppressing viral replication, are secreted in response to the triggering of Toll-like receptors (TLRs). Immune complexes may trigger this response in SLE. Type I IFNs promote the activation of autoreactive T cells and augment class switching and antibody production in B cells. IRF5 is a transcription factor involved in interferon synthesis, and IRF5 haplotypes were one of the first genetic susceptibility factors identified in SLE. Further support for the link between IFNs and SLE is provided by the fact that a small percentage of patients on IFN-α treatment develop lupus. This drug-induced SLE occurs regardless of gender and tends to clear once the IFN-α is withdrawn.

The word 'lupus' is Latin for wolf, and this word is applied to a common symptom of SLE, the butterfly rash on the face. In the 19th century, the severe scarring rash on the face was named lupus because it was said to resemble the bite of a wolf. At that time, it was not possible to distinguish lupus erythematosus from lupus vulgaris, a scarring rash caused by tuberculosis. For unknown reasons, the rash is evoked by exposure to the sun (ultraviolet light). There is a seasonal variation to the onset of SLE, which is greatest in the Northern Hemisphere between March and September, when the greatest amount of ultraviolet light penetrates the atmosphere. Antimalarials such as hydroxychloroquine seem particularly helpful in the treatment of lupus skin disease.

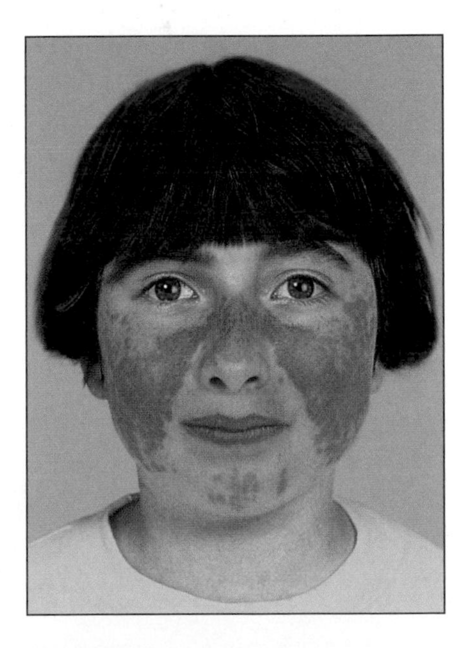

Fig. 37.3 The so-called butterfly rash typical of systemic lupus erythematosus. Photograph courtesy of M. Walport.

Questions.

1 *Why do you think Nicole's serum C3 was measured, both on her first visit to the hospital and after therapy?*

2 *What are the direct and indirect Coombs tests, and what did they tell us in this case?*

3 *Why was Nicole told to avoid direct exposure to sunlight?*

4 *Repeated analysis of Nicole's urine was negative. What does this mean?*

5 *Nicole had a serum IgG level of 2020 mg dl^{-1}. This substantially elevated level of IgG is commonly found in patients with SLE. How could you explain this? And what would you expect to find if we took a biopsy of Nicole's swollen lymph nodes?*

6 *The antigen in the immune complexes formed in SLE is often a complex antigen, such as part of a nucleosome or a ribonucleoprotein particle, which contains several different molecules. Patients often produce autoantibodies against each of these different components. What is the reason for the production of this variety of autoantibodies, and what type of failure in tolerance could be responsible for autoantibody production?*

CASE 38 Mixed Essential Cryoglobulinemia

Chronic infection can lead to immune-complex disease.

The extent and duration of antigenic stimulation is the major determinant of immunoglobulin levels in the blood and body fluids. Animals reared under germ-free conditions have low levels of immunoglobulins in the blood (hypogammaglobulinemia), whereas individuals with chronic infections have high levels (hypergammaglobulinemia). There are many chronic infectious diseases in which infection persists because of a failure of the immune response to eliminate the causative agent; examples are malaria, leishmaniasis, and hepatitis. There are other, mostly viral, infections in which the immune response fails to clear the infection because of the relative invisibility of the infectious agent to the immune system (as in infection with herpes simplex virus or hepatitis C virus).

The persistent immune response to microbial antigens can lead to the formation of immune complexes. The immune complexes may themselves become antigenic, and IgM antibodies are then formed to the IgG in the immune complexes. These anti-IgG IgM antibodies are called rheumatoid factor, because they were first detected in patients with rheumatoid arthritis (see Case 36).

The serum proteins and immunoglobulins in these immune complexes may have the peculiar property of precipitating in the cold (<37°C). Such proteins are called cryoglobulins and are not present (or are very difficult to detect) in normal blood. The detectable presence of cryoglobulins in the blood (cryoglobulinemia) is associated with the deposition of immune complexes in the kidneys, small blood vessels, and other organs, and can cause fatal immune-complex disease (Fig. 38.1).

Topics bearing on this case:
Immune-complex disease
Hepatatis C infection

Fig. 38.1 Classification of cryoglobulinemia.

Category	Immunological characterization	Associated diseases	Percentage of patients with cryoglobulinemia
I	Monoclonal immunoglobulin	Lymphoproliferative disorders (e.g., multiple myeloma, Waldenström's macroglobulinemia)	10–15
II	Polyclonal IgG and monoclonal IgM rheumatoid factor	Virtually all have chronic hepatitis C virus infection	35–40
III	Polyclonal IgG and polyclonal IgM rheumatoid factor	Chronic hepatitis C virus infection Autoimmune disorders (e.g., systemic lupus erythematosus, Sjögren's syndrome)	50

The case of Billy Budd: immunological consequences of an occult infection with hepatitis C virus.

Billy Budd was a truck driver who had been well all his life until he was 42 years old, when he complained to his wife of chronic fatigue and aching joints for some months. Raised red spots appeared on his shins, and his wife urged him to visit their family physician, Dr Melville.

On a physical examination Dr Melville found that Billy's liver was enlarged and that its edge was palpable several centimeters below the rib cage (normally the liver edge is not palpable); the spleen could not be felt. Dr Melville also noted that Billy's lower legs had multiple purpuric spots (raised lesions caused by blood leaking from damaged blood vessels; Fig. 38.2).

Various blood tests, including liver function tests, were ordered, and a week later Billy returned to see Dr Melville. His blood tests had revealed a very high level of serum alanine aminotransferase (ALT), at 831 U l^{-1} (normal <40 U l^{-1}). Other liver function tests, such as those for serum albumin and prothrombin levels, were normal. White blood cell and platelet counts were normal.

When questioned, Billy denied alcohol abuse and intravenous drug abuse, and said that his wife had been his only sexual partner for 25 years. When asked if he had ever received a blood transfusion, he said that he had had one 20 years ago, when he was 22 years old, after he fractured his tibia in a road accident. Dr Melville then ordered tests for antibodies against hepatitis B and hepatitis C viruses. Billy had a high titer of antibody against hepatitis C virus, and his serum contained large amounts of hepatitis C RNA, as determined by the polymerase chain reaction (PCR). A test for blood cryoglobulins was also positive (Fig. 38.3). A urine sample contained 4+ protein (normal 0) and innumerable red blood cells per high-power field, indicating damage due to immune complexes in the glomeruli of the kidney. Dr Melville referred Billy to a gastroenterologist.

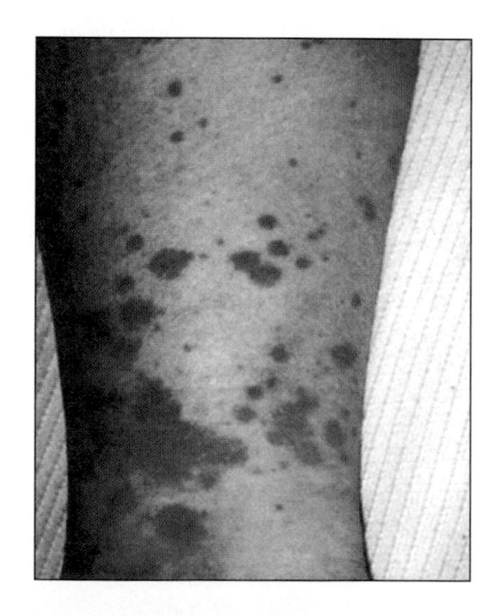

Fig. 38.2 Purpura on the leg of a patient with chronic infection with hepatitis C and cryoglobulinemia. Photograph courtesy of Sanjiv Chopra.

Before he was due to see the gastroenterologist, Billy appeared in the emergency room of the New England Deaconess Hospital with fever and severe abdominal pain, and vomiting blood. An arterial perforation was found in the small intestine, and intestinal contents had leaked from the perforation into the peritoneum, causing peritonitis. On microscopic examination of a sample of gut wall tissue, the intestinal blood vessels revealed inflammation (vasculitis), with deposition of immune complexes in the vessel walls.

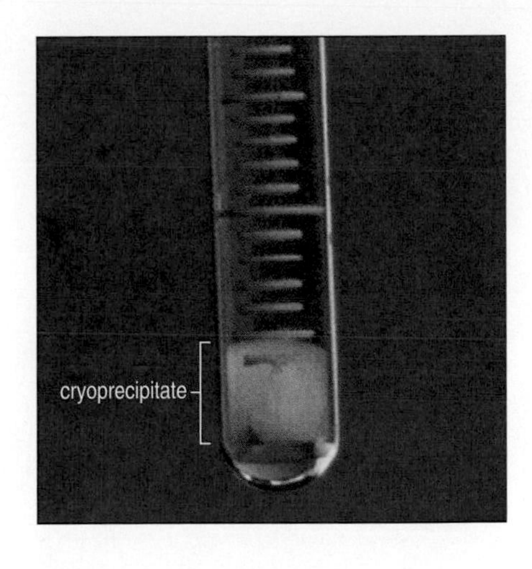

Fig. 38.3 Cryoprecipitate in Billy's blood. Billy's plasma was refrigerated overnight and then placed in a tube that is used to measure the hematocrit. The tube was centrifuged in the cold and the amount of cryoprecipitate was measured. It was 4% of the plasma volume. Photograph courtesy of Sanjiv Chopra.

Mixed essential cryoglobulinemia.

So-called cryoprecipitates—precipitates that appear when sera are stored at low temperatures and disappear at 37°C—were first noted many years ago, before it was appreciated that they were composed of immunoglobulins and other serum proteins. Cryoprecipitates are made by proteins that come out of solution, or at least partly gel, at cold temperatures and dissolve again on rewarming. This definition is based on *in vitro* behavior, however, and the true mechanism of cryoprecipitation remains obscure. It was eventually realized that the appearance of cryoprecipitates in the blood was associated with aching joints, fatigue, and purpura, especially in patients with liver disease due to chronic hepatitis C infection.

When the circulating cryoglobulin complexes are deposited in the walls of blood vessels, they activate complement. The ensuing release of the complement fragment C5a, a potent chemoattractant, causes leukocytes to infiltrate the blood vessel walls, resulting in vasculitis (Fig. 38.4). The inflammation causes the small blood vessels to burst and release blood into the skin, causing purpura. If the complexes are deposited in the synovia of joints, arthritis ensues. Vasculitis in a critical area, such as the brain or intestine, is rare, but can result in fatal bleeding.

Some monoclonal immunoglobulins produced by malignant cells in diseases such as multiple myeloma (IgG) or Waldenström's macroglobulinemia (IgM) may also behave as cryoglobulins and precipitate in the cold. The classification of the cryoglobulins is given in Fig. 38.1.

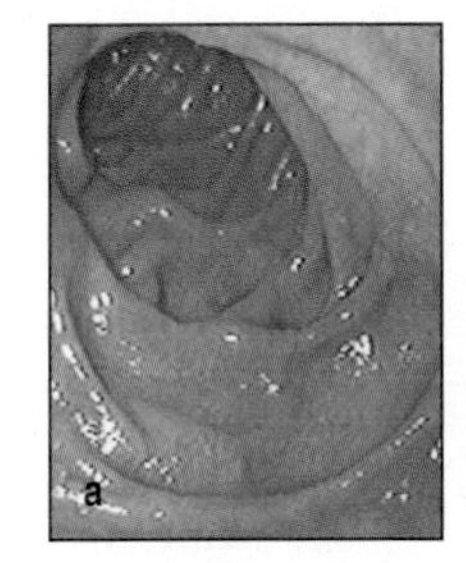

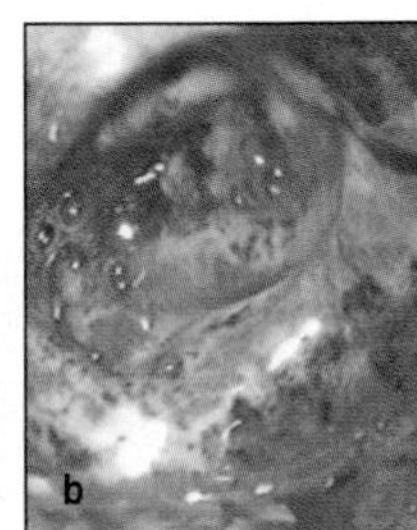

Fig. 38.4 Comparison of a healthy intestine with an intestine from a patient with cryoglobulinemia. Panel a: healthy intestine. Panel b: intestinal vasculitis from a patient with cryoglobulinemia. The photomicrographs of the gut wall were obtained by endoscopy. Photographs courtesy of Sanjiv Chopra.

Questions.

1 *What treatment do you think the gastroenterologist will suggest to Billy?*

2 *Rituximab (Rituxan), a monoclonal antibody directed against CD20, has shown some benefit in pilot clinical studies as a novel therapy for cryoglobulinemia. Why might rituximab be expected to help Billy's condition? What might be some of the risks of this therapy?*

CASE 39 Crohn's Disease

A defect in mucosal innate immunity results in inflammation of the gut.

The intestinal epithelium provides the surface essential for nutrient and vitamin absorption, water handling, and secretion of waste. The gut mucosa is also a crucial immune organ in pathogen surveillance, mucosal barrier function, and regulation of the composition of the intestinal microbial flora (the microbiota). The mucosa comprises the innermost layers of the intestinal wall (adjacent to the gut lumen) and is composed of the glandular surface epithelium overlying a layer called the lamina propria, which is separated from the submucosa by a thin muscular layer called the muscularis mucosae (Fig. 39.1). Absorptive epithelial cells (enterocytes) and interspersed goblet cells and Paneth cells perform the gut's absorptive and secretory functions, respectively. The absorptive surface area of the mucosa of the small intestine is increased by invaginations that form villi projecting into the gut lumen. Goblet cells secrete mucus into the gut lumen that protects the mucosa from the action of digestive enzymes. Paneth cells located in the bases of the intestinal crypts between the villi are specialized epithelial cells that secrete enzymes and antimicrobial peptides (defensins) that prevent the translocation of potentially pathogenic bacteria and toxins across the bowel wall. The submucosa is the second layer of the gut; it includes autonomic nerves, which control contraction of the smooth muscle of the muscularis mucosae to regulate peristalsis—the coordinated movement of gut contents. The loose connective tissue of the serosa makes up the outer layer of the bowel wall.

Cells of the gut mucosal innate and adaptive immune system are interspersed between the cells of the mucosal epithelium and throughout the lamina propria, and are also present as organized lymphoid organs (Peyer's patches) and isolated lymphoid follicles. The cells and organs of the gut mucosal immune system are collectively known as the gut-associated lymphoid tissue (GALT). The GALT is the largest lymphoid tissue in the human body, containing a significant portion of the body's 2×10^{12} lymphocytes.

Topics bearing on this case:

- NOD proteins
- Inflammatory reactions
- Gut mucosal immune system
- Host relations with the commensal microbiota

This case was prepared by Raif Geha, MD, in collaboration with Andrew Shulman, MD, PhD.

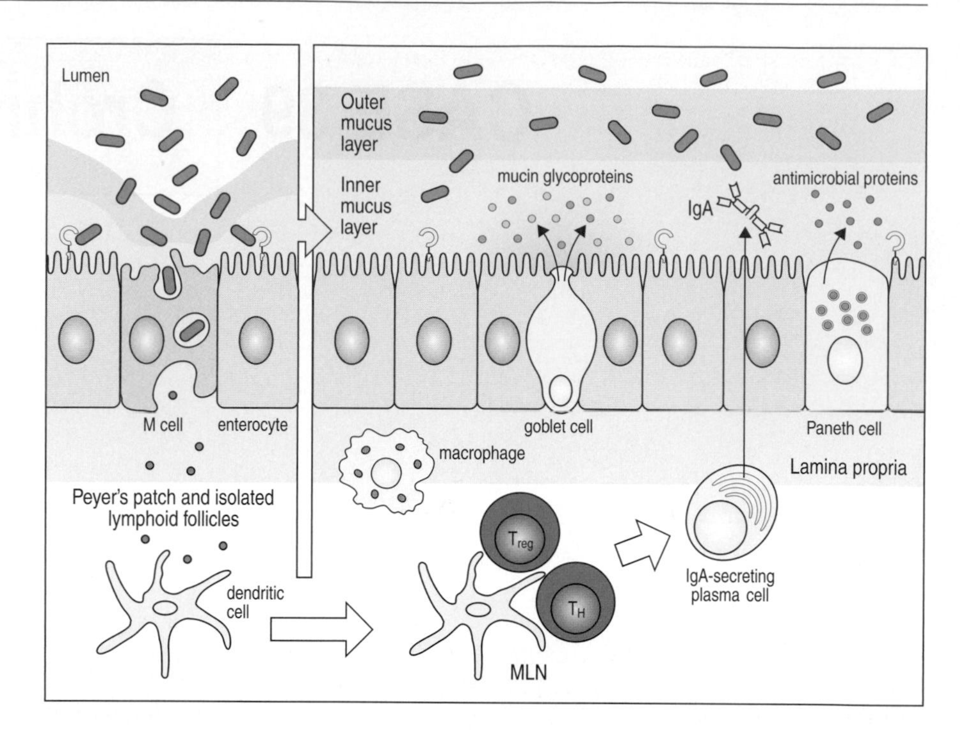

Fig. 39.1 A variety of mucosal responses keep the gut microbiota under control. The surface of the gut epithelium is protected by both innate and adaptive immune systems. A layer of mucus produced by goblet cells and antimicrobial proteins produced by Paneth cells help protect against pathogens and also keep commensal microorganisms under control, preventing them from colonizing the epithelial surface. Antigens from food, commensal microorganisms, or pathogens that enter the mucosal lymphoid tissues through M cells can initiate adaptive immune responses, including the production of IgA antibodies. In the absence of infection, dendritic cells presenting antigens (such as those derived from food and commensal microorganisms) to naive T cells in the mucosa tend to stimulate the production of regulatory T cells (T_{reg} cells), thus avoiding a damaging inflammatory response to the commensal microbiota. MLN, mesenteric lymph node.

Neutrophils, macrophages, and dendritic cells of the innate immune system express pattern recognition receptors and act as sensors for potentially pathogenic bacteria in the gut. Pattern recognition receptors recognize lipids, carbohydrates, nucleic acids, and peptides that are common features of bacteria, viruses, and fungi. One important class of such receptors is the Toll-like receptors (TLRs), which are innate immune receptors located at the cell surface and in the membranes of endosomes. Activation of TLRs by microorganisms and their products trigger innate immune pathways that mediate inflammation and host defense. NOD proteins (nucleotide-binding oligomerization domain-containing proteins) comprise another class of pattern recognition receptors that are located in the cytoplasm and bind to bacterial peptidoglycans released during infection (Fig. 39.2).

The professional antigen-presenting cells of the gut mucosa—the macrophages and dendritic cells—take up and kill bacteria via phagocytosis, and present bacterial antigens to, and activate, the resident naive T cells and B cells within the GALT. B cells within the bowel wall are activated to produce antigen-specific secretory dimeric IgA antibodies, which are transported into the gut lumen. Between 3 and 5 g of IgA is secreted into the intestinal lumen daily, where it functions to neutralize pathogenic bacterial proteins required for mucosal colonization and translocation and also helps prevent invasion by the gut's commensal microbiota. The GALT includes a full complement of T lymphocytes: CD4 helper T cells (T_H1 and T_H2) and CD8 cytotoxic T cells, and the more recently characterized lineages of CD4 effector T cells, regulatory T cells (T_{reg} cells) and T_H17 cells. The possible roles of T_H17 cells in autoimmunity and inflammation have generated intense research interest. In some contexts, T_H17 cells may have a protective role in preventing gut inflammation and maintaining homeostasis with the bacterial flora.

Given the high concentration of bacteria in the distal ileum and the colon (more than 10^{12} organisms per gram), the mucosal immune system performs a remarkable function in promoting the colonization of beneficial bacteria, which contribute to nutrient digestion, while mediating the destruction, clearance, and development of immunologic memory to pathogens and toxins. Dysregulation of the complex mechanisms responsible for mucosal immune function can result in inflammatory bowel disease (IBD): Crohn's disease and ulcerative colitis.

Fig. 39.2 Epithelial cells have a crucial role in innate defense against pathogens. TLRs are present in intracellular vesicles or on the basolateral or apical surfaces of epithelial cells, where they recognize different components of invading bacteria. NOD1 and NOD2 pattern-recognition receptors are found in the cytoplasm and recognize cell-wall peptides from bacteria. Both TLRs and NODs activate the NFκB pathway (see Fig. 23.1), leading to the generation of pro-inflammatory responses by epithelial cells. These include the production of chemokines such as CXCL8, CXCL1 (GROα), CCL1, and CCL2, which attract neutrophils and macrophages, and CCL20 and β-defensin, which attract immature dendritic cells in addition to possessing antimicrobial properties. The cytokines IL-1 and IL-6 are also produced and activate macrophages and other components of the acute inflammatory response. The epithelial cells also express MIC-A and MIC-B and other stress-related nonclassical MHC molecules, which can be recognized by cells of the innate immune system. IκB, inhibitor of NFκB.

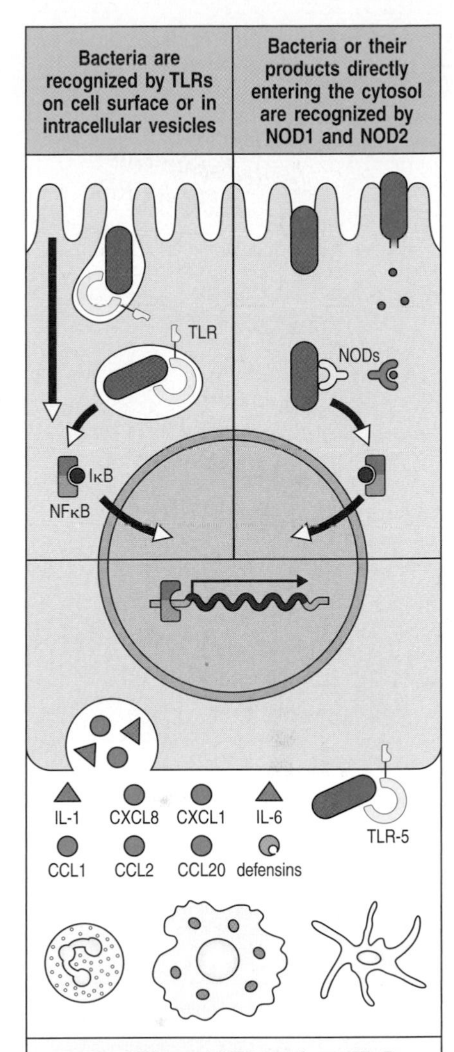

TLRs, NOD1 and NOD2 activate NFκB, inducing the epithelial cell to express inflammatory cytokines, chemokines, and other mediators. These recruit and activate neutrophils, macrophages, and dendritic cells

The case of Dorian Gray: a boy with fever, chronic abdominal pain, and weight loss.

Dorian was a healthy 8-year-old when he developed painful swelling and redness of his right first toe. The following day, he developed mouth ulcers (aphthous ulcers) that persisted for several days. He was taken to the pediatrician, who thought that the ulcers could be due to a Coxsackie virus infection. X-rays of the toe were done and showed no fracture or injury. Over the next 2 months, Dorian developed frequent poorly localized abdominal pain. Passing stools was particularly difficult, and his parents would often find Dorian crying in the bathroom. He had difficulties with constipation, passing three hard stools per week, but had no bloody stool or sensation of incomplete bowel evacuation.

The severity of Dorian's abdominal pain, toe swelling, and oral ulcers seemed to wax and wane over the next 2 months. Dorian developed daily low-grade fevers, at times as high as 39°C. His abdominal pain became more severe and he passed stools more frequently. Dorian was fatigued and listless and was unable to attend school. His appetite was very poor and his parents noticed that he had lost about 3.5 kg over the previous 2 months. He was seen again by his pediatrician, who set up outpatient appointments to investigate Dorian's illness. But when Dorian developed painful red lesions on his right shin and a pain in his jaw soon after this visit, his parents became concerned and brought him to the Children's Hospital Emergency Department.

The Grays were asked about the health of their family and told the staff that there was no history of inflammatory bowel disease or autoimmune illness. When examined, Dorian looked ill, tired, and pale. Numerous aphthous ulcers were present in his mouth. His abdominal examination showed no focal tenderness or masses. Two inflamed anal skin tags were found and the rectal exam showed no tenderness, fissures, or evidence of occult blood. Several raised, red skin lesions were present on Dorian's shins and his right first toe was swollen and warm. The skin lesions were recognized as erythema nodosum—acute nodular erythematous eruptions that typically occur on the lower extremities.

Blood tests were sent to check for evidence of infection or inflammation. The white blood cell count was elevated at 14,700 ml^{-1} (normal 5,700–9,900 ml^{-1}), and the platelet count was high at 759,000 ml^{-1} (normal 198,000–371,000 ml^{-1}). The erythrocyte sedimentation rate (ESR), an index of the levels of acute-phase reactant proteins that are synthesized by the liver in response to cytokines released during systemic inflammation, was elevated at 80 mm h^{-1} (normal 0–20 mm h^{-1}). The level of another acute-phase reactant, C-reactive protein (CRP), was also high at 2.2 mg dl^{-1} (normal less than 0.5 mg dl^{-1}). Dorian was discharged from the Emergency Department and was prescribed the proton-pump inhibitor omeprazole to reduce the secretion of gastric acid, and told to return for an upper and lower endoscopy the following week.

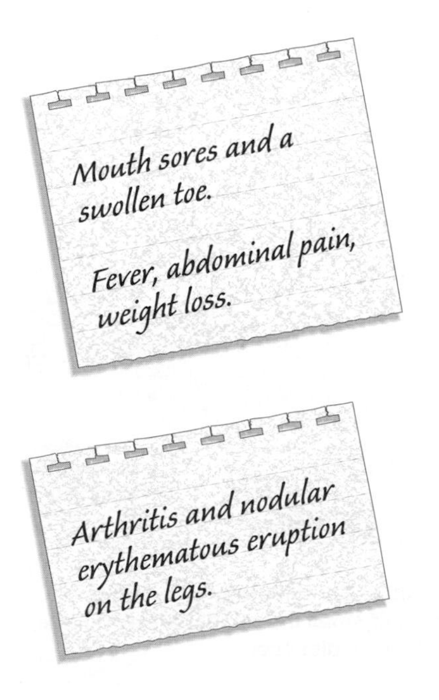

Mouth sores and a swollen toe.
Fever, abdominal pain, weight loss.
Arthritis and nodular erythematous eruption on the legs.

High leukocyte count, elevated sedimentation rate and high CRP.

Active colitis in the ileum with rare crypt abscesses.

On endoscopy, Dorian's pediatric gastroenterologist found ulceration of the esophagus and small intestine, and a perianal fistula (Fig. 39.3). Reviewing biopsy specimens from the procedure, the pathologist found neutrophilic infiltration of the esophageal and ileal lesions, active colitis with rare crypt abscesses, but no granulomatous inflammation (Fig. 39.4). The arthritis, oral aphthous ulcers, erythema nodosum lesions on the shins, and the endoscopic findings were all consistent with a diagnosis of Crohn's disease, and Dorian was started on oral corticosteroids as an initial anti-inflammatory treatment.

Over subsequent months, additional immunosuppressive agents in combination with oral steroids were tried, including 6-mercaptopurine and infusions of antibodies that inhibit the pro-inflammatory cytokine tumor necrosis factor-α (TNF-α) (infliximab and adalimumab). Unfortunately, Dorian's symptoms were not completely controlled. A year after receiving his diagnosis, Dorian developed worsening fever and abdominal pain, and an abdominal mass in the right lower quadrant could be felt. He was admitted to the Children's Hospital; abdominal ultrasound and CT scans revealed a large abscess of the distal small intestine (ileum and cecum). He was treated with intravenous antibiotics and was taken to the operating room, where the ileum and cecum were resected (ileocecectomy) and multiple smaller abscesses were drained. Dorian recovered well from surgery, continues to take weekly adalimumab injections, and has been able to discontinue steroid therapy. Although the weekly injections are difficult for Dorian and his family, the abdominal, skin, and joint symptoms of his Crohn's disease are all well controlled.

Crohn's disease.

Crohn's disease was first described in the medical literature by the Italian physician Giovanni Battista Morgagni, who presented the case of a chronically ill young man with diarrhea in 1769. American physician Burrill Crohn and colleagues published a case series of patients with "regional ileitis" in 1932. Crohn's disease is a disorder of mucosal immune dysregulation that is characterized by inflammatory lesions that can involve the entire gastrointestinal tract, from mouth to anus. In 40–50% of pediatric cases, Crohn's disease involves the ileum and colon. By contrast, the other main IBD, ulcerative colitis, typically involves only the colon and rectum. Biopsy evidence of inflammatory cell infiltrate is 'transmural' in Crohn's disease in that it involves the epithelium, the lamina propria, and adventitial layers of the bowel wall (see Fig. 39.4). As a result, fistulas and bowel abscesses, as in Dorian's case, are frequent complications. The presence of extra-intestinal inflammatory manifestations of IBDs, including skin (erythema nodosum, pyoderma gangrenosum), joints (IBD-related arthritis), and eyes (uveitis), underscores that IBDs are systemic inflammatory diseases. Gastrointestinal cancer occurs with increased frequency in individuals with Crohn's disease and ulcerative colitis, indicating that the chronic inflammation confers an increased risk of malignancy.

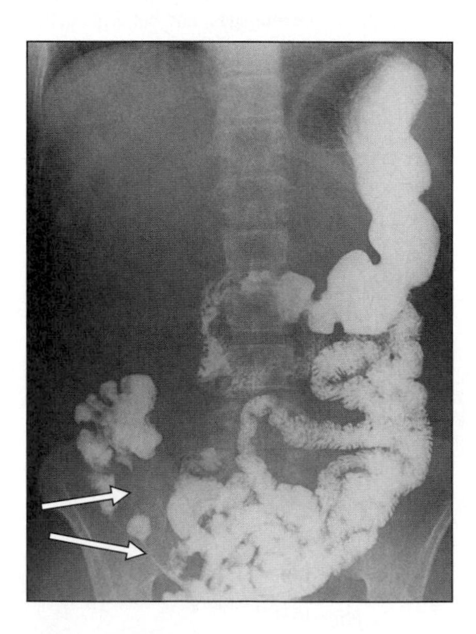

Fig. 39.3 Barium study showing two areas of small intestinal narrowing due to Crohn's disease.

Breakthroughs in understanding the pathogenesis of IBDs as a result of new human genetics tools and animal disease models have focused attention on the dysregulation of mucosal immunity to the gut commensal microbiota and a resulting impaired mucosal barrier function (see Fig. 39.1). These recent discoveries have identified proteins that participate in innate immune recognition of bacteria, cellular stress-response pathways, and host–microbiota interactions as targets of new therapies for IBD patients. The finding that humans with a deficiency in the NFκB signal pathway component NEMO (see Case 23) or in T_{reg} cells develop IBD provides evidence that the innate immune system and T_{reg} cells are essential in maintaining intestinal homeostasis. Mice

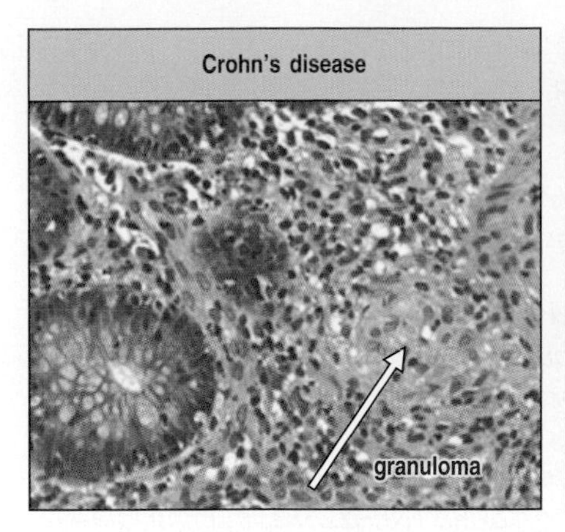

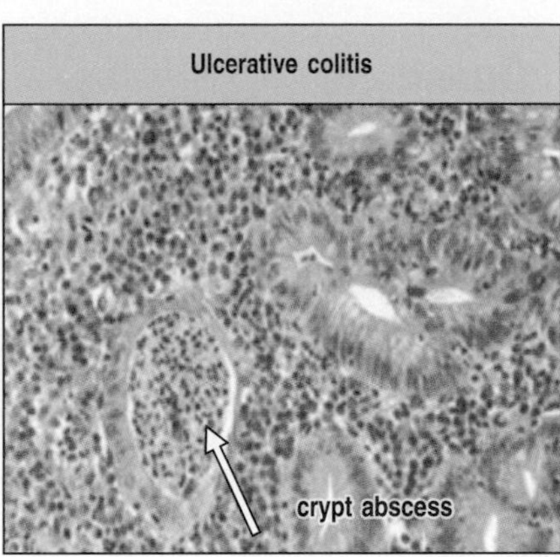

Fig. 39.4 Histology of the small intestine in Crohn's disease compared with the colon in ulcerative colitis. Left panel: Crohn's disease—biopsy from a terminal ileum with active disease. The figure illustrates a discrete granuloma composed of compact macrophages, giant cells, and epithelioid cells. Surrounding the nodule there is marked infiltration of lymphoid cells, plasma cells, and other inflammatory cells, but there is no necrosis. Right panel: ulcerative colitis—colonic mucosal biopsy taken from a patient with active disease. The crypt abscess is composed of transmigrated neutrophils and the surrounding epithelium exhibits features of acute mucosal injury. Photographs from *Nature*, 2007, **448**:427–434, with permission from Macmillan Publishers Ltd.

with genetic disruption of genes for the cytokines interleukin (IL)-2, which is essential for T_{reg}-cell function, or IL-10, an immunosuppressive cytokine, are protected from the development of intestinal inflammation when raised in germ-free conditions, revealing an essential role for host–microbiota interactions in initiating IBD pathogenesis. Isoforms of the cytokine receptor IL-23R have been shown to confer both susceptibility to and protection from Crohn's disease, indicating complex regulatory roles of T_{reg} and T_H17 cells in intestinal inflammation. The cytokine IL-23 is required for T_H17-cell maintenance and survival. T_H17 and T_{reg} cells have a reciprocal relationship, because both compete for the cytokine TGF-β for their induction, which in the case of T_H17 cells, but not T_{reg} cells, also requires IL-23R.

Genome-wide association studies, which mine genetic data from large populations of patients and control individuals to identify variants of genes that confer increased disease risk, have been particularly fruitful in the study of Crohn's disease (Fig. 39.5). One gene identified as being associated with Crohn's disease is that encoding NOD2, an intracellular innate immune receptor, expressed in macrophages and epithelial cells, that binds to muramyl dipeptides from bacterial cell walls and activates the production of inflammatory cytokines (see Fig. 39.2). Intestinal epithelial cells from individuals with a Crohn's disease-associated mutation in *NOD2* have impaired secretion of antimicrobial peptides called defensins when exposed to invasive bacteria. Blau syndrome, a rare autosomal dominant condition characterized by granulomatous inflammation of the eyes, skin, and joints, has also been found to be due to mutations in *NOD2*.

Autophagy is an evolutionarily ancient cellular stress-response pathway that delivers intracellular organelles and cytoplasmic contents to the lysosomes for degradation. Autophagy is important for walling off and eliminating bacteria that escape into the cytosol or enter it directly. *ATG16L1* and *IRGM* are two genes whose products are important for the formation of the autophagocytic vacuole. Identification of disease-related variants of the *ATG16L1* and *IRGM* genes implicate defects in autophagy in Crohn's disease, possibly because inefficient elimination of microbes from the cytosol leads to sustained cytokine secretion and Paneth cell dysfunction, resulting in intestinal inflammation. Cellular stress caused by misfolded proteins, metabolic factors, and microbes activate what is known as the 'unfolded protein response.' This response activates the transcription factor XBP1, which regulates the expression of genes important for the proper function of the immune system. Defects in XBP1 result in IBD in mouse models, and XBP1 variants have been described in patients with Crohn's disease, suggesting that abnormalities in the cellular stress response may result in intestinal inflammation and decreased survival of Paneth cells.

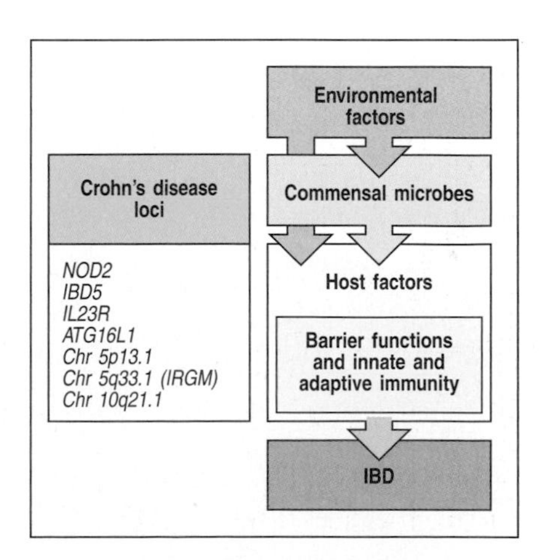

Fig. 39.5 Genome-wide association studies reveal patterns of Crohn's disease pathogenesis.

Questions.

1 *How can you account for Dorian's markedly elevated platelet count found on his initial visit to the Emergency Department?*

2 *6-Mercaptopurine was developed as a chemotherapeutic agent for the treatment of cancer. What could be the basis for its use in the treatment of inflammatory bowel disease?*

3 *Infliximab is a chimeric mouse–human monoclonal antibody generated against the inflammatory cytokine TNF-α. Adalimumab is a fully humanized anti-TNF-α antibody. How might some patients become resistant to infliximab therapy and yet still respond to adalimumab?*

4 *Natalizumab is a therapeutic monoclonal antibody that targets the cell-adhesion molecule α_4-integrin and is used as a second-line therapy for severe Crohn's disease. What might be its mechanism of action?*

CASE 40 Multiple Sclerosis

An autoimmune attack on the central nervous system.

As we saw in rheumatoid arthritis (Case 36), autoimmune disease can be caused by activated effector T cells specific for self peptides. When T cells recognize self-peptide:MHC complexes and become activated, they can cause local inflammation by activating macrophages, for example, with consequent tissue damage. Another example of a T cell-mediated autoimmune disease is the neurologic disease multiple sclerosis (MS). However, unlike the involvement of autoantibodies in some types of autoimmune diseases (see Case 42), it has been difficult to prove the involvement of T cells in MS, because T cells do not cross the placenta into the fetus and experimental T-cell transfer is not allowed in humans.

Seventy years ago, an experimental model of MS was established in mice, in which the injection of myelin in adjuvant caused the development of neurologic symptoms similar to those of MS. This disease is called experimental autoimmune encephalomyelitis (EAE). The antigens in myelin that can induce EAE are myelin basic protein (MBP), proteolipid protein (PLP), and myelin oligodendrocyte glycoprotein (MOG). The disease can be transferred to syngeneic animals by cloned antigen-specific T-cell lines derived from animals with EAE (Fig. 40.1). When the recipient animals are immunized with MBP, for example, they develop active disease. T cells specific for MBP, PLP, and MOG have been found in the blood and cerebrospinal fluid (CSF) of patients with MS.

This case was prepared by Raif Geha, MD, in collaboration with Andrew Shulman, MD, PhD.

Topics bearing on this case:
Inflammatory reactions
Interactions of co-stimulatory molecules with their receptors
The development of tolerance to self antigens
Activation of self-reactive T cells
Immunologically privileged sites
Experimental autoimmune encephalomyelitis
Induction of oral tolerance

Fig. 40.1 T cells specific for myelin basic protein mediate inflammation of the brain in experimental autoimmune encephalomyelitis (EAE). This disease is produced in experimental animals by injecting them with isolated spinal cord homogenized in complete Freund's adjuvant. EAE is due to an inflammatory reaction in the brain that causes a progressive paralysis affecting first the tail and hindlimbs before progressing to forelimb paralysis and eventual death. One of the autoantigens identified in the spinal cord homogenate is myelin basic protein (MBP). Immunization with MBP alone in complete Freund's adjuvant can also cause these disease symptoms. Inflammation of the brain and paralysis are mediated by T_H1 and/or T_H17 cells specific for MBP. Cloned MBP-specific T_H1 and/or T_H17 cells can transfer symptoms of EAE to naive recipients provided that the recipients carry the correct MHC allele. In this system it has therefore proved possible to identify the peptide:MHC complex recognized by the TH clones that transfer disease.

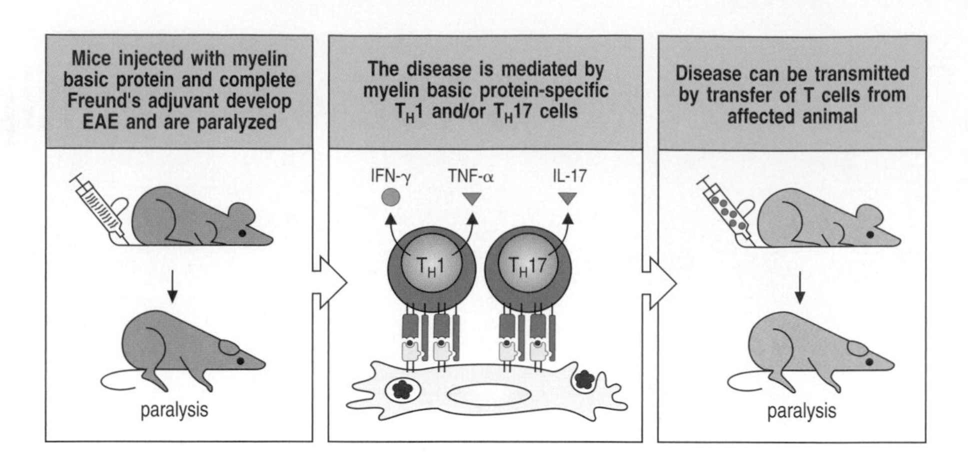

29-year-old female with sudden loss of vision in one eye.
Family history of MS; corticosteroid therapy; order MRI brain scan.

The case of Vivie Warren: an oboist who has difficulty reading a musical score.

Mrs Vivie Warren, a 29-year-old professional oboe player, was in good health until one morning she noticed a loss of vision in her left eye. Her physician referred her to a neurologist, who found that her eye movement was normal and not accompanied by any pain. The visual acuity in Vivie's left eye was 20/100 and in her right eye 20/200. Her retina was normal, and a detailed neurologic examination also proved normal. The neurologist diagnosed optic neuritis (inflammation of the optic nerve). Her family history was informative, however, in that her mother had severe MS and was permanently disabled, and a magnetic resonance imaging (MRI) brain scan was ordered.

Vivie was given a 5-day course of intravenous corticosteroids, and her vision returned to normal over the next 3 weeks. The MRI scan revealed multiple lesions in the white matter of the brain under the cortex and around the ventricles (Fig. 40.2). Intravenous injection of gadolinium, a contrast agent that leaks from blood vessels in recently inflamed tissue, showed that some of the brain lesions were probably of recent origin (Fig. 40.3). The neurologist told Vivie that she had a high probability of developing MS and advised her to return for frequent neurologic examination.

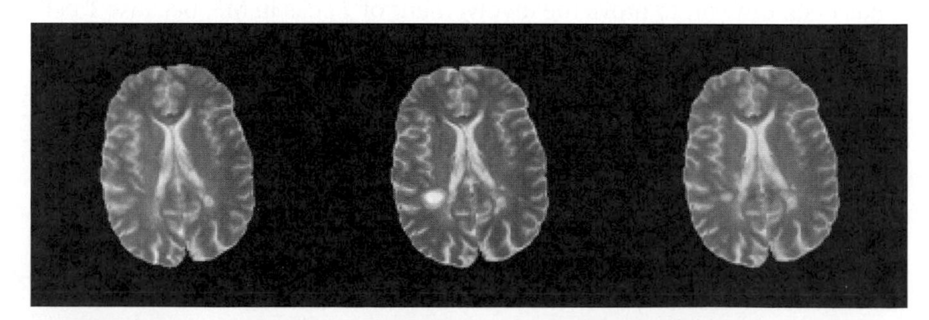

Fig. 40.2 Three-dimensional magnetic resonance images of the brain at three different time points in the course of MS. Left, early; center, during acute exacerbation; right, after therapy. The technique used causes fluid to appear white. The lateral ventricles in the middle of the brain scan and the sulci of the cerebral cortex around the edge appear white as a result of normal cerebrospinal fluid. The white spots, which are due to edema fluid and decreased myelin, are MS lesions.

Fig. 40.3 Three computer-generated images of MRI scans of the brain shown in Fig. 40.2 at the same time points. Several levels of MRI scans have been integrated by computer. The lateral and middle ventricles appear in purple. Old MS lesions appear yellow-green. A new, gadolinium-enhanced, lesion appears orange. On the right, this lesion has diminished in size after immunosuppressive therapy.

Vivie remained well for a further 3 years and then developed weakness of the muscles on the left side of her face that were innervated by the seventh cranial nerve. A repeat MRI scan with gadolinium enhancement showed new lesions in the left middle cerebellar peduncle and in the pons. CSF was obtained by lumbar puncture. It contained 28 mg dl^{-1} protein (normal) and 8 lymphocytes ml^{-1} (normal 0–3 ml^{-1}). At this point a firm diagnosis of MS was made. Despite the normal level of protein in the CSF, the IgG content was raised. On electrophoresis, discrete bands of IgG were observed, indicating clonal expansion of restricted B-cell populations in the central nervous system (CNS). Another 5-day course of corticosteroids was administered intravenously, and Vivie's symptoms improved. Weekly intramuscular injections of interferon (IFN)-β were started to prevent progression of the disease.

Vivie did well for 3 more years, after which she developed a weakness in her left leg and left hand. Her speech became slurred. She developed nystagmus (rapid uncontrolled horizontal jerking eye movements when attempting to fix the gaze on something) and ataxia (wide-based staggering gait). Vivie was given another course of corticosteroids, after which her symptoms improved, but 8 months later they recurred. The injections of IFN-β were stopped and she was put on high doses of cyclophosphamide and corticosteroids at monthly intervals. After 3 months of this therapy, the cyclophosphamide and corticosteroid injections were gradually reduced to every 12 weeks. Her neurological examination became normal and no new lesions were observed on gadolinium-enhanced MRI.

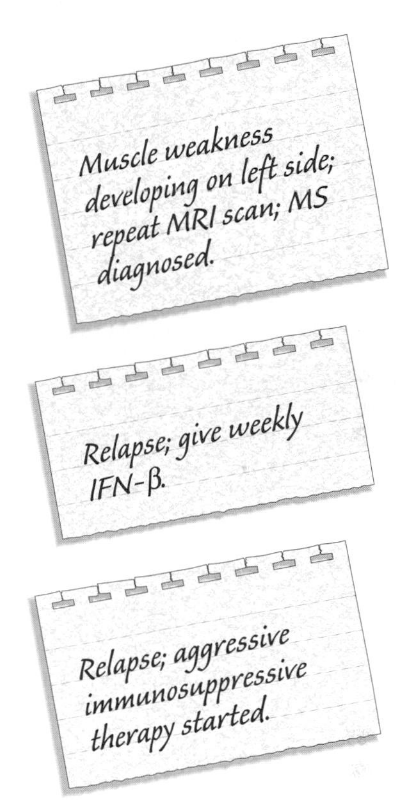

Muscle weakness developing on left side; repeat MRI scan; MS diagnosed.
Relapse; give weekly IFN-β.
Relapse; aggressive immunosuppressive therapy started.

Multiple sclerosis.

Multiple sclerosis (MS) was first described by the great French neurologist Jean-Martin Charcot in the 1860s. It was noted at autopsy that patients who died of this disease had multiple hard (sclerotic) plaques scattered throughout the white matter of the CNS. The disease is 10 times more frequent in women than in men and is associated with HLA-DR2. Those affected have a variety of nervous symptoms, such as urinary incontinence, blindness, ataxia, muscle weakness, and paralysis of limbs. The plaques characteristic of the disease show dissolution of myelin along with infiltrates of lymphocytes and macrophages, particularly along blood vessels. The inflammatory exudate causes increased vascular permeability.

The CNS is a relatively immunologically privileged site from which antigens do not normally reach the lymphoid tissues, and so there is no negative selection of T cells with the potential to react against CNS antigens. In MS, an unknown injurious event is presumed to provoke the release of CNS antigens and their presentation to lymphocytes in the peripheral lymphoid organs. This results in the expansion of clones of autoreactive T cells and their differentiation into T_H1 cells, which home to the CNS and initiate inflammation. These T_H1 cells can be readily identified in the CNS of patients with MS.

Lymphocytes and other blood cells do not normally cross the blood–brain barrier. If tissue becomes inflamed, however, activated CD4 T cells autoreactive for a brain antigen and expressing $\alpha_4:\beta_1$ integrin, which binds vascular cell adhesion molecules (VCAM) on the surface of activated venule endothelium, can migrate out of the blood into the brain. There they reencounter their specific autoantigen presented by MHC class II molecules on microglial cells and produce pro-inflammatory cytokines such as IFN-γ (Fig. 40.4). Microglia are phagocytic macrophage-like cells of the innate immune system resident in the CNS and, like macrophages, can act as antigen-presenting cells. Inflammation causes increased vascular permeability, and the site becomes heavily infiltrated by activated macrophages and T_H1 cells, which produce pro-inflammatory cytokines that exacerbate the inflammation, resulting in the further recruitment of T cells, B cells, macrophages, and dendritic cells to the site of the lesion. Autoreactive B cells produce autoantibodies against myelin antigens with help from T cells. Activated mast cells release histamine, contributing to the inflammation. In some way that is not yet fully understood, these combined activities lead to demyelination and interference with neuronal function.

Mice deficient in IFN-γ are not protected from the development of EAE, suggesting that additional T_H cells may drive CNS inflammation. T_H17 cells are a recently identified helper T-cell population that can be induced from memory T cells by a combination of cytokines including IL-6, IL-21, and transforming growth factor-β (TGF-β) and are sustained by IL-23. Although T_H17 cells do not seem to mediate inflammation in all contexts, studies in mouse EAE and human MS reveal a pathologic role for T_H17 cells and their associated cytokine, IL-17. Adoptive transfer of MOG-specific T_H1 and T_H17 cells differentiated *in vitro* induces EAE with a distinct histologic appearance, indicating that several effector T-cell types drive autoimmunity in MS. Autoreactive T_H1 cells with specificity for MS-associated myelin antigens can be found in healthy patients without MS. Investigators searching for the suppressive mechanisms that inhibit potentially autoreactive T cells in healthy individuals have focused on regulatory T cells (T_{reg} cells), a subset of naturally occurring CD4 $CD25^+$ T cells that promote peripheral tolerance and inhibit autoimmunity in multiple organs. Although patients with MS have normal numbers of T_{reg} cells, the cells have a decreased ability to suppress autoreactive T cells *in vitro*. Future research will seek to further investigate the regulation of T_H-cell subtypes and the interplay of autoreactive and suppressive T cells in MS.

Fig. 40.4 The pathogenesis of multiple sclerosis. At sites of inflammation, activated T cells autoreactive for brain antigens can cross the blood–brain barrier and enter the brain, where they reencounter their antigens on microglial cells and secrete cytokines such as IFN-γ and IL-17. The production of T-cell and macrophage cytokines exacerbates the inflammation and induces a further influx of blood cells (including macrophages, dendritic cells, and B cells) and blood proteins (such as complement) into the affected site. Mast cells also become activated. The individual roles of these components in demyelination and loss of neuronal function are still not well understood.

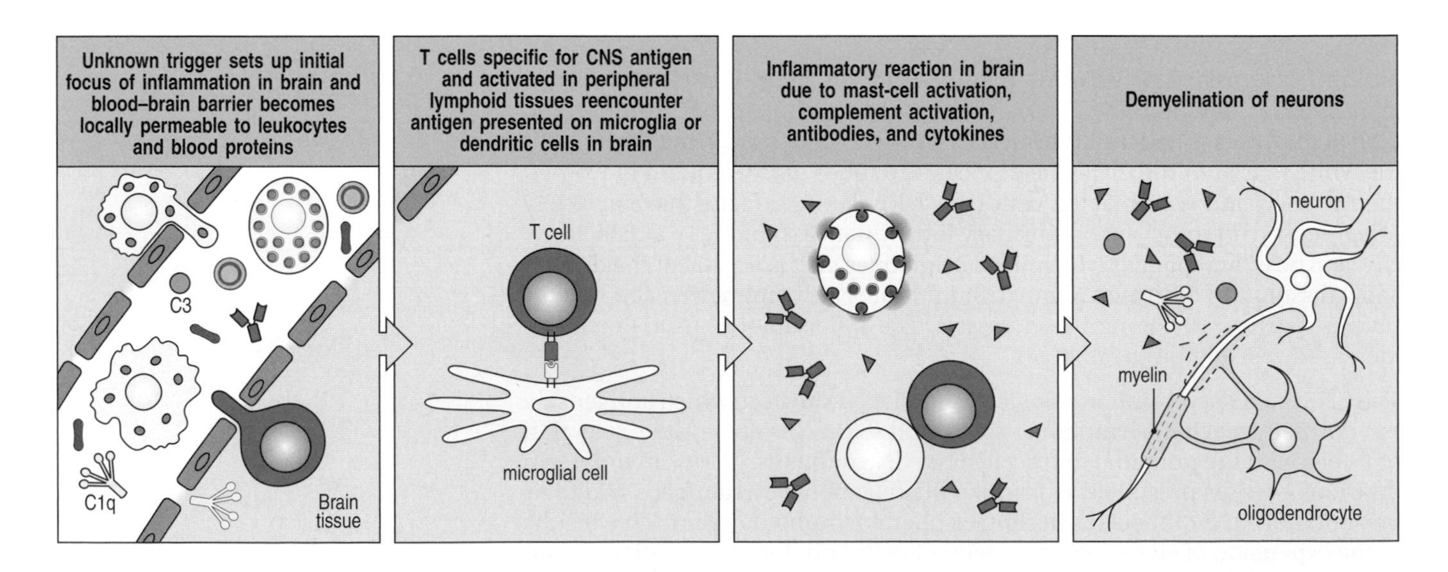

In mice, feeding with MBP before immunization with MBP in adjuvant protects against the development of EAE. MBP-specific T cells can be identified in the brain of protected mice, but they secrete TGF-β rather than IFN-γ and fail to initiate an inflammatory reaction. Furthermore, adoptive transfer of T cells from mice fed MBP can confer protection against EAE. Therapeutic attempts to treat patients with MS by oral MBP have not been successful, however. This suggests that therapeutic interventions that induce oral tolerance may not be effective in already established disease.

Questions.

1 *Oligoclonal immunoglobulins were found in Mrs Warren's central nervous system. How do you explain this?*

2 *Mrs Warren was treated with corticosteroids, cyclophosphamide, and IFN-β. What was the aim of this therapy?*

3 *An attempt has been made to treat MS patients with IFN-γ. Can you predict what the outcome was and why?*

4 *What is the rationale behind feeding MBP to mice to prevent EAE?*

5 *Can you predict whether EAE can be induced in CD28 knockout mice?*

CASE 41 Autoimmune Hemolytic Anemia

Autoimmune disease triggered by infection.

There are various ways in which an infection could induce autoimmunity: disruption of a tissue barrier might expose a normally sequestered autoantigen; the infecting microorganism might act as an adjuvant; microbial antigen might bind to self proteins and act as haptens; the microorganism might infect self-reactive lymphocytes (as in the case of infection by the Epstein–Barr virus) and induce autoantibody production; and the microorganism might share cross-reactive antigens with the host (molecular mimicry). The major histocompatibility antigens carried by an individual may also confer susceptibility to, or protection against, certain autoimmune diseases that can be incited by infection (Fig. 41.1).

There are only a few conditions in which an infection has been identified as the direct cause of the autoimmune disease. The longest established is rheumatic fever, which in the pre-antibiotic age was a frequent long-term consequence of infection with *Streptococcus pyogenes* (scarlet fever). Some strains of *S. pyogenes* lead to a type III autoimmune disease (caused by immune-complex formation; see also Case 37) of the renal glomeruli called post-streptococcal acute glomerulonephritis. These streptococcal strains are said to be nephrogenic (causing nephritis). About 3–4% of children infected with a nephrogenic strain of streptococcus will develop acute glomerulonephritis within a week or two of the onset of the streptococcal infection. What predisposes this subset of children to develop the complication is unknown.

A direct association of infection with autoimmune disease also occurs in patients with pneumonia caused by *Mycoplasma pneumoniae*, the case we shall discuss here. About 30% of patients with this infection develop a transient increase in serum antibody to a red blood cell antigen, and a small proportion of these patients develop hemolytic anemia, a type II autoimmune disease. Other type II autoimmune diseases are listed in Fig. 42.1. In this case, the decrease in the number of red blood cells (anemia) results from their immunological destruction (hemolysis) as a result of the binding of an IgM autoantibody to a carbohydrate antigen on the red cell surface. When the infection subsides, either spontaneously or after treatment, the autoimmune disorder disappears.

The IgM autoantibody agglutinates red blood cells at temperatures below 37°C. The antibodies are therefore called cold hemagglutinins or cold agglutinins, and the hemolytic disorder is known as cold agglutinin disease. Low titers of cold agglutinins (detectable at up to 1:30 dilution of serum) occur in healthy people. In about one-third of patients with mycoplasma infection there is a transient increase in the titer, but without symptoms; the rise usually goes unnoticed unless the patient happens to have a blood test, when clumping of the erythrocytes will be apparent.

The autoantibodies in the transient cold agglutinin syndrome resemble those in the chronic autoimmune disorder known as chronic cold agglutinin disease, which is of unknown etiology. This runs a protracted course in which

Topics bearing on this case:
Humoral autoimmunity
Mechanisms for breaking tolerance
Coombs test

Fig. 41.1 Association of infection with autoimmune diseases. Several autoimmune diseases occur after specific infections and are presumably triggered by the infection. The case of post-streptococcal disease is best known. Most of these post-infection autoimmune diseases also show susceptibility linked to the MHC.

Associations of infection with immune-mediated tissue damage		
Infection	**HLA association**	**Consequence**
Group A *Streptococcus*	?	Rheumatic fever (carditis, polyarthritis)
Chlamydia trachomatis	HLA-B27	Reiter's syndrome (arthritis)
Shigella flexneri, *Salmonella typhimurium*, *Salmonella enteritidis*, *Yersinia enterocolitica*, *Campylobacter jejuni*	HLA-B27	Reactive arthritis
Borrelia burgdorferi	HLA-DR2, -DR4	Chronic arthritis in Lyme disease

the autoantibodies are characteristically oligoclonal, and lymphoproliferative disorders commonly develop. Thus chronic cold agglutinin disease behaves as a variant of an IgM gammopathy called Waldenström's macroglobulinemia. A gammopathy is an abnormal monoclonal or oligoclonal spike in the immunoglobulin electrophoretic pattern.

The case of Gwendolen Fairfax: the sudden onset of fever, cough, and anemia.

Gwendolen Fairfax was a healthy, unmarried 34-year-old who worked as a manager in a bank. She had never had any illness other than minor colds and the usual childhood infections. After developing a feverish cough her symptoms got progressively worse over the next few days and she decided to seek the advice of her physician, Dr Wilde.

34-year-old female
with respiratory
infection; anemia.

Dr Wilde noted that Gwendolen was extremely pale; this was particularly noticeable in the palms of her hands, which were completely white. (Look at your palms; they have a healthy reddish pink color, unless your hemoglobin level is less than 10 g dl^{-1}.) Her temperature was raised, at 38.5°C, and her respiratory rate was 30 per minute (normal 20). On listening to her chest, the physician heard scattered rhonchi (harsh breath sounds) at the bases of both lungs.

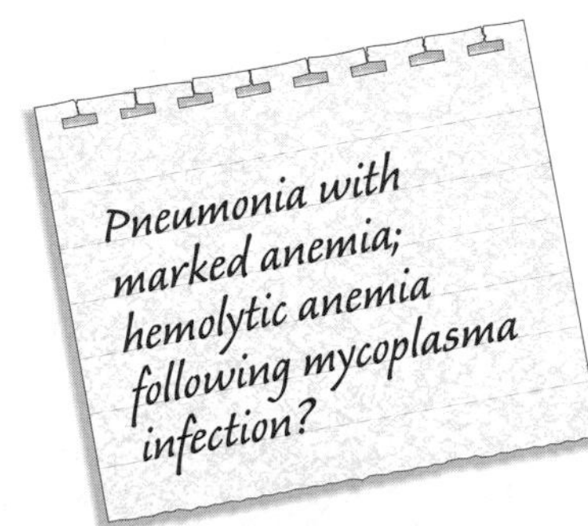

The physician ordered blood counts to be performed immediately. Gwendolen's hematocrit and hemoglobin level were low—hematocrit 26% (normal 38–46%) and hemoglobin 9.5 g dl^{-1} (normal 13–15 g dl^{-1}). Her white blood cell count was elevated at 11,300 μl^{-1}, with 70% neutrophils, 24% lymphocytes, 6% monocytes, and 1% eosinophils, revealing an increased absolute neutrophil count, and the platelet count was 180,000 μl^{-1}. The chest radiograph showed patchy infiltrates in both lower lungs (Fig. 41.2). Dr Wilde suspected a diagnosis of mycoplasma pneumonia (also known as atypical pneumonia because there was no complete opacity over a whole lobe of the lung such as typically occurs in pneumococcal and other bacterial pneumonias). He advised Gwendolen to go to Reading Hospital.

In the hospital, a sputum sample proved negative for bacteria but contained abundant neutrophils. A blood sample was tested for the presence of antibodies against the red blood cells. In the presence of sodium citrate as an anticoagulant, the sample was chilled to 4°C; the red cells became clumped and the blood looked as though it had clotted. When warmed to 37°C, however, the red cells dispersed. Thus, the red cells were reversibly agglutinated in the cold. A blood sample was spun in a warmed centrifuge, and the red cells and plasma were separated. Gwendolen's red cells were

tested using direct and indirect Coombs tests for the presence of anti-red cell IgG antibody (see Fig. 46.2); these tests proved negative. Her red cells, however, were strongly agglutinated by an antibody against the complement component C3.

When Gwendolen's plasma, or dilutions of the plasma, were incubated in the refrigerator with red cells from type O normal blood, the red cells were strongly agglutinated; when the same test was performed with red cells obtained from an umbilical cord in the delivery room of the hospital, agglutination was weaker, or negligible.

Gwendolen was started on erythromycin by intravenous administration. In 3 days her symptoms had abated; she had no fever and her cough had greatly improved. Her hemoglobin had risen to 12 g dl^{-1}. She was discharged from the hospital, and advised to continue erythromycin orally and to avoid exposure to cold weather. When she returned to Dr Wilde's office 2 weeks later for a check-up, her fever had not returned, her cough had gone, and her hemoglobin was now a normal 14.5 g dl^{-1}.

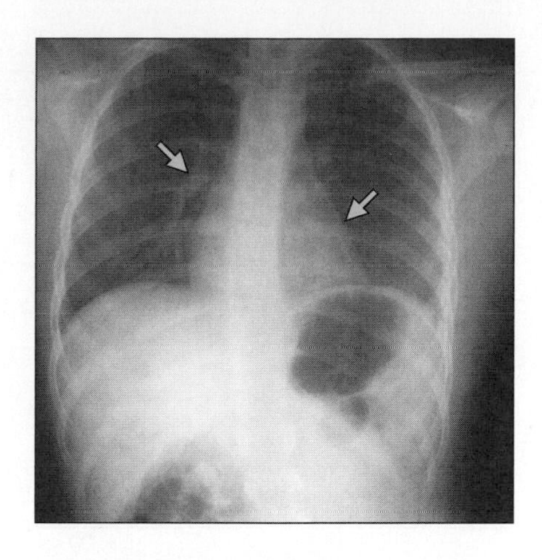

Fig. 41.2 Chest radiograph of a patient with mycoplasma pneumonia. Patchy opacities (arrowed) are evident in the lungs. Courtesy of T. Griscom.

Autoimmune hemolytic anemia.

Gwendolen's autoimmune hemolytic anemia was transient, persisting only as long as the infection persisted. In contrast, autoantibodies against red blood cells in chronic autoimmune disorders of unknown origin tend to be persistent. Transient autoantibodies most frequently develop after nonbacterial infections, such as infectious mononucleosis (see Case 45). Autoimmune hemolytic anemia can also be induced by certain chemicals and drugs.

The autoantibodies in autoimmune hemolytic anemias are of the IgM or IgG class. They bring about hemolysis of the red blood cells either by complement fixation or adherence of the red cells to the Fcγ receptors on cells of the fixed mononuclear phagocytic system—principally in the spleen, but also in the liver and other organs (Fig. 41.3).

Anti-erythrocyte antibodies can differ in their physical properties. Some bind to the red blood cells at 37°C (warm autoantibodies); some bind at lower

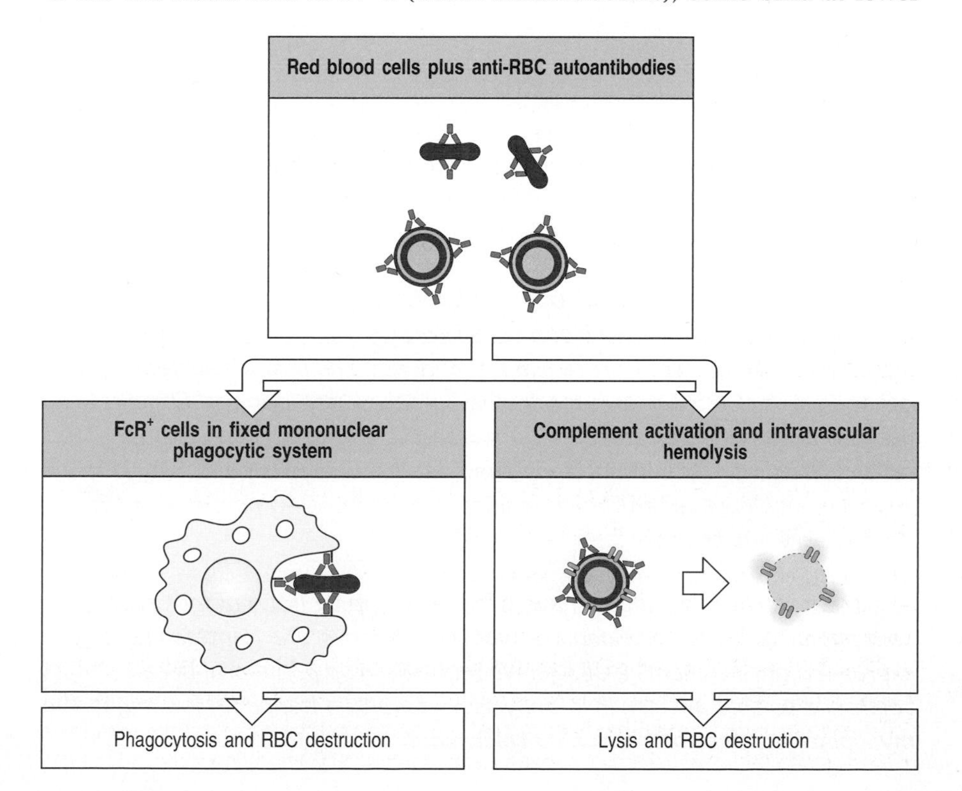

Fig. 41.3 Autoantibodies specific for the surface antigens of red blood cells can cause hemolytic anemia. Antibody-coated red blood cells (RBCs) can be destroyed in several ways. Cells coated with IgG autoantibodies are cleared predominantly by uptake by Fc receptor-bearing macrophages in the fixed mononuclear phagocytic system (left panel). Cells coated with IgM autoantibodies and complement are cleared by macrophages bearing complement receptors CR1 and CR3 in the fixed mononuclear phagocytic system (not shown). The binding of certain rare autoantibodies that fix complement extremely efficiently causes the formation of the membrane-attack complement complex on the red cells, leading to intravascular hemolysis (right panel).

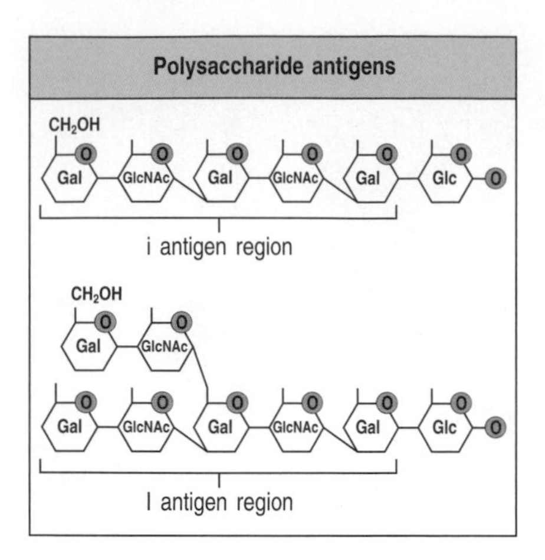

Fig. 41.4 Structure of the I and i antigen regions with which cold agglutinins react. Gal, galactose; GlcNAc, *N*-acetylglucosamine.

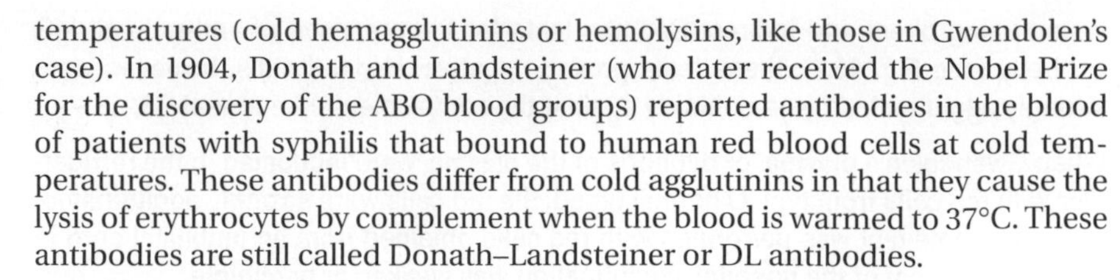
temperatures (cold hemagglutinins or hemolysins, like those in Gwendolen's case). In 1904, Donath and Landsteiner (who later received the Nobel Prize for the discovery of the ABO blood groups) reported antibodies in the blood of patients with syphilis that bound to human red blood cells at cold temperatures. These antibodies differ from cold agglutinins in that they cause the lysis of erythrocytes by complement when the blood is warmed to 37°C. These antibodies are still called Donath–Landsteiner or DL antibodies.

A transient increase in cold agglutinins is most frequently encountered during mycoplasma infection, infectious mononucleosis (the result of infection with the Epstein–Barr virus; see Case 45), and, rarely, during other viral infections. In mycoplasma pneumonia the cold autoantibodies react with a red cell surface antigen called the I antigen. This is a branched carbohydrate structure that occurs on glycoproteins and glycolipids (Fig. 41.4). Cord blood cells express only small amounts of the I antigen and that is why Gwendolen's plasma did not agglutinate them, or did so only weakly. Cord erythrocytes express a related antigen, i (see Fig. 41.4); the I antigen is not fully expressed on red blood cells until 6 months of age.

It has not been possible to find with certainty a cross-reacting I antigen on *Mycoplasma pneumoniae*, which would provide a simple explanation for this autoimmune disease. Exactly how *M. pneumoniae* triggers the transient production of an autoantibody directed to red cells is not known, although there is a strong biochemical similarity between the attachment site (ligand) for the mycoplasma on the host cell surface and the red cell antigen to which the autoantibody binds (Fig. 41.5).

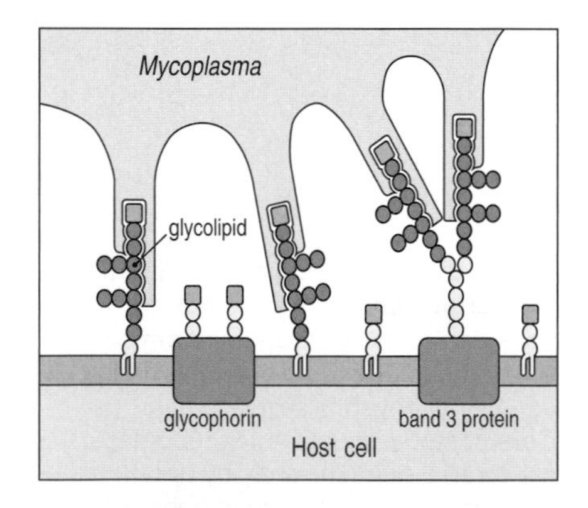

Fig. 41.5 *Mycoplasma pneumoniae* interacts with long carbohydrate chains of I antigen type on the human red cell membrane. These chains include the carbohydrates of band 3 protein and certain glycolipids, but not glycolipids and glycoproteins with short carbohydrate chains. Sequences of sugars that act as I antigen are shown as orange circles; the orange squares depict sialic acid.

Questions.

1 *The hemolytic anemia in Gwendolen Fairfax's case was not severe and it subsided rapidly. In similar cases the hemolysis can be extensive and the drastic fall in hemoglobin can be life threatening. What emergency measure might be undertaken to stop severe hemolysis?*

2 *Autoimmune hemolytic anemia is caused, in general, by autoantibodies of the IgG and IgM classes. Could antibodies of other classes such as IgA, IgD, and IgE cause autoimmune hemolytic anemia?*

3 *In the direct and indirect Coombs tests (see Case 46), agglutination of the antibody-coated red blood cells is brought about by an antibody against human IgG that cross-links the IgG on the red cell surface and causes the cells to agglutinate. In the so-called 'non-gamma' Coombs test an antibody against C3 or C4 is used to agglutinate red blood cells coated with an IgM antibody that has fixed complement. Why is it important to have an antibody against the C3d or C4d portions of C3 or C4 to carry out the test?*

4 *Most normal individuals have a low titer (up to 1:30) of IgM anti-I cold agglutinins in their blood. As mentioned above, the I antigen is little expressed until 6 months of age. We must develop peripheral B-cell tolerance to this antigen in infancy. How does this happen? How might the mycoplasma infection break this tolerance?*

CASE 42 Myasthenia Gravis

The immune response turns against the host.

The specific adaptive immune response can, in rare instances, be mounted against self antigens and cause autoimmune disease. Injury to body tissues can result from antibodies directed against cell-surface or extracellular-matrix molecules, from antibodies bound to circulating molecules that deposit as immune complexes, or from clones of T cells that react with self antigens. A special class of autoimmune disease is caused by autoantibodies against cell-surface receptors (Fig. 42.1). Graves' disease and myasthenia gravis are two well-studied examples. Graves' disease is caused by autoantibodies against the receptor on thyroid cells for thyroid-stimulating hormone (TSH), secreted by the pituitary gland. In this disease, autoantibody binds to the TSH receptor; like TSH, it stimulates the thyroid gland to produce thyroid hormones. In myasthenia gravis, the opposite effect is observed: antibodies against the acetylcholine receptor at the neuromuscular junction impede the binding of acetylcholine and stimulate internalization of the receptor, thereby blocking the transmission of nerve impulses by acetylcholine (Fig. 42.2). In addition, the presence of autoantibodies at the neuromuscular junction initiates complement-mediated lysis of the muscle endplate and damages the muscle membrane.

Myasthenia gravis means severe (*gravis*) muscle (*my*) weakness (*asthenia*). This disease was first identified as an autoimmune disease when an immunologist immunized rabbits with purified acetylcholine receptors to obtain antibodies against this receptor. He noticed that the rabbits developed floppy ears, like the droopy eyelids (ptosis) that are the most characteristic symptom of myasthenia gravis in humans. Subsequently, patients with this disease were found to have antibodies against the acetycholine receptor. In addition, pregnant women with myasthenia gravis transfer the disease to their newborn infants. As IgG is the only maternal serum protein that crosses the placenta from mother to fetus, neonatal myasthenia gravis is clear evidence that myasthenia gravis is caused by an anti-IgG antibody. More recently, patients with myasthenia gravis have been identified who have autoantibodies against muscle-specific kinase (MUSK) rather than the acetylcholine receptor. MUSK is a tyrosine kinase receptor involved in clustering acetylcholine receptors; therefore, these autoantibodies also inhibit signaling through the neuromuscular junction.

This case was prepared by Raif Geha, MD, in collaboration with Janet Chou, MD.

Topics bearing on this case:

- Humoral autoimmunity
- Transfer of maternal antibodies
- Mechanisms for breaking tolerance

Fig. 42.1 Autoimmune diseases caused by antibody against surface or matrix antigens. These are known as type II autoimmune diseases. Damage by IgE-mediated responses (type I) does not occur in autoimmune disease. In most type II diseases, autoantibodies bind to the cell surface or extracellular matrix and target them for destruction by phagocytes (often with the help of complement) and/or natural killer cells. A special class of autoimmune diseases is caused by autoantibodies that bind cellular receptors and either stimulate or block their normal function. Immune-complex disease (type III) is discussed in Case 37. T cell-mediated disease (type IV) is discussed in Cases 36, 40, and 53.

Some common type II autoimmune diseases caused by antibody against surface or matrix antigens		
Syndrome	**Autoantigen**	**Consequence**
Autoimmune hemolytic anemia (see Case 41)	Rh blood group antigens, I antigen	Destruction of red blood cells by complement and phagocytes, anemia
Autoimmune thrombocytopenic purpura	Platelet integrin GpIIb:IIIa	Abnormal bleeding
Goodpasture's syndrome	Noncollagenous domain of basement membrane collagen type IV	Glomerulonephritis Pulmonary hemorrhage
Pemphigus vulgaris (see Case 43)	Epidermal cadherin	Blistering of skin
Graves' disease	Thyroid-stimulating hormone receptor	Hyperthyroidism
Myasthenia gravis	Acetylcholine receptor	Progressive weakness
Insulin-resistant diabetes	Insulin receptor (antagonist)	Hyperglycemia, ketoacidosis
Hypoglycemia	Insulin receptor (agonist)	Hypoglycemia

The case of Mr Weld: from floppy ears to droopy eyelids.

Mr Weld, a 71-year-old retired engineer, had been in good health and active all his life. He developed double vision (diplopia). Initially, he did not want to seek medical attention because the double vision sometimes improved spontaneously. However, it gradually worsened over the course of 4 months and he finally scheduled an appointment with his physician.

On examination, the doctor noticed that Mr Weld had ptosis of both eyelids so that they covered the upper third of the irises of his eyes. When the doctor asked Mr Weld to look to the right and then to the left, he noticed limitations in the ocular movements of both eyes, as shown in Fig. 42.3.

The remainder of the neurological examination was normal. No other muscle weakness was found during the examination.

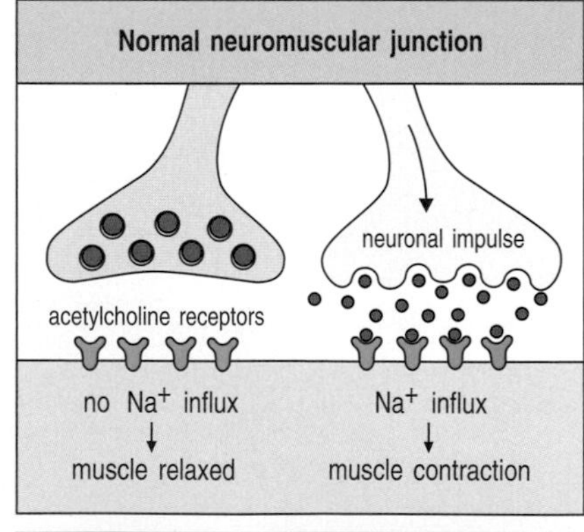

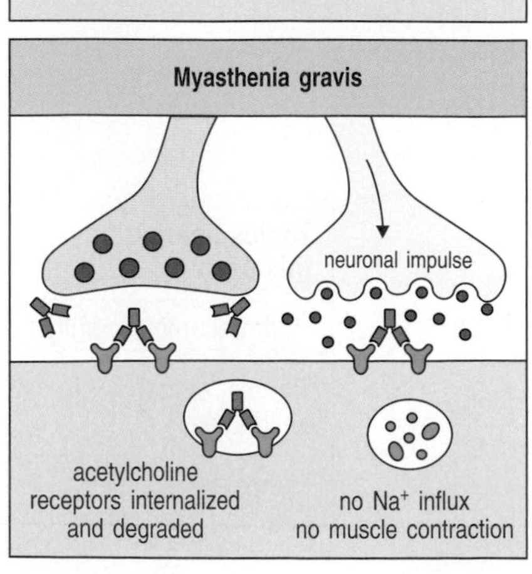

Fig. 42.2 Autoantibodies against the acetylcholine receptor weaken the reception of the signal from nerve ends that cause the muscle cell to contract. At the neuromuscular junction, acetylcholine is released from stimulated neurons and binds to acetylcholine receptors, triggering muscle contraction. The acetylcholine is destroyed rapidly by the enzyme acetylcholinesterase after release. In myasthenia gravis, autoantibodies against the acetylcholine receptor induce its endocytosis and degradation, and prevent muscles from responding to neuronal impulses.

A radiological examination of the chest was performed, and it was normal. There was no evidence in the radiograph of enlargement of the thymus gland. A blood sample was taken from Mr Weld, and his serum was tested for antibodies against the acetylcholine receptor. The serum contained 6.8 units of antibody against the acetylcholine receptor (normal less than 0.5 units). Mr Weld was told to take pyridostigmine, an inhibitor of cholinesterase. His double vision improved steadily but he developed diarrhea from the pyridostigmine, and this limited the amount he could take.

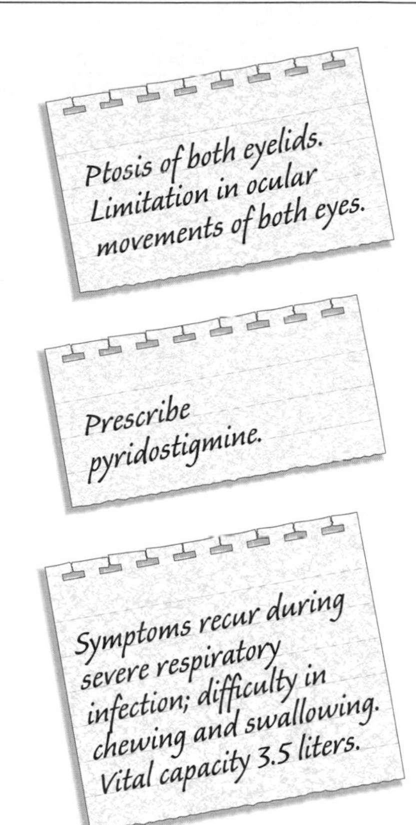

Ptosis of both eyelids. Limitation in ocular movements of both eyes.
Prescribe pyridostigmine.
Symptoms recur during severe respiratory infection; difficulty in chewing and swallowing. Vital capacity 3.5 liters.

Three years later, Mr Weld developed a severe respiratory infection. Soon afterward, his ptosis became so severe that he had to lift his eyelids by taping them with adhesive tape. His diplopia recurred and his speech became indistinct. He developed difficulty in chewing and swallowing food. He could only tolerate a diet of soft food and it would take him several hours to finish a meal.

On examination the neurologist noted that Mr Weld now had weakness of the facial muscles and the tongue, and the abnormality in ocular movements again became apparent. Because of the diarrhea Mr Weld was only able to tolerate one-quarter of the prescribed dose of pyridostigmine. He also developed difficulty in breathing. His vital capacity (the amount of air he could exhale in one deep breath) was low, at 3.5 liters.

He was admitted to hospital and treated with azathioprine. Thereafter he showed steady improvement. His ptosis and diplopia improved remarkably and he was able to eat normally. His vital capacity returned to normal and was measured to be 5.1 liters.

Myasthenia gravis.

The defining characteristic of myasthenia gravis is a fluctuating weakness that worsens with activity and improves with rest. Normally, repetitive nerve stimulation during sustained physical activity results in the release of decreased amounts of acetylcholine with each successive stimulus; however, enough acetylcholine is released to achieve the desired muscle strength in healthy individuals. In contrast, patients with myasthenia gravis have fewer functional acetylcholine receptors as a result of the presence of anti-receptor autoantibodies. During repetitive nerve stimulation, the combination of fewer functional acetylcholine receptors with the physiologic decrease in neurotransmitter release results in muscular weakness.

Mr Weld experienced a common type of myasthenia gravis, called the oculobulbar form because it primarily affects the muscles of the eye. Older patients tend to have more generalized muscle weakness as well, and often have autoantibodies against muscle proteins in addition to anti-acetylcholine receptor antibodies. In very severe cases, difficulty in swallowing can cause the aspiration of food particles into the lung and impaired breathing, which may be fatal. Plasmapheresis (the filtration and removal of plasma from whole blood) can be used to remove the autoantibodies and treat a myasthenic crisis.

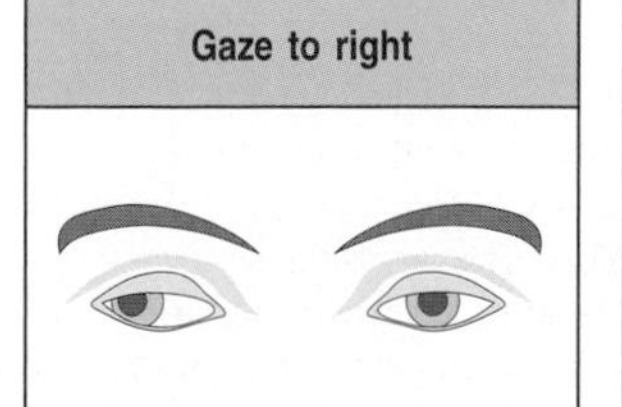

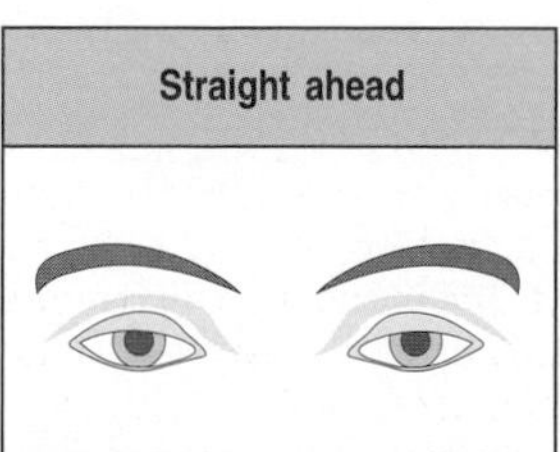

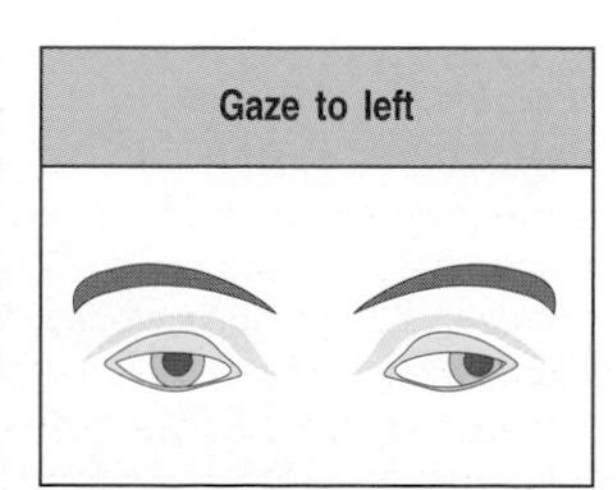

Fig. 42.3 Diagram of ocular movement limitation.

In younger people, the disease presents most often with weakness in the eye muscles. Chest radiographs of younger people with myasthenia gravis frequently reveal enlargement of the thymus gland; however, an association between myasthenia gravis and tumors of the thymus (thymomas) is more common in adults. Early removal of the thymus gland (thymectomy), particularly in those with thymomas, may lead to symptomatic improvement. Although the definitive mechanisms underlying the association between thymomas and myasthenia gravis are not yet identified, it has been hypothesized that neoplastic epithelial cells in the thymoma express selflike epitopes resembling proteins such as the acetylcholine receptor. In addition, thymomas have been found to have decreased expression of the autoimmune regulator gene (*AIRE*; see Case 17) and smaller numbers of regulatory T cells, indicating that an abnormal microenvironment within the thymomas results in impaired negative selection. However, as the occurrence of myasthenia gravis does not correlate with decreased expression of *AIRE* in the thymus, there are still unidentified factors that influence the development of myasthenia gravis in patients with thymomas.

Questions.

1 *Newborn infants of mothers with myasthenia gravis exhibit symptoms of myasthenia gravis at birth. How long would the disease be likely to last in these infants?*

2 *Pyridostigmine is an ideal drug for the treatment of myasthenia gravis. It inhibits the enzyme cholinesterase, which normally cleaves and inactivates acetylcholine. In this way, pyridostigmine prolongs the biological half-life of acetylcholine. Unfortunately, it also causes diarrhea by increasing the amount of acetylcholine in the intestine. Acetylcholine binds to the muscarinic receptors in the intestine and increases intestinal motility. Because he could not tolerate full therapeutic doses of pyridostigmine and was getting worse, Mr Weld was given azathioprine (Fig. 42.4) and showed marked improvement. What did the azathioprine do? What would concern you about prolonged use of this drug?*

3 *Mr Weld had a severe relapse in his disease after a respiratory infection. Many autoimmune diseases seem to be triggered by infection, and relapses in autoimmune diseases frequently follow an infection. Can you explain how this might happen?*

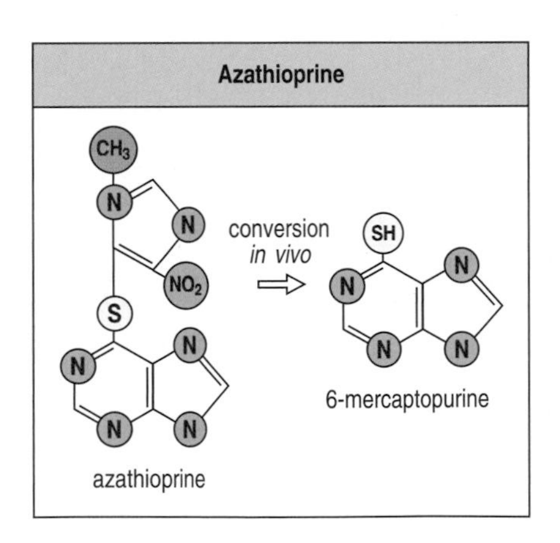

Fig. 42.4 The structure of azathioprine and its active product, mercaptopurine.

CASE 43 Pemphigus Vulgaris

Autoimmune attack on the integrity of the skin.

Autoimmune disease results when the adaptive immune response is directed against self antigens. The particular pathology of the disease depends on the nature of the self antigen and the type of immune response that is mounted against it. In the type II autoimmune diseases (see Fig. 42.1), the pathology is caused by autoantibodies that interact with self antigens in the extracellular matrix or on cell surfaces. In myasthenia gravis, for example, autoantibodies against the acetylcholine receptor on skeletal muscle block its function at the neuromuscular junction and cause paralysis (see Case 42), whereas in Graves' disease an autoantibody against the receptor for thyroid-stimulating hormone acts as an agonist, stimulating receptor activity and causing hyperthyroidism. In the type II autoimmune disease pemphigus vulgaris, the self-reactive agent is an autoantibody against a structural protein of the epidermal cells of the skin. Its actions result in the skin cells coming apart from each other; the affected skin blisters and is destroyed.

'Pemphigus' is derived from the Latin word for blister, and 'vulgaris' means common or ordinary. Patients with this disease have autoantibodies against desmoglein-3, which is a protein component of the desmosome—one of the intercellular junctions that link skin cells and other epithelial cells tightly to each other. Desmogleins are members of the cadherin family of cell adhesion molecules, proteins that effect intercellular adhesion in a calcium-dependent manner. Disruption of desmoglein causes blisters to form in the skin and on the mucous membranes; extensive sloughing of the skin may ensue (Fig. 43.1).

Topics bearing on this case:

- Peptide binding by MHC class II antigens
- Humoral autoimmunity
- Epitope spreading
- HLA associations with disease

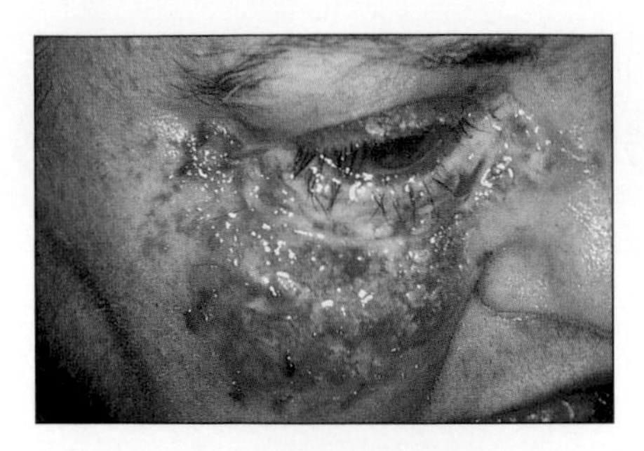

Fig. 43.1 Lesions of pemphigus vulgaris that have coalesced to produce a large plaque. The plaque has a moist surface that is oozing and has crusts of dried serum from blister fluid. Photograph courtesy of Razzaque Ahmed.

Evidence that pemphigus vulgaris is caused by an autoantibody comes from the observation that infants born to mothers with the disease have a transient period of skin blistering during the neonatal period. There is also a good animal model in which human IgG from patients with pemphigus injected into neonatal mice reproduces pemphigus.

The case of Arthur Sammler, Esq.: a hoarse lawyer with a sore mouth and blistering skin.

Arthur Sammler is a lawyer of Ashkenazi Jewish descent. He was 55 years of age when an irritation in his throat became persistent enough for him to consult a physician. The physician suggested that he gargle with warm saline, which seemed to help. A month later, however, after a long day in court he became very hoarse and the hoarseness persisted for a week. A week later he noticed a 'sore' on his right cheek. He told the physician he was under considerable stress over a custody case, and the physician diagnosed a herpes simplex infection and advised him to take oral acyclovir. The lesion did not improve.

Persistent skin sore and ulceration of mouth; resistant to usual therapies; do biopsy.

At a more thorough examination 2 weeks later, numerous erosions and ulcers in the mucosa of his mouth and gums were also revealed. These were thought to result from a yeast infection with *Candida albicans* and he was treated with a Mycostatin (nystatin) mouthwash.

However, new lesions soon appeared on the palate, uvula, and tongue, and Mr Sammler returned to the physician. This time, a biopsy of the lesions was taken by an oral surgeon. While awaiting the results of the biopsy Mr Sammler developed erosions of the skin on his scalp and neck.

Diagnosis of pemphigus vulgaris; start immunosuppressive therapy.

Microscopic examination of the biopsy revealed disruption of the epidermal layer of the skin characteristic of pemphigus vulgaris (Fig. 43.2). Mr Sammler was sent to a dermatologist, who took another biopsy for immunofluorescence studies; blood samples were taken for analysis. The immunofluorescence studies revealed deposits of IgG in an intercellular pattern within the epidermis (Fig. 43.3). His serum contained an antibody against desmoglein-3 (known commonly as 'pemphigus antibody') in a titer of 1:640.

Condition not responding to steroids; try cyclophosphamide.

Mr Sammler was treated with large doses of prednisone, 120 mg per day. After 3 weeks of this therapy, during which time no improvement occurred, new lesions appeared on his back and scalp. The dose of prednisone was increased to 180 mg per day. Two weeks later he developed a persistent cough, fever, and chest pain. A chest radiograph during an emergency room visit suggested that he had developed interstitial pneumonia due to *Pneumocystis jirovecii*. He was admitted to hospital and treated for this infection.

It was found that his pemphigus antibody titer had risen to 1:1280. He was started on cyclophosphamide at 150 mg per day, and was told to decrease his prednisone dose by 10 mg every 10 days and to rinse his mouth daily with hydrogen peroxide and elixir of Decadron (a corticosteroid). New lesions kept appearing, so the cyclophosphamide was increased to 200 mg per day.

Responding to cyclophosphamide; decrease dose gradually.

Over the next 8 weeks the pemphigus antibody titer decreased to 1:320; no new lesions appeared, and the older ones healed. The cyclophosphamide dose was decreased over an 8-month period, and Mr Sammler remained symptom-free after cyclophosphamide was stopped. In the year since then, he has noticed the occasional new lesion after damage to his skin; these new lesions are promptly injected with a corticosteroid. His pemphigus antibody titer has remained at 1:80 during this asymptomatic period.

Pemphigus vulgaris.

The case of Arthur Sammler illustrates how an autoimmune disease that was frequently fatal in the past can now be controlled by immunosuppressive drugs. The epidermis, with its cells held tightly together, forms an important barrier to the entry of pathogens into the body. Extensive blistering and sloughing of skin destroy this barrier; in the past, this often resulted in bloodstream infection (septicemia) with *Staphylococcus aureus*, a common contaminant of human skin.

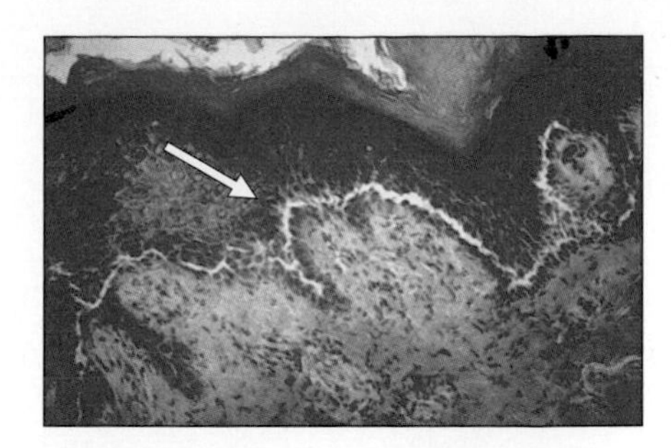

Fig. 43.2 Histopathology of an early skin lesion of pemphigus vulgaris. Stained with hematoxylin and eosin. An epidermal, suprabasal vesicle (blister) is clearly seen. Photograph courtesy of Razzaque Ahmed.

The mechanism of autoimmune tissue destruction seems to be somewhat unusual in pemphigus vulgaris. In some autoimmune diseases, the autoantibodies cause tissue destruction by stimulating the complement system, but the autoantibody in pemphigus vulgaris is of the IgG4 subclass and does not fix complement. Some other mechanism must therefore be invoked to explain the disruption of desmoglein-3 adhesion. It is thought that the binding of the autoantibody to its antigen in some way causes the upregulation of serine proteinase activity on the surface of the epidermal cells and that this results in the proteolytic digestion of desmoglein-3. The particular enzyme that digests the desmoglein-3 is as yet unknown.

Long-term study of the antibody response in some patients with pemphigus vulgaris has shown that the response evolves to involve epitopes other than the one against which antibodies are initially directed. This is known as epitope spreading and has been found to correlate with the clinical progression of disease. Antibodies present before the onset of disease are specific for epitopes in the extracellular part of the molecule nearest to the cell membrane. These antibodies do not bind to cell-surface desmoglein, nor do they transfer disease when injected into mice. With disease onset, antibodies specific for the domains farthest away from the cell surface become detectable (Fig. 43.4). These IgG antibodies bind to cell-surface desmoglein, and when injected into mice they induce disease. The process by which the immune response initially targets epitopes in one part of an antigenic molecule and then progresses to other, non-cross-reactive, epitopes of the same molecule is called intramolecular epitope spreading. In pemphigus, it is only with epitope spreading that disease-causing antibodies are made.

Pemphigus vulgaris is encountered most frequently in Ashkenazi Jews and has a strong association with an HLA haplotype found mainly in that ethnic group. This HLA haplotype includes HLA-DR4 and HLA-DQ3, which are present in virtually all Ashkenazi Jews with pemphigus vulgaris. Patients are usually heterozygous for this haplotype and only in rare cases are they homozygous. Unaffected relatives of patients with pemphigus who bear the same HLA haplotype frequently also have antibodies against desmoglein-3. However, these antibodies are of the IgG1 subclass and do not cause disease. In other ethnic groups, pemphigus vulgaris is usually associated with HLA-DR14 and HLA-DQ5.

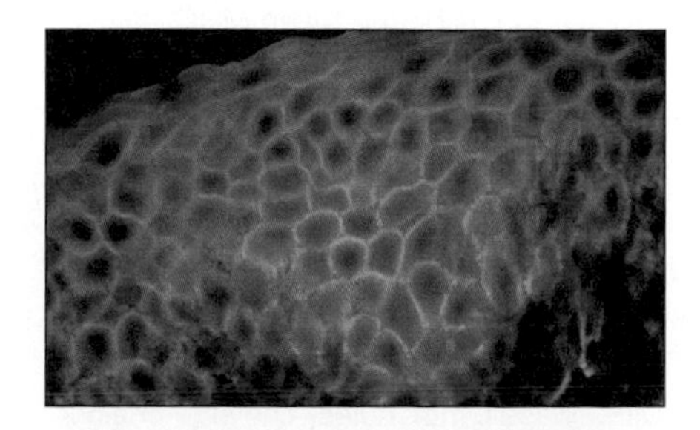

Fig. 43.3 Direct immunofluorescence study of tissue around a pemphigus vulgaris lesion. The normal-appearing tissue around the lesion contains deposits of IgG, other immunoglobulins, and complement components in the intercellular spaces of the entire stratum malpighii of the epidermis. Photograph courtesy of Razzaque Ahmed.

A molecular explanation has been discovered for the association of pemphigus vulgaris with HLA-DR4. Each major histocompatibility complex (MHC) class II variant can bind a subset of all possible peptides; the actual peptides that can be bound by any given MHC molecule are determined by the particular pattern of amino-acid binding sites in the peptide-binding cleft of the MHC molecule (Fig. 43.5). The β chain of the MHC class II molecule DR is designated DRB1; 22 known subtypes of this chain have been identified by DNA typing among those bearing DR4 (DRB1*04). Only one of these variants—DRB1*0402—is associated with pemphigus vulgaris. It differs from the other 21 variants in that the amino acid residue at position 71, which lies in one of the binding pockets in the peptide-binding groove of the MHC molecule, is negatively charged (glutamic acid), whereas all the other variants have

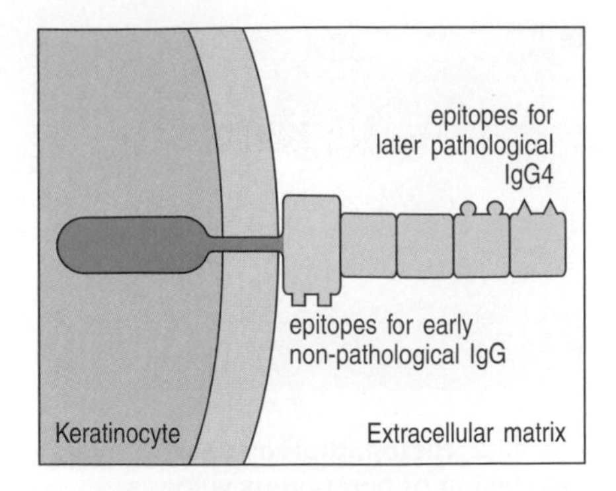

Fig. 43.4 The progressive involvement of desmoglein epitopes in pemphigus vulgaris. This skin-blistering disease is caused by antibodies that bind to desmoglein-3, an adhesion molecule in the cell junctions that hold keratinocytes together. When the autoimmune B-cell response begins, harmless antibodies are made against the extracellular region of desmoglein-3 nearest to the cell surface. In time, the response spreads and antibodies are made against the region farthest away from the cell surface. These antibodies cause disease and are of the IgG4 subclass.

positively charged amino acid residues (lysine or arginine) at this position. Only DRB1*0402 will bind the antigenic peptide derived from desmoglein-3 and present it to T lymphocytes.

Questions.

1 *The intravenous administration of gamma globulin frequently improves autoimmune disease. It has been observed that the administration of gamma globulin to patients with pemphigus vulgaris significantly lowers the titer of IgG4 anti-desmoglein-3 and improves the clinical course of the disease. How do you explain this effect?*

2 *Relatives of patients with pemphigus vulgaris who share the DRB1*0402 MHC class II type have IgG1 antibodies against desmoglein-3. These antibodies fix complement but do not cause disease. Can you speculate why this antibody does not cause disease, whereas the IgG4 antibody causes disease?*

3 *What might explain the difference between unaffected people who make only the IgG1 antibody against desmoglein-3 and the patients who make the IgG4 antibody? Can you design an experiment to test your hypothesis?*

4 *We have seen cyclophosphamide used to eradicate bone marrow cells in Cases 8 and 27. Why is it useful in the treatment of pemphigus vulgaris?*

5 *Why did Mr Sammler develop pneumonia caused by* Pneumocystis jirovecii *during his initial treatment for pemphigus vulgaris?*

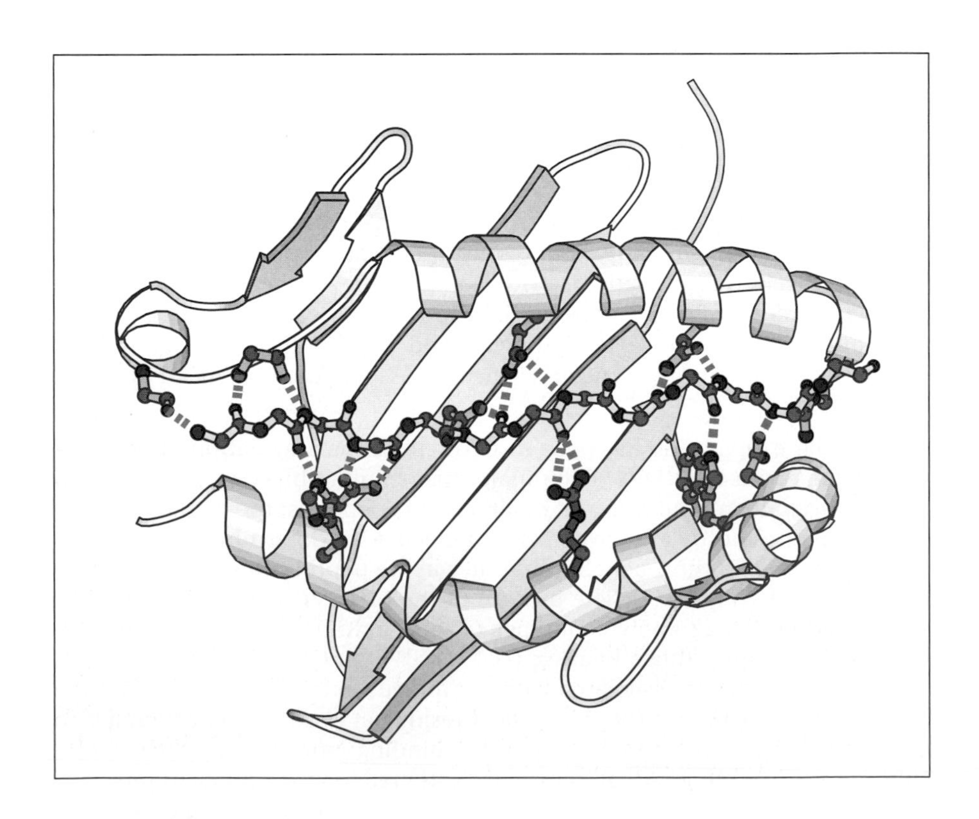

Fig. 43.5 Peptides bind to MHC class II molecules by interactions along the length of the binding groove. A peptide (yellow; shown as the peptide backbone only, with the amino terminus to the left and the carboxy terminus to the right), is bound by an MHC class II molecule through a series of hydrogen bonds (dotted blue lines) that are distributed along the length of the peptide. The hydrogen bonds toward the amino terminus of the peptide are made with the backbone of the class II polypeptide chain, whereas throughout the peptide's length, bonds are made with residues that are highly conserved in MHC class II molecules. The side chains of these residues are shown in gray on the ribbon diagram of the MHC class II binding groove. Most DR4 alleles have a positively charged arginine (shown here in purple) or lysine at position 71 of the β chain. The DRB1*0402 allele that is linked to susceptibility to pemphigus has a negatively charged glutamic acid residue instead.

CASE 44 Celiac Disease

Chronic tissue damage in the gut caused by hypersensitivity to a common food antigen.

Celiac disease (also known as gluten-sensitive enteropathy) is an immune-mediated inflammatory disease of the gastrointestinal tract caused by a permanent sensitivity to gluten, a mixture of proteins present in the grains of wheat, barley, and rye. Gluten is degraded into antigenic peptides that trigger an immunologic process in the gut that leads to diarrhea, malabsorption, and, ultimately, nutritional deficiencies and failure to thrive. There is a description of celiac disease dating back to the 1st century AD, but it was not until the 20th century that the trigger for the disease was identified. Willem Karel Dicke, a Dutch pediatrician, noted that children with celiac disease had a clear improvement in their symptoms during the Second World War, when cereals were scarce. After the war and the reinstatement of cereals in their diet, children with celiac disease relapsed, and wheat was recognized as the major culprit.

Celiac disease is perhaps the most common genetic disorder, with a prevalence of 0.5–1.0% in many populations. The principal genetic determinants of the disease have been identified as the highly variable HLA class II *DQA* and *DQB* genes located on chromosome 6 in the major histocompatibility complex. Celiac disease is singular among immunologic diseases in that the exogenous antigen, the autoantigen, and the genetic susceptibility have all been clearly described, and therefore it provides a useful model for studying the genesis of immunologically mediated diseases.

Topics bearing on this case:

Gut mucosal immune system

Allergy and hypersensitivity

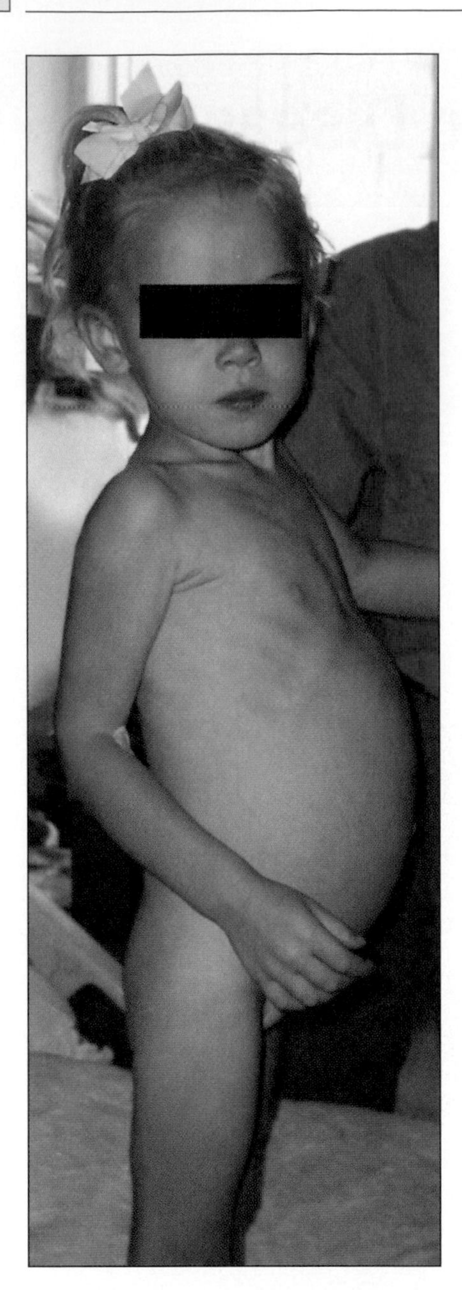

Fig. 44.1 A child with celiac disease. Note the protuberant abdomen, muscle wasting, and malnourished appearance. Photograph courtesy of Dascha Weir.

Demeter O'Reilly: a little girl who lost weight and became weak.

Demeter O'Reilly was a healthy baby who grew normally until 12 months old, when her parents noticed that she became irritable and stopped gaining weight. She was observed closely by her pediatrician for the next 3 months during which time she developed a distended abdomen, passed large, foul-smelling stools, and lost nearly 2 kg in weight. Her parents also noticed that Demeter was becoming progressively weaker.

At birth, Demeter weighed 3.8 kg, was 52 cm long, and had a head circumference of 36 cm (all within the normal range). She grew normally until nearly 15 months of age and reached all the developmental milestones appropriately.

Both of her parents were of Irish American ancestry and were healthy, as was her 3-year-old sister. There was no known history of metabolic, neurologic, or gastrointestinal disease in children in the family. There was no history of autoimmune diseases, including type 1 diabetes mellitus.

Given the progression of her symptoms and the weight loss, Demeter was referred to the Children's Hospital Emergency Room at 15 months. When she arrived she appeared pale and chronically ill. She weighed 8.5 kg (5th centile), her height was 77.5 cm (50th centile), and her head circumference was 46 cm (50th centile). The examination made Demeter fretful but she could be consoled. She was mildly dehydrated, and had a protuberant abdomen and significant muscle wasting in her buttocks, legs, and arms (Fig. 44.1). No enlargement of liver, spleen, or lymph nodes was evident.

Laboratory tests revealed anemia, with a hemoglobin level of 10 g dl^{-1} (normal 12–14 g dl^{-1}). Her white-cell count and platelets were normal. She had a mild elevation of liver enzymes (transaminitis), with aspartate aminotransferase at 48 U l^{-1} (normal range 2–40 U l^{-1}) and alanine aminotransferase at 46 U l^{-1} (normal range 3–30 U l^{-1}). Her albumin was normal at 3.7 g dl^{-1}. No evidence of metabolic disease was found, and so serologic samples for testing for celiac disease were sent to the laboratory. Demeter tested positive for anti-endomysium IgA antibodies (autoantibodies against the connective tissue that sheathes muscle—the endomysium) at a titer of 1:2560 and for anti-tissue transglutaminase (TTG) IgA antibodies at a level greater than 118 U ml^{-1} (greater than 25 U ml^{-1} is considered positive). She was positive for anti-gliadin IgA antibodies at greater than 104 U ml^{-1} (greater than 23 U ml^{-1} is considered positive) and anti-gliadin IgG antibodies at greater than 77 U ml^{-1} (greater than 28 U ml^{-1} is considered positive).

Given Demeter's symptoms and the positive serologic results, the decision was made to perform a biopsy of her gastrointestinal tract. An upper gastrointestinal endoscopy revealed edema and flattening of the mucosal folds in the duodenal bulb. Multiple biopsies were taken from the stomach and duodenum. Review of the pathology of the tissues showed subtotal to total villous atrophy and increased intraepithelial lymphocytes in the second portion of the duodenum and in the duodenal bulb (Fig. 44.2). The clinical picture, serology, and biopsy results confirmed the diagnosis of celiac disease.

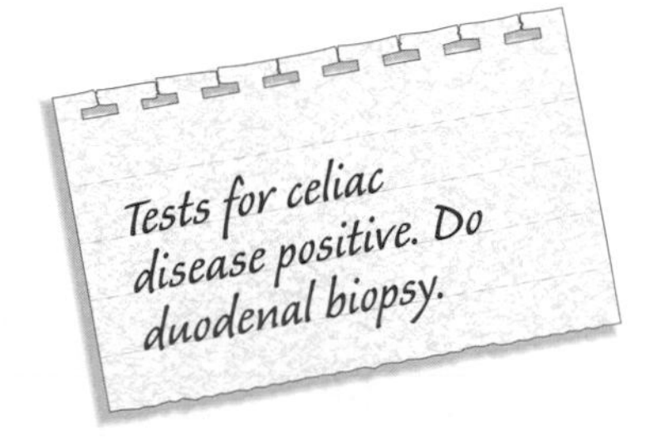

Tests for celiac
disease positive. Do
duodenal biopsy.

Demeter's family received extensive nutritional counseling and she was started on a gluten-free diet. At the 6-month follow-up, she was doing extremely well, with excellent energy level. She had normal bowel movements once a day and had no abdominal pain or vomiting. Her weight was now in the 75th centile and her height was at the 50th centile for her age. She appeared a robust and healthy toddler with a normal abdomen and muscle bulk. Serologic tests revealed no anti-endomysium and anti-TTG antibodies. Her family had availed themselves of the many sources of advice and support available to families with celiac disease, and felt comfortable with her gluten-free diet.

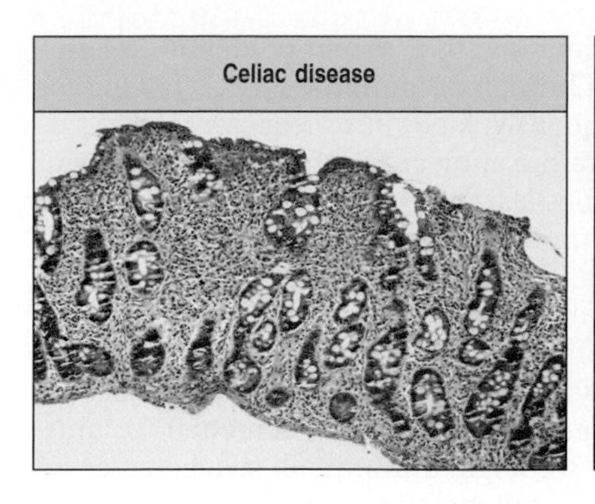

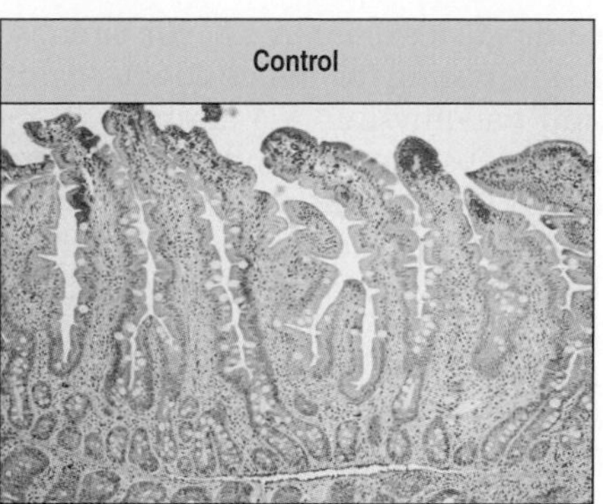

Fig. 44.2 Comparison of mucosal tissue taken by internal biopsy from a patient with celiac disease and from a healthy intestine. The tissue from the patient with celiac disease is on the left, the healthy tissue on the right. Note the flattened mucosal surface, crypt hyperplasia, and extensive inflammatory infiltrate in the celiac disease biopsy. Photograph courtesy of Donald A. Antonioli.

Celiac disease.

The prevalence of celiac disease, or gluten-sensitive enteropathy, varies by population, and it has been estimated that 0.5–1.0% of the general population in Europe and the United States are affected. However, variability in presentation and delay in onset of symptoms probably lead to the disease being under-diagnosed. Ninety to 95% of patients with celiac disease express HLA-DQ2, and the remainder express DQ8. The disease clusters in families. First-degree relatives of affected patients have a 2–5% prevalence of symptomatic celiac disease, and up to 10% may have histological evidence of the disease, even in the absence of symptoms. In addition, patients with type 1 diabetes mellitus, Down syndrome (trisomy 21), Turner syndrome (45,XO), Williams syndrome (congenital facial and cardiac abnormalities with hypercalcemia), autoimmune thyroiditis, or selective IgA deficiency are at higher risk than the rest of the population of developing celiac disease.

Children with celiac disease classically come to the physician's attention between 6 and 24 months of age with abdominal distension, diarrhea, malabsorption, and weight loss. Anemia, irritability, and muscular wasting are often observed. Rarely, children in celiac crisis are seen, with severe diarrhea, dehydration, and marked disturbance in electrolyte balance; they require immediate treatment with intravenous corticosteroid. Older children may have more subtle symptoms but will also have diarrhea, nausea, vomiting, and abdominal pain. Interestingly, celiac disease with constipation is sometimes seen. Atypical features of the disease include hypoplasia of the enamel of the permanent teeth, osteopenia/osteoporosis, short stature, and delayed puberty. Dermatitis herpetiformis, an intensely itchy rash consisting of small solid papules and water-filled vesicles, is a dermatologic manifestation of celiac disease; small-bowel biopsies in patients with dermatitis herpetiformis reveal lesions that are consistent with gluten-sensitive enteropathy. Both the rash and the gastrointestinal lesions remit if gluten is removed from the diet.

Adults can also develop celiac disease. An iron-deficiency anemia that does not respond to oral therapy is the most common presentation in adults; gastrointestinal symptoms such as diarrhea and flatulence sometimes predominate. It is also important to consider celiac disease in patients with unexplained bone fractures, transaminitis, or neurologic symptoms including peripheral neuropathy or ataxia, as well as in women with infertility.

The development of highly sensitive and specific serologic tests for celiac disease has greatly aided its diagnosis. Serologic tests are also important, to monitor compliance with a gluten-free diet, because no anti-gluten antibodies can be detected when patients avoid gluten completely. Many different serologic tests are in use; for initial screening, the current guidelines recommend testing for IgA antibody against human recombinant TTG. Testing for anti-endomysium IgA antibody is frequently used, but measurement of this test is observer-dependent and is therefore subject to error. Positive titers of anti-TTG and anti-endomysium antibodies are almost 100% successful in predicting celiac disease lesions in symptomatic patients. Anti-gliadin antibodies are less predictive, because elevated levels of anti-gliadin IgG antibodies are found in patients with inflammatory bowel disease and in healthy people. IgA deficiency occurs in 2% of children with celiac disease and can lead to misinterpretation of low antibody levels, so it is important to measure IgA levels. Patients with known IgA deficiency should be screened for anti-TTG IgG antibodies. Regardless of the serologic findings, all suspected cases should be confirmed by a biopsy of the small intestine. Characteristic pathology includes partial to total villous atrophy and lymphocyte infiltration with crypt hyperplasia in the small intestine.

The treatment for celiac disease is lifelong adherence to a gluten-free diet. This is challenging to comply with and is also more expensive than a normal diet. It is essential to give patients and their families extensive nutritional counseling as well as support in their attempts to maintain the diet.

Untreated celiac disease leads to significant illness and an increased risk of mortality. Children with untreated celiac disease are most likely to have problems with growth and decreased bone mineralization, which will be corrected on a gluten-free diet. Women with untreated celiac disease are more likely to have spontaneous abortions, babies of low birth weight, and a shorter duration of breastfeeding. Lastly, there is an increase in mortality due to cancer in adults with celiac disease, primarily enteropathy-associated T-cell lymphoma. Recent results suggest that strict compliance with a gluten-free diet may reduce the risk of cancer compared with those patients who do not comply.

The pathogenesis of celiac disease is now quite well understood. The exogenous antigen is gluten, a mixture of gliadin and glutenin polypeptides present in wheat. Ingested proteins are usually degraded into small residues by gastric, pancreatic, and intestinal enzymes during transit through the gastrointestinal tract. In most cases, these residues are not appropriate stimulators of the immune system because they are not big enough to bind HLA molecules. However, gluten contains a proline-rich peptide of 33 amino acids (the 33-mer) that survives transit through the gastrointestinal tract and arrives intact in the small bowel. This peptide passes through the gastrointestinal lining into the subepithelial space. Such trafficking is usually prevented by the competent tight intercellular junctions of the gastrointestinal epithelium. A number of factors may damage the epithelial barrier, including infection, gluten-induced upregulation of zonulin (an intestinal peptide involved in the regulation of tight junctions), or mechanical stress.

In the subepithelial space the peptides are modified by enzymes, including TTG, that increase their antigenicity, and they are picked up and presented by antigen-presenting cells to CD4 T cells, thereby initiating an inflammatory response that is self-perpetuating in the presence of gluten and damages the gastrointestinal tract. The enzyme TTG deamidates the peptide, making it more antigenic, and has also been identified as an autoantigen in the disease (Fig. 44.3). The key genetic determinants of CD have been identified as HLA-DQ2 or HLA-DQ8. The interplay of these three factors underpins the immune-mediated damage that is seen in patients with celiac disease (Fig. 44.4).

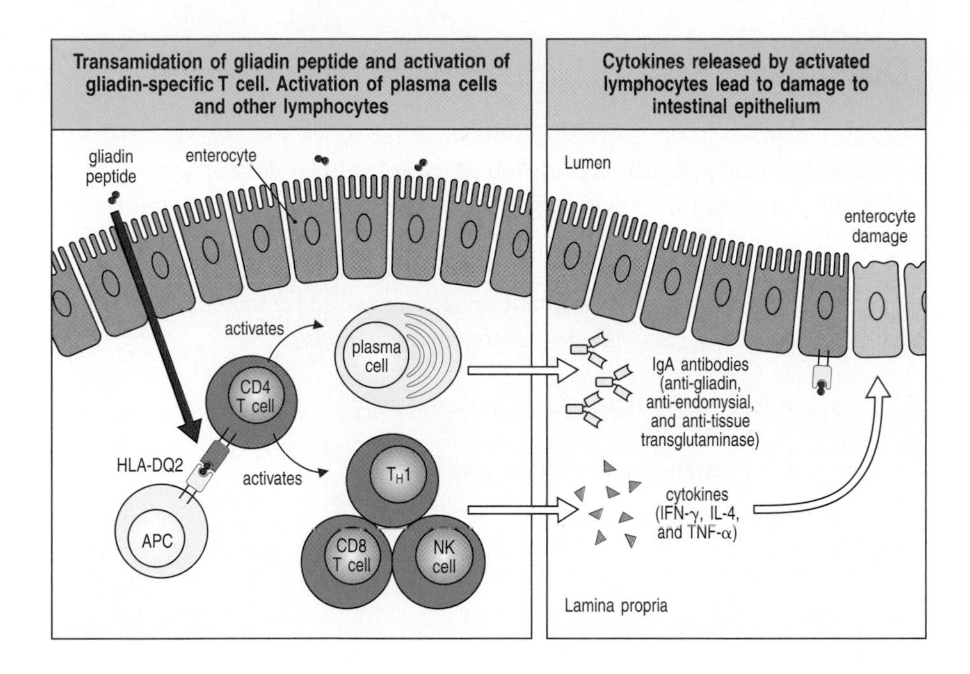

Fig. 44.3 The pathogenesis of celiac disease. Inflammation of the small intestine is caused by CD4 T cells responding to peptides derived from gluten that are deamidated by tissue transglutaminase and presented by HLA-DQ8 or HLA-DQ2 molecules. APC, antigen-presenting cell.

After presentation of the immunogenic peptides to CD4 T cells, an inflammatory response is initiated that is characterized by a CD4 T_H1 response, with interferon-γ (IFN-γ) as the predominant cytokine. We know relatively little about the effector mechanisms that give rise to the celiac lesion. It is likely that some of the intestinal damage is mediated via a bystander effect of IFN-γ, but there may be other mechanisms involved. Examination of gastrointestinal lesions in patients shows infiltration of the intestinal mucosa by both CD4 lamina propria lymphocytes and intraepithelial CD8 T cells. As described above, there is significant evidence for the role of CD4 cells, but the role of the CD8 cells remains elusive.

The anti-TTG antibodies produced in celiac disease are of clinical importance in diagnosis and in tracking disease activity, but the mechanism of their production is not yet fully understood. It has been proposed that hapten–carrier-like complexes form between TTG and complexes of gliadin, which activate helper T cells that then provide help to TTG-specific B cells. When gluten is removed from the diet, antibody levels decrease and the inflammation and injury resolve.

The innate immune system may also be stimulated by gluten. There is evidence that gluten-derived peptides induce the release of IL-15 by intestinal epithelial cells. IL-15 activates antigen-presenting dendritic cells in the lamina propria and also leads to the upregulation of expression of the cell-surface protein MIC-A by epithelial cells. Potentially cytotoxic CD8 intraepithelial lymphocytes in the mucosal epithelium can be activated via their NKG2D receptors, which recognize MIC-A, and they then kill MIC-A-expressing epithelial cells. These innate immune responses may create some intestinal damage and may also induce the co-stimulatory molecules necessary for initiating an antigen-specific CD4 T-cell response to other parts of the α-gliadin molecule. The ability of gluten to stimulate both innate and adaptive immune responses may explain its unique ability to induce celiac disease.

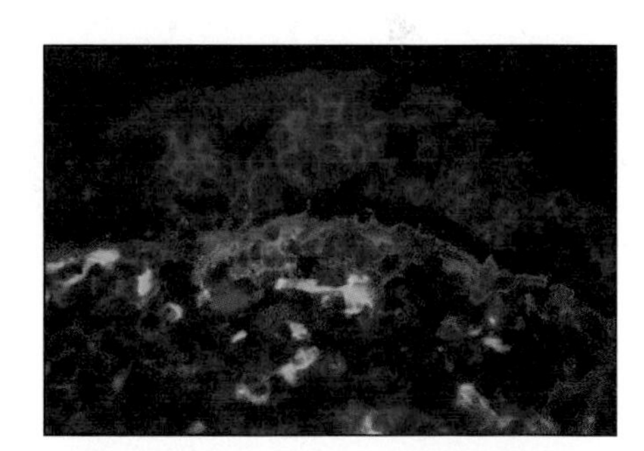

Fig. 44.4 Immunofluorescence staining of the mucosa of the small intestine from an untreated patient with celiac disease. Tissue transglutaminase (TTG, pink), HLA-D (green), and T cells (CD3, purple) have been specifically stained for. Note the close spatial relationship between TTG, the antigen-presenting cells expressing HLA-DQ, and T cells just beneath the intestinal epithelium.

Questions.

1 *Given the high prevalence of celiac disease and the difficulty in maintaining a gluten-free diet, are there any new therapies on the horizon for this condition?*

2 *What are the current recommendations regarding screening of first-degree relatives and high-risk groups for celiac disease?*

3 *How is a food determined to be gluten free?*

4 *Can patients with celiac disease eat oats?*

CASE 45 Acute Infectious Mononucleosis

Cytotoxic T cells terminate viral infection.

All viruses, and some bacteria, multiply inside infected cells; indeed, viruses are highly sophisticated parasites that do not have a complete biosynthetic or metabolic apparatus of their own and, in consequence, must replicate inside a living cell. Once inside a cell, a pathogen is not accessible to antibodies and has to be eliminated by other means.

Some intracellular bacteria live and multiply in membrane-bound phagosomes within macrophages and are killed by antibacterial agents released into these vacuoles after macrophage activation by CD4 T_H1 cells. Viruses, in contrast, together with those bacteria that live in the cytosol, can be eliminated only by destruction of the infected cell itself. This role in host defense is fulfilled by the cytotoxic CD8 T cells of adaptive immunity and the natural killer (NK) cells of innate immunity.

CD8 cytotoxic T cells kill infected cells by recognizing foreign, pathogen-derived peptides that are transported to the cell surface bound to MHC class I molecules (see Fig. 12.1). The peptides carried by MHC class I molecules come from the degradation of proteins in the cytosol, and so cytotoxic T cells act against pathogens whose proteins are found in the cytosol of the host cell at some stage in their life cycle (Fig. 45.1). The critical role of cytotoxic CD8 T cells in host defense is seen in the increased susceptibility of animals artificially depleted of cytotoxic T cells (or with inherited defects of cytotoxicity function) to many viral and intracytosolic bacterial infections. Mice and humans lacking the MHC class I molecules that present antigen to CD8 cells are also more susceptible to such infections (see Case 12).

Cytotoxic T cells kill their infected targets with great precision and neatness, by inducing apoptosis in the infected cell while sparing adjacent normal cells; this strategy minimizes tissue damage (Fig. 45.2). CD8 cytotoxic T cells release two types of preformed cytotoxins—the fragmentins or granzymes, which seem able to induce apoptosis in any type of target cell, and the protein perforin, which is thought to act as a translocator, pore-forming protein to enable granzymes to cross the membrane of the target cell (Fig. 45.3). A membrane-bound molecule, the Fas ligand, which is expressed on CD8 T cells as well as on some CD4 T cells, can also induce apoptosis by binding to Fas on a limited range of target cells. Together, these properties allow the cytotoxic T cell to attack and destroy virtually any infected cell. Cytotoxic CD8 T cells also produce the cytokine interferon (IFN)-γ; this cytokine inhibits viral replication, induces MHC class I expression, and also activates macrophages. In addition to combating infection by viruses and intracytosolic bacteria, CD8 T cells are important in controlling some protozoal infections; they are crucial, for example, in host defense against *Toxoplasma gondii*, an intracellular protozoan.

Topics bearing on this case:
Processing and presentation of cytosolic antigens
Activation of cytotoxic T cells
Cell killing by cytotoxic T cells

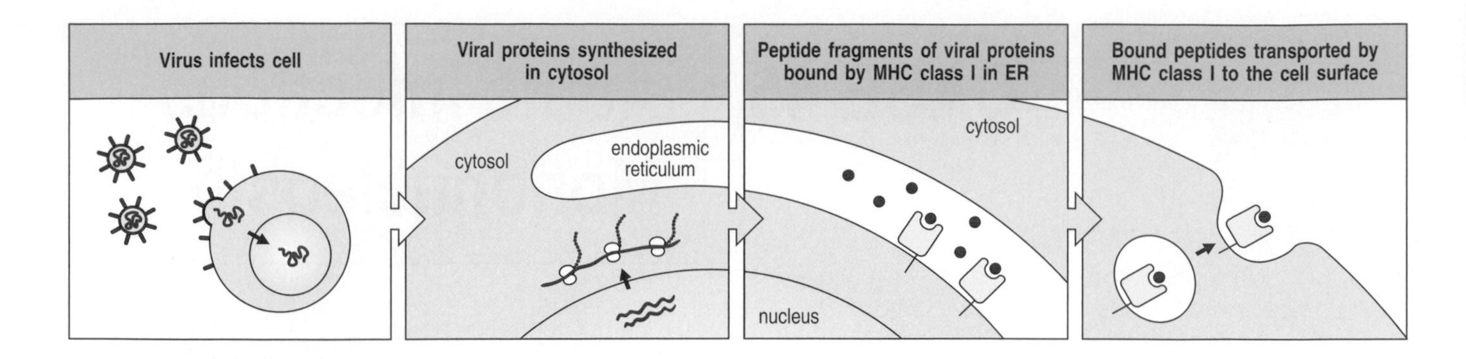

Fig. 45.1 MHC class I molecules present antigen derived from proteins in the cytosol. In cells infected with viruses, viral proteins are synthesized in the cytosol. Peptide fragments of viral proteins are transported into the endoplasmic reticulum, where they are bound by MHC class I molecules, which then deliver the peptides to the cell surface.

The importance of cytotoxic T cells in the control of viral replication is highlighted by many aspects of Epstein–Barr virus (EBV) infection, which is described in this case study. EBV (also known as human herpesvirus 4) is a member of the virus family Herpesviridae. It has a double-stranded linear DNA genome enclosed in an icosahedral capsid and a lipid envelope, and replicates its DNA genome in the host-cell nucleus. EBV infects only humans and is one of the most successful infectious agents on its obligate host. It can even be thought of as a commensal that only seldom causes injury to the host; anywhere from 60–98% of healthy adults show serological evidence of infection with EBV. The virus infects mainly B cells and epithelial cells.

The case of Emma Bovary: a bad sore throat from a B-cell infection.

15-year-old female with severe sore throat, fever and malaise.

Emma Bovary was a healthy 15-year-old when she suddenly developed a very sore throat accompanied by fever and malaise. Her throat was so swollen she had difficulty swallowing. Over the next few days the fever waxed and waned, her sore throat became worse, and she became progressively more tired and anorectic (lost her appetite). On the third day of illness her pediatrician noted severe pharyngitis and took a throat culture for β-hemolytic streptococci; the culture proved negative.

Emma's symptoms persisted, and she was unable to eat because she could hardly swallow. She said she had no difficulty breathing but that her left upper abdomen felt slightly uncomfortable. Emma's 1-year-old brother became ill at the same time, but did not have such severe symptoms. He was merely listless and felt warm. He had no particular physical symptoms, and seemed to recover completely after a few days.

Sore throat, temperature, and swollen lymph nodes in neck. Infectious mononucleosis?

On physical examination on the tenth day of illness, Emma appeared very ill. She had a high temperature (38.2°C), a pulse rate of 84 min^{-1}, a respiratory rate of 18 min^{-1}, and a blood pressure of 85/55 mmHg. Her mouth was dry and her tonsils were red and enlarged. They met in the midline, leaving a passage of only about 2 cm × 2 cm. Palatal petechiae (very small hemorrhages under the mucosa) could be seen. Her anterior and posterior cervical lymph nodes were swollen and tender (lymphadenopathy); the largest nodes were 2 cm × 2 cm. Her abdomen felt soft and the liver was enlarged, the edge being palpable 2 cm below the right costal margin. The spleen was also enlarged; the tip was easily palpable under the left costal margin.

A blood test gave a white blood cell count of 18,590 μl^{-1}, with 39% neutrophils, 27% lymphocytes, 22% atypical lymphocytes (very high), and 11% monocytes (high); her hematocrit was 45% and the platelet count 397,000 μl^{-1}. Serum electrolytes were normal. Another throat culture was obtained, and blood tests for Epstein–Barr virus (EBV) were ordered.

Fig. 45.2 Cytotoxic T cells kill target cells bearing specific antigen while sparing neighboring uninfected cells. All the cells in a tissue are susceptible to the induction of apoptosis by the cytotoxins of armed effector CD8 T cells, but only infected cells are killed. Specific recognition by the T-cell receptor identifies which target cell to kill, and the polarized release of cytotoxic granules (see Fig. 45.3) ensures that neighboring cells are spared.

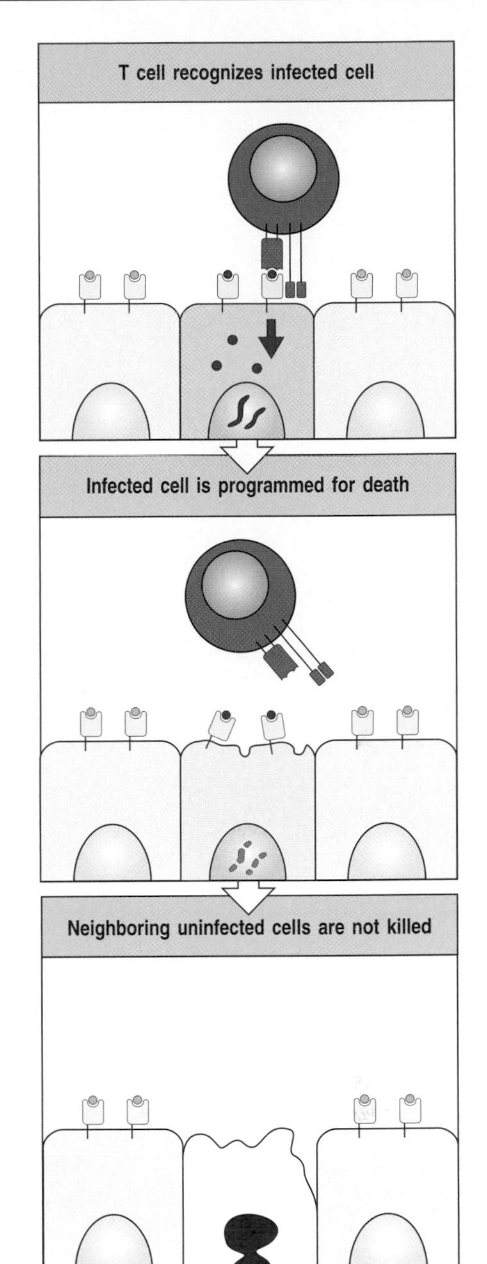

In the meantime a presumptive diagnosis of acute infectious mononucleosis was made with complications including partial pharyngeal obstruction and mild dehydration. Emma was admitted to the hospital and received 1 liter of normal saline intravenously followed by 20 mg of methylprednisolone (a corticosteroid) intravenously every 12 hours.

Her throat culture again proved negative for streptococcus but her blood serum was positive for IgM and IgG antibodies against EBV capsid antigen. Emma improved quickly with the symptomatic treatment and was discharged on the second day after admission to complete her recovery at home.

Acute infectious mononucleosis.

Emma showed many of the clinical features characteristic of acute infectious mononucleosis (IM) induced by the Epstein–Barr virus, a disease also known as glandular fever in some countries. She had severe pharyngitis with petechiae on the palate, swollen lymph nodes in the neck, enlarged liver and spleen, and large numbers of atypical lymphocytes (the mononucleosis after

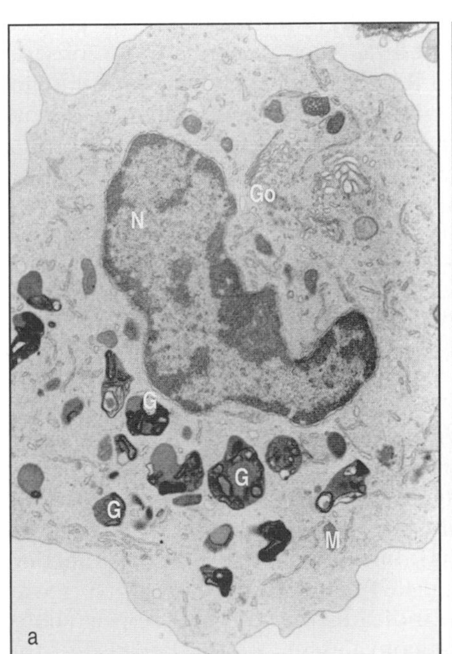

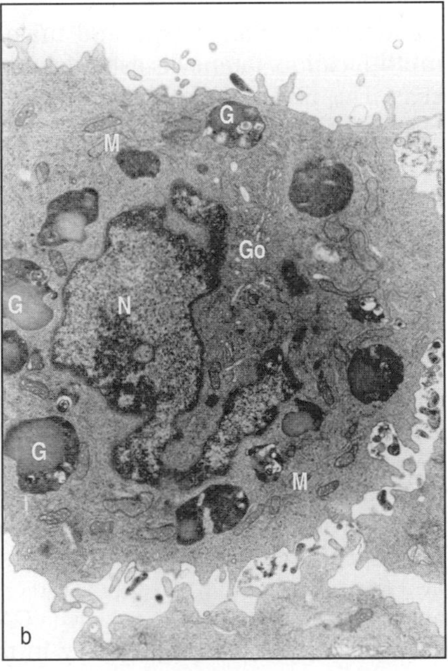

Fig. 45.3 The cytotoxic granules of cytotoxic T cells are released in a polarized fashion. Perforin molecules, as well as several other effector molecules, are contained in the granules of cytotoxic T cells (panel a: G, granules; N, nucleus; M, mitochondria; Go, Golgi apparatus). When an effector CD8 cytotoxic T cell recognizes its target, the granules are released onto the target cell (panel b, bottom right quadrant). Photographs courtesy of E. Podack.

which the disease is named) in her blood. These cells, also known as Downey–McKinlay cells, are large cells with foamy basophilic cytoplasm and fenestrated nuclei. They are mostly T cells, with a preponderance of CD8 cytotoxic T cells, and are present in 90% of patients with IM, in which they sometimes constitute the majority of blood leukocytes.

It is these cells that control the acute infection by destroying EBV-infected B cells. The atypical lymphocytosis in IM is a reflection of the increased CD8 T cell cytotoxic activity. In the vast majority of individuals, the infection is brought under control but not eradicated, because the viral genome persists latently in many B cells. In the latent phase some viral antigens are produced and peptides derived from them are presented by MHC I molecules at the surface of the infected cell, thus enabling latently infected cells to be recognized and destroyed by EBV-specific cytotoxic T cells. Some of the latently infected B lymphocytes become transformed; they are able to propagate themselves indefinitely if released from the presence of EBV-specific cytotoxic T cells, and are potentially malignant. In healthy people after infection, approximately 1 in 10^6 B cells is transformed. In patients with immune deficiency or who are immunosuppressed, infected B cells can grow unchecked. In immunodeficient patients, EBV infection can cause immunoblastic lymphoma and B-cell lymphoma.

EBV has a long incubation period: the time between primary EBV infection and the onset of illness is 30–50 days. Infection in infancy or early childhood is almost always asymptomatic, or results in mild disease, as evidenced by Emma's younger brother, who probably also had a primary EBV infection. In developed countries, primary infection is delayed in about half of the population to adolescence and early adulthood, so that 30–50% of primary EBV infection results in acute IM.

EBV enters B cells by binding to the B-cell surface molecule CD21 (also called complement receptor 2 (CR2) because it acts as a receptor for the complement fragment C3dg). This receptor is also present on a small subpopulation of T cells and on various types of epithelial cells in the nasopharynx, parotid gland duct, female cervix, and male urethra. After an active phase of viral multiplication, latency is established in B cells and, in some cases, epithelial cells. The EBV DNA is maintained during latency as an extrachromosomal DNA within the nucleus. Virus production is reactivated from time to time, and periodic shedding of infectious virus in oral secretions of healthy infected people is common and may recur multiple times during life. Adolescents like Emma often catch the disease through kissing; although Emma was not yet sexually active, she had dated several boys in her class.

Definitive diagnosis of EBV infection is best made by serological or molecular biological tests. Infected B cells are stimulated to secrete immunoglobulin, producing, among other antibodies, a so-called heterophile IgM antibody whose detection is one of the most widely used diagnostic tests for EBV infection. This heterophile antibody is not specific for EBV antigens but binds to antigens present on heterologous red blood cells (that is, those of other animals, such as sheep or goat) and agglutinates them. In addition, specific antibody responses are generated against several EBV-specific antigens, and the appearance of different antibodies is informative as to the time of infection and the pattern of virus replication (Fig. 45.4). The presence of IgM antibody against the EBV capsid antigen (VCA) indicates that the infection is acute; this antibody declines gradually in the convalescent phase. Because of the long incubation period of the virus, anti-VCA IgG antibody is also detectable at the onset of illness. The continued presence of EBV antigens in the host maintains antibody production throughout life.

Antibodies against so-called early antigens (EA) are produced mainly in the convalescent phase (1–6 months after disease onset); they then disappear.

	Viral capsid antigen; IgM	Viral capsid antigen; IgG	Early antigen	Epstein–Barr virus nuclear antigen
Never exposed	–	–	–	–
Acute infection	+	+	+/–	–
Recent infection	+/–	+	+/–	+/–
Past infection	–	+	–	+
Reactivated or chronic infection	–	+	+/–	+/–

Fig. 45.4 Serologic diagnosis of EBV infection.

The absence of EA antibody indicates that the virus is mostly quiescent and is not undergoing replication on a large scale. If the infection is reactivated, EA antibody titers rise again. Antigens expressed later in the viral life cycle, during its latent phase, are the EBNAs (Epstein–Barr nuclear antigens). Appearance of EBNA antibody indicates that the virus has been present in the body for at least a few months.

In very young, immunosuppressed, or immunodeficient patients, antibody formation can be so impaired that serologic diagnosis is impossible. In these cases EBV antigens can be detected by immunofluorescence microscopy on blood or tissue (for example lymph node) specimens or by *in situ* hybridization for small EBV RNAs (EBERs). Alternatively, viral DNA can be amplified from infected cells and tissue by the polymerase chain reaction with oligonucleotide primers specific for EBV DNA.

EBV infection is mitogenic for B cells, overriding the normal regulatory mechanisms that prevent them from dividing. Activation of B cells by virus infection also leads to immunoglobulin production, as noted above. Several EBV genes are critical for B-cell activation. One is the viral gene *BCRF-1*, which encodes the protein BCRF-1, also called vIL-10, very similar to human interleukin-10 (IL-10). The BCRF-1 protein enhances the activation and proliferation of EBV-infected cells. EBV infection also stimulates endogenous synthesis of IL-10 and IL-6, with further autocrine stimulatory effects on B cells. IL-10 also inhibits the T-cell production of IL-2 and IFN-γ, and enhances the production of B-cell stimulatory cytokines such as IL-4. IL-6 may inhibit the ability of NK cells to destroy EBV-infected cells. The polyclonal activation of B lymphocytes induced by EBV infection often results in the activation of self-reactive B lymphocytes and autoantibody production. Hence, autoimmune manifestations, and especially autoimmune cytopenias, such as autoimmune hemolytic anemia, thrombocytopenia, and/or neutropenia, are common during IM.

In most patients, IM is a self-limited disease for which supportive therapy suffices. Nucleoside analogs such as acyclovir or ganciclovir, or the DNA polymerase inhibitor foscarnet, have limited ability to inhibit the replication of EBV *in vitro*. The clinical usefulness of these drugs is so far unproven. Corticosteroids are often prescribed as a palliative measure, especially when airway obstruction is a potential concern. In the most extreme cases, when respiratory distress is present, tonsillectomy can be required. Corticosteroids reduce virus shedding and provide some symptomatic relief as a result of their anti-inflammatory effects. They do not significantly alter the course of the disease. Immunocompromised individuals are at high risk of severe lymphoproliferative disease and B-cell lymphoma after infection with EBV or reactivation of this infection. In these circumstances, monitoring of the infection is best performed by evaluating the number of viral copies in peripheral blood. If a significant viremia is detected, treatment is most often based on injection of the B-cell-depleting anti-CD20 monoclonal antibody rituximab, which destroys both infected and uninfected B lymphocytes. This treatment causes a profound B-cell lymphopenia and hypogammaglobulinemia, which requires immunoglobulin replacement therapy until circulating B lymphocytes reappear in normal numbers.

The ability of EBV to transform B cells is an extremely useful laboratory tool. When peripheral blood B cells are cultured with EBV, they become immortalized at a relatively high frequency and can be propagated indefinitely *in vitro*. This allows the general study of various aspects of B-cell biology, as well as providing material for the study of individual patients.

The virus-encoded Epstein–Barr virus nuclear antigen 2 (EBNA-2) is critical for transformation. EBNA-2 protein interacts with transcription factors, leading to the activation of several host genes, for example those encoding B-cell activating molecules such as CD21 (the EBV receptor) and CD23 (FcεRII, the low-affinity receptor for IgE), and viral genes such as LMP-1 (latent membrane protein-1, the primary oncogene of EBV). EBNA-3, -4, -5, and -6 also have a role in B-cell transformation.

Acute IM is only one of the possible outcomes of EBV infection. The Epstein–Barr virus is in fact named after two workers who studied Burkitt's lymphoma in Africa in the 1960s and first cultured the virus from these patients. This B-cell lymphoma is strongly associated with EBV infection in Africa, but not in other parts of the world; the high rate of EBV infection in early infancy together with the high incidence of malaria in Africa seem to be the predisposing factors. EBV infection is also strongly associated with nasopharyngeal carcinoma in Southeast Asia. This may perhaps be due to a particular strain of EBV that no longer possesses the epitope provoking an immunodominant response in people with a certain HLA class I allele that is present in a high proportion of people in Southeast Asia. Such people are therefore not so efficient at clearing cells infected with this EBV strain, making transformation and eventual malignancy more likely.

Questions.

1 *Patients with humoral immunodeficiency (an impaired antibody response) are susceptible to infection with some viruses such as poliomyelitis or enteric viruses, but they have no problems with EBV. Why?*

2 *There is a high risk of EBV-induced lymphoproliferative malignancy after T cell-depleted bone marrow transplantation. Why?*

3 In vitro *transformation by EBV of B cells from umbilical cord blood rarely fails. Transformation of B cells in blood cultures from some adults is difficult. Why?*

4 *Why is heterophile antibody produced during EBV infection?*

5 *Males with X-linked agammaglobulinemia (XLA) never get infected with EBV. How do you explain this?*

CASE 46 Hemolytic Disease of the Newborn

The adaptive immune response distinguishes genetic differences between individuals of the same species.

Adaptive immune responses evolved to protect vertebrate species against the world of microorganisms. However, anything discerned as 'nonself' may become a target of such responses, which can also be directed at molecular differences between individuals within a species. An antigenic determinant present in some members of a species but not in others is said to show polymorphic variation, and antibodies directed against such determinants are called alloantibodies. Perhaps the best-known alloantibodies are those that are used to determine our blood groups. Individuals who have red cells of type A have alloantibodies that react with the red blood cells of individuals who are type B and vice versa. Individuals who have red blood cells of type AB have no alloantibodies, and those with type O red blood cells have antibodies against both A and B red blood cells (Fig. 46.1). These alloantibodies arise from the fact that the capsules of Gram-negative bacteria, which inhabit our gut, bear antigens that stimulate antibodies that cross-react with the carbohydrate antigens of the ABO blood groups.

Alloantibodies induced by a fetus in the pregnant mother frequently cause serious problems. Alloimmunization most often results from Rhesus (Rh) incompatibility between mother and fetus. Approximately 13% of women are Rh-negative; that is, their red blood cells do not bear the Rh antigen. A woman who is Rh-negative has an 87% chance of marrying an Rh-positive man, and their chances of having an Rh-positive baby are very high. Not infrequently, during delivery of the newborn infant, some blood escapes from the fetal circulation into the maternal circulation and the mother develops alloantibodies against the Rh antigen as a result. During a subsequent pregnancy with an Rh-positive fetus, the maternal IgG alloantibodies cross the placenta and cause destruction of the fetal red cells, causing anemia. As we shall see, the consequences of this can be very grave and result in fetal death or severe damage to the newborn infant.

The Rh antigenic determinants are spaced very far apart on the red cell surface. As a consequence, IgG antibodies against the Rh antigen do not fix complement and therefore do not hemolyze red blood cells *in vitro*. For reasons that are less well understood, IgG antibodies against the Rh antigen do not agglutinate Rh-positive red blood cells. Because of this it was very difficult to detect Rh antibodies until Robin Coombs at the University of Cambridge devised a solution to the problem by developing antibodies against human immunoglobulin. He showed that Rh-positive red blood cells coated with IgG anti-Rh antibodies could be taken from a fetus and agglutinated by antibodies against IgG. Furthermore, he showed that when the serum of an

Topics bearing on this case:
Antibody suppression of B-cell activation
ABO blood groups
Alloantibodies
Rhesus blood group
Coombs'test
Hemagglutination assays

This case was prepared by Raif Geha, MD, in collaboration with Andrew Shulman, MD, PhD.

Fig. 46.1 Hemagglutination is used to type blood and match compatible donors and recipients for blood transfusion. Common gut bacteria bear antigens that are similar or identical to blood group antigens, and these stimulate the formation of antibodies against these antigens in individuals who do not bear the corresponding antigen on their own red blood cells (left column); thus, type O individuals, who lack A and B antigens, have both anti-A and anti-B antibodies, whereas type AB individuals have neither. The pattern of agglutination of the red blood cells of a transfusion donor or recipient with anti-A and anti-B antibodies reveals the individual's ABO blood group. Before transfusion, the serum of the recipient is also tested for antibodies that agglutinate the red blood cells of the donor, and vice versa, a procedure called a cross-match, which may detect potentially harmful antibodies against other blood groups that are not part of the ABO system.

	Red blood cells from individuals of type			
	O	A	B	AB
	Express the carbohydrate structures			
Serum from individuals of type	R–GlcNAc–Gal (Fuc)	R–GlcNAc–Gal–GalNAc (Fuc)	R–GlcNAc–Gal–Gal (Fuc)	R–GlcNAc–Gal–GalNAc (Fuc) + R–GlcNAc–Gal–Gal (Fuc)
O Anti-A and anti-B antibodies	no agglutination	agglutination	agglutination	agglutination
A Anti-B antibodies	no agglutination	no agglutination	agglutination	agglutination
B Anti-A antibodies	no agglutination	agglutination	no agglutination	agglutination
AB No antibodies against A or B	no agglutination	no agglutination	no agglutination	no agglutination

alloimmunized woman was incubated with Rh-positive red blood cells, these red blood cells could then be agglutinated by antibody against IgG (Fig. 46.2). The former is called the direct Coombs test and the latter the indirect Coombs test. This application of immunology to a vexing clinical problem led ultimately to treatment and prevention of the problem.

The case of Cynthia Waymarsh: a fetus in immunological distress.

Mrs Waymarsh was 31 years old when she became pregnant for the third time. She was known to have blood group A, Rh-negative red cells. Her husband was also type A but Rh-positive. Their first-born child, a male, was healthy. During her second pregnancy Mrs Waymarsh was noted to have an indirect Coombs titer at a 1:16 dilution of her serum. The fetus was followed closely, and the delivery of a healthy baby girl was induced at 36 weeks of gestation.

Five years later Mrs Waymarsh became pregnant again. At 14 weeks of gestation her indirect Coombs titer was 1:8, and at 18 weeks it was 1:16. Amniotic fluid was obtained at 22, 24, 27, and 29 weeks of gestation and was found to have increasing amounts of bilirubin (a pigment derived from the breakdown of heme, indicating that the fetus's red blood cells were being hemolyzed). At 29 weeks of gestation a blood sample was obtained from the umbilical vein and found to have a hematocrit of 6.2% (normal 45%). (The hematocrit is the proportion of blood that is composed of red cells, and because the volume of white cells is comparatively negligible, this is simply ascertained by centrifuging whole unclotted blood in a tube.) On finding that the fetus was profoundly anemic, 85 ml of type O, Rh-negative packed red blood cells were transfused into the umbilical vein. At 30.5 weeks of gestation another sample of

30.5 weeks: hematocrit
up to 16.3%. Further
transfusion requested.

blood from the umbilical vein was obtained; the hematocrit was 16.3%. The fetus was transfused with 75 ml of type O, Rh-negative packed red blood cells.

The fetus was examined at weekly intervals for the appearance of hydrops (see below), and none was observed. At 33.5 weeks of gestation the hematocrit of a blood sample from the umbilical vein was 21%, so 80 ml of type O, Rh-negative packed red blood cells were again transfused into the umbilical vein. At 34.5 weeks of gestation it was determined that the fetus was sufficiently mature to sustain extrauterine life without difficulty; labor was induced and a normal female infant was born. The hematocrit in the umbilical vein blood was 29%. The baby did well and no further therapeutic measures were undertaken.

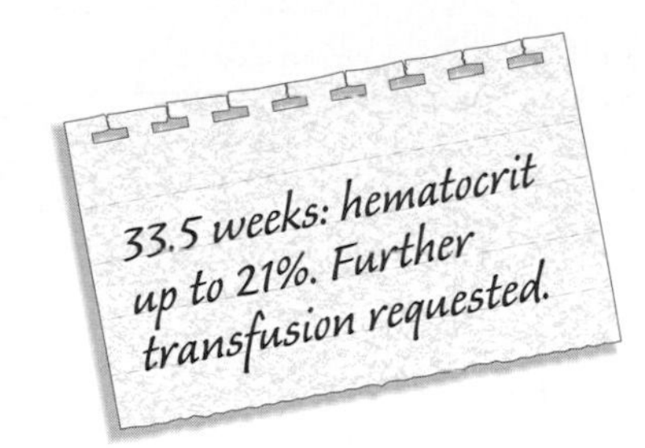

33.5 weeks: hematocrit up to 21%. Further transfusion requested.

Hemolytic disease of the newborn.

Although hemolytic disease of the newborn is most commonly the result of alloimmunization with Rh antigen, other red blood cell alloantigens, such as Lewis, Kell, Duffy, Kidd, and Lutheran, may cause alloimmunization. In any case, the maternal IgG antibodies cross the placenta in increasing amounts during the second trimester of pregnancy and hemolyze the fetal red blood cells. The resulting anemia may become so severe that, if untreated, the fetus goes into heart failure and develops massive edema; this is called hydrops fetalis and results in fetal death. The risk of fetal development of hydrops rises from 10% when the indirect Coombs titer of the mother is 1:16 to 75% when the maternal titer is 1:128. If the anemia is not so severe as to cause hydrops, the affected infant at birth is still massively hemolyzing red blood cells. Now the newborn must dispose of the heme breakdown pigments rapidly because an excessive accumulation of bilirubin results in the deposition of this pigment in the brain, and severe neurological impairments. In response to the profound anemia, the number of red blood cell precursors (erythroblasts) in the spleen, liver, and bone marrow expands rapidly; for this reason, hemolytic disease of the newborn has also been called erythroblastosis fetalis.

As we have seen in this case, the extent of hemolysis can be determined easily by obtaining amniotic fluid, into which the fetus begins to urinate by 20 weeks of gestation. The quantity of bilirubin excreted into the amniotic fluid correlates with the amount of hemolysis in the fetus. Second, the fetus can be followed by ultrasonography for the development of hydrops. Third, the degree of anemia can be ascertained directly, but with some difficulty, by obtaining a sample of blood from the umbilical vein.

It has become possible in the past few decades to eliminate hemolytic disease of the newborn to a very great extent. All Rh-negative women are given 300 mg of purified polyclonal IgG antibody derived from alloimmunized donors

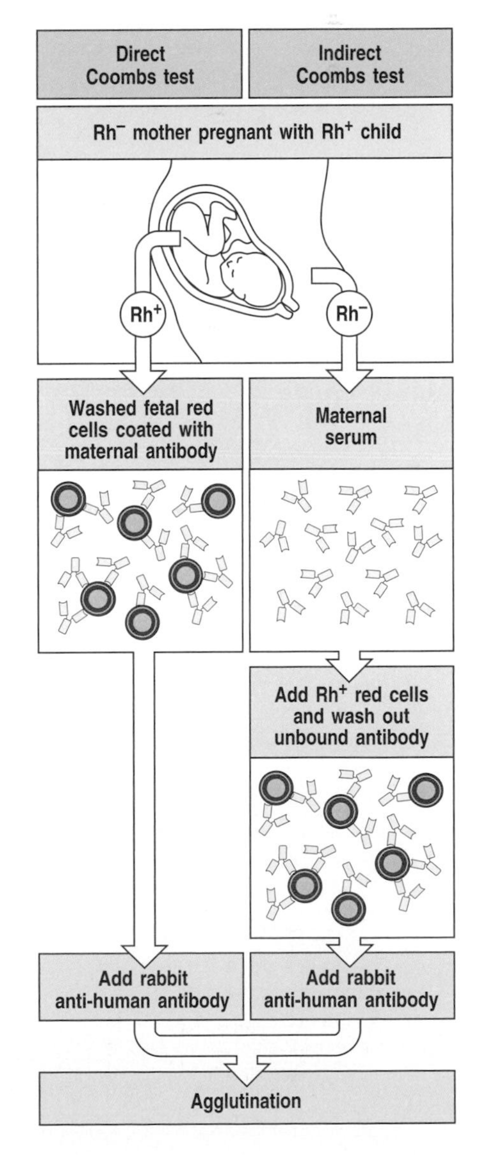

Fig. 46.2 The Coombs direct and indirect anti-globulin tests for antibody to red blood cell antigens. An Rh-negative (Rh^-) mother of an Rh-positive (Rh^+) fetus can become immunized to fetal red blood cells that enter the maternal circulation at the time of delivery. In a subsequent pregnancy with an Rh^+ fetus, IgG anti-Rh antibodies can cross the placenta and damage the fetal red blood cells. In contrast to anti-Rh antibodies, anti-ABO antibodies are of the IgM isotype; they cannot cross the placenta, and so do not cause harm. Anti-Rh antibodies do not agglutinate red blood cells, but their presence can be shown by washing the fetal red blood cells and then adding antibody against human immunoglobulin, which agglutinates the antibody-coated cells. The washing removes unrelated immunoglobulins that would otherwise react with the anti-human immunoglobulin antibody. Anti-Rh antibodies can be detected in the mother's serum in an indirect Coombs test; the serum is incubated with Rh^+ red blood cells, and once the antibody has bound, the red cells are treated as in the direct Coombs test.

specific for the Rh antigen (Rhogam) at 28 weeks of gestation and again within 72 hours of delivery. The amount of Rhogam in one vial (300 μg) is sufficient to neutralize 30 ml of fetal blood in the maternal circulation. This procedure fails to prevent alloimmunization in only 0.1% of Rh-negative women, such as Mrs Waymarsh. It must be presumed in such failures that the fetus bled more than 30 ml of blood into its mother. Removal of the Rh antigen from the maternal circulation has been presumed to be the mechanism of action of Rhogam. However, the limited efficacy of monoclonal antibodies generated against the Rh antigen in preventing alloimmunization suggests that Rhogam has important effects in addition to Rh antigen clearance. The hypothesis that Rhogam antibodies suppress the priming of maternal B cells, an early stage in B-cell activation, and prevent maternal anti-Rh antibody production is the focus of current investigation.

Questions.

1 *It was stated that the Rh antigens are so sparsely scattered on the red cell surface that IgG molecules bound to the Rh antigens are too far apart to fix C1q. Therefore, complement-mediated hemolysis cannot be invoked to explain hemolytic disease of the newborn. By what mechanism are the red cells destroyed?*

2 *When an Rh-negative woman is ABO-compatible with her husband, as Mr and Mrs Waymarsh are, the risk of Rh alloimmunization is 16%. When they are ABO incompatible the risk falls to 7%. How do you explain this difference?*

3 *Why were Rh-negative red blood cells used for the intrauterine transfusion?*

4 *Do you have concerns about administering Rhogam to women at 28 weeks of gestation?*

5 *The serum of an Rh-negative woman who is pregnant gives a negative indirect Coombs test but her serum agglutinates Rh-positive cells suspended in saline. What is your interpretation of this phenomenon and what do you do about it?*

CASE 47 Toxic Shock Syndrome

Superantigens cause excessive stimulation of T cells and macrophages.

A conventional protein antigen activates only those T lymphocytes with a T-cell receptor specific for peptides derived from the antigen. This is typically a tiny subset of the total T-cell pool. For example, only about 1 in 10,000 circulating T cells from donors immunized with tetanus toxoid proliferate when subsequently stimulated by the toxoid. In contrast, a special class of antigens known as superantigens directly activates large numbers of T cells. Superantigens are bacterial and viral proteins that can bind, as whole unprocessed proteins, simultaneously to MHC class II molecules outside the peptide-binding groove and to certain V_β chains of the T-cell receptor. By engaging all T-cell receptors that share the same V_β chains rather than a receptor that recognizes the peptide in the MHC class II cleft, superantigens can activate a sizeable fraction of T cells and generate an amplified T-cell response far greater than that caused by the typical peptide antigen.

Superantigens first bind with high affinity to MHC class II molecules on antigen-presenting cells. As the concentration of bound superantigen increases, it will bind and cross-link T-cell receptors with the appropriate V_β specificity (Fig. 47.1). Cross-linking activates both the T cell and the antigen-presenting cell through signaling events downstream of the T-cell receptor and the MHC class II molecule, respectively. Because of the high affinity of superantigens for MHC class II molecules, a very small amount of superantigen results in intense T-cell signaling. Each superantigen is capable of binding to a limited group of V_β regions. For example, the toxic shock syndrome toxin-1 (TSST-1), which is produced by certain strains of the Gram-positive bacterium *Staphylococcus aureus*, stimulates all those T cells that express the $V_\beta2$ gene segment. As there is a limited repertoire of V_β gene segments, any superantigen will stimulate between 2% and 20% of all T cells.

This mode of stimulation is not specific for the pathogen and thus does not lead to adaptive immunity. Instead, it causes excessive production of cytokines by the large number of activated CD4 T cells, the predominant responding population. This massive release of cytokines has two effects on the host: systemic toxicity, which manifests itself as toxic shock syndrome; and suppression of the adaptive immune response. Both of these effects contribute to the pathogenicity of microbes that produce superantigens.

This case was prepared by Raif Geha, MD, in collaboration with Janet Chou, MD.

Topics bearing on this case:

- Cytokines
- T-cell activation
- Superantigens
- Macrophage activation
- Toxic shock

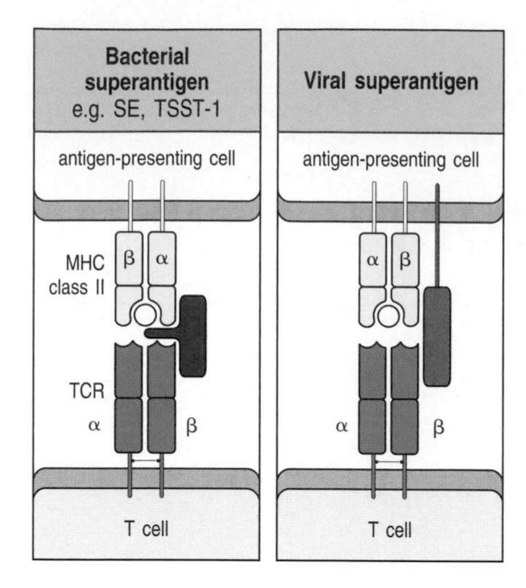

Fig. 47.1 Superantigens bind directly to T-cell receptors and MHC molecules. Superantigens interact with MHC class II molecules and T-cell receptors in a way that is quite distinct from the way in which normal peptide antigens bind. Superantigens bind independently to MHC class II molecules and to T-cell receptors, binding to the V_β domain of the T-cell receptor, away from the peptide-binding site. The T-cell receptor α chain is not directly involved in binding superantigen.

The case of Claire Bourbon: life-threatening shock from a superantigen.

Claire Bourbon was a healthy 16-year-old with a history of mild asthma and allergic rhinitis who suddenly became ill with a fever, general muscle aches, and dizziness. She felt nauseous and vomited. Claire's temperature rose to 39.8°C and her mother rushed her to the family physician. En route she briefly lost consciousness, and a red rash developed over her arms and spread rapidly to most of the body.

On arrival at the physician's she appeared quite ill and was immediately referred to the emergency department. She was alert but listless, and her general condition gave cause for concern. On examination, her temperature was 37.8°C, and heart rate and respiration rate were markedly elevated, at 140 beats per minute and 24 breaths per minute, respectively. Blood pressure was depressed—98/67 lying supine, 83/49 when seated, and 67/25 when standing—and showed evidence of significant volume depletion.

A bright red rash of flat and raised lesions was apparent on her trunk and extremities, but there were no petechiae (small subcutaneous hemorrhages) and no signs of localized infection.

Questioning revealed that Claire had not taken alcohol or drugs and had not been exposed to other ill individuals. Her last menstrual period had been 6 weeks before, and she had developed vaginal bleeding on the day previous to the onset of her illness. She had not used a tampon overnight, but had inserted one that morning, before she became ill.

Given Claire's critical status, extensive laboratory tests were conducted. Her white blood cell count was raised, at 21,000 cells μl^{-1}, with a predominance of neutrophils and band forms (immature neutrophils), indicating increased mobilization of neutrophils from the bone marrow. Serum electrolyte levels were within normal limits. The blood coagulation time was slightly prolonged and serum transaminase levels were raised; both of these signs are consistent with abnormal liver function.

Cerebrospinal fluid (CSF) was normal and showed no evidence of infection. Cultures of blood, urine, CSF, and vaginal fluid were made, and Claire was given a cephalosporin antibiotic (ceftriaxone) along with 2 liters of fluid intravenously. Her blood pressure improved and she was immediately admitted to the intensive care unit, where she developed petechiae. She was treated with intravenous fluids, two anti-staphylococcal antibiotics (oxacillin and clindamycin), cephalosporin (cefotaxime), and intravenous immunoglobulin (IVIG), and her overall condition gradually improved.

Claire's blood, urine, and CSF cultures remained sterile, but her vaginal culture was positive for abundant *S. aureus*. She was subsequently transferred to the regular in-patient ward and treated for 7 days with anti-staphylococcal antibiotics. The rash slowly faded.

Toxic shock syndrome.

Claire suffered from staphylococcal toxic shock syndrome (TSS), a striking example of the dramatic physiologic alterations caused by superantigens. TSS is a serious disease characterized by rapid onset of fever, a rash, organ failure, and shock. Most cases occur in menstruating women, typically in their teenage years, but cases do occur in all age groups. TSS is typically associated with a localized *S. aureus* infection (for example, subcutaneous abscesses, osteomyelitis, and infected wounds), staphylococcal food poisoning, or local colonization, as occurred in the vagina in this case. When kept in the vagina for a long time (more than 12 hours) tampons soaked with menstrual fluids can enhance the growth of the bacteria that are the source of superantigens. It is unlikely that the tampon played a part in Claire's illness because it was inserted less than 6 hours before the onset of symptoms. Toxigenic strains of *S. aureus* can produce enterotoxins (such as enterotoxin B) as well as TSST 1; these also act as superantigens with similar clinical consequences. In addition, microorganisms other than *S. aureus* secrete superantigens that can cause disease (Fig. 47.2).

Consistent with the $V_\beta2$ specificity of TSST-1, examination of the circulating lymphocytes from patients in the acute phase of TSS typically reveals a much higher proportion of circulating $V_\beta2$ T cells than cells using other V_β segments.

Disease	Superantigen	TCR V_β
Definite role for superantigen		
Toxic shock syndrome	TSST-1	$V_\beta2$
Staphylococcal food poisoning	SEA	$V_\beta3$, $V_\beta11$
	SEB	$V_\beta3$, $V_\beta12$, $V_\beta14$, $V_\beta15$, $V_\beta17$, $V_\beta20$
	SEC	$V_\beta5$, $V_\beta12$, $V_\beta13.1–2$, $V_\beta14$, $V_\beta15$, $V_\beta17$, $V_\beta20$
	SED	$V_\beta5$, $V_\beta12$
	SEE	$V_\beta5.1$, $V_\beta6.1–3$, $V_\beta8$, $V_\beta18$
Streptococcal toxic shock syndrome	SPE-A	$V_\beta8$, $V_\beta12$, $V_\beta14$
Scarlet fever	SPE-B	$V_\beta2$, $V_\beta8$
Mycoplasma arthritidis (rodent)	MAM	$V_\beta17$
Clostridium perfringens	Enterotoxin	$V_\beta6.9$, $V_\beta22$
Suspected role for superantigen		
HIV	CMV	$V_\beta12$
Type I diabetes mellitus	MMTV-like	$V_\beta7$
Rabies virus	Nucleocapsid	$V_\beta8$
Toxoplasmosis	?	$V_\beta5$
Mycobacterium tuberculosis	?	$V_\beta8$
Yersinia enterocolitica	?	$V_\beta3$, $V_\beta6$, $V_\beta11$
Kawasaki disease	?	$V_\beta2$, $V_\beta8$

Fig. 47.2 Superantigens, V_β usage, and disease.

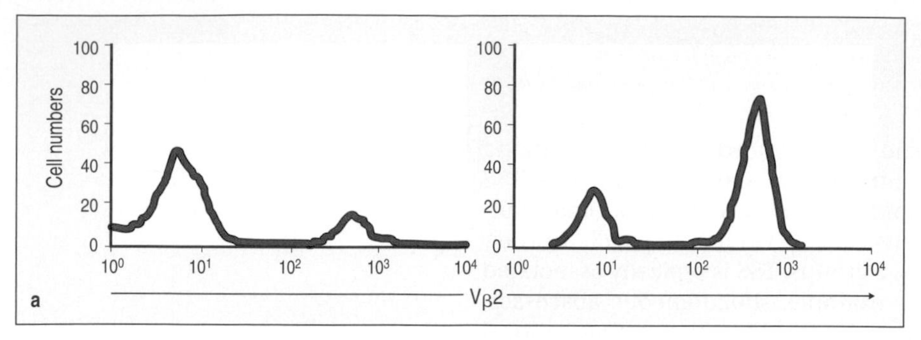

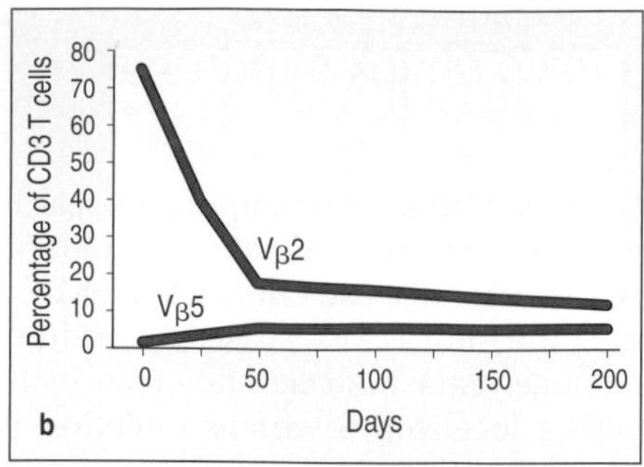

Fig. 47.3 Expansion in numbers of $V_\beta2$ T cells in toxic shock syndrome (TSS). Panel a, FACS analysis. Peripheral blood T cells from a normal control (left) and a patient with acute TSS (right). Cells are stained with anti-$V_\beta2$ monoclonal antibody with a fluorescein tag. There is an increased percentage of $V_\beta2$ T cells in the patient. The horizontal axis represents the mean fluorescence intensity. Panel b, time course of persistence of high numbers and return to normal of $V_\beta2$ T cells in a patient with TSS.

As the illness resolves, there is a gradual return to near-normal proportions. The expansion in the numbers of $V_\beta2$ cells can be measured by examining the surface expression of $V_\beta2$-containing T-cell receptors with the use of immunofluorescence (Fig. 47.3) or by semiquantitative measurement of mRNA transcripts encoding $V_\beta2$ T-cell receptor chains by using reverse transcription and the polymerase chain reaction (RT-PCR) (Fig. 47.4).

Although all the T cells activated by a given superantigen share a common V_β region, they will differ in their specificity for conventional peptide antigens. Sequencing of the T-cell receptors from superantigen-activated T cells reveals the use of different D and J gene segments by the β chains and a wide diversity of α chains. These receptors will encompass a wide variety of antigen specificities. In contrast, conventional antigens induce the clonotypic expansion of T cells—meaning that all the T cells in any given clone will have identical D and J gene segments in their β chains and identical α chains. Because the pool of V_β-restricted T cells activated by superantigen may contain autoreactive T cells, it has been postulated that superantigens could trigger autoimmune disease.

Many of the manifestations of TSS are the result of massive and unregulated cytokine production triggered by the activation of immune-system cells. TSST-1 is more effective than bacterial lipopolysaccharide in inducing the synthesis and secretion of interleukin (IL)-1 and tumor necrosis factor (TNF)-α by monocytes. TSST-1 is also a potent T-cell mitogen for those T cells whose receptors it engages; it also induces them to produce large amounts of cytokines, including IL-2 and interferon (IFN)-γ.

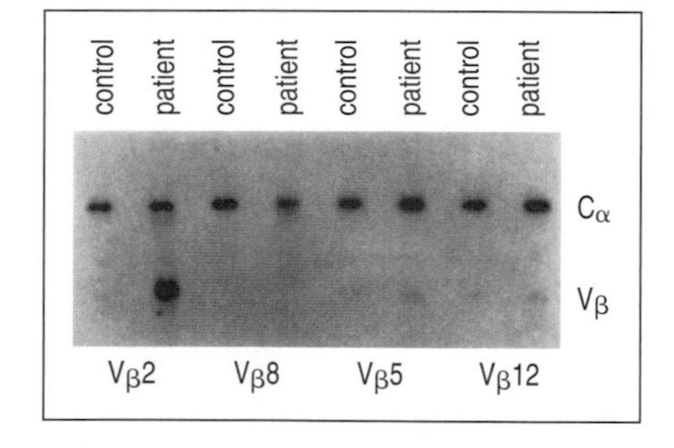

Fig. 47.4 RT-PCR analysis of T-cell receptor V_β mRNA. Autoradiograms of T-cell receptor chain transcripts amplified by reverse transcription followed by polymerase chain reaction (RT-PCR). T cells from a patient with toxic shock syndrome and a control individual were stimulated with anti-CD3 antibody and IL-2 before the extraction of RNA and generation of cDNA. Each reaction contained specific oligonucleotide primers to expand the particular V_β gene segment indicated (170–220 base pairs), as well as a C_α gene segment (about 600 base pairs) as a control to ascertain that equivalent amounts of mRNA were used. Photograph courtesy of Y. Choi.

IL-1 and TNF-α are critical in the induction of the acute-phase response, characterized by fever and the production of IL-6. IL-1 and TNF-α also activate vascular endothelium and, together with IL-2, increase vascular permeability, with the subsequent leakage of fluid from the intravascular space into the perivasculature. These effects of the massive overproduction of TNF-α result in the toxic shock: edema and intravascular volume depletion leading to hypotension, shock, and multi-organ failure.

S. aureus is a common organism that colonizes 25–50% of the population, and nearly 50% of isolates produce superantigens. However, TSS is of rare occurrence. There are several reasons for this discrepancy. The staphylococcal cell wall contains peptidoglycans that bind to the Toll-like receptor TLR-2:TLR-6 on monocytes, resulting in activation of the NFκB pathway (see Fig. 23.1). Although TLR signaling can cause a pro-inflammatory response, staphylococcal peptidoglycans induce an anti-inflammatory response mediated by IL-10 production and subsequent apoptosis of antigen-presenting cells. This aborts the amplification of the inflammatory response associated with TSS. In addition, susceptibility to TSS correlates with a poor antibody response to TSST-1. Typically, individuals are exposed to bacterial numbers that are too low to produce dangerous levels of superantigen but are high enough to generate toxin-specific antibodies. But although most healthy individuals have protective antibody titers to TSST-1, more than 80% of patients with TSS lack anti-TSST-1 antibodies during the acute illness, and most fail to develop anti-TSST-1 antibodies after convalescence. Possible explanations include an individual's inability to mount an antibody response to TSST-1 and staphylococcal enterotoxins, or a specific inhibition of such a response by the toxins. In addition, certain HLA polymorphisms, particularly in the HLA-DQ allele, are associated with increased superantigen affinity for MHC class II molecules and more severe clinical symptoms. Because antibodies against staphylococcal and streptococcal superantigens confer protection against superantigen-mediated disease, and the vast majority of donors of the plasma pools used to prepare IVIG have antibodies against TSST, IVIG is a potential treatment for patients with TSS. IVIG should be used in conjunction with antibiotics, and its efficacy in neutralizing superantigens can differ between preparations.

Questions.

1 *How do you determine whether a protein behaves as a superantigen?*

2 *Explain the rapid progression of clinical symptoms after the introduction of superantigen compared with the delay in apparent responses to conventional antigen.*

3 *What are the potential mechanisms of liver injury in TSS?*

4 *Is Claire susceptible to another bout of TSS?*

CASE 48 Lepromatous Leprosy

T_H1 versus T_H2 responses in the outcome of infection.

Mature naive CD4 T cells emerging from the thymus can differentiate into effector CD4 T cells of several different phenotypes, each with different functions in the immune response. Two of these subsets are called T_H1 and T_H2 (Fig. 48.1). They develop from naive T cells activated by pathogen antigens in the presence of different signals provided by antigen-presenting cells and the local environment. These two types of effector T cells are distinguished chiefly by the cytokines that they secrete when they encounter their target cell. T_H1 cells secrete interleukin-2 (IL-2), interferon-γ (IFN-γ), and lymphotoxin (LT); T_H2 cells secrete IL-4, IL-5, and IL-10 (Fig. 48.2). CD4 T cells can also develop into effector T_H17 cells, which secrete IL-17, or into induced regulatory T (T_{reg}) cells, each of which has distinct functional properties (see Cases 20 and 18, respectively).

The consequences of the decision to differentiate into T_H1 or T_H2 cells are profound. Selective production of T_H1 cells enables the immune response to activate macrophages and cell-mediated immunity, whereas selective production of T_H2 biases the response toward antibody production only. The decision as to which pathway a naive T cell will follow is made during its first encounter with antigen and is dependent on cytokines. T_H1 differentiation is dependent on IL-12 and IFN-γ, whereas T_H2 differentiation is dependent on IL-4 (see Fig. 48.2). These cytokines trigger pathways of signal transduction; for example, mice deficient in the intracellular signaling molecule STAT6, which is induced by IL-4, fail to develop T_H2 cells.

Other factors influencing the T-cell phenotype are the amount of antigen present and thus which cells are most likely to present it. Large amounts of antigen are usually presented by dendritic cells, which produce IL-12 and therefore favor T_H1 differentiation. A limited amount of antigen leads to preferential presentation by antigen-specific B cells that take up antigen more avidly; they induce T_H2 differentiation. The co-stimulatory molecules (B7.1 versus B7.2) expressed by the antigen-presenting cells also influence the maturation process, in that B7.1 (CD80) is more likely to provoke T_H1 development, and B7.2 (CD86) to provoke T_H2 development.

Topics bearing on this case:

- Cytokine production in innate immunity
- Differentiation of T_H1 versus T_H2 CD4 T cells
- Role of cytokines in T-cell differentiation
- Functions of T_H1 and T_H2 cells in immune responses
- Responses to mycobacteria

Fig. 48.1 CD4 T cells can differentiate into a variety of effector cells after encounter with antigen. Naive CD4 T cells first respond to their antigens (peptide:MHC class II complexes) by making IL-2 and proliferating. These cells then differentiate into a cell type sometimes known as T_H0. The T_H0 cell has the potential to differentiate into distinct functional subsets (T_H1, T_H2, T_H17, and T_{reg} cells).

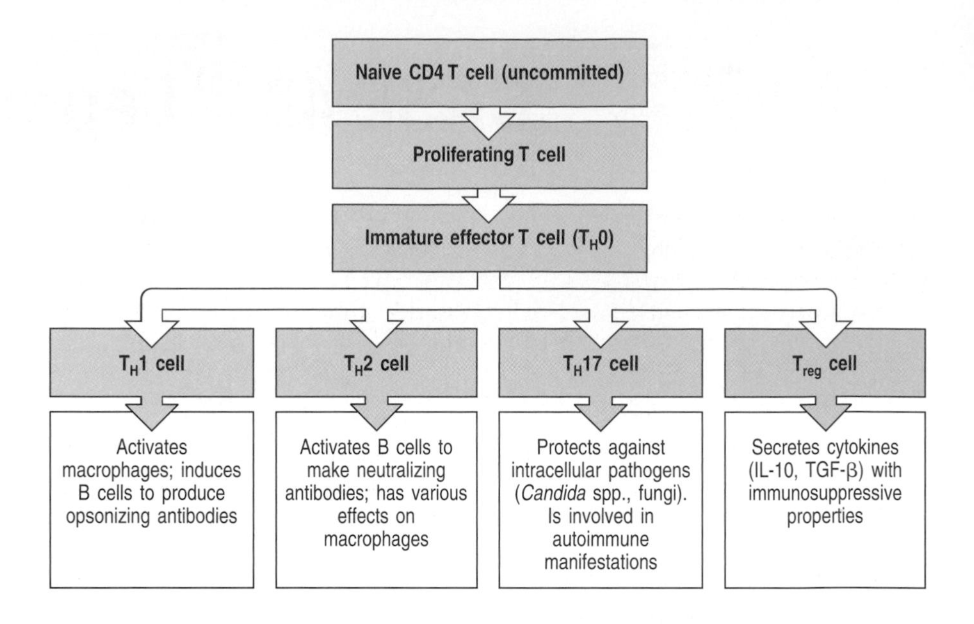

Because the decision to differentiate into T_H1 versus T_H2 cells occurs early in an adaptive immune response, the ability of pathogens to stimulate cytokine production by cells of the innate immune system has an important role in determining the subsequent course of the response. Infectious agents that invade or nonspecifically activate macrophages and NK cells, as do most viruses and intracellular bacteria such as mycobacteria, induce cells to secrete IL-12, thus favoring the differentiation of IFN-γ-secreting T_H1 cells. This loop is amplified because IFN-γ in turn favors T_H1 development and blocks the development of T_H2 cells (see Fig. 48.2). IL-12 also enhances the proliferation of T_H1 cells but has no effect on T_H2 cells because they do not express the β chain of the IL-12 receptor and therefore do not respond to the mitogenic effects of IL-12.

The differentiation of T_H2 cells is favored by pathogens, such as parasites, that elicit IL-4 production from specialized subsets of cells that include mast cells, eosinophils, and thymus-derived invariant NKT cells (iNKT cells), which express both the NK1.1 marker and a T-cell receptor of restricted V_β and invariant V_α chain usage. This loop is amplified by the cytokines produced by T_H2 cells—IL-4 and IL-10. IL-4 promotes the development of T_H2 cells, and IL-10 blocks T_H1 development (see Fig. 48.2).

Once one T_H phenotype becomes dominant in the course of a response, it is difficult to shift the antigen-specific response to the other. One reason for this is that the cytokine products of T_H1 and T_H2 cells are reciprocally inhibitory (see Fig. 48.2). The outcome of certain infections is greatly influenced by the type of T-cell response elicited. As we see in this case, infection with *Mycobacterium leprae*, the leprosy bacillus, is a good example.

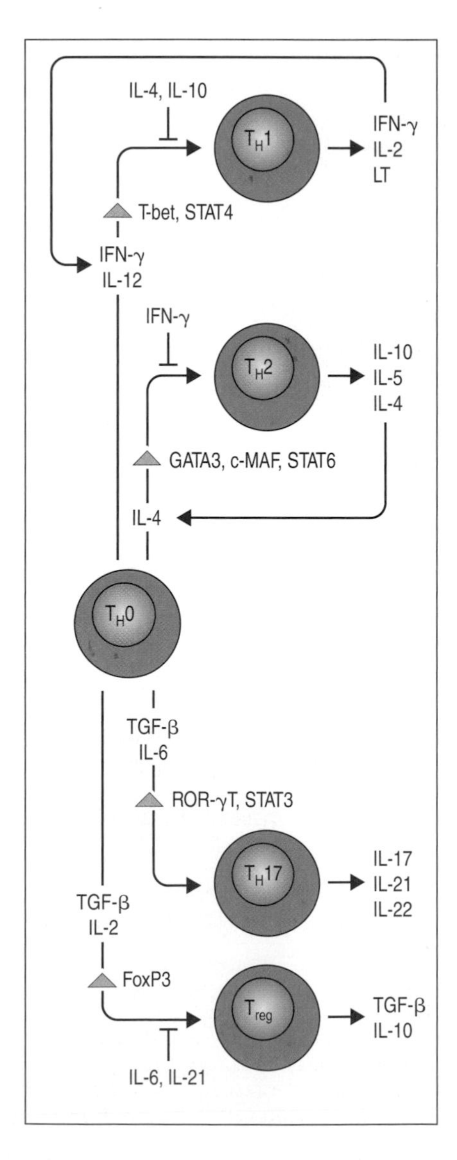

Fig. 48.2 Differentiation of T_H0 cells into T_H1, T_H2, T_H17 or T_{reg} cells depends on the cytokines present. IL-4 induces T_H2 differentiation, whereas IL-12 and IFN-γ induce differentiation into T_H1 cells. TGF-β and IL-6 drive the differentiation of T_H17 cells, whereas TGF-β and IL-2 favor the generation of regulatory T (T_{reg}) cells. IL-4, which is secreted by T_H2 cells, also inhibits T_H1 development. Similarly, cytokines produced by T_H1 cells, for example IFN-γ, inhibit T_H2 differentiation. Development of T_H1, T_H2, T_H17, and T_{reg} cells is driven by transcription factors induced by the cytokines. In particular, induction of the transcription factors T-bet and STAT4 in naive T cells leads to T_H1 development, and induction of GATA3 and STAT6 leads to T_H2 development, whereas induction of ROR-γT and STAT3 favors the differentiation of T_H17 cells, and expression of FoxP3 leads to the development of T_{reg} cells.

The case of Ursula Iguaran: a T_H2 response to the leprosy bacillus has severe consequences.

Ursula first sought medical advice when she was 18 years old, having left her home in Colombia to attend Harvard University on a scholarship. From the age of 16 years, she had started to notice a gradual loss of sensation on the backs of her hands and had developed hypopigmented lesions over both arms. The lesions progressively became worse and she noticed that she was losing her eyelashes and hairs from her eyebrows. She also experienced recurrent nosebleeds. A month after first noticing the hair loss she decided to seek medical help.

18-year-old female;
light-colored lesions
on skin.
Acid-fast bacilli
(? mycobacteria) in
skin lesions; leprosy?

On examination at the physician's office, Ursula seemed to be well apart from her immediate symptoms. She reported a history of mild asthma, which required treatment with inhaled β_2-adrenergic agents on an as-needed basis. Multiple hypopigmented macules (coin-sized raised lesions with ill-defined borders) were evident on her skin, along with cutaneous nodules 1 cm in diameter. These lesions were predominantly on her elbows, wrists, and hands (Fig. 48.3) and showed traces of dried blood; she also had similar lesions on her knees, ears, and buttocks. The absence of eyelashes and the ends of her eyebrows was obvious.

Cardiovascular and abdominal examinations were normal; a neurological examination was negative except for a decreased response to pinprick on the outer edges of the right and left hand and the right fourth and fifth fingers. There was a flexion contracture of the fourth and fifth fingers of both hands so that she could not straighten these fingers completely.

On blood test, Ursula's hematocrit was 35.1%; her white blood count was 7100 μl^{-1}, with 68% neutrophils, 23% lymphocytes, 5% monocytes, and 4% eosinophils (all normal values). Serum electrolytes were normal. Because her symptoms were becoming more severe, Ursula was referred to a dermatologist. She told him that she had grown up in a small village on the Caribbean coast of Colombia where many people, including her mother, had leprosy. The dermatologist performed a biopsy of the lesions on her left arm and right forearm, which revealed numerous acid-fast bacilli in clumps. A routine hematoxylin and eosin stain of lesion tissue showed up numerous Virchow's cells (highly vacuolated cells of the macrophage lineage, also known as foam cells) and few lymphocytes (Fig. 48.4). Cultures for acid-fast bacilli were negative.

The suspected diagnosis of lepromatous leprosy led to a more extensive immunologic work-up. Delayed hypersensitivity skin tests with intradermal injections of candida, mumps, and tuberculin antigens showed no reactions when the injection sites were inspected 48 and 72 hours later. Ursula's serum IgG was mildly elevated at 1800 mg dl^{-1} (normal 600–1100 mg dl^{-1}); her IgA and IgM levels were normal.

A diagnosis of lepromatous leprosy was made on the basis of the presence of acid-fast bacilli in the biopsy and Ursula's progressive neurologic symptoms. She was placed on a multiple drug regime consisting of dapsone, clofazamine, and rifampin, drugs that kill *M. leprae*. Her skin lesions gradually flattened and improved.

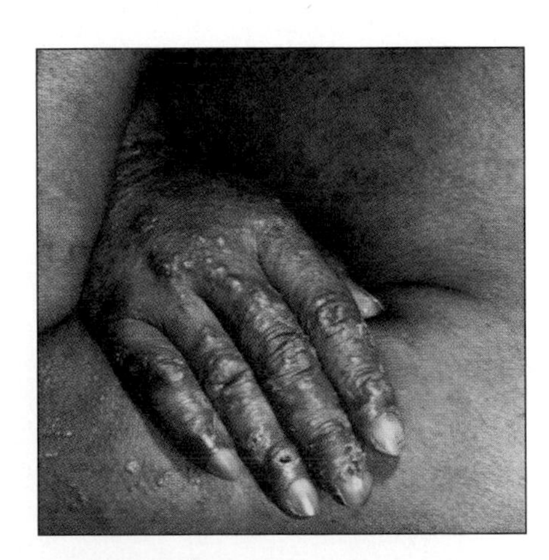

Fig. 48.3 Cutaneous nodules in lepromatous leprosy. Patients with lepromatous leprosy have multiple skin lesions. This photograph shows subcutaneous nodules on the hand. Photograph courtesy of E. Gonzalez.

Lepromatous leprosy.

The classical clinical feature in patients with leprosy is the association of cutaneous lesions, neuropathologic changes, and deformities. Leprosy is caused by *Mycobacterium leprae*, which colonizes macrophages and other host cells and multiplies within them. Mycobacteria within macrophages are protected

Fig. 48.4 Responses to *M. leprae* are sharply differentiated in lepromatous and tuberculoid leprosy. The photographs show sections of lesion biopsies stained with hematoxylin and eosin. Infection with *M. leprae* bacilli, which can be seen in the right-hand photograph as numerous small dark red dots inside macrophages, can lead to two very different forms of the disease. In tuberculoid leprosy (left), growth of the microorganism is well controlled by T_H1-like cells that activate infected macrophages. The tuberculoid lesion contains granulomas (see Case 26) and is inflamed, but the inflammation is localized and causes only local peripheral nerve damage. In lepromatous leprosy (right), infection is widely disseminated and the bacilli grow uncontrolled in macrophages. In the late stages there is severe damage to connective tissues and to the peripheral nervous system. There are several intermediate stages between these two polar forms. Photographs courtesy of G. Kaplan.

Infection with *Mycobacterium leprae* can result in different clinical forms of leprosy

There are two polar forms, tuberculoid and lepromatous leprosy, but several intermediate forms also exist

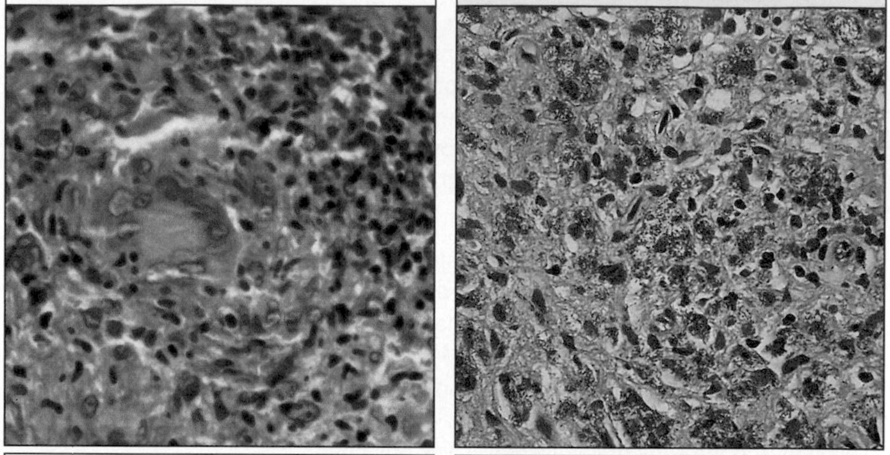

Tuberculoid leprosy	Lepromatous leprosy
Organisms present at low to undetectable levels	Organisms show florid growth in macrophages
Low infectivity	High infectivity
Granulomas and local inflammation. Peripheral nerve damage	Disseminated infection. Bone, cartilage, and diffuse nerve damage
Normal serum immunoglobulin levels	Hypergammaglobulinemia
Normal T-cell responsiveness. Specific response to *M. leprae* antigens	Low or absent T-cell responsiveness. No response to *M. leprae* antigens

from attack by antibody and can be eliminated only when their host macrophages are activated and produce increased amounts of nitric oxide, oxygen radicals, and other microbicidal molecules. *M. leprae* grows best at 30°C and therefore lesions tend to appear on the extremities—the colder areas of the body—for example the hands, ears, and buttocks as in Ursula's case. Unlike *M. tuberculosis, M. leprae* does not grow in culture.

The clinical symptoms of leprosy vary depending on the type of immune response to the mycobacteria. The clinical spectrum is typically divided into two polar forms, tuberculoid and lepromatous leprosy, although intermediate forms exist. Tuberculoid leprosy is associated with a vigorous cell-mediated (T_H1) response against the bacillus. This results in macrophage activation with efficient killing of intracellular mycobacteria, localized tissue damage, and usually a milder clinical picture. In the lepromatous form, the T_H1 cell-mediated response is defective and a T_H2 response predominates; this leads to a vigorous but ineffective antibody response against *M. leprae* and dissemination of the bacilli to other sites in the body, which results in further tissue destruction and aggravation of the symptoms. The importance of T_H1-derived IFN-γ in containing mycobacterial infections is further illustrated by the observation that infants with genetic defects in the IFN-γ receptor die from disseminated mycobacterial infections (see Case 24).

Infection with *M. leprae* illustrates a situation in which the same microorganism, in different individuals, can trigger either a T_H1 or a T_H2 response. A T_H1

response predominates in tuberculoid leprosy, in which the mycobacteria are contained within well-circumscribed granulomas and propagate poorly, usually accompanied by subsequent minimal tissue damage. In contrast, a T_H2 response predominates in lepromatous leprosy, in which the mycobacteria propagate rapidly, with resulting extensive tissue damage. Analysis of mRNA isolated from lesions of patients with lepromatous and tuberculoid leprosy illustrates the cytokine patterns in the two forms of the disease. T_H2 cytokines (IL-4, IL-5, and IL-10) dominate in the lepromatous form, whereas T_H1 cytokines (IL-2, IFN-γ, and LT) dominate in the tuberculoid form (Fig. 48.5).

The neurologic damage in leprosy has two main causes. It can arise from bacterial multiplication within Schwann cells—the cells that form the insulating myelin sheath around some nerve cell axons. Disruption of the myelin sheath interferes with the normal conduction of nerve impulses along the axon. In the tuberculoid form, nerve damage also arises from the formation of granulomas and inflammation of the tissue surrounding the nerve. The nerve damage results in dysfunctional nerve terminals, causing decreased sensation and eventually a loss of motor function.

The nosebleeds that Ursula experienced are common in leprosy. They are due to large numbers of *M. leprae* in the nasal tissue with extensive involvement of the nasal mucosa, leading to congestion and breakage of blood vessels.

The T_H2 response can influence the course of the infection in various other ways. By binding to mycobacterial antigens displayed on the surface of infected cells, antibodies against the leprosy bacillus can interfere with the action of cytotoxic CD8 T cells. CD8 T cells can, in addition to their cytolytic function, also respond to antigen by secreting cytokines. Patients with lepromatous leprosy have CD8 T cells that suppress the T_H1 response by making IL-10 and LT. IL-10 inhibits the development of T_H1 cells and inhibits both cytokine release from macrophages and their capacity to kill internalized microorganisms. LT also inhibits the intracellular killing capacity of macrophages. Inhibition of macrophages leads to decreased production of IL-12, fewer T_H1 cells, and more T_H2 cells. In contrast, patients with the less destructive tuberculoid leprosy lack suppressor CD8 T cells and thus make a vigorous T_H1 response, leading to macrophage activation and the destruction of the leprosy bacilli.

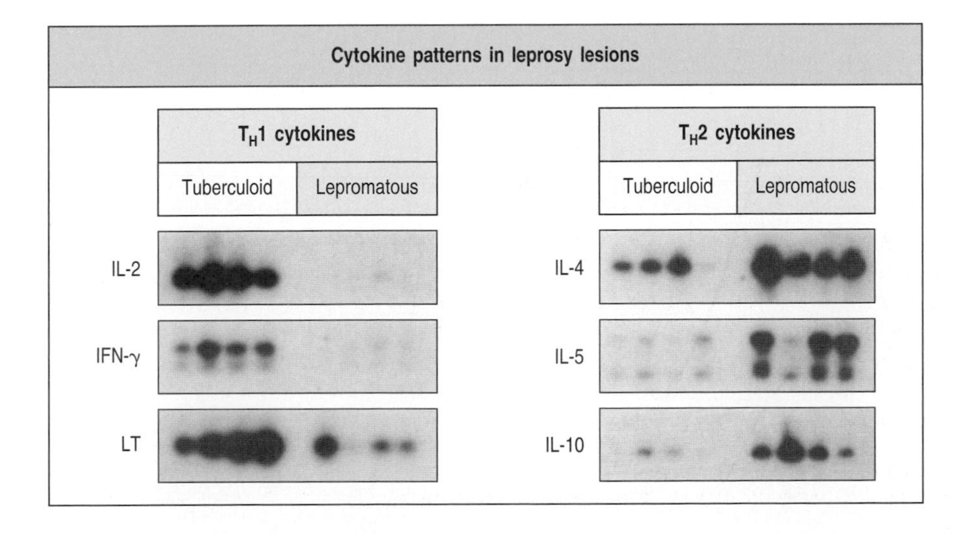

Fig. 48.5 Cytokine patterns in leprosy lesions. The cytokine patterns in the two polar forms of leprosy are distinctly different, as shown by Northern blot analysis of the mRNA from lesions of three patients with lepromatous leprosy and three patients with tuberculoid leprosy. Cytokine mRNAs typically produced by T_H2 cells predominate in the lepromatous form, whereas cytokines produced by T_H1 cells predominate in the tuberculoid form. Cytokine blots courtesy of R.L. Modlin.

Questions.

1 *Ursula did not respond to candida and mumps antigens, which are common recall antigens, with a delayed-type hypersensitivity reaction. Give a possible explanation.*

2 *Which cytokine might be beneficial to a patient with lepromatous leprosy?*

3 *Describe the mechanism for Ursula's hypergammaglobulinemia.*

4 *Why would Ursula be prone to asthma?*

CASE 49 Acute Systemic Anaphylaxis

A life-threatening immediate hypersensitivity reaction to peanuts.

Adaptive immune responses can be elicited by antigens that are not associated with infectious agents. Inappropriate immune responses to otherwise innocuous foreign antigens result in allergic or hypersensitivity reactions, and these unwanted responses can be serious. Allergic reactions occur when an already sensitized individual is reexposed to the same innocuous foreign substance, or allergen. The first exposure generates allergen-specific antibodies and/or T cells; reexposure to the same allergen, usually by the same route, leads to an allergic reaction.

Acute systemic anaphylaxis is a type I IgE-mediated hypersensitivity reaction (Fig. 49.1) that is rapid in onset and can cause death. There is typically involvement of at least two organ systems, including the skin, respiratory, gastrointestinal, cardiovascular, or central nervous systems. As with any type I hypersensitivity reaction, the first exposure to the allergen generates allergen-specific IgE antibodies, which become bound to Fc receptors (FcεRI) on the surface of mast cells. On repeat exposure to allergen, cross-linking of IgE bound to FcεRI on mast cells and basophils leads to degranulation, with the release of preformed mediators such as histamine and tryptase and the synthesis and release of other mediators such as prostaglandins and leukotrienes. Histamine is a major mediator of the immediate effects of anaphylaxis, causing multiple symptoms including increased permeability of blood vessels, which can cause life-threatening hypotension. The mast-cell mediators important for anaphylaxis, and the clinical consequences of their release, are illustrated in Fig. 49.2.

Allergens introduced systemically are most likely to cause a serious anaphylactic reaction through the activation of sensitized connective tissue mast cells. The disseminated effects on the circulation and on the respiratory system are the most dangerous, and localized swelling of the upper airway can cause suffocation. Ingested antigens cause a variety of symptoms through their action on mucosal mast cells.

Any protein allergen can provoke an anaphylactic reaction, but those that most commonly cause acute systemic anaphylaxis are foods, medications, and insect venoms (Fig. 49.3). Proteins in food, most commonly milk, soy beans, eggs, wheat, peanuts, tree nuts, and shellfish, can also cause systemic anaphylaxis. Contact with protein antigens found in latex, a common

Topics bearing on this case:

- Class I hypersensitivity reactions
- Allergic reactions to food
- Mast-cell activation via IgE

This case was prepared by Raif Geha, MD, in collaboration with Lisa Bartnikas, MD.

Type I immune-mediated tissue damage	
Immune reactant	IgE antibody
Antigen	Soluble antigen
Effector mechanism	Mast-cell activation antigen
Example of hyper-sensitivity reaction	Allergic rhinitis, allergic asthma, systemic anaphylaxis

Fig. 49.1 Type I immunological hypersensitivity reactions. Type I hypersensitivity reactions involve IgE antibodies and the activation of mast cells (see also Case 50).

constituent of rubber gloves, is also known to cause anaphylaxis. In addition, small-molecule antibiotics such as penicillin can act as haptens, binding to host proteins.

22-month-old child, unconscious, swollen face, difficulty in breathing. Give epinephrine immediately.

Type I allergic responses are characterized by the activation of allergen-specific CD4 helper cells (T_H2 cells) and the production of allergen-specific IgE antibody. The allergen is captured by B cells through their antigen-specific surface IgM and is processed so that its peptides are presented by MHC class II molecules to T-cell receptors of antigen-specific T_H2 cells. The interleukins IL-4 and/or IL-13 produced by the activated T_H2 cells induce a switch to the production of IgE, rather than IgG, by the B cell (see Fig. 2.4). However, allergen-specific IgE antibodies can exist without the occurrence of anaphylaxis, suggesting that factors other than IgE may be required.

This case concerns a child who suffered from life-threatening systemic anaphylaxis caused by an allergy to peanuts.

The case of John Mason: a life-threatening immune reaction.

John was healthy until the age of 22 months, when he developed swollen lips while eating cookies containing peanut butter. The symptoms disappeared in about an hour. A month later, while eating the same type of cookies, he started to vomit, became hoarse, had great difficulty in breathing, started to wheeze and developed a swollen face. He was taken immediately to the emergency room of the Children's Hospital, but on the way there he became lethargic and lost consciousness.

On arrival at hospital, his blood pressure was catastrophically low at 40/0 mmHg (normal 80/60 mmHg). His pulse was 185 beats min^{-1} (normal 80–90 beats min^{-1}), and his respiratory rate was 76 min^{-1} (normal 20 min^{-1}). His breathing was labored. An anaphylactic reaction was diagnosed and John was immediately given an intramuscular injection of 0.15 ml of a 1:1000 dilution of epinephrine (adrenaline). An intravenous solution of normal saline was infused as a bolus. The antihistamine Benadryl (diphenhydramine hydrochloride) and the anti-inflammatory corticosteroid Solu-Medrol (methylprednisolone) were also administered intravenously. A blood sample was taken to test for histamine and the enzyme tryptase.

Mediators of anaphylaxis		
Mediator	Action	Signs/symptoms
Histamine	Vasodilation, bronchoconstriction	Pruritus, swelling, hypotension, diarrhea, wheezing
Leukotrienes	Bronchoconstriction	Wheezing
Platelet-activating factor*	Bronchoconstriction, vasodilation	Wheezing, hypotension
Tryptase	Proteolysis	Unknown

Fig. 49.2 Mediators released by mast cells during anaphylaxis and their clinical consequences. *Platelet-activating factor is not released by mast cells but by neutrophils, basophils, platelets, and endothelial cells.

IgE-mediated allergic reactions			
Syndrome	**Common allergens**	**Route of entry**	**Response**
Systemic anaphylaxis	Drugs Serum Venoms	Intravenous (either directly or following oral absorption into the blood)	Edema Vasodilation Tracheal occlusion Circulatory collapse Death
Acute urticaria (wheal-and-flare)	Insect bites Allergy testing	Subcutaneous	Local increase in blood flow and vascular permeability
Allergic rhinitis (hay fever)	Pollens (ragweed, timothy, birch) Dust-mite feces	Inhaled	Edema of nasal mucosa Irritation of nasal mucosa
Allergic asthma	Danders (cat) Pollens Dust-mite feces	Inhaled	Bronchial constriction Increased mucus production Airway inflammation
Food allergy	Shellfish Milk Eggs Fish Wheat	Oral	Vomiting Diarrhea Pruritus itching Urticaria (hives) Anaphylaxis (rarely)

Fig. 49.3 IgE-mediated reactions to extrinsic antigens. All IgE-mediated responses involve mast-cell degranulation, but the symptoms experienced by the patient can be very different depending on whether the allergen is injected, inhaled, or eaten, and depending on the dose of the allergen.

Within minutes of the epinephrine injection, John's hoarseness improved, the wheezing diminished, and his breathing became less labored (Fig. 49.4). His blood pressure rose to 50/30 mmHg, the pulse decreased to 145 beats min^{-1} and his breathing to 61 min^{-1}. Thirty minutes later, the hoarseness and wheezing got worse again and his blood pressure dropped to 40/20 mmHg, his pulse increased to 170 beats min^{-1} and his respiratory rate to 70 min^{-1}.

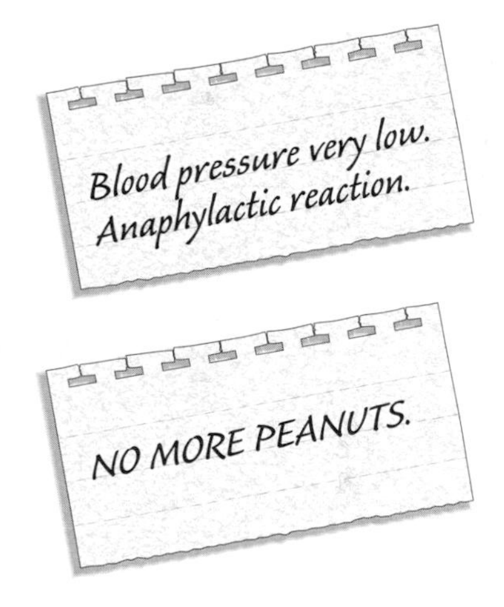

John was given another intramuscular injection of epinephrine and was made to inhale nebulized albuterol (a β_2-adrenergic agent). This treatment was repeated once more after 30 minutes. One hour later, he was fully responsive, his blood pressure was 70/50 mmHg, his pulse was 116 beats min^{-1} and his respiratory rate had fallen to 46 min^{-1}. John was admitted to the hospital for further observation.

Treatment with Benadryl and methylprednisolone intravenously every 6 hours was continued for 24 hours, by which time the facial swelling had subsided and John's blood pressure, respiratory rate, and pulse were normal. He had stopped wheezing and when the doctor listened to his chest with a stethoscope it was clear.

He remained well and was discharged home with an Epi-Pen. His parents were instructed to avoid giving him foods containing peanuts in any form, and were asked to bring him to the Allergy Clinic for further tests.

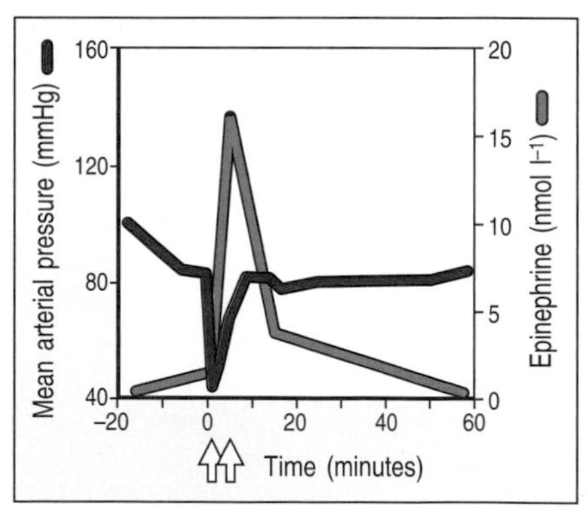

Fig. 49.4 Mean arterial pressure and epinephrine levels in a representative patient with insect-sting anaphylactic shock. Time 0 indicates the onset of the anaphylactic reaction as reported by the patient. The arrows indicate administration of antihistamines and epinephrine.

Acute systemic anaphylaxis.

Anaphylaxis presents a medical emergency and is the most urgent of clinical immunologic events; it requires immediate therapy. It results from the generation and release of a variety of potent biologically active mediators and their concerted effects on a number of target organs. John showed classic rapid-onset symptoms of anaphylaxis, starting with vomiting and swelling of the face and throat, and constriction of the bronchial smooth muscle, which led to his difficulty in breathing. This was soon followed by a catastrophic loss of blood pressure, due to leakage of fluid from the blood vessels. Anaphylaxis can also cause urticaria (hives), heart arrhythmias, and myocardial ischemia, and gastrointestinal symptoms such as nausea, vomiting, and diarrhea. All these signs and symptoms can occur singly or in combination.

Fatal allergic reactions to the venoms in bee and wasp stings have been recognized for at least 4500 years and account today for roughly 40 deaths each year in the United States. In 1902, Portier and Richet reported that a second injection of a protein from a sea anemone caused a fatal systemic reaction in dogs that had been injected previously with this protein. Because this form of immunity was fatal rather than protective, it was termed 'anaphylaxis' to distinguish it from the 'prophylaxis' (protection) generated by immunization.

Anaphylaxis requires a latent period for sensitization after the first introduction of antigen followed by reexposure to the sensitizing agent, which can be any foreign protein or a hapten. In the early part of the 20th century, the most frequent cause of systemic anaphylaxis was horse serum, which was used as a source of antibodies to treat infectious diseases.

In many cases, the presentation of food allergy occurs on the first known ingestion, suggesting that routes other than the oral one may be important in sensitization. For example, epidemiologic data suggest that sensitization to peanut protein may occur in children through the application of peanut oil to inflamed skin. A recent study demonstrated that the incidence of peanut allergy in children who avoided peanut ingestion correlated with the level of peanut consumption in their homes, which is consistent with the skin's being an important route of allergen sensitization. At present there is no cure for food allergy. Current therapy relies on allergen avoidance and the treatment of severe reactions with epinephrine.

Anaphylaxis is increasing in prevalence and is a frequent cause of visits to the emergency room, with 50–2,000 episodes per 100,000 persons, or a lifetime prevalence of 0.05–2.0%. The rate of fatal anaphylaxis from any cause is estimated at 0.4 cases per million individuals per year. Although in John's case the reaction was brought on by eating a food, an antigen administered by subcutaneous, intramuscular, or intravenous injection is more likely to induce a clinical anaphylactic reaction than one that enters by the oral or respiratory route.

Questions.

1 *Anaphylaxis results in the release of a variety of chemical mediators from mast cells, such as histamine and leukotrienes. Angioedema (localized swelling caused by an increase in vascular permeability and leakage of fluid into tissues) is one of the symptoms of anaphylaxis. With the above in mind, why did John get hoarse and why did he wheeze?*

2 *When his parents brought John back to the Allergy Clinic, a nurse performed several skin tests by pricking the epidermis of his forearm with a shallow plastic needle containing peanut antigens. John was also tested in a similar fashion with antigens from nuts as well as from eggs, milk, soy, and wheat. Within 5 minutes John developed a wheal, 10 mm × 12 mm*

in size, surrounded by a red flare, 25 mm × 30 mm (see Fig. 50.5), at the site of application of the peanut antigen. No reactions were noted to the other antigens. A radioallergosorbent test (RAST) was performed on a blood sample to examine for the presence of IgE antibodies against peanut antigens. It was positive. What would you advise John's parents to do?

3 *Why was John treated first with epinephrine in the emergency room?*

4 *Why was John given a blood test for histamine and the enzyme tryptase?*

5 *Why was the skin testing for peanuts not done in the hospital immediately after John had recovered, instead being done at a later visit?*

6 *The incidence of peanut allergy is increasing. Why?*

7 *John's parents want to know whether there are therapies that might cure him of his peanut allergy. What do you tell them?*

CASE 50 Allergic Asthma

Chronic allergic disease caused by an adaptive immune response to inhaled antigen.

Chronic allergic reactions are much more common than the acute systemic anaphylaxis reaction discussed in Case 49. Among these are allergic reactions to inhaled antigens (Fig. 50.1), which range in severity from a mild allergic rhinitis (hay fever) to potentially life-threatening allergic asthma, the disease discussed in this case. Once an individual has been sensitized, the allergic reaction becomes worse with each subsequent exposure to allergen, which not only produces allergic symptoms but also increases the levels of antibody and T cells reactive to the allergen.

Allergic asthma is an example of a type I hypersensitivity reaction. Type I reactions involve the activation of helper CD4 T_H2 cells, IgE antibody formation, mast-cell sensitization, and the recruitment of eosinophils. The allergen-specific IgE antibodies formed in sensitized individuals bind to and occupy high-affinity Fcε receptors (FcεRI) on the surfaces of tissue mast cells and basophils (Fig. 50.2). When the antigen is encountered again, it cross-links these bound IgE molecules, which triggers the immediate release of mast-cell granule contents, in particular histamine and various enzymes that increase blood flow and vascular permeability. This is the early phase of an immediate allergic reaction.

Within 12 hours of contact with antigen, a late-phase reaction occurs (Fig. 50.3). Arachidonic acid metabolism in the mast cell generates prostaglandins and leukotrienes, which further increase blood flow and vascular permeability. Cytokines such as interleukin-3 (IL-3), IL-4, IL-5, and tumor necrosis factor-α (TNF-α) are produced by both activated mast cells and helper T cells, and these further prolong the allergic reaction. The mediators and cytokines released by mast cells and helper T cells cause an influx of monocytes, more

This case was prepared by Raif Geha, MD, in collaboration with Lisa Bartnikas, MD.

Topics bearing on this case:

- Inflammatory reactions
- iNKT cells
- Differential activation of T_H1 and T_H2 cells
- IgE-mediated hypersensitivity
- Skin tests for hypersensitivity
- Radioimmunoassay
- Tests for immune function

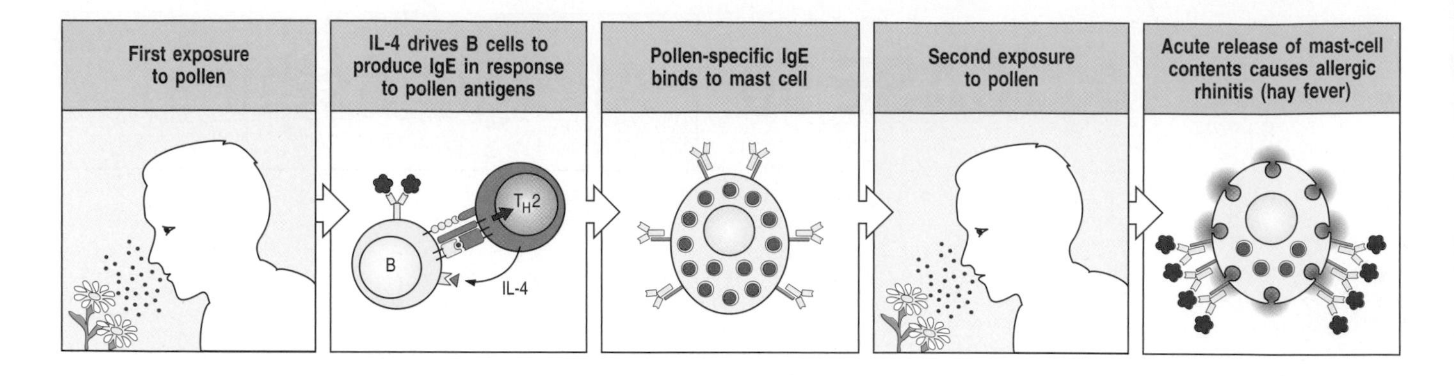
First exposure to pollen
IL-4 drives B cells to produce IgE in response to pollen antigens
Pollen-specific IgE binds to mast cell
Second exposure to pollen
Acute release of mast-cell contents causes allergic rhinitis (hay fever)
T_H2
B
IL-4

Fig. 50.1 Allergic reactions require previous exposure to the allergen. In this example, the first exposure to pollen induces the production of IgE anti-pollen antibodies, driven by the production of IL-4 by helper T cells (T_H2). The IgE binds to mast cells via FcεRI. Once enough IgE antibody is present on mast cells, exposure to the same pollen induces mast-cell activation and an acute allergic reaction, here allergic rhinitis (hay fever). Allergic reactions require an initial sensitization to the antigen (allergen), and several exposures may be needed before the allergic reaction is initiated.

T cells, and eosinophils into the site of allergen entry. The late-phase reaction is dominated by this cellular infiltrate. The cells of the infiltrate, particularly the eosinophils, make a variety of products that are thought to be responsible for much of the tissue damage and mucus production that is associated with chronic allergic reactions. Cytokine-producing NKT cells have also been implicated in allergic asthma.

Approximately 15% of the population suffers from IgE-mediated allergic diseases. Many common allergies are caused by inhaled particles containing foreign proteins (or allergens) and result in allergic rhinitis, asthma, and allergic conjunctivitis. In asthma, the allergic inflammatory response increases the hypersensitivity of the airway not only to allergen reexposure but also to nonspecific agents such as exercise, pollutants, and cold air.

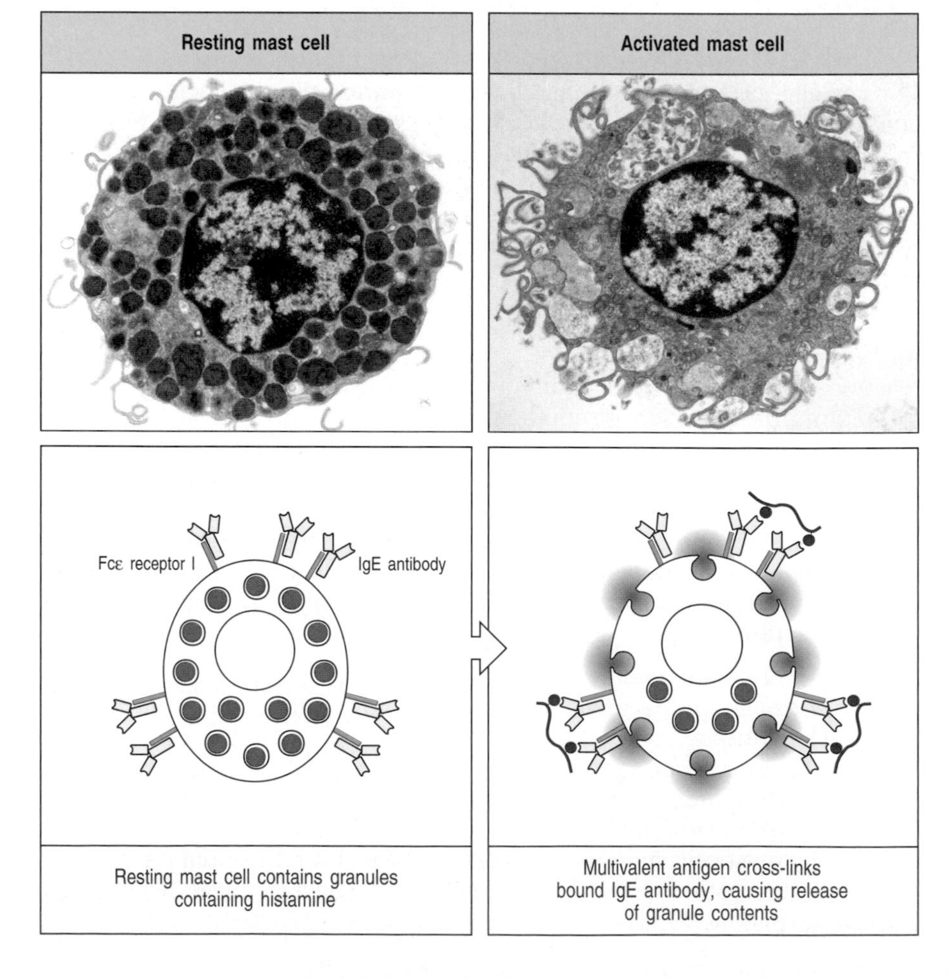
Resting mast cell
Activated mast cell
Fcε receptor I
IgE antibody
Resting mast cell contains granules containing histamine
Multivalent antigen cross-links bound IgE antibody, causing release of granule contents

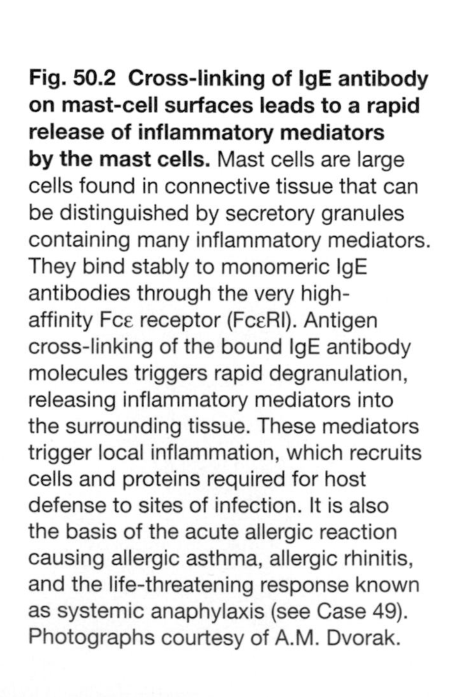

Fig. 50.2 Cross-linking of IgE antibody on mast-cell surfaces leads to a rapid release of inflammatory mediators by the mast cells. Mast cells are large cells found in connective tissue that can be distinguished by secretory granules containing many inflammatory mediators. They bind stably to monomeric IgE antibodies through the very high-affinity Fcε receptor (FcεRI). Antigen cross-linking of the bound IgE antibody molecules triggers rapid degranulation, releasing inflammatory mediators into the surrounding tissue. These mediators trigger local inflammation, which recruits cells and proteins required for host defense to sites of infection. It is also the basis of the acute allergic reaction causing allergic asthma, allergic rhinitis, and the life-threatening response known as systemic anaphylaxis (see Case 49). Photographs courtesy of A.M. Dvorak.

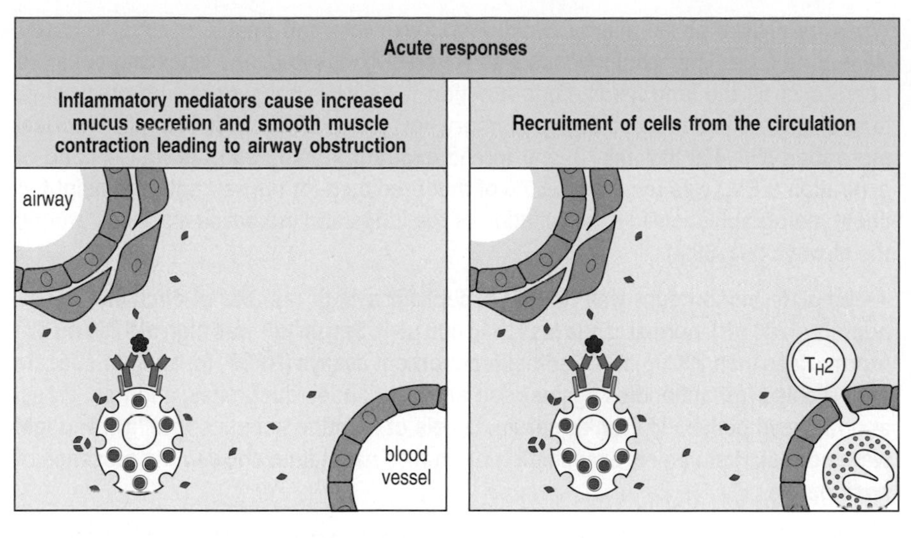

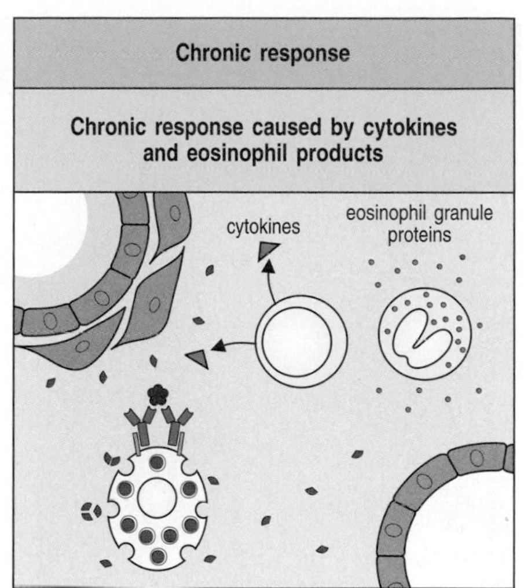

Fig. 50.3 The acute response in allergic asthma leads to T_H2-mediated chronic inflammation of the airways. In sensitized individuals, cross-linking of specific IgE on the surface of mast cells by inhaled allergen triggers them to secrete inflammatory mediators, causing bronchial smooth muscle contraction and an influx of inflammatory cells, including eosinophils and T_H2 lymphocytes. Activated mast cells and T_H2 cells secrete cytokines that also augment eosinophil activation, which causes further tissue injury and influx of inflammatory cells. The end result is chronic inflammation, which may then cause irreversible damage to the airways.

The case of Frank Morgan: a 14-year-old boy with chronic asthma and rhinitis.

Frank Morgan was referred by his pediatrician to the allergy clinic at 14 years of age because of persistent wheezing for 2 weeks. His symptoms had not responded to frequent inhalation treatment (every 2–3 hours) with a bronchodilator, the β_2-adrenergic agonist albuterol.

This was not the first time that Frank had experienced respiratory problems. His first attack of wheezing occurred when he was 3 years old, after a visit to his grandparents who had recently acquired a dog. He had similar attacks of varying severity on subsequent visits to his grandparents. Beginning at age 4 years, he had attacks of coughing and wheezing every spring (April and May) and toward the end of the summer (second half of August and September). A sweat test at age 5 years to rule out cystic fibrosis, a possible cause of chronic respiratory problems, was within the normal range.

As Frank got older, gym classes, basketball, and soccer games, and just going outside during the cold winter months could bring on coughing and sometimes wheezing. He had been able to avoid wheezing induced by exercise by inhaling albuterol 15–20 minutes before exercise. Frank had frequently suffered from a night-time cough, and his colds had often been complicated by wheezing.

Frank's chest symptoms had been treated as needed with inhaled albuterol. During the previous 10 years, Frank had been admitted to hospital three times for treatment of his asthma with inhaled bronchodilators and intravenous steroids. He had also been to the Emergency Room many times with severe asthma attacks. He had maxillary sinusitis at least three times, and each episode was associated with green nasal discharge and exacerbation of his asthma.

Since he was 4 years old, Frank had also suffered from intermittent sneezing, nasal itching, and nasal congestion (rhinitis), which always worsened on exposure to cats and dogs and in the spring and late summer. The nasal symptoms had been treated as needed with oral antihistamines with moderate success. Frank had had eczema as a baby, but this cleared up by the time he was 5 years old.

Family history revealed that Frank's 10-year-old sister, his mother, and his maternal grandfather had asthma. Frank's mother, father, and paternal grandfather suffered from allergic rhinitis.

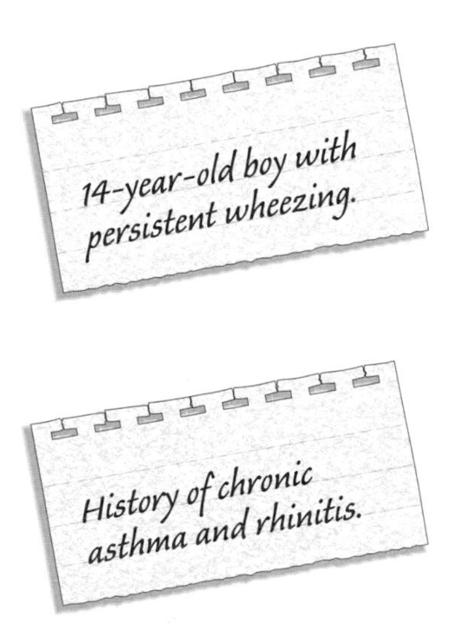

14-year-old boy with persistent wheezing.
History of chronic asthma and rhinitis.

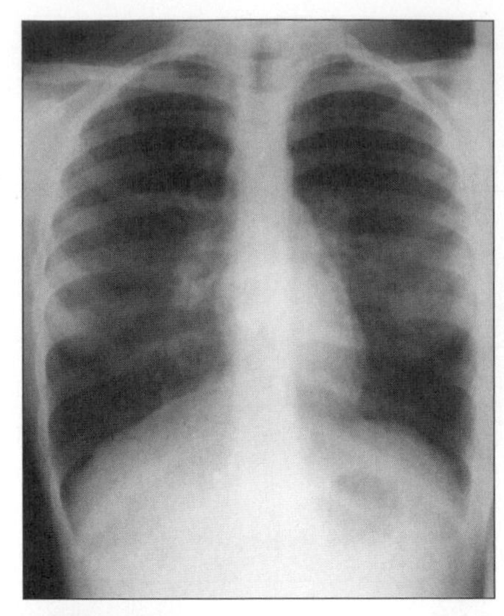

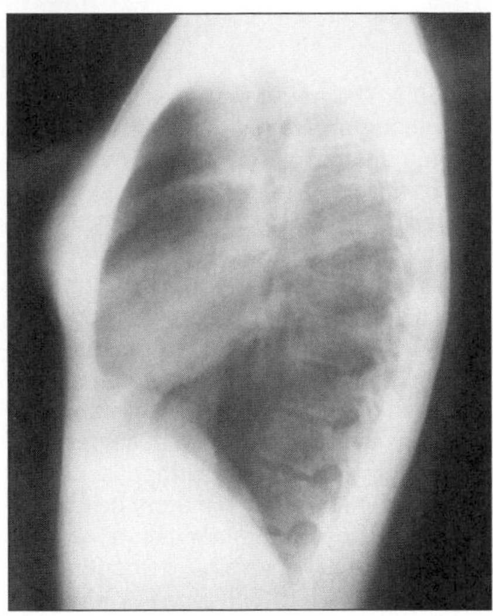

Fig. 50.4 Chest radiographs of a patient with asthma. Top: anteroposterior (A–P) view. Bottom: lateral view. The volume occupied by the lungs spans eight to nine rib spaces instead of the normal seven in the A–P view and indicates hyperinflation. The lateral view shows an increased A–P dimension, also reflecting hyperinflation. Hyperinflation indicates air trapping, which is a feature of the obstructive physiology seen in asthma. The bronchial markings are accentuated and can be seen to extend beyond one-third of the lung fields. This indicates inflammation of the airways.

When he arrived at the allergy clinic, Frank was thin and unable to breathe easily. He had no fever. The nasal mucosa was severely congested, and wheezing could be heard over all the lung fields. Lung function tests were consistent with obstructive lung disease with a reduced peak expiratory flow rate (PEFR) of 180 liter min^{-1} (normal more than 350–400 liter min^{-1}), and forced expiratory volume in the first second of expiration (FEV_1) was reduced to 50% of that predicted for his sex, age, and height. A chest radiograph showed hyperinflation of the lungs and increased markings around the airways (Fig. 50.4).

A complete blood count was normal except for a high number of circulating eosinophils (1200 μl^{-1}; normal range less than 400 μl^{-1}). Serum IgE was high at 1750 ng dl^{-1} (normal less than 200 ng dl^{-1}). Radioallergosorbent assays (RAST) for antigen-specific IgE revealed IgE antibodies against dog and cat dander, dust mites, and tree, grass, and ragweed pollens in Frank's serum. Levels of immunoglobulins IgG, IgA, and IgM were normal. Histological examination of Frank's nasal fluid showed the presence of eosinophils.

Frank was promptly given albuterol nebulizer treatment in the clinic, after which he felt better, his PEFR rose to 400 liter min^{-1}, and his FEV_1 rose to 65% of predicted. He was sent home on a 1-week course of the oral corticosteroid prednisone. He was told to inhale albuterol every 4 hours for the next 2–3 days, and then to resume taking albuterol every 4–6 hours as needed for chest tightness or wheezing. He was also started on fluticasone propionate (Flovent), an inhaled corticosteroid, and montelukast (Singulair), a leukotriene receptor antagonist for long-term control of his asthma. To relieve his nasal congestion, Frank was given the steroid fluticasone furoate (Flonase) to inhale through the nose, and was advised to use an oral antihistamine as needed. He was asked to return to the clinic 2 weeks later for follow-up, and for immediate hypersensitivity skin tests to try to detect which antigens he was allergic to (Fig. 50.5).

On the next visit Frank had no symptoms except for a continually stuffy nose. His PEFR and FEV_1 were normal. Skin tests for type I hypersensitivity were positive for multiple tree and grass pollens, dust mites, and dog and cat dander. He was advised to avoid contact with cats and dogs. To reduce his exposure to dust mites the pillows and mattresses in his room were covered with zippered covers. Rugs, stuffed toys, and books were removed from his bedroom. He was also started on immunotherapy with injections of grass, tree, and ragweed pollens, cat, dog, and house dust mite antigens, to try to reduce his sensitivity to these antigens.

A year and a half later, Frank's asthma continues to be stable with occasional use of albuterol during infections of the upper respiratory tract and in the spring. His rhinitis and nasal congestion now require much less medication.

Allergic asthma.

Like Frank, millions of adults and children suffer from allergic asthma. Asthma is the most common chronic inflammatory disorder of the airways and is characterized by reversible inflammation and obstruction of the small airways. Asthma has become an epidemic; the prevalence in the United States is increasing by 5% per year, with more than 500,000 new cases diagnosed annually. It is the most common cause of hospitalization and days lost from school in children. About 70% of patients with asthma have a family history of allergy. This genetic predisposition to the development of allergic diseases is called atopy. Wheezing and coughing are the main symptoms of asthma, and both are due to the forced expiration of air through airways that have become temporarily narrowed by the constriction of smooth muscle as a result of the

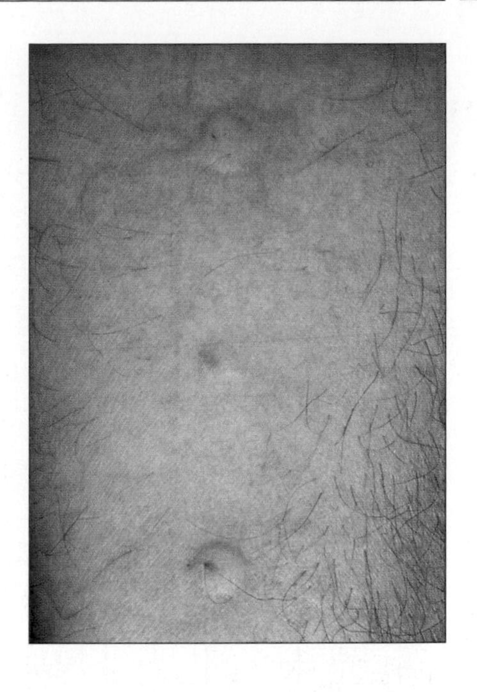

Fig. 50.5 An intradermal skin test. The photograph was taken 20 minutes after intradermal injections had been made with ragweed antigen (top), saline (middle), and histamine (bottom). A central wheal (raised swelling), reflecting increased vascular permeability, surrounded by a flare (red area), reflecting increased blood flow, is observed at the sites where the ragweed antigen and the positive histamine control were introduced. The small wheal at the site of saline injection is due to the volume of fluid injected into the dermis.

allergic reaction. As a consequence of the narrowed airways, air gets trapped in the lung, and the lung volume is increased during an attack of asthma (Fig. 50.6).

Once asthma is established, an asthma attack can be triggered not only by the allergen but by viral infection, cold air, exercise, or pollutants. This is due to a general hyperirritability or hyperresponsiveness of the airways, leading to constriction in response to nonspecific stimuli, thus reducing the air flow. The degree of hyperresponsiveness can be measured by determining the threshold dose of inhaled methacholine (a cholinergic agent) that results in a 20% reduction in airway flow. Airway irritability correlates positively with eosinophilia and serum IgE levels.

CD4 T cells are the central effector cells of airway inflammation in asthma. During asthma exacerbations, secretion of the T_H2-specific cytokines IL-4, IL-5, IL-9, and IL-13 is increased. Clinical improvement in asthma is associated with decreased T cells in the airways. Mast cells are also important effector cells in asthma and, after stimulation by allergen, release preformed and newly generated mediators, contributing to acute and chronic mucosal inflammation. Cysteinyl leukotrienes, a product of arachidonic metabolism, are also key inflammatory mediators in asthma (Fig. 50.7). Cysteinyl leukotriene receptors include at least three types of transmembrane receptors. Activation of the cysteinyl leukotriene receptor 1 ($CysLT_1$) leads to bronchial smooth muscle constriction and muscle-cell proliferation, plasma leakage,

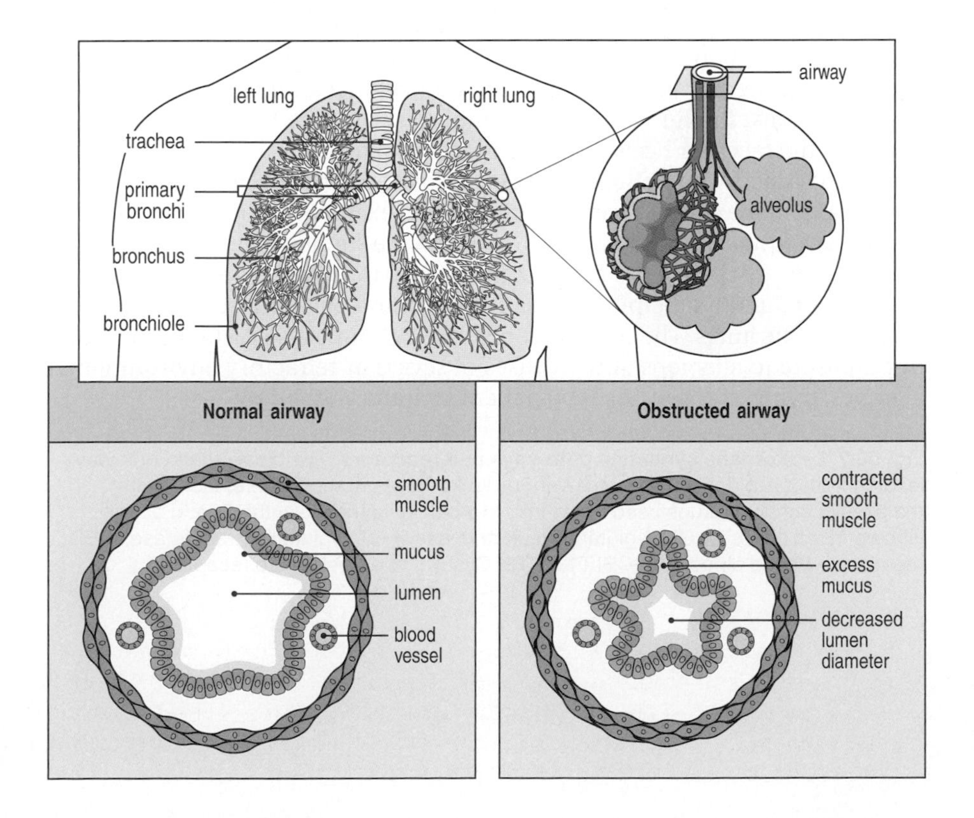

Fig. 50.6 Obstruction of the airways in chronic asthma. The top panels show the general anatomy of the lungs. Asthma is a chronic inflammatory disorder of the small airways—the bronchi and the bronchioles. In susceptible individuals, inflammation leads to recurrent wheezing, shortness of breath, chest tightness, and coughing. In between asthma attacks, patients are often asymptomatic, with normal physical exams and breathing tests. The bottom panels show schematic diagrams of sections through a normal airway (left) and an obstructed airway as a result of chronic asthma (right). During an asthma attack, there is infiltration of blood vessels of the small airways with immune cells (T_H2 lymphocytes and eosinophils), hypersecretion of mucus, and constriction and proliferation of bronchial smooth muscle. This leads to a decreased diameter of the airway lumen, resulting in wheezing and difficulty in breathing. In patients with severe asthma, there may be permanent airway remodeling.

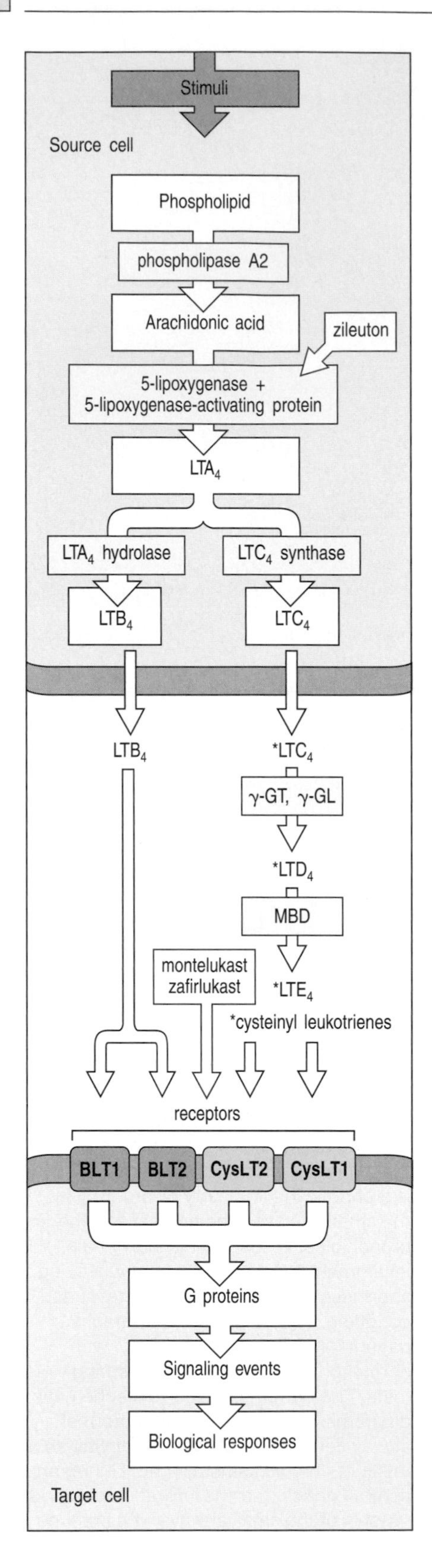

hypersecretion of mucus, and eosinophil migration. The role of neutrophils in asthma is less clear. Elevated neutrophil numbers are more frequently seen in non-allergic asthma, steroid-unresponsive asthma, and in fatal asthma, suggesting that neutrophil-dominated asthma may represent a distinct asthma phenotype. Elevated neutrophil numbers in asthmatic lungs are associated with increased expression of IL-17.

The subset of T cells called invariant NKT cells (iNKT cells) is also elevated in asthmatic airways, suggesting that they may be important in human asthma. iNKT cells are a subpopulation of thymus-derived T cells that express markers of both T cells (such as the T-cell receptor:CD3 complex) and NK cells (such as NK1.1 and Ly-49). In humans, iNKT cells express an invariant antigen receptor with a variable region composed of $V_{\alpha}24$–$J_{\alpha}15$ paired with $V_{\beta}11$. Unlike conventional T cells, iNKT cells can recognize glycolipid antigens bound and presented by the major histocompatibility complex (MHC) class Ib molecule CD1d. On activation, iNKT cells rapidly produce large amounts of the T_H1-type cytokine IFN-γ, the T_H2-type cytokines IL-4 and IL-13, and TNF-α and IL-2. The trigger for their activation in people with asthma could be glycolipids derived from microbes colonizing asthmatic airways.

Although asthma is a reversible disease, severe uncontrolled asthma can lead to airway remodeling, and a severe attack can be fatal. The mortality from asthma has been rising alarmingly in recent years. Risk factors for fatal asthma include frequent use of β_2-agonist therapy, poor perception of asthma severity, membership in a minority group, low socioeconomic status, adolescence, and male gender.

Several classes of drugs are commonly used to treat asthma, including corticosteroids, leukotriene antagonists, anti-IgE antibodies, anticholinergics, and β_2-adrenergic agonists. Corticosteroids (oral prednisone and inhaled fluticasone) inhibit the transcription of allergic and pro-inflammatory cytokines and can also activate the transcription of anti-inflammatory cytokines. This leads to a decrease in the numbers of mast cells, eosinophils, and T lymphocytes in the bronchial mucosa. Leukotriene antagonists (zileuton, montelukast, and zafirlukast) inhibit the synthesis of leukotrienes (which are products of arachidonic acid metabolism) or their receptor binding (see Fig. 50.7). Leukotriene modifiers have both mild bronchodilator and anti-inflammatory properties. Anti-IgE therapy uses a humanized monoclonal antibody (omalizumab) directed against the IgE that forms complexes with free IgE and prevents its biding to the receptor FcεRI on the surfaces of mast cells and basophils. This results in a decrease in circulating free IgE and the downregulation of FcεRI expression on the cell surfaces. β_2-agonists (for example albuterol) bind to the β_2-adrenergic receptor, which is expressed on the surface of bronchial smooth muscle cells. β_2-agonists relax smooth muscle, thus rapidly relieving airway constriction, and are helpful in treating the immediate phase of the allergic reaction in the lungs. The treatment of allergic asthma also includes minimizing exposure to allergens and, in cases of severe or refractory environmental allergies, trying to desensitize the patient by immunotherapy.

Fig. 50.7 Leukotriene synthesis pathways and receptors. The biosynthetic pathway leading from arachidonic acid to the various leukotrienes is shown here, along with the sites of action of drugs used in asthma to block leukotriene synthesis and action (shown in red boxes). γ-GL, γ-glutamyl leukotrienase; γ-GT, γ-glutamyl transferase; MBD, membrane-bound dipeptidase. BLT1, BLT2, CysLT2, and CysLT1 are receptors.

Questions.

1 *Explain the basis of Frank's chest tightness and the radiograph findings.*

2 *Explain the failure of Frank's asthma to improve despite the frequent use of bronchodilators, and his response to steroid therapy.*

3 *Eosinophilia is often detected in the blood and in the nasal and bronchial secretions of patients with allergic rhinitis and asthma. What is the basis for this finding?*

4 *What is the basis of the wheal-and-flare reaction that appeared 20 minutes after Frank had had a skin test for hypersensitivity to ragweed pollen?*

5 *Frank called 24 hours after his skin test to report that redness and swelling had recurred at several of the skin test sites. Explain this observation.*

6 *Frank developed wheezing on several occasions after taking the nonsteroidal anti-inflammatory drugs (NSAIDs) aspirin and ibuprofen (Motrin). Explain the basis for these symptoms.*

7 *How would the immunotherapy that Frank received help to alleviate his allergies?*

8 *Although atopic children are repeatedly immunized with protein antigens such as tetanus toxoid, they almost never develop allergic reactions to these antigens. Explain.*

CASE 51 Atopic Dermatitis

Skin as a target organ for allergic reactions.

As we saw in Cases 49 and 50, allergic or hypersensitivity reactions to otherwise innocuous antigens occur in certain individuals. The site of such reactions and the symptoms they produce vary depending on the type of allergen and the route by which it enters the body. Here we consider reactions to allergens entering the skin.

The main function of skin is to provide a physical barrier to the entry of foreign materials—including irritants, allergens, and pathogens—and to regulate the loss of water. The skin is composed of three layers—the epidermis (the outermost layer), the dermis, and the hypodermis—and the barrier function is performed by the epidermis. The epidermis consists mainly of keratin-synthesizing stratified epithelial cells—keratinocytes—in various stages of differentiation, interspersed with antigen-presenting Langerhans cells, a type of immature dendritic cell. The outermost cornified layer of the epidermis—the stratum corneum—consists of dead keratinocytes (corneocytes) filled with fibrous proteins such as keratin and involucrin. The dermis is a vascularized layer consisting of fibroblasts and dense connective tissue with collagen and elastic fibers, populated by dendritic cells, mast cells, macrophages, and few lymphocytes. The hypodermis is a layer of fat cells and loose connective tissue in contact with the underlying muscle.

After an antigen has been introduced into the skin, it is captured by Langerhans cells and other dendritic cells; these then migrate to local draining lymph nodes, where they present the antigen to recirculating naive T cells. Activated antigen-specific T cells proliferate and differentiate into memory or effector cells that express skin-homing receptors such as cutaneous lymphocyte antigen (CLA), and the chemokine receptors CCR4 and CCR10. CLA, which is an inducible carbohydrate modification of P-selectin glycoprotein ligand-1, is almost absent on naive T cells but is expressed by 30% of circulating memory T cells and by approximately 90% of infiltrating T cells in inflamed skin. CLA^+ T cells coexpress CCR4, whose ligands are the chemokines CCL17 (TARC) and CCL22 (MDC). Chemokines are small polypeptides that are synthesized by many cells, including keratinocytes (skin epidermal cells), fibroblasts, effector T cells, eosinophils, and macrophages (Fig. 51.1). They act through receptors that are members of the G-protein-coupled seven-span receptor family.

Topics bearing on this case:

- Cytokines and chemokines
- Inflammatory responses
- Bacterial superantigens
- T_H2 cell responses
- Migration and homing of lymphocytes
- IgE and allergic reactions

Fig. 51.1 Properties of selected chemokines. Chemokines fall into three related but distinct structural subclasses: CC chemokines have two adjacent cysteines at a particular point in the amino acid sequence; CXC chemokines have the same two cysteine residues separated by another amino acid; and C chemokines (not shown) have only one cysteine residue at this site.

Chemokine class	Chemokine	Produced by	Receptors	Major effects
CXC	CXCL8 (IL-8)	Monocytes Macrophages Fibroblasts Keratinocytes Endothelial cells	CXCR1 CXCR2	Mobilizes, activates, and degranulates neutrophils Angiogenesis
CC	CCL3 (MIP-1α)	Monocytes T cells Mast cells Fibroblasts	CCR1, 3, 5	Competes with HIV-1 Anti-viral defense Promotes T_H1 immunity
	CCL4 (MIP-1β)	Monocytes Macrophages Neutrophils Endothelium	CCR1, 3, 5	Competes with HIV-1
	CCL2 (MCP-1)	Monocytes Macrophages Fibroblasts Keratinocytes	CCR2B	Activates macrophages Basophil histamine release Promotes T_H2 immunity
	CCL5 (RANTES)	T cells Endothelium Platelets	CCR1, 3, 5	Degranulates basophils Activates T cells Chronic inflammation
	CCL11 (Eotaxin)	Endothelium Monocytes Epithelium T cells	CCR3	Role in allergy

Antigen-specific effector CD4 T cells leave the draining lymph nodes and are returned to the bloodstream, from which they reenter the skin via their receptors for skin-specific adhesion molecules and chemokines. In the skin, they can be activated locally by specific antigen to proliferate further and secrete effector cytokines.

When an allergen enters skin whose normal barrier function is disrupted in some way, it elicits an immune response dominated by allergen-specific T_H2 cells secreting their characteristic cytokines (for example interleukin-4 (IL-4)), and an IgE antibody response. This T_H2-dominated immune response results in a chronic skin inflammation and localized tissue destruction called atopic dermatitis or atopic eczema—to distinguish it from cases of dermatitis that do not have an allergic basis.

A chronic inflammatory reaction is sustained by the lymphocytes, eosinophils, and other inflammatory cells that are attracted out of the blood vessels at the site of inflammation. The first step in the process of lymphocyte homing to skin is the reversible binding (rolling) of lymphocytes to the vascular endothelium through interactions between CLA on the lymphocyte with E-selectin on the endothelial cells. The rolling cells are brought into contact with chemokines retained on heparan sulfate proteoglycans on the endothelial cell surface. Chemokine signaling activates the lymphocyte integrin lymphocyte function-associated antigen-1 (LFA-1), leading to firm adherence to intercellular cell adhesion molecule-1 (ICAM-1) on the endothelium followed by extravasation, or departure from the blood vessel, into the skin (see Fig. 27.1). Eosinophils migrate into tissues in a similar way, via an interaction between the integrin very late antigen-4 (VLA-4) on the eosinophil and vascular cell adhesion molecule-1 (VCAM-1) on the vascular endothelium. Once the lymphocytes and other cells have crossed the endothelium into the dermis, their migration to the focus of inflammation is directed by a gradient of chemokine molecules bound to the extracellular matrix.

If effector T cells are activated by their specific antigen once they have reentered the skin, they produce chemokines such as CCL5 and cytokines such as TNF-α (which activates endothelial cells to express E-selectin, VCAM-1, and ICAM-1). The chemokines produced by the effector T cells—and, under their influence, by keratinocytes—act on other T cells to upregulate their adhesion molecules, thus increasing recruitment of T cells into the affected tissue. At the same time, monocytes and polymorphonuclear leukocytes are recruited to these sites by adhesion to E-selectin. The TNF-α and interferon (IFN)-γ released by the activated T cells also act synergistically to change the shape of endothelial cells, resulting in increased blood flow, increased vascular permeability, and increased immigration of leukocytes, fluid, and protein into the site of inflammation. Thus, a few allergen-specific T cells encountering antigen in a tissue can initiate and amplify a potent local inflammatory response that recruits both antigen-specific and accessory cells to the site.

The case of Tom Joad: the itch that rashes.

Tom was admitted to hospital when he was 2 years old because of his worsening eczema. In the week before admission he had developed many open skin lesions (erosions), increased itching (pruritis), redness (erythema), and swelling (edema) of the skin. The lesions oozed a clear fluid, which formed crusts around them.

2-year-old boy with
severe eczema and
family history of
allergy.

Tom had suffered from eczema since the age of 2 months, when he developed a scaly red rash over his cheeks and over his knees and elbows (Fig. 51.2). He was breast-fed until 3 months old, when he was given a cow's milk-based formula. After 24 hours on the formula he started to vomit and to scratch his skin. A casein hydrolyzate formula was substituted for milk and he tolerated this well, but as new foods were added to his diet the eczema worsened. At 9 months old he developed a wheeze and was treated with bronchodilators. At 2 years old he had hives after eating peanut butter.

Tom's mother suffers from hay fever and his father has atopic dermatitis. The family lives in an old house with 20-year-old carpeting. Tom slept on a 10-year-old mattress surrounded by lots of stuffed animals. There were no pets, and his parents did not smoke.

During physical examination Tom was evidently uncomfortable and scratched continuously at his skin. His temperature was 37.9°C, pulse 96 beats min^{-1}, respiratory rate 24 min^{-1}, blood pressure 98/58 mmHg, weight 12 kg (10th centile), and height 90 cm (25th centile). His skin was very red, with large scales, and with scratched and infected lesions on his face, trunk, and extremities. Pustules were present on his arms and legs, and there were thick scales in his scalp. Thickened plaques of skin (lichenification) with a deep criss-cross pattern were seen around the creases on the insides of his elbows and knees, and on the backs of his hands and feet.

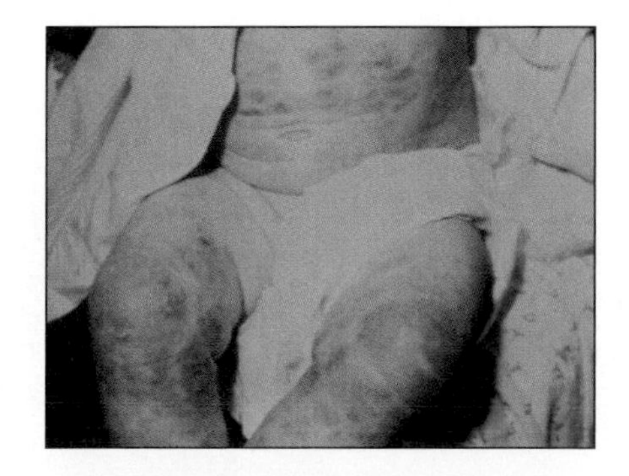

Fig. 51.2 An infant with severe eczema due to atopic dermatitis. Note the reddened and broken skin, especially in places, such as over the knees, where it is subjected to continual stretching and stress. Photograph courtesy of S. Gellis.

A skin culture was positive for *Staphylococcus aureus* and *Streptococcus pyogenes* Group A. Tom was treated with intravenous oxacillin (an antibiotic), antihistamines, topical steroids, and skin emollients such as coal tar. The infection resolved and his skin healed. Laboratory studies during hospitalization showed a white blood cell count of 9600 μl^{-1}, with 41% polymorphonuclear leukocytes, 26% lymphocytes, and 25% eosinophils (normal 0–5%), 13.1 g dl^{-1} hemoglobin, and a hematocrit of 37.2%. The absolute eosinophil count was elevated at 2400 μl^{-1} (normal 0–500 μl^{-1}). Serum IgE was also much elevated at 32,400 IU ml^{-1} (normal 0–200 IU ml^{-1}).

After discharge from hospital, Tom attended the allergy clinic, where he was tested for sensitivity to a range of allergens. He showed a positive type I allergic skin response (see Fig. 50.5) to numerous inhaled allergens including dust mites, mold spores, animal dander, and a variety of pollens. Tom also tested positive in a type I skin response to milk, cod, wheat, egg white, peanut, and tree nuts (cashew, almond, pecan, walnut, Brazil nut, and hazel nut) but had no reaction to rice or soybean. To determine whether any of these foods could be causing his atopic dermatitis, double-blind placebo-controlled food challenges were performed. He developed hives and wheezing after eating 1 g of egg white, and hives and eczema after drinking 2 g of powdered milk, but had no reaction to wheat or to cod. Four hours after the milk challenge, his serum tryptase level was raised, indicating IgE-induced mast-cell degranulation.

Reduce exposure to allergens.

His parents were advised to cover his mattress and pillows with a plastic covering and to remove the carpet and stuffed animals from his bedroom to decrease exposure to dust and mite allergens. Tom had a history of allergic reaction to peanuts, so continued avoidance of peanuts and tree nuts was recommended; he was therefore placed on a diet that excluded milk, eggs, peanuts, and tree nuts.

Tom's eczema improved significantly in response to these environmental and dietary control measures, together with the use of emollients and low-potency topical steroids on his skin. The family reduced the risk of skin irritation by avoiding all perfumed soaps and lotions, double-rinsing Tom's laundry, and dressing him in cotton clothes. He did well on this regimen until he was 12 years old, when he awoke one day with itchy vesicles on his lower left leg and ankle. The lesions progressed to become painful punched-out erosions, and were diagnosed as herpes simplex infection. He was given the antiviral drug acyclovir and the lesions resolved. Tom is now 15 years old.

Atopic dermatitis.

Atopic dermatitis, or atopic eczema, is a common pruritic (itching) inflammatory skin disease often associated with a family and/or personal history of allergy. Its prevalence is currently about 17% in children in the United States, and it is on the rise all over the world but particularly in Western and industrialized societies. Atopic dermatitis almost uniformly starts in infancy and although it tends to resolve, or improve remarkably, by the age of 5 years, it can persist into adult life in about 15% of cases. Many children with atopic dermatitis develop other indications of atopy—a predisposition to develop allergies—such as food allergies, asthma, and allergic rhinitis (hay fever).

The hallmark of atopic dermatitis is skin barrier dysfunction, which results in dry itchy skin. This leads to scratching, which inflicts mechanical injury and allows access to environmental antigens, resulting in sensitization and allergic skin inflammation (Fig. 51.3). The barrier to water permeation even in normal skin is not absolute, and the movement of water through the stratum corneum into the atmosphere is known as transepidermal water loss (TEWL). Almost all patients with atopic dermatitis have increased TEWL, an indicator of disrupted barrier function.

The terminal differentiation of keratinocytes from granular cells to corneocytes is a critical step in the maintenance of skin barrier function. The formation of the stratum corneum involves the sequential expression of several major structural proteins. Many of the proteins involved in skin cornification are encoded in a locus containing about 70 genes on chromosome 1q21, termed the epidermal differentiation complex (EDC). EDC genes are expressed during the late stages of terminal keratinocyte differentiation and encode proteins such as loricrin, involucrin, and the S100 calcium-binding

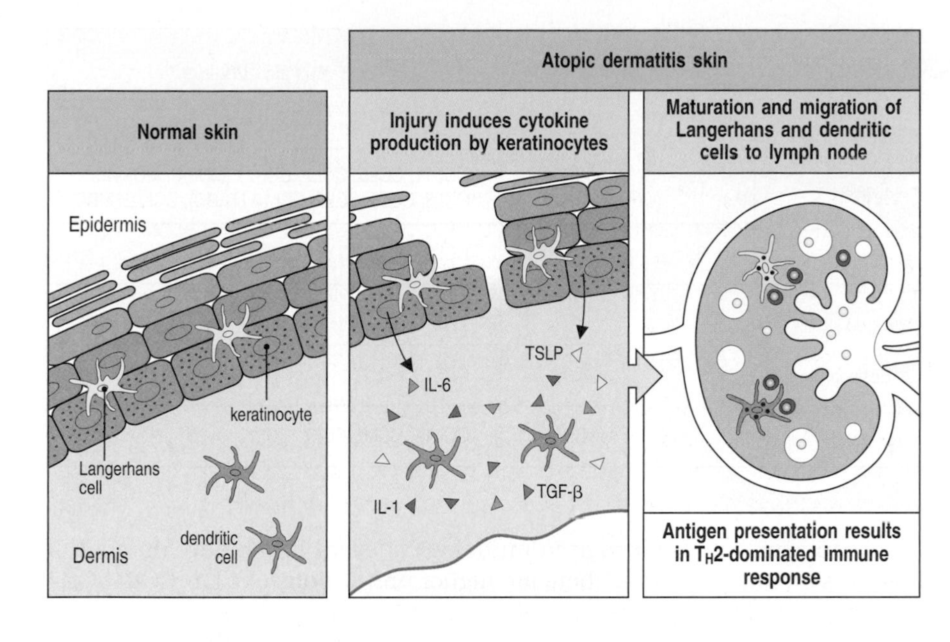

Fig. 51.3 Skin barrier dysfunction in atopic dermatitis. An intact skin barrier (left panel) prevents allergens from entering normal skin. Damage to this barrier (right panels) allows allergens to penetrate into the subepidermal layer and interact with antigen-presenting cells, and induces keratinocytes to produce cytokines that include TSLP, IL-1, IL-6, and TGF-β. This leads to maturation and migration of the antigen-presenting cells to the draining lymph nodes, where they present antigens to naive T cells, resulting in a T_H2-dominated immune response.

protein filaggrin. During corneocyte differentiation, the giant inactive precursor polypeptide profilaggrin is dephosphorylated and cleaved by serine proteases into multiple filaggrin polypeptides, which are further degraded to hydrophilic amino acids that have a critical role in the hydration of the skin.

The filaggrin gene is mutated in about 20% of patients with atopic dermatitis, and this association suggests that the skin-barrier defect precedes the development of the disease. Evidence for a role of defective skin barrier in atopic dermatitis is provided by the observation that application of the protein ovalbumin to mouse skin that has been disrupted by tape-stripping results in an allergic inflammation of the skin that has features of atopic dermatitis. Because *filaggrin* mutations are present in only a fraction of patients with atopic dermatitis, however, genetic variants of other genes involved in the skin's barrier function are also likely to be important in the pathogenesis of the disease. Indeed, mutations in two other genes involved in skin barrier function have been associated with atopic dermatitis, namely *SCCE* and *SPINK5*.

It is not clear why the introduction of antigen through a disrupted skin barrier results in a T_H2-dominated allergic response. However, it is known that mechanical injury causes keratinocytes to release the epithelial cytokine thymic stromal lymphopoietin (TSLP). TSLP acts on dendritic cells to promote their ability to skew naive T cells toward differentiation to T_H2 cells. TSLP levels are increased in skin lesions of atopic dermatitis, and expression of a TSLP transgene in keratinocytes in mice results in allergic skin inflammation. TSLP also acts on skin-infiltrating effector T helper cells to promote T_H2 cytokine production.

The skin lesions in atopic dermatitis contain a mononuclear cell infiltrate that is predominantly located in the dermis (Fig. 51.4) and is composed of activated memory CD4 T cells and macrophages. The T cells involved in the skin lesions of acute atopic dermatitis are principally skin-homing CLA^+ $CCR4^+$ T_H2 cells. The T_H2-cell cytokines IL-4, IL-13, and IL-5 in particular have key roles in the condition. IL-4 and IL-13 cause skin-resident keratinocytes and dermal fibroblasts to secrete the chemokines CCL5 (RANTES), CCL11, and CCL22, ligands for CCR3 and CCR4 expressed on skin-homing T cells; CCL11 (eotaxin), which attracts eosinophils; and CCL13 (MCP-4), which attracts macrophages. There is therefore an influx of T cells, eosinophils, and macrophages into the skin. The chemokines and receptors known to be involved in atopic dermatitis are listed in Fig. 51.5. The selective homing of memory or effector CD4

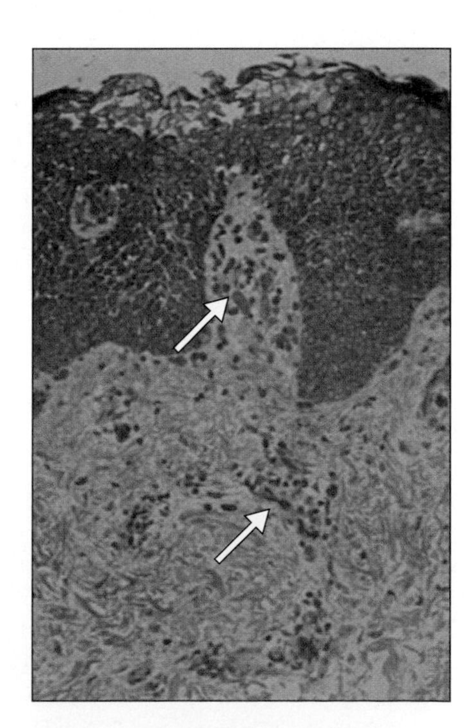

Fig. 51.4 A section through an acute skin lesion from a patient with atopic dermatitis. The section is stained with hematoxylin and eosin. The dermis has been infiltrated by mononuclear cells (arrowed), which are mostly T cells.

Fig. 51.5 Chemokines that act on T cells and eosinophils.

Target cell		Receptor	Chemokine
T cells	T_H1	CXCR3	CXCL10 (IP-10), CXCL9 (MIG)
	T_H2	CCR3 CCR4	CCL11, CCL5, CCL7, CCL13 (MCP-3, MCP-4), CCL5, CCL3, CCL2, CCL17 (TARC), CCL22 (MDC)
	Both	CCR10	CCL27 (CTACK), CCL28
Eosinophils		CCR1	CCL7
		CCR2	CCL13
		CCR3	CCL11, CCL5, CCL7, CCL13

T cells to skin is an important immunological event in the development of allergic skin inflammation. There are higher proportions of CLA^+ $CCR4^+$ CD4 T cells circulating in the peripheral blood of patients with atopic dermatitis compared with unaffected individuals, and these cells are abundant in the skin lesions.

IL-4 and IL-13 also increase IgE synthesis by promoting B-cell isotype switching to IgE, and IL-4 stimulates the preferential differentiation of T_H2 cells from naive CD4 T cells after antigen encounter. IL-5 promotes the differentiation and survival of eosinophils, which secrete a range of inflammatory mediators. In addition, the T_H2 cytokine IL-31 induces itching, and Fas ligand and TNF-α expressed by activated T_H2 cells induce keratinocyte damage. T_H2 cytokines also downregulate the expression of antimicrobial peptides by keratinocytes, which may underlie the increased susceptibility of patients with atopic dermatitis to cutaneous bacterial and viral infections that include *S. aureus*, herpes simplex type 1, molluscum contagiosum, and vaccinia virus.

In addition to infiltration by T_H2 cells, there is also evidence of a modest infiltration of T_H17 and T_H22 cells, which secrete IL-17 and IL-22, respectively, although these cytokines are much more prominent in psoriasis, another inflammatory skin disease. In chronic atopic dermatitis lesions there is a mixture of infiltrating T_H1 and T_H2 cells. The reason for the switch from T_H2 cell-dominated infiltrates in acute atopic dermatitis lesions to a mixed infiltrate in chronic lesions is not known. Microbial products that act via Toll-like receptors such as TLR-9 may promote a T_H1 response to microbial antigens in chronic atopic dermatitis.

The Langerhans cells and macrophages that infiltrate the skin lesions have IgE bound to their surface through CD23, a low-affinity receptor for IgE. Langerhans cells and monocytes can present antigen to naive T cells and activate them, and the bound IgE on the surface of the antigen-presenting cells allows them to concentrate the antigen, rendering them more efficient at antigen presentation. IgE is also bound to mast cells in the tissues through the high-affinity IgE receptor FcεRI. Signaling through this receptor after the cross-linking of IgE by antigen leads to the production and secretion of IL-4 and IL-5 by the mast cells, thus further biasing the T-cell response to a T_H2 phenotype. In chronic atopic dermatitis, the dermis is infiltrated by dendritic cells that bear IgE bound to the high-affinity IgE receptor on their surface. These cells secrete large amounts of IL-12 upon IgE cross-linking and therefore may be critical in inducing infiltrating T cells to secrete IFN-γ, a hallmark of chronic atopic dermatitis lesions.

Treatment of atopic dermatitis is aimed at softening the underlying dry skin and reducing inflammation. Emollients are the first line of topical therapy for the condition because they improve the skin's barrier function. Avoidance of irritants such as soaps and synthetic fabrics, which disrupt this barrier, is crucial in controlling atopic dermatitis. Acute flare-ups require treatment with topical corticosteroid and/or calcineurin inhibitor ointments or creams. Antihistamines can be helpful, especially at night, to control the itching.

Questions.

1 *Topical steroids are effective in reducing the eczema associated with atopic dermatitis. Why?*

2 *What other immunomodulatory agents might be effective in atopic dermatitis?*

3 *Why did Tom develop an extensive herpesvirus infection?*

4 *Atopic dermatitis is described as the 'itch that rashes.' If patients are prevented from scratching, no rash occurs. What is the relationship of scratching to the rash?*

5 *Skin infections with staphylococci and other bacteria exacerbate atopic dermatitis. Can you suggest a possible explanation for this?*

6 *Many patients with atopic dermatitis have associated asthma and/or food allergy. Why?*

CASE 52 Drug-Induced Serum Sickness

An adverse immune reaction to an antibiotic.

The intravenous administration of a large dose of antigen can evoke in some individuals a type III hypersensitivity reaction or immune-complex disease (Fig. 52.1). Antigen administration produces a rapid IgG response and the formation of antigen:antibody complexes (immune complexes) that can activate complement.

As a result of the large amount of antigen present and the rapid IgG response, small immune complexes begin to be formed in conditions of antigen excess (Fig. 52.2). Unlike the large immune complexes that are formed in conditions of antibody excess, which are rapidly ingested by phagocytic cells and cleared from the system, the smaller immune complexes are taken up by endothelial cells in various parts of the body and become deposited in tissues. Local activation of the complement system by these immune complexes provokes localized inflammatory responses.

The experimental model for immune-complex disease is the Arthus reaction, in which the subcutaneous injection of large doses of antigen evokes a brisk IgG response. The activation of complement by the IgG:antigen complexes generates the complement component C3a, a potent stimulator of histamine release from mast cells, and C5a, one of the most active chemokines produced by the body. The local endothelial cells are activated by the interactions in blood vessels between the immune complexes, complement and circulating leukocytes and platelets. They upregulate their expression of adhesion molecules such as selectins and integrins, which facilitates the emigration of white blood cells from the bloodstream and the initiation of a local inflammatory reaction (Fig. 52.3). Platelets also accumulate at the site, causing blood clotting; the small blood vessels become plugged with clots and burst, producing hemorrhage in the skin (Fig. 52.4).

When an antigen is injected intravenously, the immune complexes formed can be deposited at a wide range of sites. When deposited in synovial tissue, the resulting inflammation of the joints produces arthritis; in the kidney glomeruli they cause glomerulonephritis; and in the endothelium of the blood vessels of the skin and other organs they provoke vasculitis (Fig. 52.5).

Topics bearing on this case:

- Inflammatory reactions
- Properties of IgG antibodies
- Activation of complement by antigen:antibody complexes
- Type III immunological hypersensitivity reactions
- Immune-complex formation

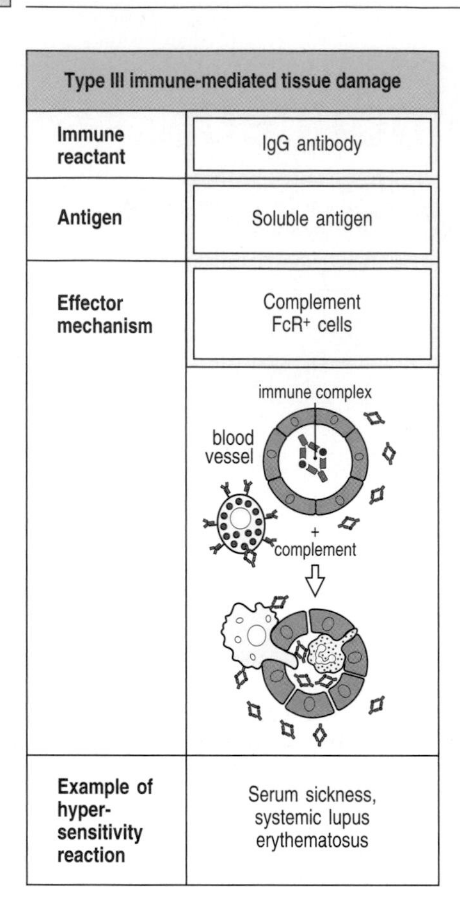

Type III immune-mediated tissue damage
Immune reactant
IgG antibody
Antigen
Soluble antigen
Effector mechanism
Complement
FcR+ cells
immune complex
blood vessel
+ complement
Example of hyper-sensitivity reaction
Serum sickness, systemic lupus erythematosus

Fig. 52.1 Type III hypersensitivity reactions. These can be caused by large intravenous doses of soluble antigens (serum sickness) or by an autoimmune reaction against some types of self antigen (as in systemic lupus erythematosus, see Case 37). The IgG antibodies produced form small immune complexes with the antigen in excess. The tissue damage involved is caused by complement activation and the subsequent inflammatory responses, which are triggered by immune complexes deposited in tissues.

In the early years of the twentieth century, the most common cause of immune-complex disease was the administration of horse serum, which was used as a source of antibodies to treat infectious diseases, and so this type of hypersensitivity reaction to large doses of intravenous antigen is still known as serum sickness.

This case describes a 12-year-old boy who received massive intravenous injections of penicillin and of ampicillin (one of its analogues) to treat pneumonia. He developed a serum-sickness reaction to the antibiotics.

The case of Gregory Barnes: serum sickness precipitated by penicillin.

When Gregory was brought to the Children's Hospital Emergency Room, his parents told the physicians that for 2 days he had had high fever (more than 39.5°C), a cough, and shortness of breath. Before then he had enjoyed excellent health. On physical examination he was pale, looked dehydrated, and was breathing rapidly with flaring nostrils. His respiratory rate was 62 min^{-1} (normal 20 min^{-1}), his pulse was 120 beats min^{-1} (normal 60–80 beats min^{-1}), and his blood pressure was 90/60 mmHg (normal). When his chest was examined with a stethoscope the emergency room doctors heard crackles (bubbly sounds) over the lower left lobe of his lungs. A chest radiograph revealed an opaque area over the entire lower lobe of the left lung. A diagnosis of lobar pneumonia was made.

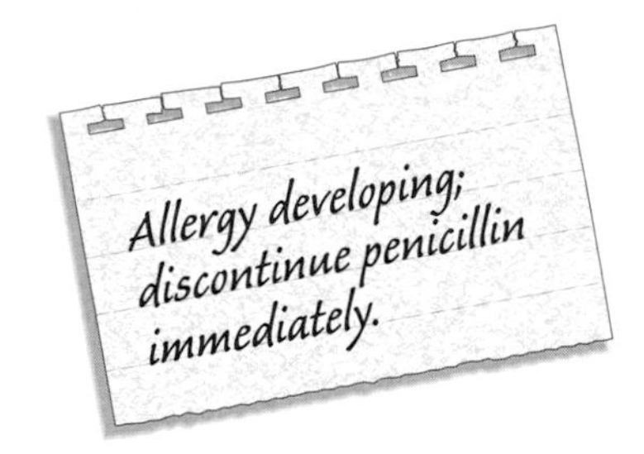

Allergy developing; discontinue penicillin immediately.

A white blood count revealed 19,000 cells μl^{-1} (normal 4000–7000 cells μl^{-1}) with an increase in the percentage of neutrophils to 87% of total white blood cells (normal 60%) and the abnormal presence of immature forms of neutrophils. A Gram stain of Gregory's sputum revealed Gram-positive cocci. Sputum and blood cultures grew *Streptococcus pneumoniae* (the pneumococcus).

Gregory was admitted to the hospital and treated with intravenous ampicillin at a dose of 1 g every 6 hours. Gregory gave no history of allergy to penicillin so ampicillin was used, to cover both Gram-positive and Gram-negative bacteria. On the fourth day of treatment, he felt remarkably better, his respiratory rate had decreased to 40 min^{-1}, and his temperature was 37.5°C. His white cell count had decreased to 9000 μl^{-1}. Because the *S. pneumoniae* grown from his sputum and blood was sensitive to penicillin, the ampicillin was replaced by penicillin. On his ninth day in hospital, Gregory had no fever, his white cell count was 7000 μl^{-1}, and his chest radiograph had improved. Plans for discharge from hospital were made for the following day.

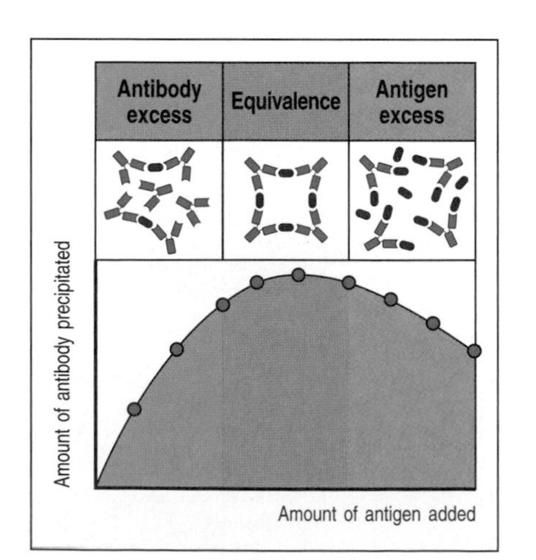

Fig. 52.2 Antibody can precipitate soluble antigen in the form of immune complexes. *In vitro*, the precipitation of immune complexes formed by antibody cross-linking the antigen molecules can be measured and used to define zones of antibody excess, equivalence, and antigen excess. In the zone of antigen excess, some immune complexes are too small to precipitate. When this happens *in vivo*, such soluble immune complexes can produce pathological damage to blood vessels.

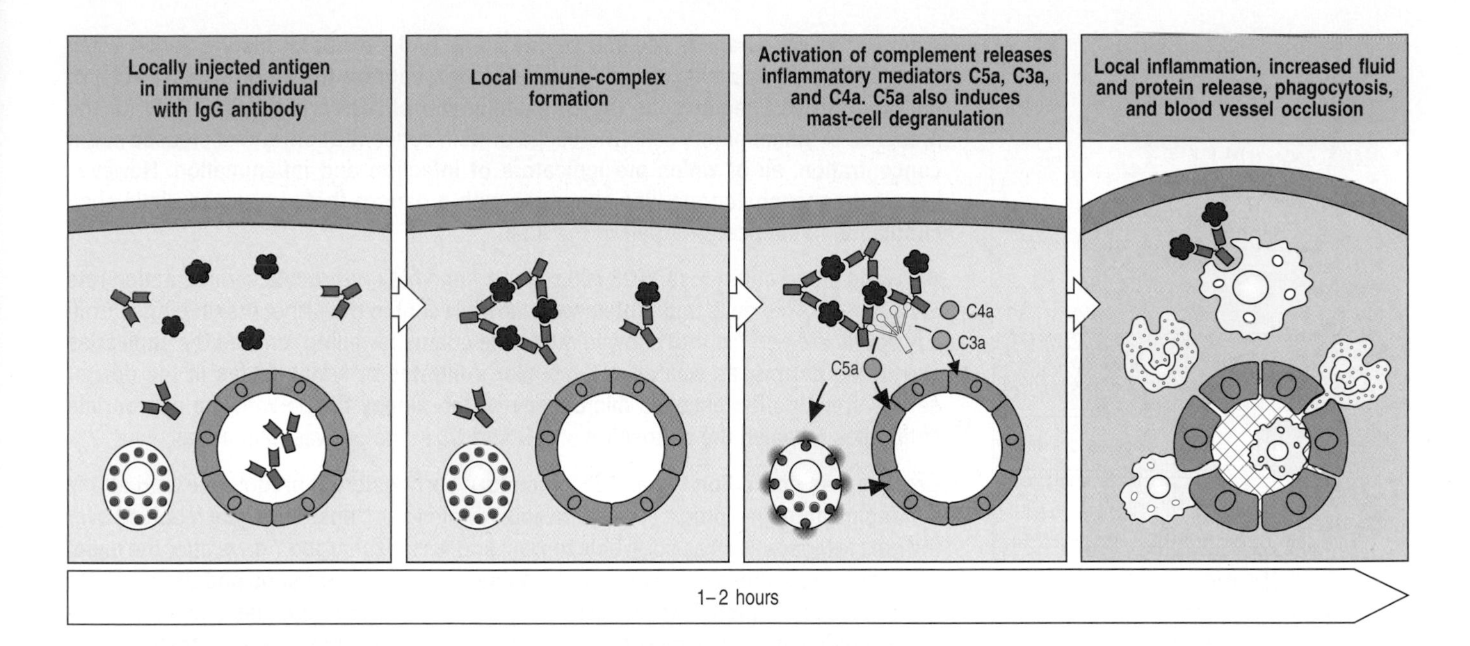
Locally injected antigen in immune individual with IgG antibody
Local immune-complex formation
Activation of complement releases inflammatory mediators C5a, C3a, and C4a. C5a also induces mast-cell degranulation
Local inflammation, increased fluid and protein release, phagocytosis, and blood vessel occlusion
C4a
C3a
C5a
1–2 hours

Fig. 52.3 The deposition of immune complexes in local tissues causes a local inflammatory response known as an Arthus reaction. In individuals who have already made IgG antibodies against an allergen, the same allergen injected into the skin forms immune complexes with IgG antibody that has diffused out of the capillaries. Because the dose of antigen is low, the immune complexes are only formed close to the site of injection, where they activate complement, releasing inflammatory mediators such as C5a, which in turn can activate mast cells to release inflammatory mediators. As a result inflammatory cells invade the site, and blood vessel permeability and blood flow are increased. Platelets also accumulate at the site, ultimately leading to occlusion of the small blood vessels, hemorrhage, and the appearance of purpura.

The next morning, Gregory had puffy eyes, and welts resembling large hives on his abdomen. He was given the antihistamine Benadryl (diphenhydramine hydrochloride) orally, and penicillin was discontinued. Two hours later he developed a tight feeling in the throat, a swollen face, and widespread urticaria (hives). With a stethoscope, wheezing could be heard all over his lungs. The wheezing responded to inhalation of the β_2-adrenergic agent albuterol. That evening Gregory developed a fever (a temperature of 39°C) and swollen and painful ankles, and his urticarial rash became more generalized. He appeared once again acutely ill.

The rash spread over his trunk, back, neck, and face, and in places became confluent (Fig. 52.6). Gregory also had reddened eyes owing to inflamed conjunctivae, and had swelling around the mouth. The anterior cervical, axillary, and inguinal lymph nodes on both sides were enlarged, measuring 2 cm by 1 cm. The spleen was also enlarged, with its tip palpable 3 cm below the rib margin. Ankles and knee joints were swollen and tender to palpation, and were too painful to move very far. The child was alert and his neurologic examination was normal.

Laboratory analysis of a blood sample revealed a raised white blood cell count (19,800 μl^{-1}) in which the predominant cells were lymphocytes (72%, in contrast with the normal 30%). Plasma cells were detected in a blood smear, although plasma cells are normally not present in blood. The erythrocyte sedimentation rate, an indicator of the presence of acute-phase reactants in the blood, was elevated at 30 mm h^{-1} (normal less than 20 mm h^{-1}). His total serum complement level and his serum C1q and C3 levels were decreased.

A presumptive diagnosis of serum sickness was made, and Gregory was given Benadryl and Naprosyn (naproxen), a nonsteroidal anti-inflammatory agent. On the following day, the rash and joint swellings were worse and the child complained of abdominal pain. There were also purpuric lesions, caused by hemorrhaging of small blood vessels under the skin, on his feet and around his ankles. There was no blood in his stool.

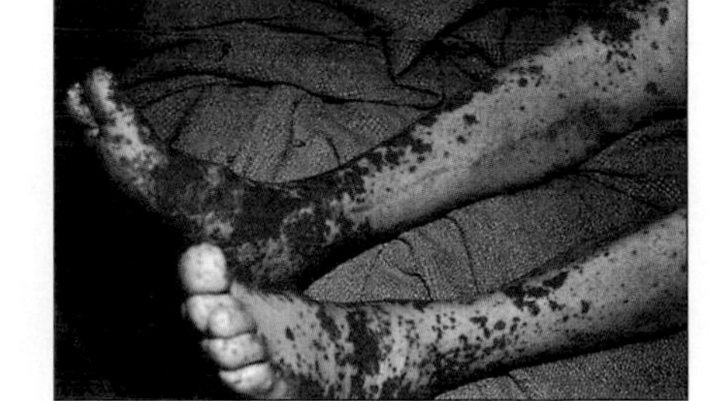

Fig. 52.4 Hemorrhaging of the skin in the course of a serum-sickness reaction.

Later in the day, Gregory became agitated, and had periods of disorientation when his speech was unintelligible and he could not recognize his parents. A CT scan of his brain proved negative, as did an examination of his cerebrospinal fluid for the presence of inflammatory cells, increased protein concentration, and decreased sugar concentration, all of which are indicators of infection and inflammation. However, his electroencephalogram was abnormal, with a pattern that suggested diminished circulation in the posterior part of the brain.

His white blood count rose to 23,700 cells μl^{-1} and his erythrocyte sedimentation rate to 54 mm h^{-1}. Red cells and protein were now present in the urine. A skin biopsy from a purpuric area on his foot showed moderate edema (swelling) around the capillaries and in the dermis, as well as perivascular infiltrates of lymphocytes in the deeper dermis. Immunofluorescence microscopy of the biopsy tissue with the appropriate antibodies revealed the deposition of IgG and C3 in the perivascular areas.

Gregory was started on the anti-inflammatory corticosteroid prednisone, and all his symptoms improved progressively; the joint swelling and splenomegaly resolved over the next few days. He was soon able to walk and was discharged 7 days after the onset of his serum sickness on a slowly decreasing course of prednisone and Benadryl. On follow-up examination 2 weeks later, Gregory had no IgE antibodies against penicillin or ampicillin, as detected by both immediate hypersensitivity skin tests and by an *in vitro* radioallergosorbent test (RAST). His parents were instructed that Gregory should never be given any penicillin, penicillin derivatives, or cephalosporins.

Route	Resulting disease	Site of immune complex deposition
Intravenous (high dose)	Vasculitis	Blood vessel walls
	Nephritis	Renal glomeruli
	Arthritis	Joint spaces
Subcutaneous	Arthus reaction	Perivascular area
Inhaled	Farmer's lung	Alveolar/capillary interface

Fig. 52.5 The dose and route of antigen delivery determine the pathology observed in type III allergic reactions.

Serum sickness.

The classic symptoms of serum sickness that Gregory showed were first described in great detail by Clemens von Pirquet and Bela Schick in a famous monograph entitled *Die Serumkrankheit* (serum sickness), published in 1905. Schick subsequently translated this monograph into English and it was reissued by Williams and Wilkins in 1951. It is fascinating to read this short work in the light of current knowledge. In the 1890s it had become common practice to treat diphtheria with horse serum containing antibodies taken from horses that had been immunized with diphtheria toxin. Immune horse serum was also used to treat scarlet fever, which was then a life-threatening illness. Von Pirquet and Schick made systematic observations on dozens of children who developed the symptoms and signs of serum sickness at the St Anna's Children's Hospital in Vienna and described the classic symptoms of the disease. They correctly surmised that serum sickness was due to an immunologic reaction to horse serum proteins in their patients (Fig. 52.7).

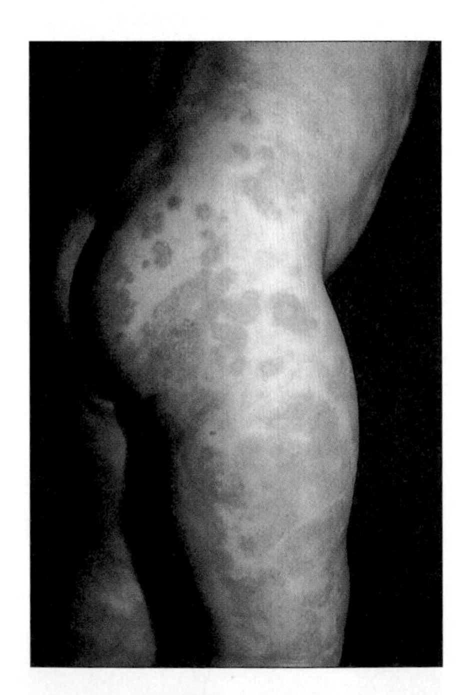

Fig. 52.6 Urticarial rash as a consequence of a serum-sickness reaction.

Experimental models of serum sickness were developed in the 1950s by Hawn, Janeway, and Dixon, who injected rabbits with large amounts of bovine serum albumin or bovine gamma globulin. They noted that the rabbits developed glomerulonephritis just at the time when antibody against the foreign protein first appeared in the rabbit serum, accompanied by a profound and transitory fall in the serum complement level. By this time, immunochemistry had advanced to the point where it was possible to show that the disease was caused by the formation and deposition of small immune complexes.

Although horse serum is no longer used in therapy, other foreign proteins are still administered to patients. Antitoxins to snake venom are produced in various animal species, and mouse monoclonal antibodies are used in clinical practice. However, the commonest causes of serum sickness today are antibiotics, particularly penicillin and its derivatives, which act as haptens. These drugs bind to host proteins that serve as carriers and thus can elicit a rapid and strong IgG antibody response.

Serum sickness, although very unpleasant, is a self-limited disease that terminates as the immune response of the host moves into the zone of antibody excess. It can prove fatal if it provokes kidney shutdown or bleeding in a critical area such as the brain. Its course can be ameliorated by anti-inflammatory drugs such as prednisone and antihistamines. It is also unlike the other types of hypersensitivity in that a reaction can appear on first encounter with the antigen, if that is long-lived and given in a sufficiently large dose. This seems to have been the case for Gregory.

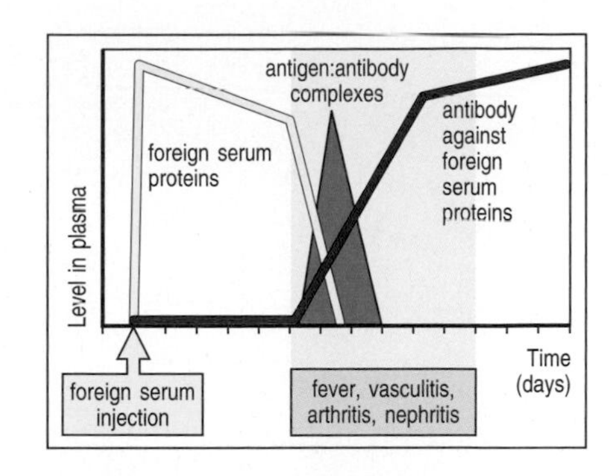

Fig. 52.7 Serum sickness is the classic example of a transient syndrome mediated by immune complexes. An injection of large amounts of foreign proteins, in this case derived from horse serum, leads to an antibody response. These antibodies form immune complexes with the circulating foreign proteins. These complexes activate complement and phagocytes, inducing fever, and are deposited in small blood vessels, inducing the symptoms of vasculitis, glomerulonephritis, and arthritis. All these effects are transient and resolve when the foreign protein is cleared from the system.

Questions.

1 *Hives (urticaria) and edema about the mouth and eyelids were the first symptoms of serum sickness developed by Gregory. What caused these early symptoms?*

2 *At one point Gregory became confused and disoriented and did not recognize his parents. His cerebrospinal fluid was normal and a CT scan of his brain was normal. However, an electroencephalogram displayed an abnormal pattern of brain waves. What produced these clinical and laboratory findings?*

3 *What other manifestations of vasculitis were noted in Gregory?*

4 *Gregory had enlarged lymph nodes everywhere and his spleen was also enlarged. If you had a biopsy of a node what would you expect to see?*

5 *Gregory had a brisk 'acute-phase response'. What is this, and what causes it?*

6 *Penicillin can cause more than one type of hypersensitivity reaction. What laboratory test gave the best evidence that Gregory was suffering from a disease caused by immune complexes?*

7 *When Gregory returned for a follow-up clinic visit, a skin test for immediate hypersensitivity was performed by intradermal injection of penicillin. He did not respond. Does this mean that an incorrect diagnosis was made and that he did not have serum sickness due to penicillin?*

CASE 53 Contact Sensitivity to Poison Ivy

A delayed hypersensitivity reaction to a hapten.

Allergic or hypersensitivity reactions can be elicited by antigens that are not associated with infectious agents, for example pollen, dust, food, and chemicals in the environment. They do not usually occur on the first encounter with the antigen, but a second or subsequent exposure of a sensitized individual causes an allergic reaction. Allergic symptoms will depend on the type of antigen, the route by which it enters the body, and the cells involved in the immune response. These unwanted responses can cause distressing symptoms, tissue damage, and even death. These are the same reactions that would be provoked by a pathogenic antigen, had it been introduced and presented in the same way. When they are not helping to combat an infection, however, these damaging side-effects are clearly unwanted.

There are four main types of immunological hypersensitivity reactions, which are distinguished by the type of immune cells and antibodies involved, and the pathologies produced. The one discussed here is an example of a type IV (delayed hypersensitivity) reaction (Fig. 53.1). Many allergic reactions occur within minutes or a few hours of encounter with the antigen, but some take a day or two to appear (Fig. 53.2). The latter are the delayed hypersensitivity reactions. Delayed hypersensitivity reactions are mediated by T cells only, either T_H1 CD4 T cells or cytotoxic CD8 T cells, or sometimes both. Antibodies are not involved. The reactions can be triggered by foreign proteins or by self proteins that have become modified by the attachment of a hapten, such as a small organic molecule or metal ion. A common type of delayed hypersensitivity reaction is allergic contact dermatitis, a skin rash caused by direct contact with the antigen.

Delayed hypersensitivity reactions fall into two classes (see Fig. 53.1). In the first, the damage is due to an inflammatory response and tissue destruction by T_H1 cells and the macrophages they activate. In the second class of delayed

This case was prepared by Raif Geha, MD, in collaboration with Lisa Bartnikas, MD.

Topics bearing on this case:
T-cell priming (sensitization)
Apoptosis
Preferential activation of T_H1 cells
Allergic reactions
Inflammatory response

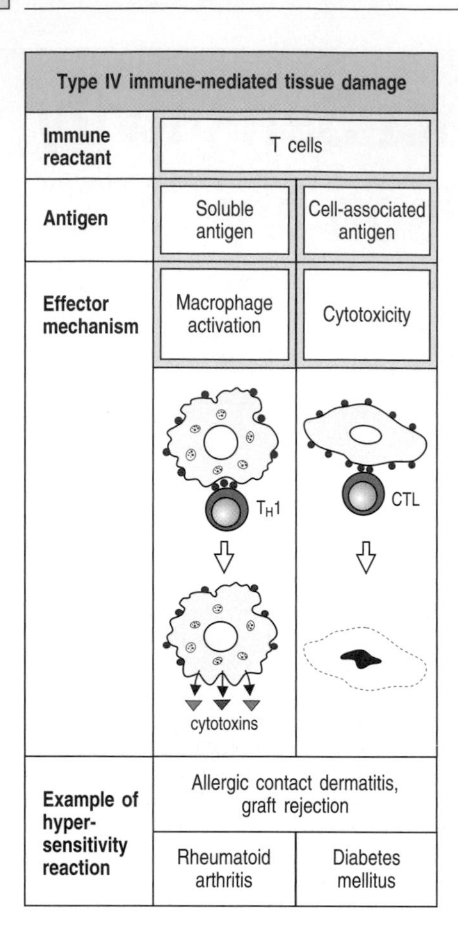

Type IV immune-mediated tissue damage
Immune reactant
T cells
Antigen
Soluble antigen
Cell-associated antigen
Effector mechanism
Macrophage activation
Cytotoxicity
TH1
CTL
cytotoxins
Example of hyper-sensitivity reaction
Allergic contact dermatitis, graft rejection
Rheumatoid arthritis
Diabetes mellitus

Fig. 53.1 Type IV hypersensitivity reactions. There are four types of immune-mediated tissue damage. Types I–III are antibody-mediated and are distinguished by the different types of antigens recognized and the different classes of antibodies involved. We have seen examples of type I responses in Case 49 and Case 50 and of type III in Case 52. Type IV hypersensitivity reactions are T cell-mediated and can be subdivided into two classes. In the first class, tissue damage is caused by T_H1 cells, which activate macrophages, leading to an inflammatory response. On encounter with antigen, effector T_H1 cells secrete cytokines, such as interferon-γ, that activate macrophages, and to a lesser extent mast cells, to release cytokines and inflammatory mediators that cause the symptoms. In the second class of type IV reactions, damage is caused directly by cytotoxic T lymphocytes (CTLs) that attack tissue cells presenting the sensitizing antigen on their surface. The delay in the appearance of a type IV hypersensitivity reaction is due to the time it takes to recruit antigen-specific T cells and other cells to the site of antigen localization and to develop the inflammatory response. Because a delayed hypersensitivity response involves antigen processing and presentation to achieve T-cell activation, quite large amounts of antigen need to be present at the site of contact. The amount of antigen required is two or three orders of magnitude greater than that required to initiate an antibody-mediated immediate hypersensitivity reaction.

hypersensitivity reactions, tissue damage is caused mainly by the direct action of antigen-specific cytotoxic CD8 T cells on target cells displaying the foreign antigen. Some antigens may cause a combination of both types of reactions.

This case describes the most frequently encountered delayed hypersensitivity reaction in the United States—allergic contact dermatitis due to the woodland plant poison ivy.

The case of Paul Stein: a sudden appearance of a severe rash.

Paul Stein was 7 years old and had enjoyed perfect health until 2 days after he returned from a hike with his summer camp group, when itchy red skin eruptions appeared all along his right arm. Within a day or two, the rash had spread to his trunk, face, and genitals. His mother gave him the antihistamine Benadryl (diphenhydramine

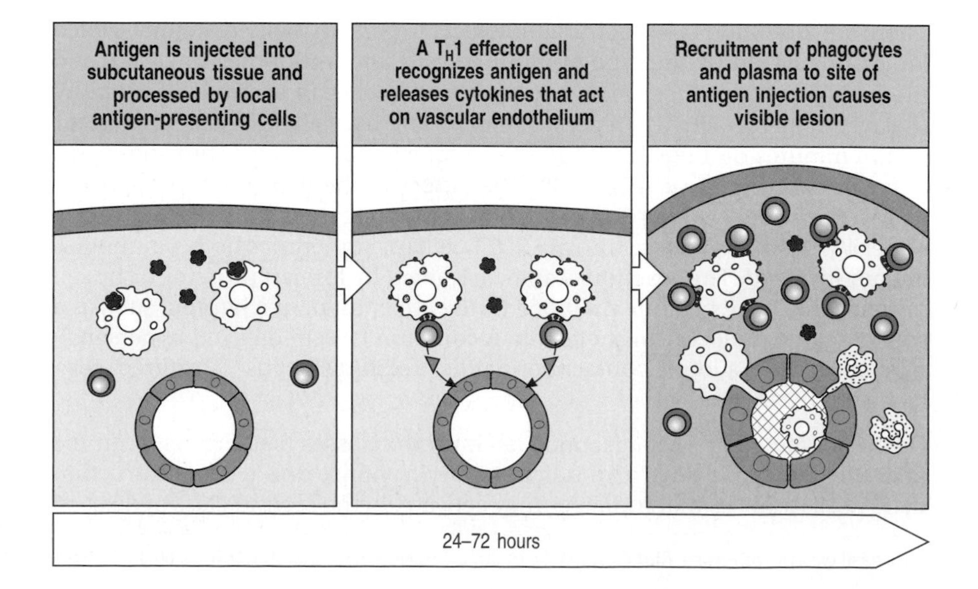

Recruitment of phagocytes and plasma to site of antigen injection causes visible lesion
24–72 hours

Fig. 53.2 The time course of a delayed-type hypersensitivity reaction. The first phase involves the uptake, processing, and presentation of the antigen by local antigen-presenting cells. In the second phase, T_H1 cells that were primed by a previous exposure to the antigen migrate into the site of injection and become activated. Because these specific cells are rare, and because there is no inflammation to attract cells into the site, it may take several hours for a T cell of the correct specificity to arrive. These cells release mediators that activate local endothelial cells, recruiting an inflammatory cell infiltrate dominated by macrophages and causing the accumulation of fluid and protein. At this point, the lesion becomes apparent.

hydrochloride) orally to suppress the itching, but this gave only partial relief. The rash did not improve, and a week after it first appeared he attended the Dermatology Clinic at the Children's Hospital.

Physical examination revealed large patches of raised, red, elongated blisters, oozing scant clear fluid, on his body and extremities (Fig. 53.3). Paul also had swollen eyelids and a swollen penis. There was no history of fever, fatigue, or any other symptom. A contact sensitivity reaction to poison ivy was diagnosed.

He was given a corticosteroid-containing cream to apply to the skin lesions three times a day, and Benadryl to take orally three times a day. He was asked to shampoo his hair, wash his body thoroughly with soap and water, and cut his nails short.

Two days later, his parents reported that, although no new eruptions had appeared, the old lesions were not significantly better. Paul was then given the corticosteroid prednisone orally, which was gradually decreased over a period of 2 weeks. The topical steroid cream was discontinued.

Within a week, the rash had almost disappeared. Upon stopping the prednisone there was a mild flare-up of some lesions, and this was controlled by application of topical steroid for a few days. Paul was shown how to identify poison ivy in order to avoid further contact with it, and told to wear long pants and shirts with long sleeves on any future hikes in the woods.

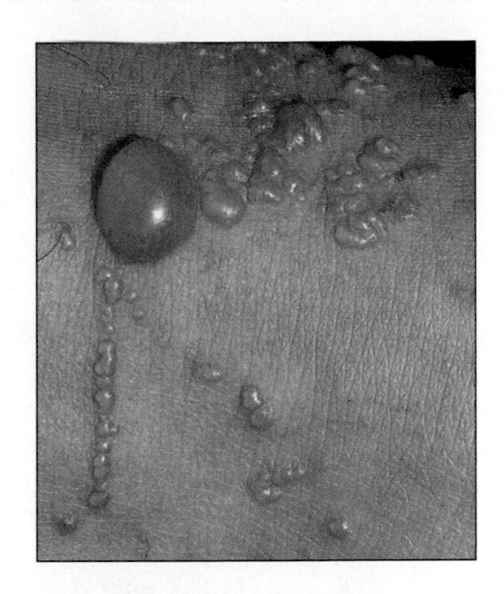

Fig. 53.3 Blistering skin lesions of patient with poison ivy contact dermatitis. Note the linear pattern of blisters in several areas. This is called the Koebner phenomenon and is due to exposure to the hapten along a line, possibly due to initial wiping or scratching. A rash only occurs on the initial areas of skin contact. Once the hapten is cleaned from the skin, no additional areas of skin will become involved. Therefore, touching skin lesions or blister fluid will not result in additional spread.

Contact sensitivity to poison ivy.

The reaction to poison ivy is the most commonly seen delayed hypersensitivity reaction in those parts of the United States where the plant grows wild. The absence of fever or general malaise accompanying the rash, and Paul's otherwise excellent health except for the skin lesions, point to a contact sensitivity reaction rather than to a viral or bacterial infection or some other underlying long-term illness. The appearance of the rash just 2 days after Paul returned from a hiking trip where he could easily have been in contact with poison ivy virtually clinches the diagnosis.

Allergic contact dermatitis due to poison ivy is caused by a T-cell response to a chemical in the leaf called pentadecacatechol (Fig. 53.4). On contact with the skin, this small, highly reactive, lipid-like molecule penetrates the outer layers, and binds covalently and nonspecifically to proteins on the surfaces of skin cells, in which form it functions as a hapten. Most people are susceptible, and sensitivity, once acquired, is lifelong.

The generation of a delayed-type hypersensitivity reaction requires the completion of both an 'afferent' and an 'efferent' response. In the afferent part of the response, the hapten enters the epidermis, and haptenated self proteins are ingested by specialized phagocytic cells in the epidermis (Langerhans cells) and dermis (dendritic cells) into intracellular vesicles, where they are cleaved into peptides. Some of these peptides will have hapten attached. The peptides bind to MHC class II molecules in the vesicles and are presented as peptide:MHC complexes on the Langerhans cell surface. Over the next 12–48 hours, some of these Langerhans cells and dermal dendritic cells migrate to a regional lymph node, where they become antigen-presenting cells that can activate naive hapten-specific T cells to become recirculating hapten-specific effector CD4 T_H1 cells and effector cytotoxic CD8 T cells that express skin-homing receptors such as E-selectin and CCR4. The efferent part of the response involves homing by these sensitized T_H1 and cytotoxic T cells to the site of contact with the plant. There, the activated effector T cell can react with haptenated peptides presented by Langerhans cells and dermal

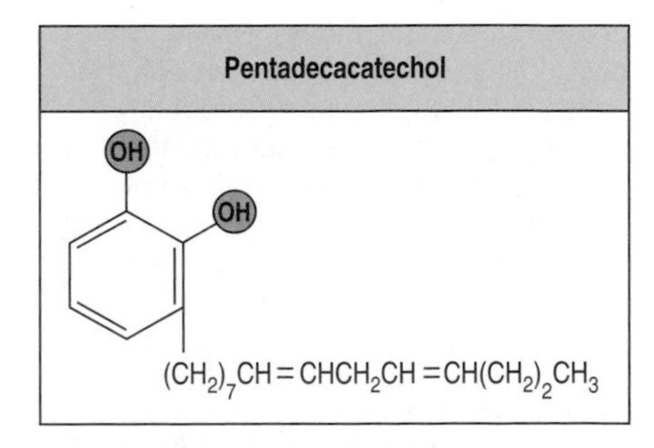

Fig. 53.4 The chemical formula of pentadecacatechol, the causative agent of contact sensitivity to poison ivy.

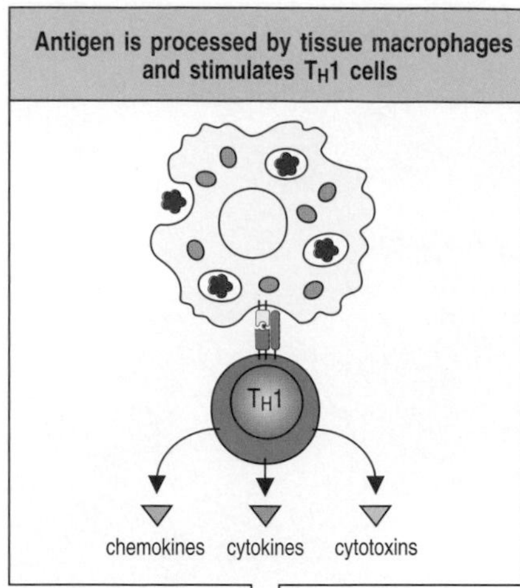
Antigen is processed by tissue macrophages and stimulates T_H1 cells
T_H1
chemokines
cytokines
cytotoxins

IFN-γ	TNF-α and/or LT
Induces expression of vascular adhesion molecules. Activates macrophages, increasing release of inflammatory mediators	Local tissue destruction. Increased expression of adhesion molecules on local blood vessels
Chemokines	**IL-3 and/or GM-CSF**
Macrophage recruitment to site of antigen	Monocyte production by bone marrow stem cells

Fig. 53.5 The delayed-type (type IV) hypersensitivity response is directed by cytokines released by T_H1 cells stimulated by antigen. Antigen in the local tissues is processed by antigen-presenting cells and presented on MHC class II molecules. Antigen-specific T_H1 cells can recognize the antigen locally at the site of injection, and release chemokines and cytokines that recruit macrophages to the site of antigen deposition. Antigen presentation by the newly recruited macrophages then amplifies the response. T cells may also affect local blood vessels through the release of TNF-α and LT (lymphotoxin) and stimulate the production of macrophages through the release of IL-3 and GM-CSF (granulocyte–macrophage colony-stimulating factor). Finally, T_H1 cells activate macrophages through the release of IFN-γ, and kill macrophages and other sensitive cells through the release of LT or by the expression of Fas ligand.

dendritic cells, leading to a release of inflammatory mediators by the T_H1 cells and cytotoxic molecules, such as perforin, by the cytotoxic T cells, thereby initiating a local delayed hypersensitivity reaction in the skin. With each subsequent exposure to antigen, the period of latency from contact to appearance of a rash is shortened (anamnesis).

The appearance of Paul's rash 2 days after his suspected exposure to poison ivy is typical of a delayed hypersensitivity reaction. The haptenated self peptides presented on skin macrophages and Langerhans cells at the site of contact with poison ivy are initially recognized by the small number of activated hapten-specific T_H1 cells within the pool of recirculating T cells. On encounter with the haptenated peptides, these T_H1 cells release chemokines, cytokines, and cytotoxins that initiate an inflammatory reaction and also kill cells directly (Fig. 53.5).

One of the cytokines produced by the T_H1 cells is interferon-γ (IFN-γ), whose main effect in this context is to activate macrophages. The subsequent macrophage activity causes many of the symptoms of the delayed hypersensitivity reaction. Macrophages activated by IFN-γ release cytokines and inflammatory mediators such as interleukins, prostaglandins, nitric oxide (NO), and leukotrienes. The combined effects of T-cell and macrophage activity cause a local inflammatory response and tissue damage at the site of the contact with poison ivy.

The red, raised, blistering skin lesions of poison ivy dermatitis are due to the infiltration of large numbers of blood cells into the tissue at the site of contact, combined with the localized death of tissue cells and the destruction of the extracellular matrix that holds the layers of skin together. One of the first actions of effector T_H1 cells on contact with their antigen is to release the cytokines TNF-α and LT (lymphotoxin), and chemokines such as CCL5 (formerly known as RANTES). TNF-α in particular increases the expression of adhesion molecules on the endothelium lining postcapillary venules and increases vascular permeability so that macrophages and other leukocytes adhere to the sides of the blood vessel. This aids their migration from the bloodstream into the tissues in response to the secreted chemokines.

Once at the contact site, macrophages are activated and themselves release cytokines and other inflammatory mediators, which attract more monocytes, T cells, and other leukocytes to the site, thus helping to amplify and maintain the inflammatory reaction. The blood vessels also dilate, which causes the redness associated with the rash. The inflammatory mediators also act on mast cells to cause degranulation and the release of histamine, which is the main cause of the itching that accompanies the reaction.

Tissue destruction, which is a feature of delayed hypersensitivity reactions, is caused both by cytokines and by direct cell–cell interactions. The TNF-α and LT released by T_H1 cells and macrophages act at the same TNF receptors, which are expressed on virtually all types of cells, including skin cells. Stimulation of these receptors induces a 'suicide' pathway in the cells—apoptosis—which causes their death. Activated CD4 T cells also express the Fas ligand (FasL) in their plasma membrane, which interacts with the ubiquitously expressed cell-surface molecule Fas to cause the death of the target cell by apoptosis. Other mediators released by activated T cells, such as the enzyme stromelysin, degrade the proteins of the extracellular matrix, which maintain the integrity of the skin.

Lipid-like haptens such as pentadecacatechol can also cause the priming and activation of cytotoxic CD8 T cells, as small fat-soluble molecules can enter the cytosol of skin cells directly by diffusing through the plasma membrane. Once inside, pentadecacatechol binds to intracellular proteins. Peptides generated from the haptenated proteins in the cytosol are delivered to the cell surface associated with MHC class I molecules. These target cells are recognized and attacked by antigen-specific cytotoxic CD8 T cells, which have become primed and activated on a previous encounter with the antigen. The outcome of all these reactions is the raised, red, weeping blisters characteristic of sensitivity to poison ivy.

Corticosteroids are the standard treatment for hapten-mediated contact sensitivity, because they inhibit the inflammatory response by inhibiting the production of many of the cytokines and chemokines. Corticosteroids are lipid-like molecules that can diffuse freely across plasma membranes. Once inside the cell, they bind to receptor proteins in the cytoplasm. The receptor:steroid complex enters the nucleus, where it controls the expression of several genes. Of relevance here is the fact that it induces the production of an inhibitor of the transcription factors required to switch on transcription of the cytokine and chemokine genes. In mild cases of poison ivy dermatitis, topical steroids applied locally are sufficient. In more severe cases such as Paul's, oral steroids are needed to achieve a concentration necessary to inhibit the inflammatory response.

Paul was given antihistamines to block the histamine receptors and counteract the action of the histamine released from mast cells. Antihistamines also counteract itching caused by substances other than histamine (for example prostaglandins released from macrophages).

Questions.

1 *Paul had lesions not only on the exposed skin but also on areas that would be covered, like the trunk and penis and in areas that were not in obvious contact with poison ivy leaves. How do you explain that?*

2 *How do you explain the recurrence of the lesions after discontinuation of the corticosteroids?*

3 *Paul must take great care to avoid poison ivy all his life, because subsequent reactions to it could be even more severe. Why would this be?*

4 *How would you confirm that Paul's contact dermatitis was caused by poison ivy rather than by another chemical such as the one found in the leaves of poison sumac (Toxicodendron vernix), another plant that gives rise to contact dermatitis?*

5 *One of Paul's friends, Brian, has X-linked agammaglobulinemia. What is the likelihood that this boy will develop poison ivy sensitivity?*

6 *Delayed hypersensitivity reactions are a rapid, inexpensive, and easy measure of T-cell function. What antigens might you use to test people for T-cell function in this way?*

7 *What are some common causes of contact sensitivity?*

Answers

Case 1

Answer 1

In every somatic cell of a female, one of the two X chromosomes is inactivated. Which of the X chromosomes is inactivated is a random process, so each is normally active in 50% of the cells on average. However, if the normal X chromosome is inactivated in a pre-B cell of Bill's mother or grandmother, that cell has no normal *BTK* gene product and cannot mature. All of their B cells therefore have the normal X chromosome active (Fig. A1.1). This makes it seem that in the B cells of Bill's mother and grandmother the inactivation of the X chromosome has been nonrandom. If we have a marker that allows us to distinguish between the two X chromosomes of Bill's aunts we can determine whether their B cells exhibit random or nonrandom X inactivation. In fact it turned out that one of Bill's aunts was a carrier and the other was not.

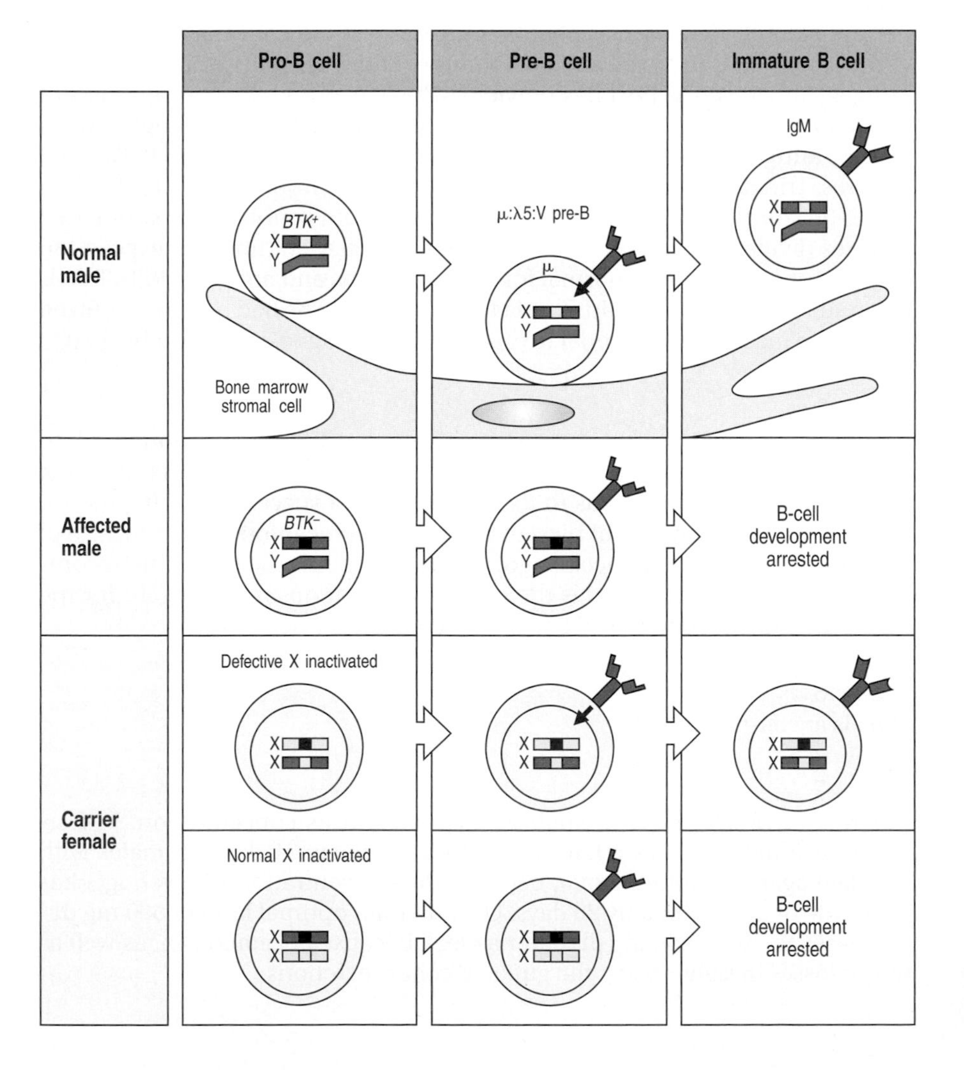

Fig. A1.1 Arrested B-cell development in X-linked agammaglobulinemia is responsible for apparently nonrandom X inactivation in the B cells of female carriers. In normal individuals, B-cell development proceeds through a stage in which the pre-B-cell receptor consisting of μ:λ5:Vpre-B tranduces a signal via Bruton's tryosine kinase (BTK), triggering further B-cell development. In males with XLA, no signal can be transduced and, although the pre-B-cell receptor is expressed, the B cells develop no further. In females, one of the two X chromosomes in each cell is permanently inactivated early in development. Because the choice of which chromosome to inactivate is random, half of the pre-B cells in a carrier female express a normal *BTK*, and half express the defective gene (*BTK*$^-$). None of the B cells that express *BTK*$^-$ from the defective chromosome can develop into mature B cells. Therefore, in the carrier, mature B cells always have the nondefective X chromosome active. This is in sharp contrast to all other cell types, which express the nondefective chromosome in only half of the population. Apparently nonrandom X chromosome inactivation in a particular cell lineage is a clear indication that the product of the X-linked gene is required for the development of cells of that lineage. It is also sometimes possible to identify the stage at which the gene product is required, by detecting the point in development at which X-chromosome inactivation develops bias. Using this kind of analysis, one can identify carriers of traits like XLA without needing to know the nature of the gene.

Answer 2
Maternal IgG crossed the placenta into Bill's circulation during fetal life. He was protected passively by the maternal IgG for 10 months.

Answer 3
Live polio vaccines are made from viruses with a disabling mutation in the gene that allows the virus to enter the motor nerve cells and cause paralysis. When a normal infant is given live attenuated poliovirus orally, the poliovirus establishes a harmless infection in the gut. Within 2 weeks the infant makes IgG and IgA antibodies that neutralize the poliovirus and prevent the infection from spreading, so that as the infected gut cells die the infection is terminated. When male infants with X-linked agammaglobulinemia are given this same vaccine, they are incapable of making any antibodies and the infection can persist. They continue to excrete the poliovirus from their gut. After a time there may be a mutation in some of the viruses that causes them to reacquire the ability to enter nerve cells (neurotropism). These so-called revertant viruses disseminate through the bloodstream and infect the neurons in the spinal cord, thus causing paralytic poliomyelitis. Another example of a virus that may disseminate from the gastrointestinal tract in males with X-linked agammaglobulinemia is the echoviruses. These are cleared readily by normal individuals; however, without the benefit of either IgA or IgG antibodies against the infecting serotype, the virus can disseminate to the central nervous system and cause meningoencephalitis.

Answer 4
Certain plant lectins, such as phytohemagglutinin and concanavalin A, cause virtually all T cells to divide and are therefore known as nonspecific mitogens. Antigens to which the host has previously been exposed also cause T cells to divide *in vitro*. After 72 hours exposure either to nonspecific mitogens or to specific antigens, ^{3}H-thymidine (tritiated thymidine) was added to Bill's T-cell cultures. Tritiated thymidine becomes incorporated into the DNA of dividing cells. The stimulation indices (the number of tritium counts in the stimulated cultures divided by the number of counts in similar cultures not exposed to mitogen or antigen) were normal for both mitogens and antigens. Bill's T cells responded to tetanus and diphtheria toxoids because he had been immunized with these inactivated bacterial toxins before the diagnosis was established.

Answer 5
Hereditary deficiency of complement component C3. The fixation of C3 to the bacterial surface, either by the classical or by the alternative pathways of complement activation, leads to its cleavage into a succession of fragments, two of which (C3b and iC3b) bind to complement receptors on the surface of phagocytic cells and enhance phagocytosis. iC3b binds the most potent complement receptor (CR3), and is the most important opsonizing agent for the ingestion and phagocytosis of encapsulated bacteria (see Case 32).

Answer 6
Tonsils are 80–90% B cells.

Answer 7
The rate at which IgG is catabolized depends on its concentration. In normal individuals, IgG has a half-life of approximately 21 days. In males with X-linked agammaglobulinemia, because the concentration is lower, IgG has a half-life of approximately 28 days. Overall, Bill's optimal level of 600 mg dl^{-1} decreases to about 450 mg dl^{-1} after a week because of catabolism, as well as minor losses in saliva, tears, the gut, and other secretions.

Bill weighs 75 kg.

The vascular volume is 8% of body weight or 0.08 liters kg^{-1}.

Bill's blood volume is therefore 75 × 0.08 = 6 liters.

Bill's hematocrit (the portion of the blood composed of cells) is 45%.

Therefore his plasma is 55% of his blood volume or 6 × 0.55 = 3.3 liters (3300 ml).

Bill injects himself with 10 g of gamma globulin or 10,000 mg. He thereby raises his IgG level by 10,000/3300 or roughly 3 mg ml^{-1}, or 300 mg dl^{-1}.

Half of the IgG equilibrates into the extravascular pool, so that Bill has really raised his plasma level by only 150 mg dl^{-1}.

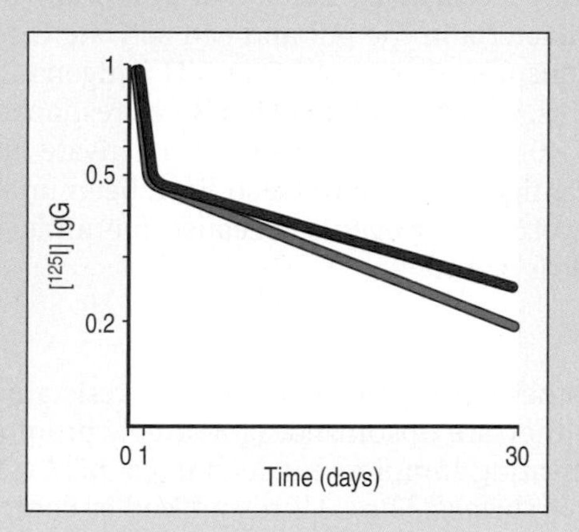

Fig. A1.7 IgG was radiolabeled with ^{125}I and then injected intravenously. After 10 minutes to allow for complete mixing of the radioactive dose in the blood, a plasma sample was obtained and the radioactivity was assayed. The amount of radioactivity at that time point was considered to be 100%. Plasma samples were obtained subsequently at frequent intervals and the percentage of residual radioactivity was determined and plotted on semi-logarithmic graph paper. The blue line shows the rate of disappearance of IgG in a normal person and the red line the rate in a male with X-linked agammaglobulinemia. From inspection of the curve it can be determined how long it takes for the radioactivity to decrease by 50%, say from 50% to 25%.

IgG, like all plasma proteins, distributes into the extravascular space: half the body IgG is in the blood and the other half is in the extravascular space. A dose of gamma globulin administered intravenously would equilibrate with the extravascular fluid in 24 hours. The dose of 10 g is arrived at as shown in Fig. A1.7.

Answer 8

Signaling via the pre-B-cell receptor is critical for B-cell development. Therefore, a deficiency of any one of the nonredundant components of pre-B-cell receptor signaling can cause failure of B-cell development. Defects transmitted as an autosomal recessive trait have been described in the μ heavy chain, the surrogate light chain λ5, the signaling components Igα and Igβ, and the B-cell linker protein BLINK. Another form of agammaglobulinemia is associated with a defect in the protein leucine-rich repeat containing 8 (LRRC8), namely a translocation that affects one *LRRC8* allele and results in a truncated LRRC8 protein. The role of LRRC8 in B-cell development is unknown.

Case 2

Answer 1

There is no DNA switch region 5′ to the Cδ gene. A single transcript of VDJCμCδ is alternatively spliced to yield either the μ or the δ heavy chain (see Fig. 2.4). In contrast, there are DNA switch regions 5′ to all the other heavy-chain C genes, and isotype switching must occur before functional transcripts of these genes are made. Isotype switching requires the enzyme activation-induced cytidine deaminase, which is expressed in B lymphocytes in response to signals from T cells.

Answer 2

These experiments tell us that the activation of macrophages against this opportunistic microorganism requires binding of CD40 on the macrophage surface to the CD40L on the surface of activated T cells.

Answer 3

Patients with CD40L deficiency have impaired antibody responses to T-cell dependent antigens, but can make IgM antibodies to antigens that can stimulate a B-cell response without T-cell help. The blood group antigens are sugars that are also found on bacteria in the gut and can activate B cells in the absence of T-cell help. Tetanus toxoid, typhoid O and H antigens, and streptolysin, in contrast, are proteins, which cannot elicit a B-cell response in the absence of T cells. Without CD40L, Dennis's T cells cannot activate his B cells to respond to these protein antigens. He would also have been unable to make antibodies against *Streptococcus pyogenes,* because the antigenic component of the bacterial capsule is a protein.

Answer 4

The polysaccharide capsules of these pyogenic bacteria are resistant to destruction by phagocytes unless they are opsonized. IgM largely promotes the phagocytosis of bacteria by activating complement, leading to the deposition of C3 fragments on the bacterial surface. The C3 is recognized by complement receptors on the phagocytic cells. IgG, however, is more efficient than IgM in promoting the phagocytosis of most bacteria. In addition, there is a range of Fc receptors for IgG isotypes on phagocytes, and IgG1, IgG2, and IgG3 antibodies all promote complement activation on bacterial surfaces. This means that bacteria coated with IgG stimulate phagocytosis through two different classes of receptor, Fc and C3. This results in much more efficient phagocytosis than stimulation through a single class of receptor.

Answer 5

Both neonates and people taking immunosuppressive drugs such as cyclosporin A exhibit increased susceptibility to both pyogenic and opportunistic infections.

A second reason for the increased susceptibility of neonates to some pyogenic infections is the immaturity of many of their B cells. Neonates are normally protected by preexisting maternal IgG until their lymphocytes mature.

Cyclosporin A also inhibits transcription of the *IL2* gene, thereby preventing the expansion of T-cell clones activated by antigen. This means that all T cell-mediated immune responses, including cytotoxic T-cell responses, are suppressed by cyclosporin A.

Case 3

Answer 1

Although the most notable features of hyper IgM syndrome caused by CD40L deficiency (Case 2) are elevated IgM and a lack of other immunoglobulin isotypes, the defect is actually of T cells, not B cells. Lack of expression of CD40L by T cells does not only affect B-cell function: it also results in a failure to activate monocytes/macrophages and dendritic cells via CD40. Interleukin-12 is not synthesized. In the absence of CD40 signaling, pulmonary macrophages are inefficient in killing *Pneumocystis jirovecii*, leading to pneumonia, and liver macrophages may be similarly deficient in killing *Cryptosporidium*, leading to chronic inflammation of the bile ducts (cholangitis). Lack of activation of dendritic cells via CD40 may impair their ability to elicit robust T-cell responses. This may contribute to the susceptibility to opportunistic infections in CD40L deficiency. In AID deficiency, the defect is solely in the B cells, and only affects antibody production, resulting in a phenotype similar to other B-cell deficiencies such as X-linked agammaglobulinemia (Case 1). There is susceptibility to pyogenic bacteria in particular, leading to frequent bacterial infections of the ears, lungs, and sinuses, but no increased risk of opportunistic infections.

Answer 2
Because the syndrome in this patient resembles that seen in CD40 ligand deficiency, we might hypothesize that she has a defect in CD40, the receptor for CD40 ligand. In fact, several cases of hyper IgM syndrome secondary to defects in CD40 have now been reported. As might be expected, these patients are clinically indistinguishable from those with CD40 ligand deficiency, except in the pattern of inheritance.

Answer 3
In this case it is likely that the defect is in another of the proteins required in the B cell for class switching. One of the genes involved in class switching is that for uracil-DNA glycosylase (*UNG*), which removes uracil from uridine. Defects in *UNG* have been described in a few patients with hyper IgM syndrome.

Answer 4
B cells from patients deficient in CD40L undergo normal class-switch recombination in response to CD40 ligation and cytokine action, whereas B cells from AID patients do not. Measurement of IgE synthesis after stimulation of cultured blood lymphocytes with anti-CD40 antibody plus IL-4 is an excellent indicator of class switching, because blood lymphocytes normally contain very few or no B cells that have already switched to IgE. B cells from patients deficient in CD40L make normal amounts of IgE in response to anti-CD40 plus IL-4. In contrast, B cells from patients with AID deficiency, or CD40 deficiency or defects in genes downstream of CD40, fail to secrete IgE in response to anti-CD40 and IL-4.

Answer 5
Enlargement of lymph nodes in children is usually due to expansion in the number and size of germinal centers. The interaction of CD40 with its ligand is required for germinal center formation, and patients with CD40L deficiency or CD40 deficiency have no germinal centers. In contrast, AID is not required for germinal center formation. Mitotic stimuli result in normal cell proliferative responses in these patients, and infections result in enlarged lymph nodes, as observed in Daisy.

Case 4

Answer 1
TACI, like other members of the TNFR family, such as TNFR-I and Fas, might require ligand-induced trimerization for signaling. The TRAFs that associate with TACI cytoplasmically have been shown to have a higher affinity for trimeric receptors. Therefore, in the case of mutations in the cytoplasmic domain of TACI, the recruitment of mutant and normal TACI subunits into a trimeric complex will compromise the binding of downstream signaling molecules and thus interfere with signaling. In this case, the mutant TACI will be acting as a dominant negative.

Answer 2
There is considerable variation in the severity of clinical symptoms in both related and unrelated individuals with the same *TACI* mutations. As seen in this case, in the same family, the *TACI* mutation could be associated with CVID and IgA deficiency. This suggests that the penetrance of the *TACI* mutation is determined by the particular genotype of the individual. In addition, environmental modifiers could also be involved. Indeed, some family members of patients with CVID carry the same *TACI* mutation but have few or no symptoms. Furthermore, some of the common mutations in *TACI* found in CVID have been identified in a small percentage of 'normal' subjects on blood screening.

Answer 3
A deficiency in CD19, which is a component of the B-cell co-receptor complex, mutations in the co-stimulatory protein ICOS, and possibly a mutation in BAFF-R have been found.

Case 5

Answer 1
His mother had nonrandom inactivation of the X chromosome in her T cells, thereby demonstrating that she carried an X-linked gene required for the normal maturation of T cells.

Answer 2
From what we know of their functions, defects in the receptors for IL-2, IL-4, IL-9, IL-15, and IL-21 do not seem relevant to the early block in T-cell development seen in X-linked SCID, because all of them activate mature lymphocytes or other effector cells. However, the receptor for IL-7 is thought to be important for pre-T-cell growth in humans and mice, and also for pre-B-cell growth in mice (which are deficient in both B and T cells when they lack the γ_c chain). Mice with a defect in the IL-7 receptor alone suffer from blocks in T- and B-cell development that resemble those seen in mice lacking the γ_c chain. A loss of IL-7 receptor function is therefore likely to be the most important loss of function responsible for X-linked SCID.

Answer 3
To prevent graft-versus-host disease. The mother's bone marrow donation also contained mature T cells capable of reacting with the paternal HLA antigens inherited by Martin from his father. Recognition of the paternal HLA antigens in Martin by the T-cell antigen receptors of the maternal T cells would incite graft-versus-host disease. The CD34 antigen is expressed by hematopoietic stem cells and hematopoietic progenitors; therefore, positive selection of these cells depletes the mature T lymphocytes. Alternatively, several strategies can be used to deplete T lymphocytes contained in the bone marrow, including *in vitro* treatment of the bone marrow with monoclonal antibodies directed against T lymphocytes in the presence of complement (to lyse T lymphocytes).

Answer 4
The *P. jirovecii* organisms are present in lung fluid and in the pulmonary macrophages. They do not incite an inflammatory response until the infant has T cells bearing a T-cell antigen receptor for *P. jirovecii* antigens. After a successful transplant, the infant is rendered chimeric. The transplanted T cells recognize *P. jirovecii* antigens and incite an inflammatory response, which makes the pneumonia more severe.

Answer 5
They may develop disseminated BCG infection and even die because of this complication. This ordinarily harmless attenuated bacillus is a pathogen in individuals with compromised cell-mediated immunity.

Answer 6
They developed a progressive vaccinia infection, which spread contiguously in the skin from the site of inoculation—so-called vaccinia gangrenosa. It proved invariably fatal in these infants. No live vaccines of any kind should be given to children or adults with significant T- or B-cell immunodeficiency.

Answer 7
Martin was treated with bone marrow transplantation without chemotherapy: he could not reject maternal cells because he lacked mature T lymphocytes. This strategy is often used in patients with SCID. However, with this strategy, autologous B lymphocytes persist, if present. B lymphocytes from patients with X-linked SCID or with JAK3 deficiency are impaired in their response to IL-4 and IL-21, two cytokines that are important in the maturation of B-cell responses. In particular, IL-4 is important for class-switch recombination, whereas IL-21 is produced by follicular helper T cells and promotes germinal center B-cell reaction and terminal differentiation of B lymphocytes.

Answer 8
Gene therapy. The gene encoding γ_c is introduced by retroviral transfer into bone marrow stem cells of patients with X-linked SCID, and the cells are injected into the blood. This therapy has resulted in full immune reconstitution in patients with X-linked SCID. Unfortunately, some of these patients developed leukemia, prompting the development of novel and safer vectors that are currently being tested in clinical trials.

Case 6

Answer 1
The thymus gland can shrink (involute) as the result of many kinds of stress, particularly infection.

Answer 2
The T cells of a patient with SCID are incapable of responding to a mitogenic stimulus. Because Roberta's cells could not respond to John's it was not necessary to add mitomycin to the mixed lymphocyte reaction.

Answer 3
The metabolic abnormalities of ADA deficiency are present already in prenatal life, and continue after birth until detoxification is achieved with treatment. Accumulation of toxic metabolites of adenosine is particularly detrimental to the thymus, and may cause irreversible damage, so that full restoration of T-cell generation may not be achieved, even after detoxification.

Case 7

Answer 1
The few T cells produced and activated must have had a T_H2 phenotype and secreted large amounts of interleukin-4 (IL-4) and interleukin-5 (IL-5). IL-4 is required for switching to IgE synthesis and IL-5 for the recruitment

of eosinophils. The few B cells in the patient (which were below the limit of detection) must have been induced to switch immunoglobulin class to IgE.

Answer 2

The few clones of T cells that were able to mature were activated, as shown by their surface expression of CD45R0 and MHC class II molecules. The activated clones would have expanded within the lymph node.

Answer 3

Neither of Ricardo's parents had Omenn syndrome, but Ricardo had an affected brother and an affected sister. This indicates Mendelian autosomal recessive inheritance of the defect. If the defect had been X-linked recessive, a female would not have been affected; if it were dominant, one of the parents would have had to have been affected.

Answer 4

The T cells that are present are activated, as shown by their expression of CD45R0 and MHC class II molecules, and express homing receptors for the skin. In the skin, the activated T cells secrete chemokines that attract other inflammatory cells, such as monocytes and eosinophils, into the skin. The perivascular inflammation in the skin causes the blood vessels to dilate, and this appears as a bright red rash.

Case 8

Answer 1

The maturation of CD4 T cells in the thymus depends on the interaction of thymocytes with MHC class II molecules on thymic epithelial cells. When the MHC class II genes are deleted genetically in mice, the mice also exhibit a deficiency of CD4 T lymphocytes.

Answer 2

The polyclonal expansion of B lymphocytes and their maturation to immunoglobulin-secreting plasma cells requires helper cytokines, such as IL-4, from CD4 T cells. Helen's hypogammaglobulinemia is thus a consequence of her deficiency of CD4 T lymphocytes.

Answer 3

Helen's T cells, although decreased in number, are normal and are not affected by the defect. They are capable of normal responses to nonspecific mitogens and to an allogeneic stimulus in which the antigen is presented by the MHC molecules on the surface of the (nondefective) allogeneic cells and thus does not require to be processed and presented by the defective cells. However, the failure of her lymphocytes to respond to tetanus toxoid *in vitro* resulted from the fact that, in this situation, there were no cells that could present antigen on MHC class II molecules to the CD4 T cells.

Answer 4

Yes. Helen's T cells would be capable of recognizing the foreign MHC molecules on the grafted skin cells and would reject the graft.

Case 9

Answer 1

Bone marrow transplantation is not the treatment of choice in patients with

DiGeorge syndrome or with *FOXN1* deficiency, because in these patients the defect lies in the thymic epithelium, not in cells of hematopoietic origin. However, bone marrow transplantation has been successfully performed in patients with DiGeorge syndrome who have received a transplant from HLA-identical siblings. In this case, the bone marrow is not manipulated. The mature T lymphocytes contained in the graft expand in the recipient and provide immune reconstitution, but no new T cells are generated in the thymus.

Answer 2

The idea that T cells develop across the HLA barrier in patients with DiGeorge syndrome treated with unrelated thymic transplantation is indeed against the dogma that positive selection requires recognition of self-HLA molecules. Although this remains puzzling, one possibility is that dendritic cells from the host home to the thymus and mediate the positive selection of newly generated thymocytes.

Answer 3

In most patients with DiGeorge syndrome, some thymic (and parathyroid) tissue is present, even if it cannot be seen. These tissues undergo expansion after birth, leading to progressive correction of the hypocalcemia and some significant T-cell development.

Answer 4

The thymus has a crucial role in the establishment of tolerance. In particular, expression of the transcription factor Aire by medullary thymic epithelial cells induces the expression of peripheral tissue antigens in the thymus. These antigens can be presented to newly generated self-reactive thymocytes that are deleted. Severe defects of the thymic stroma (as in DiGeorge syndrome) may lead to a reduced expression of Aire and of peripheral tissue antigens, hence allowing the survival of self-reactive T cells. Furthermore, the thymus is also the tissue in which natural regulatory T (nT_{reg}) cells are generated. These cells suppress autoimmune manifestations in the periphery. The thymic defect in DiGeorge syndrome may also lead to a reduction of nT_{reg} cells and hence contribute to autoimmunity.

Case 10

Answer 1

The lymph node would exhibit marked, if not exuberant, follicular hyperplasia, indicating an ongoing B-cell response to the virus. In SCID and XLA the lymph nodes are very small. In SCID the lymph node would contain no or very, very few lymphoid cells. In XLA the lymph node would have no follicles, no germinal centers, and no B cells or plasma cells. T cells would be present but not in an organized array (Fig. A10.1).

Answer 2

In infants, the HIV infection typically runs a more rapid course, and infants often die before they reach 1 year old. Mr Thomas, however, had been infected for many years before symptoms of immunodeficiency started to appear. This difference is probably due to the fact that infants are immunologically immature and naive, whereas an infected adult has a functionally mature immune system and decades of acquired adaptive immunity to many different antigens. This has consequences for both the response to HIV itself and the susceptibility to other infections. We have seen that T cells in newborn infants are not fully 'turned on.' For example, in hyper IgM the CD40 ligand is not readily expressed when T cells of newborns are activated (see Case 2). Their T cells do not synthesize interferon-γ in normal amounts, nor are their cytotoxic

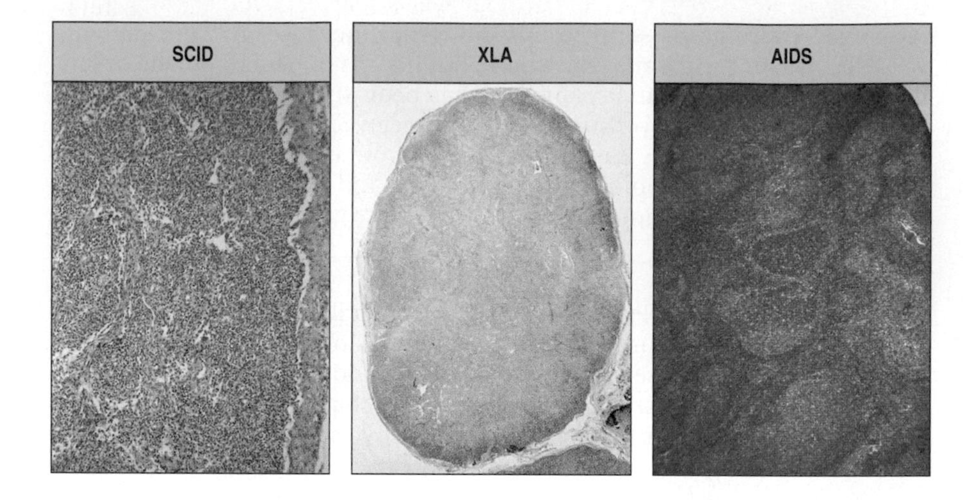

Fig. A10.1 Lymph-node sections from patients with severe combined immunodeficiency (SCID), X-linked agammaglobulinemia (XLA), and AIDS.

T lymphocytes readily activated. This functional immaturity is probably why young infants have difficulty confining and walling off infections, particularly those that require adaptive immunity mediated by T cells. Tuberculosis offers a clear example of this. It is a fast-spreading, highly lethal infection in young infants, whereas in older children and immunologically normal adults, the infection is usually confined to the lung, or more rarely to other organs.

HIV infection in infants occurs before they have had an opportunity to develop any adaptive immunity to common infections, and this means that they are prone to infections not seen in adult AIDS. An adult will already have antibodies to the common pyogenic bacteria and will not be particularly susceptible to pyogenic infections, whereas these are frequently observed in affected infants. Adults will also have been exposed to common viruses, such as Epstein–Barr virus (EBV). This is normally encountered early in life and contained as a latent infection. Virtually all adults have been infected with EBV by the end of the second or third decade of life, and primary EBV infection is therefore not a threat to HIV-infected adults. For an HIV-infected infant, however, a first encounter with this virus causes bizarre manifestations such as parotitis (inflammation of the parotid gland, like mumps) and a form of pneumonia characterized by pulmonary lymphoid hyperplasia.

Answer 3

The answer to this question is not known precisely. The immune response to the virus is illustrated in Fig. A10.3. Antibody against HIV seems to have a minor role in resisting the progress of infection. HIV-specific cytotoxic CD8

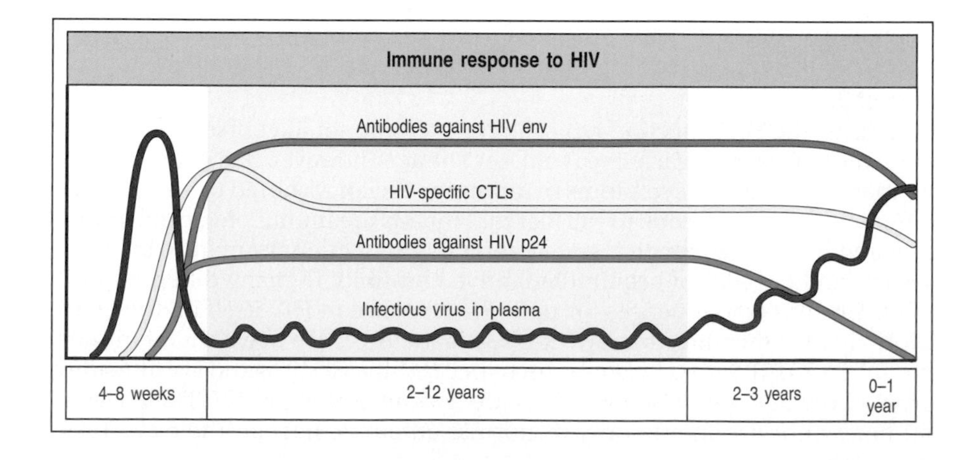

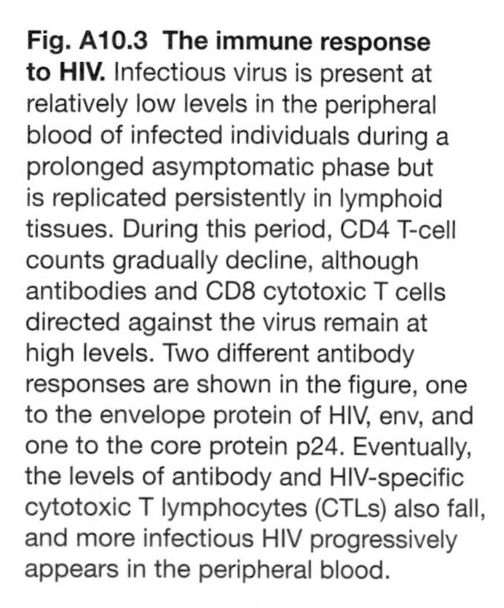

Fig. A10.3 The immune response to HIV. Infectious virus is present at relatively low levels in the peripheral blood of infected individuals during a prolonged asymptomatic phase but is replicated persistently in lymphoid tissues. During this period, CD4 T-cell counts gradually decline, although antibodies and CD8 cytotoxic T cells directed against the virus remain at high levels. Two different antibody responses are shown in the figure, one to the envelope protein of HIV, env, and one to the core protein p24. Eventually, the levels of antibody and HIV-specific cytotoxic T lymphocytes (CTLs) also fall, and more infectious HIV progressively appears in the peripheral blood.

cells arise as virus levels decline from the peak associated with primary infection, and seem to have a more important role in containing the infection. Rare individuals with mutations in the co-receptors for HIV (CCR5 and CXCR4) are resistant to HIV infection.

Answer 4

The virus burden in these individuals is very low; in some cases, the only detectable viruses carry mutations in genes such as *nef* or *tat*, which are vital to HIV replication in the infected host. However, viruses able to replicate in culture can be isolated from most of these so-called 'long-term non-progressors.' These patients seem to be able to contain replication-competent virus, most probably by continuing to maintain a successful cytotoxic CD8 T-cell response.

Answer 5

They inhibit the reverse transcriptase of HIV. Zidovudine (AZT; 3′-azido,2′,3′-dideoxythymidine) is a nucleoside analog that is phosphorylated inside the cell and used as a substrate by the reverse transcriptase of HIV (Fig. A10.5). HIV reverse transcriptase synthesizes a DNA complement of the viral RNA, at the start of a new round of virus replication in a newly infected cell. The incorporation of zidovudine blocks further extension of this DNA strand and thereby stops replication of the virus. Two other nucleoside analogs, ddI (dideoxyinosine) and ddC (dideoxycytosine) inhibit HIV replication by a similar mechanism and are also used to treat HIV infection. Lamivudine (also known as 3TC) is an enantiomer of a dideoxy analog of cytidine. Efavirenz (also called Sustiva) is a nonnucleoside inhibitor of HIV reverse transcriptase. Unfortunately, mutation allows the virus to acquire resistance to all these drugs. Replication of HIV (and other known retroviruses) is error-prone, and the virus mutates as it replicates in the infected host. Resistance to zidovudine requires multiple mutations but can arise in only a few months. Combining HIV protease inhibitors with zidovudine or other nucleoside analogs has markedly improved the survival of HIV-infected patients.

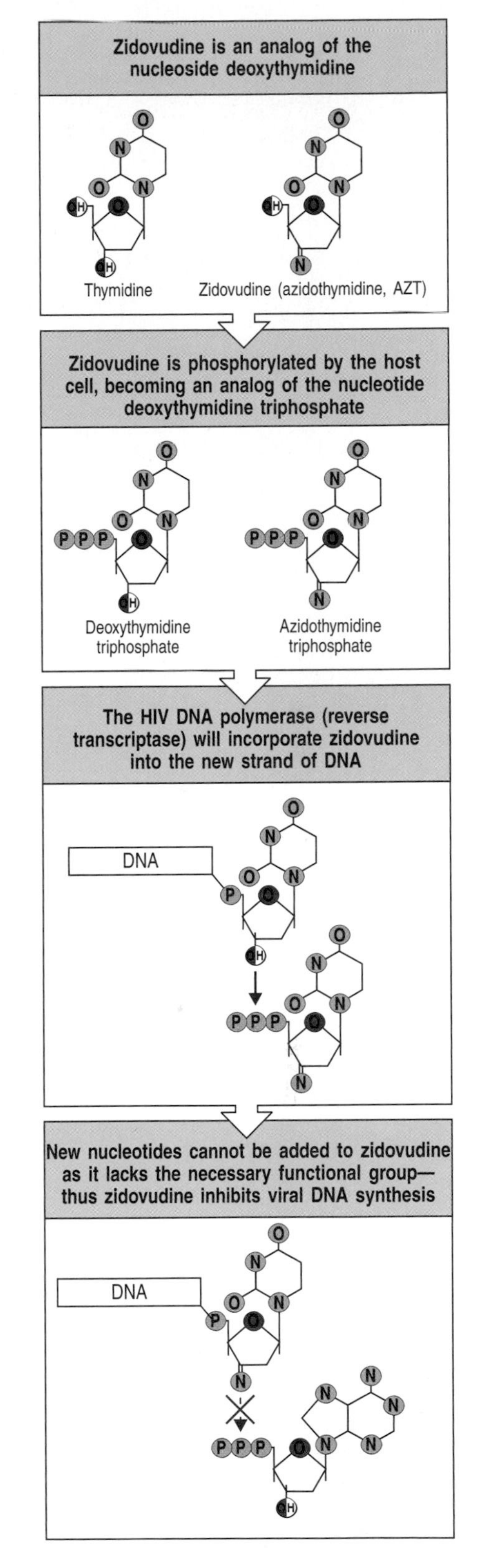

Fig. A10.5 The mechanism of action of the antiretroviral drug zidovudine. Zidovudine is azidothymidine (AZT), an analog of the nucleotide deoxythymidine.

Answer 6

HIV produces an aspartyl protease, structurally related to pepsin and renin, which has two aspartic-acid residues in the active site. This protease is required to splice the HIV Gag proteins after they are synthesized and before the packaging of the virus into its coat and its budding from the cell surface. Inhibitors of this HIV protease have been designed and are very effective in halting HIV replication. If the HIV in Mr Thomas were to become resistant to the reverse transcriptase inhibitors, adding a protease inhibitor to his drug regimen might keep the virus in check for some time longer.

Answer 7

The precise answer to this question is unknown. It is clear that the cells producing the virus are killed, either by cytotoxic T lymphocytes or by direct cytotoxic effects of the virus. It is also possible that the death of uninfected 'bystander' cells contributes to CD4 T-cell depletion. HIV is known to have a cytotoxic effect on CD4 T cells in culture. The viral capsular gp120 binds and cross-links the CD4 molecule, which depresses T-cell function and may induce apoptosis, even when the CD4 cells are not themselves infected by HIV. If, as seems likely, the early killing of HIV-infected CD4 cells by cytotoxic T cells serves to contain the virus and prevent greater CD4 T-cell depletion in the next round of HIV infection and replication, a declining ability to mount cytotoxic responses, especially to new viral variants, could be very important. Patients infected with HIV show impairment of T-cell function, especially memory cell responses, even during the asymptomatic phase. They are hypergammaglobulinemic, and humoral responses seem favored at the expense of cell-mediated immunity. This immunoregulatory bias impedes inflammatory responses, and possibly also cytotoxic responses to HIV-infected cells.

Answer 8
The CD4 T-cell count is, statistically speaking, the best indicator of the time-course of progression to AIDS. Other factors such as lifestyle and the incidence of intercurrent infections do not seem statistically significant.

Answer 9
Tumor necrosis factor-α (TNF-α). It causes loss of appetite (anorexia) and increased expenditure of body heat (thermogenesis).

Case 11

Answer 1
The engrafted T cells seem to recognize allogeneic antigens on the recipient's hematopoietic cells and thus will attack the leukemic cells. One such antigen, HB-1, which is a B-cell lineage marker, is expressed by acute lymphoblastic leukemia cells, which are B-lineage cells, and by B lymphocytes transformed by Epstein–Barr virus (EBV).

Answer 2
IFN-γ induces the expression of MHC molecules on cells; this makes GVHD worse, because it provides more targets for the donor T cells.

Answer 3
In vivo T-cell depletion can also be achieved with the intravenous injection of monoclonal antibodies, such as anti-CD3. This approach is also used to treat initial graft rejection in recipients of solid organ transplantation, but it is associated with a higher risk of lymphoproliferative disease induced by the Epstein–Barr virus. Monoclonal antibodies targeting activated, but not resting, T lymphocytes have been also used in the treatment of GVHD, and include anti-CD25 and anti-CD40L monoclonal antibodies.

Answer 4
One reason could be that the skin and intestine express a higher level of MHC molecules than other tissues. The intestinal tract is also likely to be damaged by the preparative cytotoxic treatments given to destroy the recipient's bone marrow. The damage induces the production of cytokines; as well as inducing MHC molecules, these can also drive GVHD and make the tissue susceptible to immunological attack.

Case 12

Answer 1
The intrathymic maturation of CD8 T cells depends on the expression of MHC class I molecules on thymic epithelial cells. Conversely, maturation as CD4 T cells requires interactions with the MHC class II molecules also present on thymic epithelium. Because Tatiana and Alexander do not lack MHC class II molecules, their CD4 T cells are normal and they have normal humoral immunity.

Answer 2
The maturation of CD8 T cells bearing γ:δ chains occurs after these cells emigrate from the thymus and is independent of MHC class I expression, whereas

the maturation of α:β CD8 T cells occurs in the thymus and is dependent on the expression of MHC class I molecules.

Answer 3
No. Delayed-type hypersensitivity reactions are provoked by antigen-specific CD4 T cells (see Case 8).

Answer 4
The factors that help B cells to mature and secrete immunoglobulins are derived from activated CD4 T cells. These cells were normal in Tatiana and Alexander. Factors that suppress B-cell responses are secreted by CD8 T cells, of which the children had very few. They were therefore not very efficient at terminating B cell-mediated humoral immune reactions and tended to overproduce antibody. In Case 8, we saw the opposite phenomenon in patients with MHC class II deficiency, who have a deficiency of CD4 T cells and very low levels of serum immunoglobulins.

Answer 5
The patients with TAP1 deficiency clinically resembled those with TAP2 deficiency.

Case 13

Answer 1
In retrospect, the family history and the patient's past medical history provided clues that he might be affected by XLP. The patient himself had low IgG and recurrent otitis. His uncle and grandfather were affected by some of the manifestations of XLP, including recurrent lymphoma and aplastic anemia.

Answer 2
In some patients, family history as well as a classic clinical presentation (fulminant infectious mononucleosis) provide a strong basis for a clinical diagnosis. Often, however, there is no clear family history and the severe lymphoproliferation cannot easily be distinguished from other forms of malignant lymphohistiocytosis with hemophagocytosis. The molecular defect responsible for most cases of XLP has been identified as a mutation or deletion of the gene encoding SAP (*SH2D1A*). In difficult cases, a molecular diagnosis is desirable, and PCR analysis of the four exons encoding SAP can be performed. Flow cytometry and Western blotting can be used to identify patients in whom *SH2D1A* mutations result in a lack of SAP protein expression. If family history, clinical data, and laboratory data are consistent with a diagnosis of XLP but no defects in SAP can be identified, mutation analysis should be targeted to the *BIRC4* (*XIAP*) gene, using PCR.

Answer 3
In a girl, mutation of an X-chromosome gene such as *SH2D1A* would be very unlikely to account for her problem. An autosomal disorder is more probable. If the parents were related, suspicion of an autosomal defect would be heightened. Autosomal recessive defects that may cause an XLP-like phenotype include ITK deficiency but also, and more commonly, defects in the cytolytic machinery that result in hemophagocytic lymphohistiocytosis (see Case 14).

Answer 4
Destruction of Alexander's bone marrow by radiation or chemotherapy (with busulfan and cytoxan) followed by administration of HLA-matched marrow from a normal sibling or unrelated donor would replace all lymphoid

precursors with normal SAP-expressing cells of donor origin. This approach has been used successfully in the treatment of XLP.

Case 14

Answer 1
FHL is characterized by impaired cytotoxic activity of CD8 T cells and NK cells, which are therefore unable to kill virus-infected cells. However, the ability of CD8 T cells to recognize virus-derived peptides in association with HLA class I molecules on the surface of infected cells is intact. Therefore, antigen-specific CD8 T cells continue to respond to the viral infection by becoming activated and proliferating. As part of this response, they secrete large amounts of IFN-γ, which drives the production of the pro-inflammatory cytokines IL-6 and TNF-α by macrophages.

Answer 2
The liver and spleen enlargement that Jude developed, which is typical of the accelerated phase of HLH, is the result of the marked expansion of CD8 T cells and the accumulation of activated macrophages.

Answer 3
The bone marrow is a target organ in HLH. Activated macrophages often engulf red cells, myeloid cells, lymphoid cells, and platelets, leading to bone marrow hypoplasia.

Case 15

Answer 1
The nitro blue tetrazolium (NBT) test measures the capacity of the lysosomes in a phagocyte to produce superoxide and other oxygen free radicals. The test is done by adding yellow NBT dye to phagocytes that are then stimulated with a cell activator, typically phorbol myristate acetate (PMA). Activation of the NADPH oxidase enzyme complex in the lysosome by PMA leads to the production of the reactive intermediates of oxygen that modify NBT into formazan, which has a deep blue color. Chediak–Higashi syndrome (CHS) affects the normal formation and traffic of vesicles in the cell, but does not affect function of the NADPH oxidase. Thus, Shweta's neutrophils could readily reduce NBT to a blue color.

Answer 2
An easy and non-invasive test with which to confirm CHS is microscopic observation of a hair shaft. The hair of patients with CHS has abnormal pigmentation: when observed under the microscope, the hair shaft is seen as speckled with clumps of pigment instead of the normal homogeneous distribution (see Fig. 15.5). When the hair bulb of CHS patients is examined under electron microscopy, melanocytes showing enlarged melanosomes with variable amounts of melanin pigment are seen.

Answer 3
The accelerated phase of CHS is thought to be due to impaired lymphocyte cytotoxicity, although the precise pathogenic mechanism is so far unclear. A similar process is observed in other diseases with impaired cytotoxicity such as familial hemophagocytic lymphohistiocytosis (FHL)(see Case 14). Many cases of FHL are due to mutations in *perforin 1*, a gene that encodes a lytic

enzyme that is contained in the granules of cytotoxic lymphocytes and is essential for cytotoxicity. Evidence suggests that the accelerated phase of CHS may typically occur after infection with Epstein–Barr virus. In this setting, the immune system is unsuccessful in killing the virus-infected cells; in attempting to control the infection, lymphocytes proliferate without restraint.

Answer 4

Bone marrow transplantation is only able to correct defects that are due to cells of hematopoietic origin. Oculocutaneous albinism and the neurological defects seen in patients with CHS are not due to dysfunction of hematopoietic cells, so these abnormalities cannot be corrected by a bone marrow transplant. At one time, it was believed that the neurological disease was due to persistent lymphocyte infiltration of the CNS, but later studies have shown intrinsic defects in the neurons and glia of patients with CHS, which have disproved that hypothesis. The oculocutaneous albinism is due to a defect in the melanocytes that produce pigment in the skin, eyes, and hair.

Case 16

Answer 1

Signals that direct the maturation of hematopoietic stem cells into the various lineages are transmitted by their contact with stromal cells in the bone marrow. Presumably, this interaction, like T-cell–B-cell interaction, requires cytoskeletal reorientation, and thus will be impaired in cells containing an active affected X chromosome. The stem cells bearing an active normal X chromosome thus have a survival advantage. An alternative hypothesis, based on studies in mice, is that fetal liver hematopoietic stem cells that express WASP have a significant advantage over WASP-negative cells in reaching the bone marrow.

Answer 2

You might try to give antibody against the B-cell cell-surface protein CD40 along with the immunogen. Ligation of CD40 by the CD40 ligand borne by activated T cells is a signal for a resting B cell to start dividing and to undergo isotype switching. The antibody should act like the CD40 ligand and induce isotype switching in B cells (see Case 2).

Answer 3

Measurement of mean platelet volume (MPV) may help diagnose XLT. WAS and XLT are the only conditions that are typically associated with low MPV. In contrast, the MPV is normal or even elevated in patients with chronic ITP.

Case 17

Answer 1

The loss of adrenal cortical hormones as a result of the autoimmune destruction of his adrenal cortex caused Robert's pituitary gland to secrete greatly increased amounts of ACTH. ACTH is composed of 39 amino acids, the amino-terminal 14 amino acids of which can be cleaved off by trypsin-like enzymes. This 14-residue peptide is called melanocortin and stimulates melanocytes in the skin to produce the brown pigment melanin. The receptor for melanocortin also binds intact ACTH, albeit at a lower affinity. The increased amounts of ACTH, and probably of melanocortin, were what led to the increased pigmentation of Robert's scrotum and the tissue around his nipples.

Answer 2
They had twice the normal number of CD4 and CD8 effector/memory cells in their lymph nodes. This apparently resulted from a lack of negative selection of autoreactive cells in the thymus.

Answer 3
This experiment implies, but does not prove, that *Aire* acts in the thymus, and that the expression of *Aire* in peripheral organs is less important. The lymphocytes transferred from the normal mice had left the thymus with self-reactive clones deleted, whereas this had not occurred in the *Aire*-deficient lymphocytes. To show definitively that the thymus is responsible for the disease, the investigators transplanted either knockout or control thymuses into *nude* mice (which do not have a thymus). Only those mice that received a knockout thymus developed autoimmune disease.

Answer 4
Autoantibodies against ovarian cells are common in females with APECED and may cause primary ovarian failure.

Case 18

Answer 1
In Foxp3-deficient mice, infusion of relatively small numbers of T_{reg} cells controls the disease symptoms. So it is likely that, in humans also, a small number of natural T_{reg} cells is sufficient for immune regulation, and Billy can make sufficient Treg $_{cells}$ that derive from his sister's bone marrow stem cells.

Answer 2
The 'conditioning' leading up to a bone marrow transplant comprises treatment with cytotoxic drugs (see Cases 8 and 27) that kill all rapidly dividing cells, including the CD4 effector T cells responsible for the uncontrolled inflammatory response in IPEX. As these cells are destroyed, the autoimmune response that they have produced will be dampened. Immunosuppressant drugs have the same effect in reducing the activation and proliferation of T cells, and this is why they are used to treat IPEX and other autoimmune disorders.

Answer 3
Because IPEX patients are unable to downregulate the immune activation triggered by infections, their disease frequently flares up on exposure to pathogens and even after vaccination. IVIG is useful in forestalling infections or ameliorating their impact when they occur. Vaccination is contraindicated in IPEX patients because of the risk of disease flare-up.

Answer 4
Venous infusion of immunocompetent purified naive CD4 T cells into mice with severe immunodeficiency (for example, *scid* mice or Rag-deficient mice (see Case 7)) results in colitis. Infusion of CD4 CD25 T_{reg} cells into these recipients can both prevent and reverse the colitis. This suggests that CD4 CD25 T_{reg} cells may have therapeutic potential in human autoimmune diseases.

Answer 5
Patients with IL-2Rα deficiency present a clinical picture with similarities to IPEX. IL-2 signaling is essential for the maintenance of T_{reg} cells. Therefore, T_{reg} activity is deficient in the absence of IL-2, IL-2Rα (CD25), or the transcription factor STAT5, which is important for transducing the IL-2 signal. This may explain the similarity of the symptoms of IL-2Rα deficiency to those of IPEX.

Case 19

Answer 1

Fas and FasL are homotrimeric signaling complexes (see Fig. 19.2). If one element of the trimer is mutant, the trimer is rendered ineffective and cannot deliver the signal to downstream elements of the pathway that ultimately cause cell death. This type of effect is called a dominant-negative effect.

Answer 2

Some family members, even though they show no clinical evidence of ALPS, will show impaired lymphocyte apoptosis *in vitro*. It is clear that environmental and/or other genetic factors have a role in the full expression of the ALPS phenotypes as in other genetically inherited diseases. This is called variable expressivity.

Answer 3

Vaccinia (the virus used for smallpox vaccination) expresses a protein, Crm A, that inhibits caspases. Herpes simplex has two genes, *Us5* and *Us3*, which encode proteins that also inhibit caspases. In contrast, Epstein–Barr virus, which causes acute infectious mononucleosis (see Case 45), produces a protein that resembles Bcl-2, which prevents apoptosis and renders infected cells resistant to killing by cytotoxic T cells (Fig. A19.3).

Answer 4

Yes, it would be. The lethality in mice of knocking out caspase-8 points to the importance of this enzyme in fetal tissue remodeling, but there are known differences between species. Moreover, a point mutation in human caspase-8, at the site where it interacts with the Fas complex, might interfere with its function in Fas-induced apoptosis but not in other, Fas-independent, processes. Thus a missense mutation in caspase-8 could conceivably cause ALPS.

Case 20

Answer 1

Bone marrow transplantation was attempted in a 7-year-old female patient with clinically diagnosed HIES in 2000. Despite good evidence of engraftment (full donor chimerism), within 4 years she again developed recurrent staphylococcal skin abscesses and her serum IgE levels rose to pre-transplant levels.

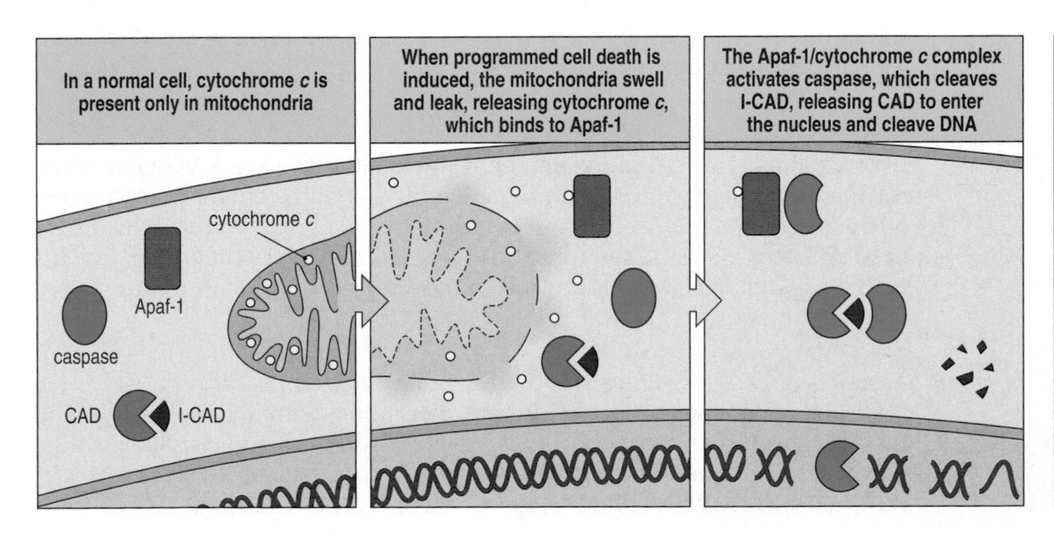

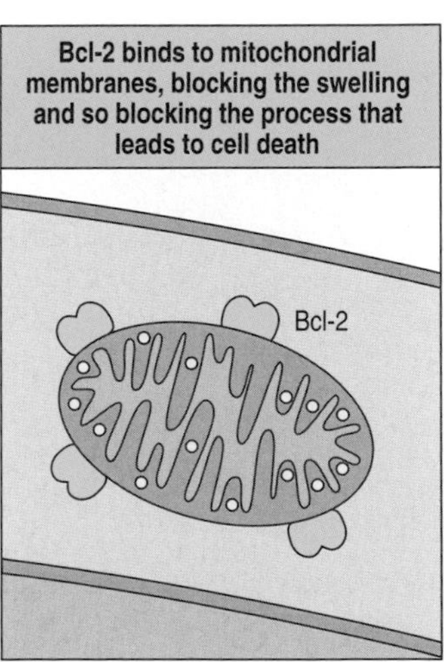

Fig. A19.3 Bcl-2 inhibits the processes that lead to programmed cell death. In normal cells, cytochrome *c* is confined to the mitochondria (first panel). However, during apoptosis the mitochondria swell, allowing the cytochrome *c* to leak out into the cytosol (second panel). There it interacts with the protein Apaf-1, forming a cytochrome *c*:Apaf-1 complex that can activate caspases. An activated caspase cleaves the I-CAD (see Fig. 19.2), which leads to DNA fragmentation (third panel). Bcl-2 interacts with the mitochondrial outer membrane and blocks the mitochondrial swelling that leads to cytochrome *c* release (last panel).

Although the experience with bone marrow transplant in HIES is limited, it seems that defects apart from those in the hematopoietic-cell precursors are important in mediating the immunologic abnormalities.

Answer 2

STAT3 forms homodimers, which translocate to the nucleus to induce the expression of RORγT. A fourfold decrease in RORγT expression is consistent with the fact that STAT3 activity depends on intact function of both molecules in the dimer. A single mutant allele of *STAT3* (heterozygous mutation) would render 75% of the STAT3 homodimers nonfunctional, which is consistent with a dominant-negative effect. Having only 25% of the functional STAT3 homodimers would lead to a roughly equivalent decrease in RORγT expression.

Answer 3

The elevation of IgE levels in HIES is likely to reflect IgE production by activated B cells and plasma cells in which T_H2-dependent class switching to IgE has already occurred. IgE synthesis by these cells is not subject to inhibition by IFN-γ, which only counteracts T_H2 cytokine induction of IgE switching in naive B cells.

Answer 4

IL-21 normally inhibits IgE synthesis in mice, and mice with a knockout of the IL-21 receptor (IL-21R) have elevated serum IgE. Because IL-21R signals via STAT3, decreased IL-21R signaling would be expected to result in increased IgE levels. In humans, however, IL-21 has been reported to increase IgE synthesis, so at present the cause of the elevation in IgE in patients with autosomal dominant HIES is not clear.

Case 21

Answer 1

Loss-of-function mutations in the genes encoding Artemis, DNA ligase IV, Cernunnos, or DNA-PKcs can lead to a radiosensitive T^- B^- NK^+ SCID phenotype. All of these proteins are essential for DNA repair in V(D)J recombination and general DNA repair via NHEJ.

Answer 2

Differences in modifier genes can affect the clinical presentation of a monogenic disease. In the mouse model of ataxia telangiectasia, hypermorphic mutations in the *RAD50* gene, which encodes a component of the MRN complex, can partly compensate for ATM deficiency.

Answer 3

Because there are ATM-independent mechanisms of DNA repair, even complete ATM deficiency results in the faulty repair of only a small fraction of double-strand DNA breaks, and the repair of single-strand DNA breaks does not require ATM. Therefore, the gradual clinical presentation in patients with ataxia telangiectasia is a result of the slow accumulation of unrepaired double-strand DNA breaks.

Answer 4

No. ATM is expressed at very low levels in peripheral blood lymphocytes, and Western blotting is not always reliable, especially if the sample size is less than 10 ml. It is therefore best if the diagnosis of ataxia telangiectasia is made by using a combination of clinical, laboratory, and genetic testing.

Case 22

Answer 1

Chronic neutropenia and hypogammaglobulinemia are also typical of CD40 ligand or CD40 deficiency (see Case 2). However, these conditions have X-linked and autosomal recessive inheritance respectively, whereas WHIM syndrome is typically autosomal dominant. Furthermore, the bone marrow of patients with CD40 ligand or CD40 deficiency shows an arrest in myeloid differentiation, whereas an accumulation of mature neutrophils is seen in patients with WHIM syndrome. Neutropenia may also be seen in other conditions with hypogammaglobulinemia. Autoimmune neutropenia may be observed in patients with common variable immunodeficiency; severe neutropenia may be transiently seen in patients with agammaglobulinemia (see Case 1) during acute infections.

Answer 2

There are two possible explanations for this. Sue's WHIM syndrome could be caused by a de novo mutation that arose in the paternal or maternal germline. Alternatively, it is possible that the disease was not clinically evident in one of the two parents, in spite of the presence of the mutation. Forms of WHIM syndrome that show only some of the symptoms of the disease have been reported.

Answer 3

CXCR4 mutations not only cause retention of neutrophils in the bone marrow but also interfere with the trafficking of other leukocytes. In particular, impaired migration of effector T cells, NK cells, and antigen-presenting cells (dendritic cells) might contribute to the increased susceptibility of patients with WHIM syndrome to viral cutaneous infections.

Answer 4

The receptor function of CXCR4 can be antagonized by competitive ligands, such as the drug AMD-3100 (plerixafor). Because CXCR4 action is important in retaining hematopoietic stem cells in the bone marrow, AMD-3100 is sometimes used to help mobilize stem cells from the bone marrow to the periphery to facilitate the collection of hematopoietic stem cells from blood (for example, for transplantation).

Case 23

Answer 1

One important defense against mycobacteria is the activation of macrophages to synthesize cyokines such as IL-18 and TNF-α after recognition of microorganisms via Toll-like receptors (TLRs). These cytokines amplify immune responses, activate macrophages' intracellular killing activity and induce nitric oxide, which is important in the destruction of the intracellular mycobacteria. In patients with *NEMO* mutations, these cytokines are not made in such quantity because TLRs signal via the activation of NFκB by the IKK complex, which contains NEMO. TNF-α and IL-18 themselves also signal through the NFκB pathway, so their effect is severely reduced. This in turn reduces the synthesis of the important cytokine interferon-γ (IFN-γ) by T cells, which is normally induced by IL-12 acting synergistically with IL-18. Thus, NEMO deficiency also results in decreased IFN-γ production. Because IFN-γ also activates macrophages' ability to kill intracellular bacteria, a lack of NEMO leads to an inability of macrophages to kill mycobacteria that have been taken up by phagocytosis and have become resident in their endosomes. The persistently infected macrophages continue to activate T cells, leading to granuloma formation. Patients with NEMO mutations thus have some similarities to children with mutations in IL-12, the IL-12 receptor, and the IFN-γ receptor in terms of their susceptibility to mycobacteria (see Case 24).

Answer 2

The killer activity of natural killer (NK) cells is deficient in most patients with *NEMO* mutations. Patients deficient in NK cells are known to suffer from recurrent CMV and herpesvirus infections. This suggests that activation of NK cells via their invariant activating receptors in response to viruses is dependent on intact IKK and NFκB activation.

Answer 3

Class switching can be induced by various members of the TNF family (to which CD40 ligand belongs), which are expressed on the surface of activated dendritic cells and engage receptors on B cells. This engagement activates NFκB in a NEMO-independent fashion by selective activation of IKKα dimers by the kinase NIK, which results in the processing of the auto-inhibited NFκB subunit p100 into its active form, p52. This small amount of B-cell activation might also explain the presence of lymph nodes in patients with *NEMO* mutations.

Answer 4

IKK is the major means of phosphorylating IκB, but other kinases can also perform this function, thereby circumventing the block to some extent. When IκB cannot be phosphorylated at all because of a mutation at the phosphorylation site, the pathway is completely blocked and a more severe immunodeficiency results.

Answer 5

It is still possible that the patients have NEMO deficiency despite a normal coding sequence. Mutations in the 5′ untranslated region of a gene may result in severely reduced mRNA levels and protein levels with normal coding sequence. NEMO deficiency due to this type of mutation has been observed, with a fivefold reduction of NEMO protein levels, impaired TLR function, and impaired antibody responses to several vaccines.

Case 24

Answer 1

Mycobacteria, particularly atypical mycobacteria, are ubiquitous in the environment, and any infection is normally contained by T-cell action. AIDS, however, is characterized by a marked reduction in the number of CD4 T_H1 cells. Hence the production of IFN-γ is compromised, and patients with AIDS have difficulty in activating their macrophages and clearing mycobacteria.

Answer 2

The delayed-type hypersensitivity reaction is provoked by a few T cells that are specific for tuberculin. After binding antigen, these T cells secrete chemokines that nonspecifically recruit macrophages and other inflammatory cells. The secretion and action of these chemokines are not dependent on IFN-γ. It is therefore not surprising that Clarissa and her cousins developed positive tuberculin skin tests.

Answer 3

Granulomas form where there is local persistence of antigen, antigen-specific T cells, and activated macrophages. These children could not activate their macrophages, so it is not surprising that they did not form granulomas. Granulomas can be beneficial in that they wall off and prevent the spread of microorganisms. These children could not do that and we find the mycobacteria spreading via the bloodstream—a highly unusual finding in mycobacterial disease. However, granuloma formation is preserved in patients with autosomal recessive partial IFN-γR1 deficiency and in patients with IL-12Rβ1 deficiency.

Answer 4
Infection with pyogenic bacteria such as pneumococci is controlled and terminated by antibody and complement. In fact Clarissa had a very high level of IgG (1750 mg dl^{-1}), probably as a result of increased IL-6 production induced by the chronic mycobacterial infection. Most viral infections, such as chickenpox, are terminated by cytotoxic CD8 cells. Activation of these cells is not dependent on IFN-γ. However, there is evidence from mice that a lack of the IFN-γ receptor increases susceptibility to certain viruses, including vaccinia virus and lymphocytic choriomeningitis virus. Salmonellae take up residence as an intracellular infection of macrophages and become inaccessible to antibody and complement; they can be destroyed in this site only when macrophages are activated by IFN-γ.

Case 25

Answer 1
If the neutropenia is due to increased peripheral destruction of neutrophils, the ability to produce neutrophils in the bone marrow is likely to be normal. In such cases, a bone marrow aspirate will demonstrate the presence of myeloid cells, including neutrophils, in all stages of differentiation. In contrast, neutropenia due to a decreased production of neutrophils is associated with a reduction of myeloid cells in the bone marrow. This reduction may involve cells at all stages of differentiation (as seen in patients with leukemia, whose bone marrow is occupied by cancer cells), or it may affect more mature myeloid cells only, with a normal presence of more immature and progenitor cells, as is typically the case in patients with SCN.

Answer 2
Neutrophils have a very rapid turnover (their half-life being about 8 hours). Therefore, almost all of the neutrophils infused intravenously will die within a day. Furthermore, such transfusions are associated with a significant risk of inflammatory reactions. These observations significantly limit the therapeutic use of transfusions of granulocytes generally. Nonetheless, this approach could be considered in patients with severe numerical or functional defects of neutrophils (SCN and chronic granulomatous disease, respectively) who have life-threatening infections that fail to respond to conventional treatment.

Answer 3
The binding of cytokines to their specific receptors elicits intracellular signaling and may promote cellular activation, leading to proliferation and/or differentiation. Negative regulation of cytokine-mediated signaling may involve internalization of the cytokine receptor, or modification of its intracytoplasmic tail (such as ubiquitination) followed by proteasome-mediated degradation. The intracytoplasmic tail of the G-CSF receptor (G-CSFR) includes a ubiquitination site. Somatic mutations of the G-CSFR that result in truncation of the intracytoplasmic tail impede G-CSFR ubiquitination and cause increased G-CSF-mediated signaling and cellular hyperactivation, thus contributing to leukemic transformation.

Case 26

Answer 1

There has been random inactivation of the X chromosomes in Randy's mother's neutrophils. Therefore 50% of her neutrophils have a normal X chromosome and 50% have an X chromosome bearing the CGD defect. The cells bearing the CGD-defective X chromosome have not been selected against. That half does not reduce NBT, whereas the half bearing the normal X chromosome does.

Answer 2

Streptococcus pneumoniae, the pneumococcus, does not produce catalase, an enzyme that converts hydrogen peroxide (H_2O_2) to water and oxygen, and is thus far less resistant to intracellular killing than are microorganisms that are catalase producers.

Answer 3

Because of persistent antigenic stimulation, Randy is making more immunoglobulins (antibodies) than a normal person. In fact, all chronic infections, such as malaria, result in hypergammaglobulinemia.

Answer 4

Rac2 is important for both the function of the membrane cytochrome b_{558} complex and for the chemotaxis of many cells including neutrophils. These two defects coexist both in humans and in the mouse model of Rac2 deficiency, but not in defects that affect other genes that encode other subunits of the cytochrome b_{558} complex.

Case 27

Answer 1

It is autosomal recessive. The parents are both healthy; they have had an affected male and an affected female child. These facts lead to the conclusion that the leukocyte adhesion deficiency (LAD) is inherited as an autosomal recessive trait. In fact, the gene encoding CD18 has been mapped to the long arm of chromosome 21 at position 21q22.

Answer 2

Monoclonal antibodies against CD18 can induce a mild phenotype of LAD. Such monoclonal antibodies have been used to prevent graft rejection in recipients of kidney grafts and, when administered before bone marrow transplantation, can prevent graft-versus-host disease.

Answer 3

T cells express the β_1 integrin VLA-4 (Fig. A27.3), and its interaction with VCAM-1 on endothelia seems to be sufficient to enable T cells to home and function normally. VLA-4 is not expressed on neutrophils and macrophages, which are much more dependent on β_2 integrins for their adhesion to other cells. B cells also home normally in LAD, and this is probably due to the integrin α_4:β_7, which is not defective in this disease.

Answer 4

Luisa's brother had delayed separation of the umbilical cord and Luisa developed an infection at the site of umbilical cord separation. The role of neutrophils and macrophages in wound healing is not well understood.

Nevertheless, it is apparent from cases of LAD that the movement of white blood cells into wounds is vital to normal tissue repair.

Answer 5

Fucose is a defining determinant of the sialyl-Lewis^x element, which is the ligand whereby white blood cells bind to selectins on endothelial cells (see Fig. CS6-0301/27.1). In patients with LAD type 2, impaired transport of fucose into the Golgi apparatus prevents the fucosylation of newly synthesized glycoproteins, such as sialyl-Lewis^x. Infants with LAD type 2 have very high white blood cell counts because their leukocytes cannot roll on the endothelium to begin the process of leukocyte emigration from the bloodstream.

Answer 6

After the transplant, her leukocytes were found to express CD18.

Answer 7

This was done to destroy her abnormal cells and create 'space' for the transplanted cells. Such preparation can be avoided in the case of severe combined immunodeficiency because the lymphoid compartment is already devoid of T cells and thus space is available for the transplanted cells without any treatment.

Answer 8

PHA is a so-called nonspecific T-cell mitogen. The mitogenic response requires cell–cell interactions that depend on the interaction of LFA-1 with ICAM-1, as well as that of VLA-4 with VCAM-1. In LAD, one of these interactions is missing.

Answer 9

In LAD, there is a defect in the mobility of the leukocytes so that their emigration from the circulation to sites of infection is impeded, whereas in chronic granulomatous disease (CGD), the microbicidal capacity of phagocytes (neutrophils and macrophages) is impaired. The microbicidal capacity of phagocytes in LAD is normal. Children with LAD therefore never develop the inflammatory lesions characteristic of CGD, which are caused by activated phagocytes. In contrast, the leukocytes in CGD have normal mobility and can reach sites of infection in a normal manner but cannot efficiently destroy the infecting bacteria when they reach them. Hence the development of abscesses and granulomas. Unlike individuals with CGD, children with LAD are no more susceptible to pneumococcal infections than they are to other bacterial infections.

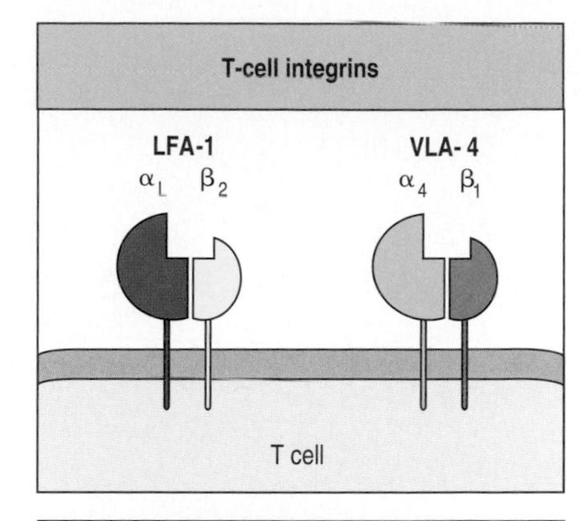

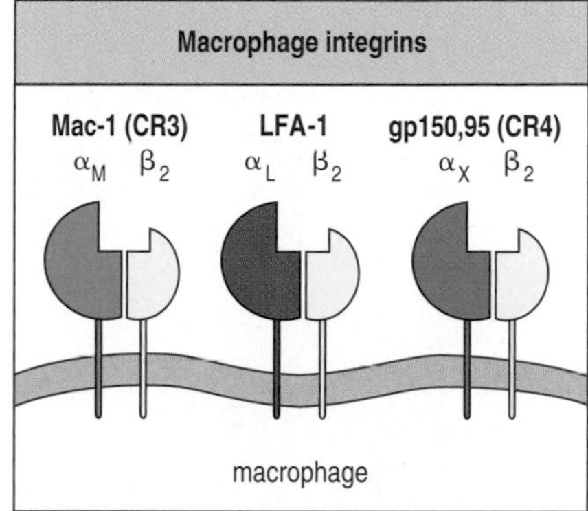

Fig. A27.3 T cells and phagocytes express different integrins. Integrins are heterodimeric proteins containing a β chain, which defines the class of integrin, and an α chain, which defines the different integrins within a class. The α chain is larger than the β chain and contains binding sites for divalent cations that may be important in signaling. Upper panel: LFA-1, a β_2 integrin, and VLA-4, a β_1 integrin, are expressed on T cells and are important in the migration and activation of these cells. Lower panel: macrophages and neutrophils express all three members of the β_2 integrin family: LFA-1, Mac-1 (also known as CR3), and gp150,95 (also known as CR4). Like LFA-1, Mac-1/CR3 binds the immunoglobulin-superfamily molecules ICAM-1, ICAM-2, and ICAM-3, but in addition it is a complement receptor (for the fragment iC3b). gp150,95/CR4 also binds complement and stimulates phagocytosis.

Case 28

Answer 1

The reasons for the central nervous system (CNS)-specific phenotype of patients with TLR-3 signaling defects remain poorly defined. However, the CNS-restricted phenotype suggests that TLR-3 signaling is redundant for the systemic control of viral infections. Indeed, leukocytes may respond to the TLR-3 agonist poly(I:C) in a TLR-3-independent manner, which indicates that there is redundancy in the mechanisms of recognition and response to double-stranded RNA (dsRNA) viral intermediates within the immune system.

Answer 2

Identification of the molecular basis of HSE could have important therapeutic implications. In particular, administration of type 1 interferons might be beneficial in patients with mutations of *TLR3*, *UNC93B*, and *TRAF3*, who are otherwise unable to produce these interferons in response to HSV-1 infection in the CNS.

Case 29

Answer 1

Several immunodeficiencies result in susceptibility to pneumococcal infection. These include defects of innate immunity, including congenital asplenia (see Case 30) and defects within the complement pathways (see Cases 31–33). Other defects in the NFκB activation pathway that lies downstream of TLRs and other cell-surface receptors include NEMO deficiency (see Case 23) and mutations in IκB that prevent its degradation and release of NFκB. In addition, defects of adaptive immunity that result in impaired antibody production, such as X-linked agammaglobulinemia (see Case 1) and common variable immunodeficiency (see Case 4), result in increased susceptibility to Gram-positive bacteria, such as pneumococci and staphylococci.

Answer 2

These infections are usually associated with fever. IRAK4 deficiency, however, leads to an early block in the TLR/IL-1R signaling pathways and barely detectable or no TLR-induced production of pro-inflammatory cytokines. The virtual absence of pro-inflammatory cytokines and the inability of IRAK4-deficient patients to respond to what little IL-1 might be produced results in an impaired febrile response (see Case 34). Thus, a history of little or no fever associated with recurrent pyogenic infections supports a possible diagnosis of IRAK4 deficiency.

Answer 3

So far, there are too few IRAK4-deficient patients for us to be sure. More cases of IRAK4 deficiency need to be identified and followed throughout their lives before conclusions can be drawn. The 18 patients identified so far all show an increased susceptibility to invasive infections with pyogenic bacteria in childhood. As they mature into adolescence, however, susceptibility to such infections becomes variable, and many no longer show significantly increased susceptibility.

Answer 4

The antibody response to polysaccharides seems to require initial signaling via TLR-2 (which recognizes lipoteichoic acid of Gram-positive bacterial cell walls and lipoproteins of Gram-negative bacteria) and TLR-4 (which recognizes LPS) in dendritic cells. This may be because stimulation of dendritic cells via these TLRs is essential for their ability to induce the differentiation of T cells (especially T_H1) that help B cells to make the antipolysaccharide antibodies. The response to Pneumovax, for example, seems to depend on the presence of such TLR ligands in the vaccine. It has been shown that depletion of endotoxin from the vaccine renders it unable to elicit an antibody response in mice.

Answer 5

The answer to this important question is not clear. The production of type I interferons in response to the ligation of TLRs 7, 8, or 9 is markedly diminished in IRAK4-deficient patients, and interferon synthesis is variably affected in response to TLR-3 ligation by long double-stranded RNA. The apparent integrity of the antiviral defenses in these patients is therefore surprising.

One explanation could be that humans can make relatively intact adaptive immune response to viruses as a result of responses by cytotoxic T cells, which kill the infected cells, and through antiviral antibodies produced by B cells. There is probably redundancy between the adaptive and innate immune responses, with (innate system) NK cells cytotoxic for virus-infected cells being activated by T-cell-derived IL-2 and IFN-γ. Other intracellular antiviral immune responses are produced via activation of the RNA-dependent protein kinase pathway and via activation of the cytoplasmic protein RIG1 by double-stranded RNA, which leads to the synthesis of type I interferons.

Answer 6

Experience in managing these patients is limited. However, they are maintained in good health into their adolescence by a combination of prophylactic antibiotics and regular infusions of intravenous immunoglobulin.

Case 30

Answer 1

First you find out that Nicholas has had all his routine immunizations. He received DPT (diphtheria, pertussis, and tetanus antigens) and oral live poliovirus vaccine at ages 3, 4, and 5 months, and a booster of both before entering kindergarten. He was also given MMR (mumps, measles, and rubella live vaccines) at 9 months of age. At the same time, he was given Hib vaccine (the conjugated capsular polysaccharide of *Haemophilus influenzae*, type b; Fig. A30.1). His growth and development have been normal. He suffered a middle ear infection (otitis media) at age 24 months. Other than that he has had no other illnesses, except for a common cold each winter. You feel comfortable that he is protected against infection with *H. influenzae* from the Hib vaccine. However, your concern about the possibility of pneumococcal infection leads you to advise the surgeon to immunize Nicholas against pneumococcal capsular poysaccharides by giving him conjugated pneumococcal polysaccharide vaccine. You also advise prophylactic antibiotics, to be taken at a low dose daily but at higher doses when Nicholas has any dental work done, or any invasive surgical procedure.

Answer 2

The typhoid vaccine was given subcutaneously and a response was mounted in a regional lymph node. The sheep red blood cells were given intravenously, and, in the absence of a spleen, failed to enter any peripheral lymphoid tissue where an immune response to them could occur.

Answer 3

The defect is inherited as an autosomal recessive. The parents are normal but each carries this recessive gene. Furthermore, they are consanguineous, a setting in which autosomal recessive disease is encountered more frequently

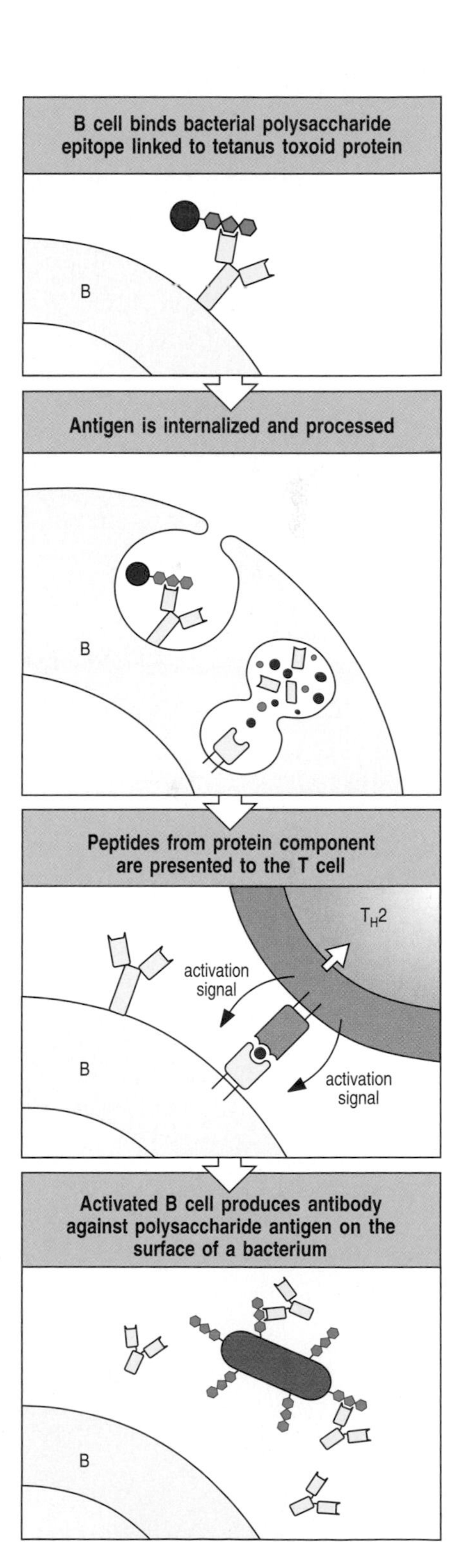

Fig. A30.1 *Haemophilus influenzae* b vaccine is a conjugate of bacterial polysaccharide with the tetanus toxoid protein, which enhances the immune response by allowing a polysaccharide-specific B cell to recruit T-cell help. The B cell recognizes and binds the polysaccharide, internalizes and degrades the toxoid protein to which it is attached, and displays the peptides derived from it on surface MHC class II molecules. Helper T cells generated in response to the protein moiety of the toxoid recognize the complex on the B-cell surface and activate the B cell to produce antibody against the polysaccharide. This antibody can then protect against infection with *H. influenzae*, type b.

than when parents are unrelated. Chance would predict that one in four (that is, two) of their eight children would be affected. Each pregnancy provides a one in four chance that the fetus will inherit the abnormal gene from both parents. As it turned out, this happened in five of Mrs Vanderveer's eight pregnancies. Because Betsy married a normal man, all her children are heterozygous for the defect, like their maternal grandparents, and have normal spleens.

Case 31

Answer 1

Histamine release on complement activation is caused by C3a (the small cleavage fragment of C3), and the main chemokine is C5a (the small cleavage fragment of C5). These are both generated by the C3/C5 convertase, which in the classical pathway is formed from C4b and C2a. In HAE, C4b and C2a are both generated free in plasma. C4b is rapidly inactivated if it does not bind immediately to a cell surface; for that reason, and because the concentrations of C4b and C2a are relatively low, no C3/C5 convertase is formed, C3 and C5 are not cleaved, and C3a and C5a are not generated.

The edema in HAE is caused not by the potent inflammatory mediators of the late events in complement activation, but by C2b generated during the early events, and by bradykinin generated through the uninhibited activation of the kinin system.

Answer 2

The only other complement component that should be decreased is C2, which is also cleaved by C1. C1 plays no part in the alternative pathway of complement activation, so complement activation by the alternative pathway is not affected. The terminal components are not affected either. The unregulated activation of the early complement components does not lead to the formation of the C3/C5 convertase (see Question 1), so the terminal components are not abnormally activated. The depletion of the early components of the classical pathway does not affect the response to the normal activation of complement by bound antibody because the amplification of the response through the alternative pathway compensates for the deficiency in C4 and C2.

Answer 3

This is not hard to explain; as we have already remarked, the alternative pathway of complement activation is intact and thus, although the classical pathway is affected by deficiencies in C2 and C4, these are compensated for by the potent amplification step from the alternative pathway.

Answer 4

Stanozolol is a well-known anabolic androgen that has been used illegally by Olympic competitors. For unknown reasons, anabolic androgens suppress the symptoms of HAE, and that is why stanozolol was prescribed to Richard. Patients, especially females, do not like to take these compounds because they cause weight gain, acne, and sometimes amenorrhea. Preparations of purified C1INH are now available, and intravenous injection of C1INH prepared from human donors is safe and very effective in halting the symptoms of the disease.

Answer 5

In practice, you would administer epinephrine immediately in any case, because most such emergencies are due to anaphylactic reactions and because epinephrine is a harmless drug. If the laryngeal edema is anaphylactic, it will respond to the epinephrine. If it is due to hereditary angioedema, it will not. Anaphylactic edema is also likely to be accompanied by urticaria and itching, and the patient may have been exposed to a known allergen. Most patients know if they are allergic or have a hereditary disease, and they should be asked whether they have had a similar problem before.

Answer 6

HAE does not skip generations: it is therefore likely that its effects are dominant. It clearly affects both males and females, so it cannot be sex-linked. If the gene has a dominant phenotype, and Richard's two children are normal, then it follows they cannot have inherited the defective gene from their father, and their children cannot inherit the disease from them.

Richard has inherited his abnormal *C1INH* gene from his mother. Because he has a normal *C1INH* gene from his father, you might expect that he would have 50% of the normal level of C1INH. However, the tests performed by his immunologist revealed 16% of the normal level. In general, functional C1INH tests in HAE patients reveal between 5% and 30% of normal activity. How could this be explained? There are two possibilities: decreased synthesis (that is, less than 50% synthesis from only one gene); or increased consumption of C1INH as a result of increased C1 activation. Both explanations have been shown to be correct. Patients with HAE synthesize about 37–40% of the normal amount of C1INH, and C1INH catabolism is 50% greater than in normal controls.

Case 32

Answer 1

As Morris lives longer, his adaptive immunity against these common bacteria becomes stronger and he has come to rely less on innate immune mechanisms for protection against infection. Bacteria coated with antibodies can be phagocytosed independently of complement via the Fc receptors on phagocytes.

Answer 2

A similar result. The factor B would be rapidly destroyed and its rate of synthesis could turn out to be normal. The overproduction of C3b in the absence of factor I leads to an increased binding of factor B to C3b and its subsequent cleavage by factor D. Thus, factor B is being consumed excessively as a result of the deficiency in factor I. The lack of factor B, like the C3 deficiency, is secondary to the basic defect in factor I.

Answer 3

His serum levels of C3 and factor B rose to normal. C3b disappeared from his serum. The effect lasted for about 10 days.

Answer 4

Deficiency in factor H. Because factor H is needed for the cleavage of C3b by factor I in the blood, factor H deficiency should result in clinical symptoms identical with those of factor I deficiency. In fact this is true: several families with factor H deficiency have been studied, and they show symptoms indistinguishable from factor I deficiency.

Answer 5

He is producing large amounts of C3b, which binds to complement receptor 1 (CR1) on red blood cells (see Case 33) and leads to their agglutination by anti-C3.

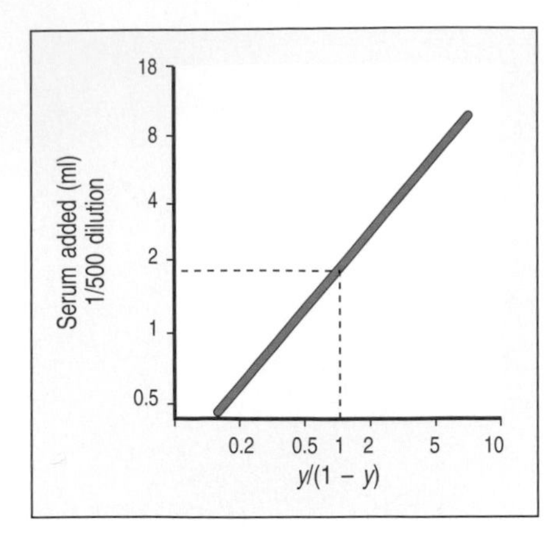

Fig. A33.1 Determination of CH_{50}. This is the quantity of complement required for 50% lysis of 5×10^8 optimally sensitized red blood cells in 1 hour at 37°C. Complement titers are expressed as the number of CH_{50} contained in 1 ml of undiluted serum. This is usually determined by plotting $\log_{10} y/(1-y)$ (where *y* is percentage lysis) against the logarithm of the amount of serum. This plot is linear near $y/(1-y) = 1$ (50% lysis). In the example shown, the serum contains 250 CH_{50} ml^{-1}.

Case 33

Answer 1

The CH_{50} is the quantity of complement required for 50% lysis of 5×10^8 optimally sensitized sheep red blood cells in 1 hour at 37°C. (The sheep red blood cells are usually sensitized with a rabbit antibody against the sheep cells.) The complement titer is expressed as the number of CH_{50} contained in 1 ml of undiluted serum from the patient. This is determined by plotting $\log y/(1-y)$ (where *y* is the percentage lysis) against the logarithm of the amount of serum. The plot is linear near $y/(1-y) = 1$ (50% lysis) (Fig. A33.1). Because this assay measures the functional integrity of all the complement components comprising the classical pathway, it is the single best screening test when a general diagnosis of complete complement deficiency is being considered.

Answer 2

The bacteria are temporarily vulnerable to killing by complement when they divide. At this time the bacterial membrane is exposed and is vulnerable to attack by the membrane-attack complex. The association between genetic deficiencies of membrane-attack complex proteins and neisserial infections illustrates that an important aspect of host defense against these infections is the killing of extracellular bacteria by complement-mediated lysis. *Neisseria* that escape killing enter a variety of cell types and establish an intracellular infection.

Answer 3

In humans, clearance of immune complexes from the blood is largely effected by their attachment to complement receptor 1 (CR1) on the surface of red blood cells. One of the ligands for CR1 is C4b (the other is C3b). All the C1 components are required for the formation of C4b, and if C4b is part of an immune complex, the binding of the complex by CR1 is facilitated (Fig. A33.3). Immune complexes bound to the surface of the red cell are transported to the liver and spleen, where C4b (or C3b) is converted by factor I to iC4b (or iC3b); this facilitates uptake of the immune complexes by phagocytes via complement receptor CR3 and others, and their destruction. In the absence of C4b, immune complexes are less efficiently attached to red cells and are therefore less efficiently cleared from the blood.

Case 34

Answer 1

Anakinra is a recombinant protein that competitively blocks the binding of IL-1β to its receptor, IL-1R, and would thus be expected to mitigate the effects of the excess IL-1β produced in the cryopyrinopathies. Pretreatment with anakinra before exposure to cold has been shown to prevent the development of symptoms and the elevation of acute-phase reactants in patients with familial cold autoinflammatory syndrome, and it also improves symptoms and corrects biochemical abnormalities in Muckle–Wells syndrome. Children with NOMID/CINCA have shown similarly promising results, with resolution of fever, rash, and uveitis (inflammation of the middle layer of the eyeball, the uvea, comprising the choroid, iris, and cilary body), and the relief of excess pressure exerted by the CSF.

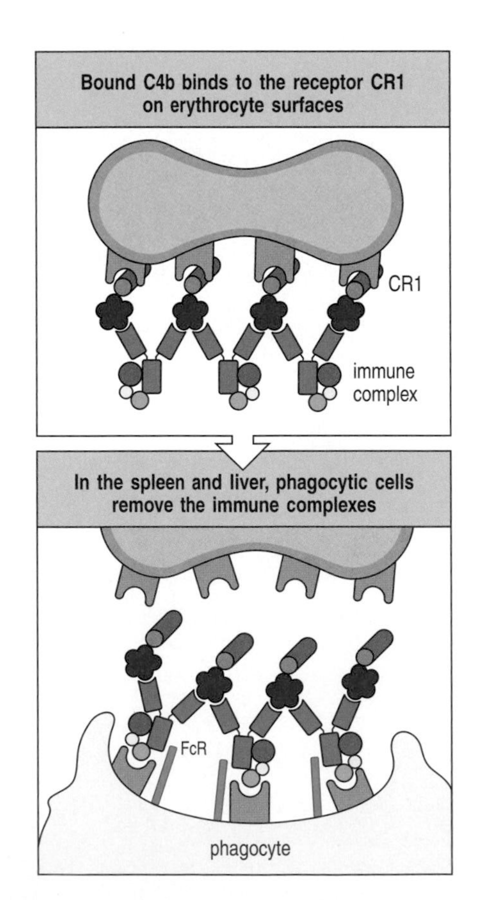

Fig. A33.3 Erythrocyte complement receptor CR1 helps to clear immune complexes from the circulation. Immune complexes bind to CR1 on erythrocytes, which transport them to the liver and spleen, where they are removed by macrophages expressing receptors for both Fc and bound complement components.

Answer 2

Colchicine inhibits the assembly of microtubules in cells by binding to β-tubulin subunits. In the context of autoinflammatory disease it is thought to act mainly by inhibiting the neutrophil response. It inhibits microtubule-dependent processes in the cell, including the division of neutrophil precursors and the secretion of pro-inflammatory mediators by mature neutrophils. In addition, colchicine has been shown to modulate the production of chemokines and pro-inflammatory prostanoids (prostaglandins and leukotrienes) by neutrophils, and it is also thought to inhibit neutrophil adhesion to the endothelium, a necessary step in the migration of neutrophils out of the blood and into tissues. Interestingly, it has also recently been shown to block IL-1β processing and secretion. Together, all of these actions are anti-inflammatory.

Answer 3

Caspase 1 is required to process the precursor forms of IL-1 and IL-18 proteolytically to produce mature active cytokines that can be secreted. It is now known that the cytokine IL-33 also requires processing by caspase 1. IL-33 is an IL-1-like cytokine that signals through the IL-1 receptor-related protein ST 2. IL-33, via ST 2, activates the transcription factor NFκB and mitogen-activated protein kinases (MAPKs), and drives the production of the T_H2-associated cytokines IL-4, IL-5, and IL-13. In addition to IL-33, one might expect that blood levels of cytokines that are induced by the action of IL-1β, such as IL-6, would be increased in response to caspase 1 activation.

Case 35

Answer 1

Patients with sJIA have an increased secretion of IL-1β (unlike in other forms of arthritis), and some show remarkable improvement when treated with anakinra. Unlike the periodic fever syndromes discussed in Case 34, however, the causative gene mutation in sJIA has not yet been identified.

Answer 2

Anti-IL-1 therapies have been on the market for the past few years. Anakinra is an IL-1 receptor antagonist. It is used for the treatment of rheumatoid arthritis and more recently for pediatric rheumatologic conditions such as sJIA. Because IL-1 is important in the normal response to infection, higher rates of serious infections or infectious complications in patients on anakinra would not be unexpected. Meta-analysis of four large anakinra trials showed overall a similar rate of infection compared with placebo; however, there was a modest increase in the infection rate for patients on the highest anakinra doses.

Tocilizumab was recently approved by the US Food and Drug Administration for rheumatoid arthritis, and studies using tocilizumab to treat sJIA look promising. Tocilizumab blocks the effects of IL-6 by binding to its receptor. Again, an increased risk of infections is the primary theoretical concern with tocilizumab. Trials of tocilizumab for patients with rheumatoid arthritis showed a small but significant increase in the infection rate, especially in patients on higher doses. Elevation of serum cholesterol has also been observed in individuals on tocilizumab. Large cohorts of patients on IL-1 and IL-6 inhibitors will need to be followed for years to ascertain their effects fully.

Answer 3

Understanding of the autoinflammatory disorders has advanced considerably over the past few years. Many of these disorders involve the inflammasome and subsequent IL-1 signaling. NLRP3, a component of the inflammasome, was found to be mutated in the cryopyrin-associated periodic fever syndromes

(CAPS) (see Case 34). The CAPS are a spectrum of disorders, with the mildest being familial cold autoinflammatory syndrome (FCAS). Patients with Muckle–Wells syndrome have an intermediate phenotype, and individuals with NOMID/CINCA (neonatal-onset multisystem inflammatory disease) develop rash, fever, poor growth, and characteristic facial features at or shortly after birth (see Case 34). These NLRP3 mutations result in an increased processing and secretion of IL-1/beta. More recently, the gene encoding the IL-1 receptor antagonist (*IL1RN*) was found to be mutated in DIRA (deficiency of IL-1 receptor antagonist). Patients with DIRA present early in life with severe rashes and bony deformities. Fortunately, patients with CAPS or DIRA respond to treatment with anakinra and other IL-1 antagonists.

Answer 4

It is still a mystery exactly why the fevers with sJIA occur daily or twice daily. The fever-inducing cytokine IL-1 drives IL-6 production. IL-1 has a very short half-life in serum; however, IL-6 levels can be measured and they correlate with fluctuations in the temperature curve. The rash may be affected by some of the same inflammatory mediators as well as by the fever itself. The rash tends to be more prominent with heat, included elevated room temperature. It can also be brought out after rubbing the skin or after trauma to the area.

Case 36

Answer 1

Rheumatoid factors can be found in the serum of patients with other immune-complex diseases. In Case 38 we encounter an example in the case of mixed essential cryoglobulinemia. Patients with hypergammaglobulinemia and chronic infection can also have circulating immune complexes and rheumatoid factor.

Answer 2

Mast cells express the IgG receptor FcγRIII. They use this receptor to take up IgG:antigen complexes. This causes the immediate release from mast cells of preformed TNF-α and IL-1. Mast cells are the only cells in which such cytokines are already preformed.

Answer 3

They upregulate the expression of the integrin CD11:CD18 (LFA-1) on the leukocytes, and this increase in integrin expression promotes the binding of the leukocytes to the blood vessel wall and their emigration from the blood vessels. The soluble IL-1 receptor antagonist anakinra (Kineret) has been used successfully in the treatment of rheumatoid arthritis.

Answer 4

The monoclonal antibody infliximab is a chimeric human–mouse immunoglobulin against which patients may develop antibodies. These antibodies would render the therapy useless and might even cause anaphylactic reactions. Better anti-TNF monoclonal antibodies are those that are completely humanized, such as adalimumab, and therefore do not elicit an immune response. A more general risk is that of the reactivation of a preexisting infection, such as tuberculosis, because TNF-α is normally important in containing infections. This risk applies to all classes of anti-TNF agents. Patients about to receive infliximab or another anti-TNF agent should have a tuberculin skin test to ensure they are free of tuberculosis. The activation of tuberculosis in the absence of TNF-α suggests that this cytokine is critical in activating macrophages to contain the latent infection.

Case 37

Answer 1
The serum levels of complement proteins C3 and C4 are lowered in SLE by the large number of immune complexes binding C3 and C4, triggering their cleavage. The depletion of these proteins is therefore proportional to the severity of the disease. Successful immunosuppressive therapy is reflected in an increase in the serum levels of C3 and C4. Measurement of either C3 or C4 is sufficient; it is not necessary to measure both, and C3 is most usually measured.

Answer 2
The objective of these tests was to establish whether Nicole had autoimmune hemolytic anemia, which occurs in SLE when there are antibodies against erythrocytes. Nicole did not have hemolytic anemia (see Case 41).

Answer 3
Because ultraviolet light provokes the onset of SLE and causes relapses.

Answer 4
She had not developed glomerulonephritis. If she had, her urine would have contained protein and red blood cells.

Answer 5
As a result of the constant stimulation of their B cells by autoantigens, patients with SLE have a greatly expanded B-cell population and consequently an increased number of plasma cells secreting immunoglobulin. A lymph node biopsy from Nicole would have exhibited follicular hyperplasia in the cortex and increased numbers of plasma cells in the medulla.

Answer 6
In the first place, a large multimolecular complex such as a nucleosome carries many separate epitopes, each of which can stimulate antibody production by a B cell specific for that epitope. Any of these antibodies can bind the nucleosome particle to form an immune complex. Such potentially autoreactive B cells probably exist normally in the circulation but, provided that T-cell tolerance is intact, they are never activated because this requires T cells to be reactive against the same autoantigen. SLE is probably caused by a failure of T-cell tolerance. T cells for each of the components of the complex antigen will not be needed to induce antibodies against its individual components. As Fig. A37.6 shows, a T cell that is specific for one protein component of a nucleosome could activate B cells specific for both protein and DNA components.

Case 38

Answer 1
The goal of treatment for symptomatic cryoglobulinemia associated with hepatitis C virus (HCV) infection is to eradicate the HCV or at least reduce the level of viral infection. There are no specific inhibitors of HCV replication currently available; however, interferon-α (IFN-α) is effective in treating HCV infection in some patients. Although the mechanism of action remains uncertain, there is evidence that in patients with chronic HCV, IFN-α clears the virus by preventing new infection of liver cells rather than by inhibiting viral replication within cells. There is also some evidence that IFN-α may tip the balance between the T_H1 and the T_H2 immune response in favor of a T_H1 response. Although IFN-α has been shown to improve many of the features

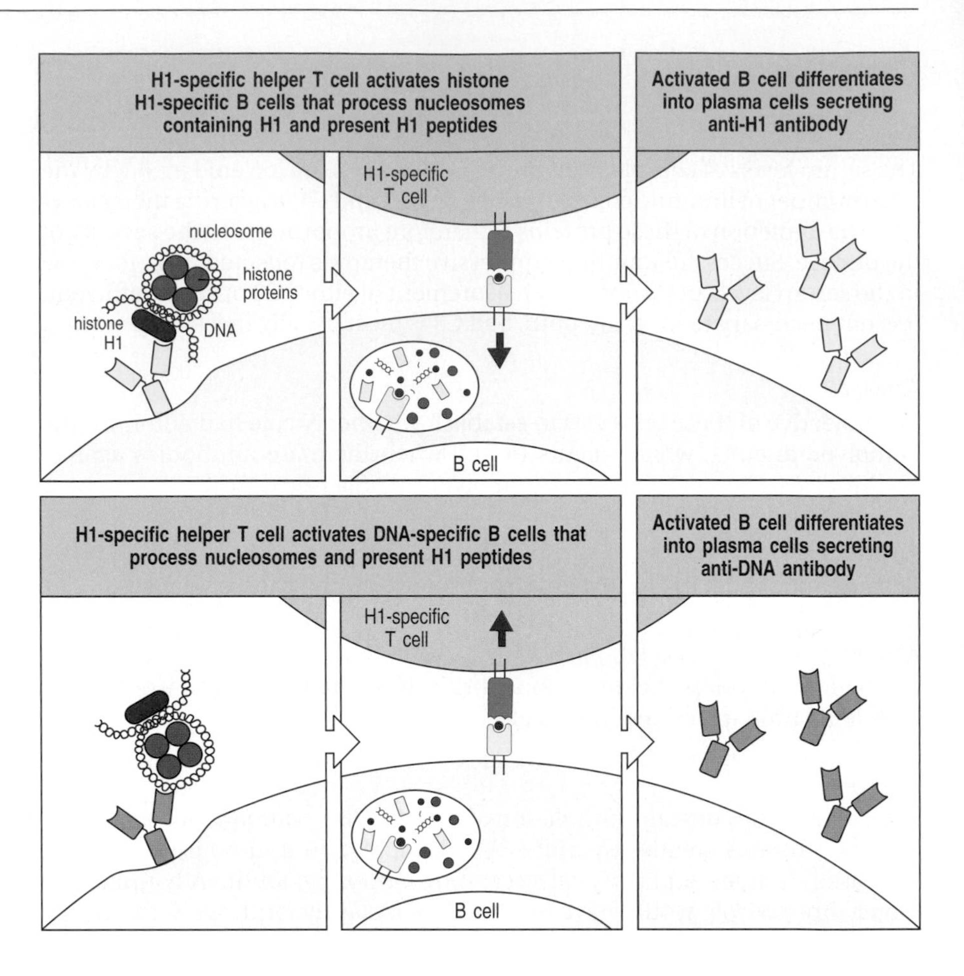

Fig. A37.6 Autoantibodies against various components of a complex antigen can be stimulated by an autoreactive helper T cell of a single specificity. In SLE, patients often produce autoantibodies against all of the components of a nucleosome, or of some other complex antigen. The most likely explanation is that all the autoreactive B cells have been activated by a single clone of autoreactive T cells specific for a peptide of one of the proteins in the complex. A B cell binding to any component of the complex through its surface immunoglobulin can internalize the complex, degrade it, and return peptides derived from the relevant protein to the cell surface bound to class II MHC molecules, where they stimulate helper T cells. These, in turn, activate the B cells. The figure illustrates this scheme for a T cell specific for the H1 histone protein of the DNA:protein complex comprising the nucleosome, and two B cells specific for the histone protein and double-stranded DNA, respectively.

of HCV-induced cryoglobulinemia, cessation of therapy often, unfortunately, results in a relapse of the disease. The combination of ribavirin with IFN-α improves both biochemical and virologic responses in chronic HCV infection and has been beneficial in symptomatic cryoglobulinemia. Ribavirin is a guanosine nucleoside analog that is active against many RNA and DNA viruses.

For severe systemic illness such as the intestinal vasculitis, plasmapheresis is an option for rapidly lowering the circulating titer of autoantibodies and immune complexes. Plasmapheresis involves withdrawal of blood from the patient's vein, separation of the plasma from blood cells, either by centrifugation or membrane separation, and reinfusion of the patient's blood cells, which are resuspended in donor plasma or another replacement solution (such as albumin, normal saline, or lactated Ringer's solution).

Answer 2

Rituximab is a chimeric anti-CD20 monoclonal antibody consisting of human IgG1 κ-chain constant regions and murine variable regions. The CD20 antigen is expressed constitutively on B-lineage cells. Rituximab bound to CD20-positive cells leads to the very efficient depletion of circulating B lymphocytes by a range of mechanisms. These include complement fixation, which results in formation of the membrane-attack complex, and antibody-dependent cell-mediated cytotoxicity (ADCC). The initial clinical application of rituximab was in the treatment of patients with B-cell lymphoma that was refractory to standard therapy; however, its use has now been extended to treat a range of autoimmune diseases, including autoimmune hemolytic anemia and cold agglutinin disease (see Case 41).

In Billy's case, rituximab might help his immune complex-mediated disease in several ways. The most straightforward interpretation is that rituximab causes

B-cell depletion, leading to decreased antibody production and a decrease in the level of circulating immune complexes. However, B cells are important not only in generating antibody responses but also in their role as antigen-presenting cells and cytokine-producing cells. Removal of B cells by rituximab may have additional benefits in symptomatic cryoglobulinemia.

A major concern about the use of rituximab is the risk of infections. In patients with HCV-related cryoglobulinemia it is possible that a decrease in anti-HCV antibody levels after B-cell depletion might allow uncontrolled replication of the HCV. Further studies need to be performed to determine whether this theoretical risk is important in patients.

Case 39

Answer 1
Platelets are an acute-phase reactant and are elevated as part of a systemic inflammatory response. Just as inflammatory cytokines increase the production of other markers of inflammation such as C-reactive protein, cytokines lead to increased platelet release and production in the bone marrow.

Answer 2
6-Mercaptopurine (6-MP) is metabolized to 6-thioguanine, which inhibits the synthesis of purine nucleotides required for DNA and RNA synthesis. 6-MP is toxic to rapidly dividing cells, which require nucleic acid synthesis. Like rapidly dividing cancer cells, lymphocytes mediating autoimmune and inflammatory responses are proliferating rapidly and can be targeted by antimetabolite drugs.

Answer 3
Some patients treated with infliximab generate neutralizing antibodies against the mouse portion of the chimeric monoclonal antibody, recognizing the drug as a foreign antigen. Because adalimumab is a fully humanized protein, the problem of neutralizing antibodies is greatly diminished.

Answer 4
Lymphocytes use cell-adhesion molecules such as α_4 integrin to bind to its ligand, VCAM-1, expressed on endothelial cells. This causes lymphocyte arrest and enables them to migrate from the vasculature to sites of inflammation, to which they are attracted by chemokines. Inhibiting T-cell homing to the gut in Crohn's disease reduces the extent of inflammation and diminishes disease symptoms.

Case 40

Answer 1
The oligoclonality of the immunoglobulins in the cerebrospinal fluid reflects the activation of a limited number of B-cell clones that have gained entry into the central nervous system after the breakdown of the blood–brain barrier. Only those B cells that recognize antigen via their surface immunoglobulin receptor and receive a stimulatory signal from an activated T cell will proceed to synthesize and secrete immunoglobulins.

Answer 2
Corticosteroids and cyclophosphamide (a powerful cytotoxic drug) inhibit T-cell proliferation and thus interfere with the secretion of cytokines that

drive the inflammation and further T-cell activation. The mechanism of action of IFN-β is not known. More recently, a monoclonal antibody, natalizumab (Tysabri), that targets the α_4 integrin subunit has been reapproved in the United States for a restricted subset of patients with MS, after being withdrawn from the market in 2005 because three patients developed progressive multifocal leukoencephalopathy due to the JC virus. This drug is aimed at blocking the movement of leukocytes from the blood into sites of inflammation.

Answer 3

The patients got markedly worse. IFN-γ upregulates the expression of MHC class II molecules and thus enhances antigen presentation. In addition, it drives the differentiation of T_H1 cells, which are involved in the pathogenesis of MS.

Answer 4

Proteins eaten as part of food have long been known not to elicit routine immune responses. The reason seems to be that there are antigen-specific mechanisms in the gut for suppressing peripheral immune responses to antigens delivered by mouth. One is that when T cells in gut-associated lymphoid tissues are presented with orally delivered protein antigens in the absence of an infection, a lack of co-stimulatory signals induces the T cells to become anergic. Another involves the development of regulatory T cells (see Case 18), which can actively suppress antigen-specific responses after rechallenge with antigen. Such cells produce cytokines, including interleukin-4 (IL-4), IL-10, and TGF-β, which inhibit the development of T_H1 responses and are associated with low levels of antibody and virtually absent inflammatory T-cell responses. However, attempts to treat MS in humans by feeding the MBP antigen have proved unsuccessful.

Answer 5

EAE cannot be induced in mice lacking CD28. CD28 on T cells is the receptor for the B7 co-stimulatory molecules, which are essential for the activation of naive antigen-specific T cells, including T cells that recognize MBP. In contrast, mice in which the cell-surface protein CTLA-4, another receptor for B7 molecules, has been knocked out develop EAE more readily than their normal littermates. This is because CTLA-4 binds B7 molecules about 20 times more strongly than does CD28, and normally delivers an inhibitory signal to the activated T cell. In CTLA-4 knockout mice, this inhibitory signal is missing and so the T cells are more readily activated.

Case 41

Answer 1

A plasma exchange (plasmapheresis). In this procedure the patient's blood is repeatedly removed 300–500 ml at a time and the blood is centrifuged (in this case at 37°C, at which temperature the IgM antibodies would be eluted from the red blood cells). The cells are resuspended in normal plasma and infused back into the patient. This is a relatively efficient procedure for removing IgM antibodies because 70% of IgM is in the plasma compartment and only 30% of IgM is in the extravascular compartment. In contrast, only 50% of IgG is in the vascular compartment and 50% is in the extravascular space.

Answer 2

They usually do not cause anemia because they do not fix complement (C1q) and there are no Fc receptors for these immunoglobulin classes on cells of the macrophage lineage, which bear only Fcγ receptors. Fcε receptors are expressed on mast cells and B cells, and their engagement would not result in the destruction of red blood cells.

Answer 3

The internal thiol ester that binds covalently to the IgM autoantibody is situated in the C3d and C4d fragments. The complement components C3b and C4b that are bound to the autoantibody may be digested by Factor I to C3c + C3d and C4c + C4d. C3c and C4c would be released from the antigen:antibody complex and thus antibodies against them would not agglutinate the red blood cells.

Answer 4

There is almost certainly no T-cell antigen receptor for this carbohydrate antigen, so that help to move the B cells into follicles would not occur and specific B cells would undergo apoptosis in the T-cell zone (41.4). Alternatively, the B cells might be rendered anergic by soluble antigen or undergo programmed cell death through the interaction of Fas and Fas ligand.

Although the precise mechanism of autoimmunization awaits elucidation, the molecular characterization of the autoantigen I and of the host cell adherence receptors for *Mycoplasma pneumoniae* have indicated collectively that the disorder has its origin in the interaction of the infective agent with carbohydrate attachment sites on host cells. The mycoplasma adheres to the ciliated bronchial epithelium, and also to red cells and a variety of other cells via ligands that consist of long carbohydrate chains of I antigen type that are capped with sialic acid (see Fig. 41.5). When bound to the carbohydrate chain that contains the I-antigen sequence, the mycoplasma may act as a carrier and the I antigen may act as a hapten.

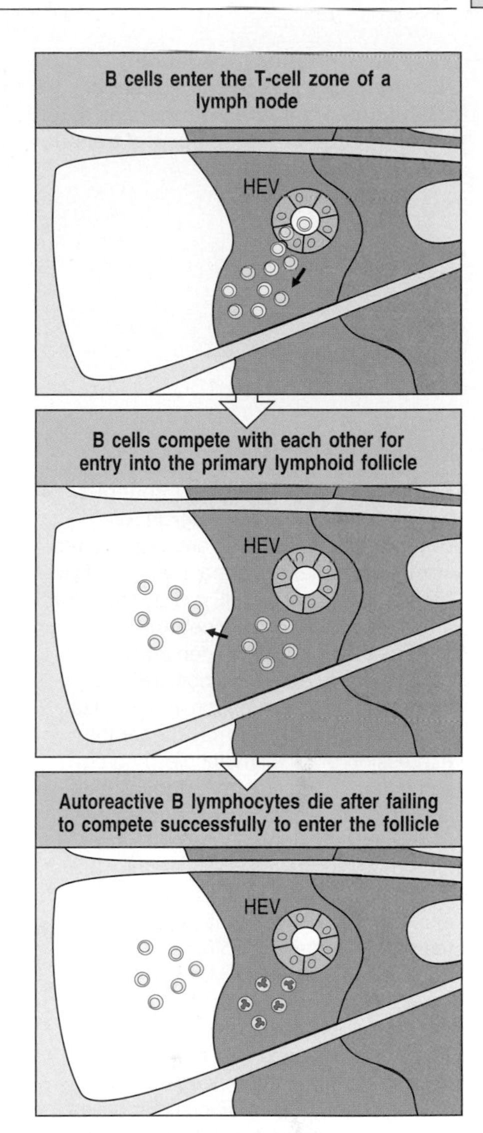

Fig. A41.4 Autoreactive B cells do not compete effectively to enter primary lymphoid follicles in peripheral lymphoid tissue. In the top panel, B cells are seen entering the T-cell zone of a lymph node through high endothelial venules (HEVs). Those with reactivity to foreign antigens are shown in yellow, and autoreactive cells are shown in gray. The autoreactive cells fail to compete with B cells specific for foreign antigens for exit from the T-cell zone and entry into primary follicles (middle panel). This is because B cells reactive to foreign antigens receive signals from antigen-specific T cells that promote their activation and survival. In contrast, the autoreactive B cells fail to receive survival signals and undergo apoptosis in the T-cell zone (bottom panel).

Case 42

Answer 1

It would be likely to last 1–2 weeks. The infant has the disease because maternal IgG antibodies against the acetylcholine receptor have crossed the placenta from the maternal circulation to the fetal circulation. The infant is not synthesizing these autoantibodies; he or she has acquired the disease passively by transfer of the antibodies (Fig. A42.1). The maternal IgG antibodies bind to the acetylcholine receptors in the baby, and the complex of the receptor with bound IgG antibodies is internalized into the cell and degraded. Within 10–15 days all the maternal IgG antibodies against the acetylcholine receptor are adsorbed from the babies' blood and the symptoms abate.

Answer 2

Azathioprine is an immunosuppressive agent. It is converted in the liver to 6-mercaptopurine, which inhibits DNA synthesis. Thus the growth of rapidly dividing cells, such as B cells and T cells, is inhibited and the immune response is suppressed. Unfortunately the effects of azathioprine are not specific. It suppresses not only the formation of antibodies against the acetylcholine receptor but also all other immune responses. Patients taking azathioprine become susceptible to infections. If used for very prolonged periods it is associated with the development of lymphomas. The reasons for this are not well understood.

Answer 3

Fig. A42.3 summarizes ways in which infectious diseases can break self tolerance and induce, or worsen, an autoimmune disease. An infectious agent may

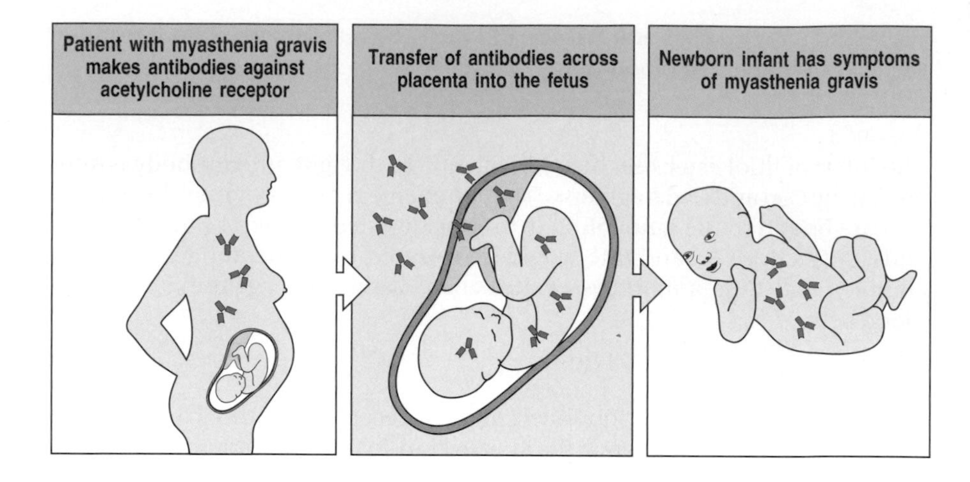
Patient with myasthenia gravis makes antibodies against acetylcholine receptor
Transfer of antibodies across placenta into the fetus
Newborn infant has symptoms of myasthenia gravis

Fig. A42.1 Antibody-mediated autoimmune disease can appear in the infants of affected mothers as a consequence of transplacental transfer of IgG autoantibodies.

expose hidden antigens, or may increase the expression of MHC molecules and co-stimulators on tissue cells so as to induce an autoimmune response. B ells already primed to make an autoantibody may receive help from nearby T cells activated by an infection, especially if pathogens become attached to self molecules. Pathogens may induce responses that cross-react with self molecules. Bacterial and viral superantigens can overcome clonal anergy and break tolerance to self antigens. At present, relatively little is known about the induction of human autoimmune disease, and there are only a few examples in which the evidence for any one of these mechanisms is strong.

Case 43

Answer 1

The monomeric IgG in the gamma globulin administered intravenously binds to high-affinity Fc receptors on macrophages and causes the release of immunosuppressive cytokines such as transforming growth factor-β (TGF-β), interleukin (IL)-10 and the IL-1 receptor antagonist. In patients with pemphigus vulgaris, clinical improvement after intravenously administered gamma globulin is noted in about 6 weeks, and decreased antibody titers are found after 6 months.

Answer 2

The reasons for this difference are not understood. We know that the IgG1 antibody reacts with a different epitope of desmoglein-3 from that targeted by the IgG4 antibody, but this in itself does not provide an explanation. IgG1

Mechanism	Disruption of cell or tissue barrier	Infection of antigen-presenting cell	Binding of pathogen to self protein	Molecular mimicry	Superantigen
Effect	Release of sequestered self antigen; activation of nontolerized cells	Induction of co-stimulatory activity on antigen-presenting cells	Pathogen acts as carrier to allow anti-self response	Production of cross-reactive antibodies or T cells	Polyclonal activation of autoreactive T cells
Example	Sympathetic ophthalmia	Effect of adjuvants in induction of EAE	? Interstitial nephritis	Rheumatic fever ? Diabetes ? Multiple sclerosis	? Rheumatoid arthritis

Fig. A42.3 Ways in which infectious agents can break self tolerance. Because some antigens are sequestered from the circulation, either behind a tissue barrier or within the cell, it is possible that an infection that breaks cell and tissue barriers may expose hidden antigens (first column). A second possibility is that the local inflammation in response to an infectious agent may trigger the expression of MHC molecules and co-stimulators on tissue cells, inducing an autoimmune response (second column). In some cases, infectious agents may bind to self proteins. Because the infectious agent induces a helper T-cell response, any B cell that recognizes the self protein will also receive help. Such responses should be self-limiting once the infectious agent has been eliminated, because at this point the T-cell help will no longer be provided (third column). Infectious agents may induce either T-cell or B-cell responses that can cross-react with self antigens. This is termed molecular mimicry (fourth column). T-cell polyclonal activation by a bacterial superantigen could overcome clonal anergy, allowing an autoimmune process to begin (fifth column). There is little evidence for most of these mechanisms in human autoimmune disease. EAE, experimental autoimmune encephalomyelitis.

fixes complement, and thus we must infer that the proteinases generated by complement activation do not digest desmoglein-3.

Answer 3
Isotype switching is required to make the IgG4 antibodies in addition to or in lieu of the IgG1 antibodies. Switching to IgG4 is stimulated by the cytokine IL-4 released by activated T_H2 cells. If we take T cells from asymptomatic individuals who make IgG1 antibodies and T cells from patients who make IgG4 antibodies and stimulate them with desmoglein-3, we might find that the patients' T cells make more IL-4 (and thus undergo more isotype switching) than those of the asymptomatic individuals. Higher levels of IL-4 production in patients' T cells are indeed found in such an experiment.

Answer 4
Cyclophosphamide (Fig. A43.4) is an alkylating agent that interferes with DNA synthesis and therefore stops cell division. Although it has many bad side effects such as anemia, thrombocytopenia, and hair loss, it is effective in halting lymphocyte cell division, and thus suppresses immune reactions.

Answer 5
Pneumocystis jirovecii is an opportunistic pathogen that is a frequent cause of pneumonia in immunosuppressed patients. In this case, the high-dose corticosteroid treatment suppressed T-cell functions and trafficking, causing increased susceptibility to this potentially serious infectious agent.

Case 44

Answer 1
The 33-mer peptide has a predominance of proline residues, and this structure is crucial to its resistance to digestion by gastrointestinal enzymes. A bacterial propyl endopeptidase could potentially catalyze the breakdown of this peptide, thereby preventing it from interacting with TTG. Removing the 33-mer and its antigenic epitopes from the subepithelial space would eliminate the interaction of HLA DQ2-bound peptides with T cells and would halt the inflammatory process.

Answer 2
Testing is recommended in asymptomatic children who are in high-risk groups (type 1 diabetes mellitus, autoimmune thyroiditis, Down syndrome, Turner syndrome, Williams syndrome, IgA deficiency, and first-degree relatives of patients with celiac disease) starting at age 3 years, as long as they have not been put on a gluten-free diet. The reasoning behind testing asymptomatic children is that celiac disease can be clinically difficult to detect—but there is reason to believe that ongoing inflammation in the gut has deleterious effects and so it should be diagnosed and treated as early as possible.

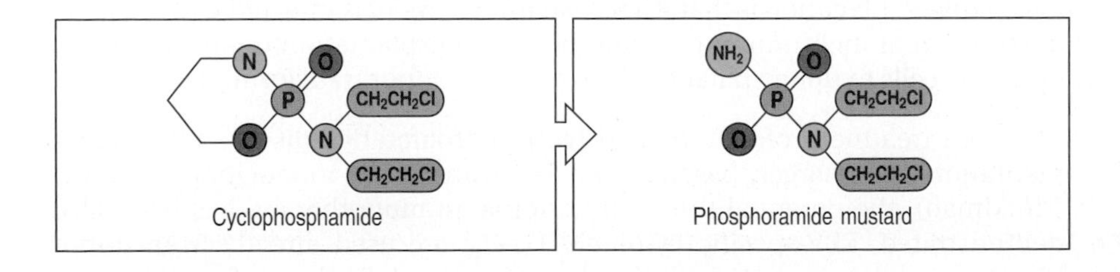

Fig. A43.4 The structure and metabolism of the cytotoxic drug cyclophosphamide. Cyclophosphamide is administered as a stable 'pro-drug', which is transformed enzymatically in the body to phosphoramide mustard, a powerful and unstable DNA alkylating agent.

An initial negative screen for celiac disease does not preclude its emergence later in some people. Repeat screening for type 1 diabetics and patients with Down syndrome and first-degree relatives with celiac disease should be performed periodically. The optimal frequency of such testing has not yet been clarified. An alternative to interval testing is to HLA type the high-risk individuals. In the absence of HLA DQ2 or DQ8, celiac disease is extremely rare.

Answer 3
A level of 220 p.p.m. gluten had been arbitrarily designated as 'gluten free,' but there is emerging evidence that even very small amounts of gluten are toxic to people with celiac disease and a new limit of 20 p.p.m. is being considered. The amount of gluten in foods is not the only challenge to patients; unclear labeling also poses a major obstacle to maintaining a gluten-free diet.

Answer 4
Oats do not contain gluten, but they can be cross-contaminated with wheat as a result of storage in silos that hold both types of grains. Patients newly diagnosed with celiac disease should not eat oats, but the grain can be reintroduced in small amounts once the disease is in remission.

Case 45

Answer 1
Cell-mediated cytotoxic activity against EBV-infected cells is the main method of controlling EBV replication. Although EBV-specific antibodies are produced in a normal infected host, they do not seem to have a major role in controlling the virus. Patients who lack the ability to produce specific antibody, but who have intact T-cell cytotoxic responses, are in most instances able to fight the infection effectively.

Answer 2
Bone marrow transplantation is most often performed in patients whose immune systems have either been destroyed with high doses of chemotherapy or in patients with primary immune deficiency. A major complication that can arise after a bone marrow transplant from a donor that is not fully HLA-matched to the recipient is graft-versus-host disease (GVHD; see Case 11). In GVHD, mature T lymphocytes in the bone marrow graft are activated and begin to attack the host's tissues. To prevent this, bone marrow grafts are often treated to remove mature T cells. If a bone marrow donor has been infected with EBV, they will have a small number of transformed B cells carrying the virus (about one in a million B cells). In the donor, these B cells are being 'held in check' by cytotoxic cells; there is an equilibrium between cell division and death of EBV-infected B cells. If mature T cells are removed from a marrow graft, but B cells are left, then the transformed B cells will escape from surveillance by cytotoxic T cells and might begin to proliferate at a high rate. Removing both mature B and T cells from donor bone marrow results in a lower rate of EBV lymphoproliferative disease after transplantation. An alternative explanation is that if a transplant recipient is infected with EBV, the destruction of their immune system before transplantation removes enough cytotoxic cells to tip the balance in favor of the donor transformed B cells.

Although treatment of EBV-related lymphoproliferative disease after transplantation is now largely based on the use of anti-CD20 monoclonal antibody (rituximab), the potential utility of adoptive immunotherapy has been also demonstrated. EBV-specific cytotoxic T cells are used directly from donor blood or can be expanded *in vitro* by culture with EBV-transformed donor cells. The cytotoxic T cells are then transfused into the transplant recipient with lymphoproliferative disease and can kill the dividing donor B cells. This treatment has achieved remission of EBV lymphoproliferative disease in up to 90% of patients in various studies so far. The major complication of this treatment is acute or chronic GVHD.

Answer 3

Many adults who have been exposed to EBV in childhood have circulating in their blood effector cytotoxic T cells specific for EBV. When a blood sample is cultured and then infected with EBV, the B cells will become infected and the effector cytotoxic T cells will then immediately destroy the infected cells. Transformed B cells can be prepared from these individuals either by removing T cells from the blood sample before culture or by adding inhibitors of T-cell activation such as cyclosporin A. It is exceedingly unlikely that a fetus will have been infected by EBV, and so their blood sample is unlikely to contain any activated EBV-specific cytotoxic T cells.

Answer 4

EBV activates B cells polyclonally; that is, without respect to the antigen specificities of the infected cells. A significant percentage (5–10%) of circulating B lymphocytes bear antigen receptors of low affinity for several cross-reacting carbohydrate, nucleotide, or glycoprotein antigens. If infected with EBV and activated, these cells will begin to secrete polyspecific IgM antibodies. Many of these antibodies will bind relatively nonspecifically to erythrocytes of other species such as horse, ox, cow, or sheep and are thus called heterophile ('other-loving') antibodies. They are found in approximately 90% of patients with EBV.

Answer 5

These patients have no B cells, the cellular host of EBV.

Case 46

Answer 1

Most IgG antibody against the Rh antigen is of the IgG3 or IgG1 subclass; these are the IgG subclasses that bind most tightly to the high-affinity Fcγ receptor (FcγRI; CD64). Red blood cells coated with anti-Rh antibody adhere tightly to the Fc receptors of macrophages in the red pulp of the spleen, the Kupffer cells of the liver, and elsewhere (Fig. A46.1). The macrophages destroy the antibody-coated red cells (Fig. A46.2).

Answer 2

If the mother and father are ABO identical, the fetus has a high likelihood of also being ABO compatible with its mother. If fetal blood enters the maternal circulation, it is likely to last much longer if the mother has no ABO alloantibodies against the fetal cells. For example, if the fetal red blood cells were of type B, they would be quickly hemolyzed if the mother were of red blood cell type A; she would have anti-B alloantibodies. Rapid destruction of the fetal red blood cells by hemolysis would impede alloimmunization of the mother.

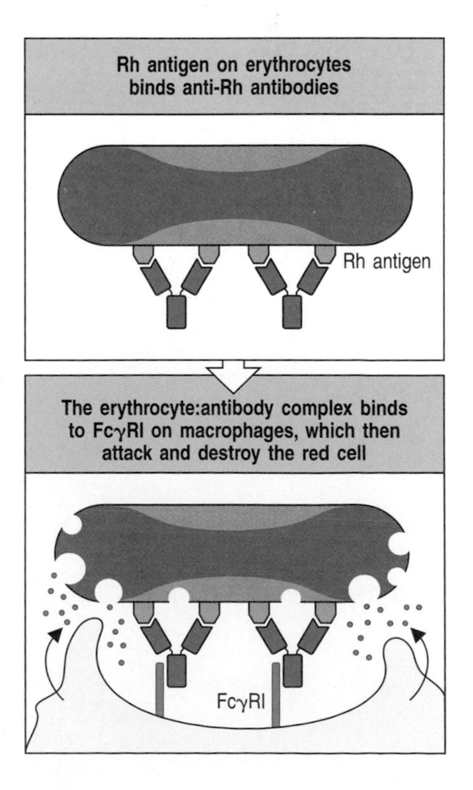

Fig. A46.1 IgG antibodies in a complex with their antigen bind via their Fc portions to the high-affinity Fcγ receptor on the surface of phagocytic cells. The Fc portions of IgG antibodies, including subclasses IgG3 and IgG1, bind the high-affinity Fc receptor FcγRI on macrophages when the antibody is complexed with an antigen (in this case a red blood cell). The binding of the Fc receptors helps to activate the macrophage, which then attacks the bound red cell.

Fig. A46.2 A rosette of Rh-positive (Rh^+) red cells. The red cells are coated with anti-Rh antibodies, which adhere to the Fc receptors on a macrophage (central cell). The macrophage is pitting the red blood cells and destroying them. Photograph courtesy of J. Jandl.

Answer 3
Rh-negative cells were used because they cannot be hemolyzed by the IgG Rh antibodies that have crossed the placenta.

Answer 4
You shouldn't, because the amount given (300 μg of IgG) is insufficient to cause the fetus harm. This amount of antibody raises the maternal titer in the indirect Coombs test to less than 1:4, a titer of Rh antibody that cannot cause significant hemolysis in the fetus. If the maternal titer is more than 1:4 she has been alloimmunized by her fetus.

Answer 5
She has IgM anti-Rh antibodies; they agglutinate Rh-positive cells in saline, unlike IgG anti-Rh antibodies. Because IgM antibodies do not cross the placenta, you would have no immediate concern about her pregnancy. However, she should be tested repeatedly to be sure that she is not developing a positive Coombs test, signifying the presence of IgG antibodies.

Case 47

Answer 1
Superantigens, but not conventional antigens, can activate naive T cells. Superantigens will thus induce the proliferation of lymphocytes from neonates and from the thymus, because previous exposure to the antigen and expansion of the number of antigen-reactive cells is not required. Superantigens do not require processing by accessory cells and are thus able to induce the proliferation of purified T cells in the presence of paraformaldehyde-treated monocytes, which lack the capacity to process antigen. Direct binding of a labeled protein to cells positive for MHC class II, or its co-precipitation with MHC class II molecules, confirms it as a superantigen.

Answer 2
During the evolution of an adaptive immune response to conventional antigen, a cascade of events must occur over a relatively long period. The antigen has to be internalized, processed, and presented as peptide:MHC complexes by antigen-presenting cells. The complexes are recognized only by those T cells bearing a T-cell receptor specific for the antigen-derived peptides—a fraction of a percent (less than 0.1%) of the entire T-cell pool. These few antigen-specific T cells must then proliferate and bystander cells must be recruited before an effective response can be mounted. In contrast, superantigen-induced immune activation is independent of antigen processing, thus bypassing the first step, and immediately activates a sizeable fraction of T cells. A very small number of superantigen molecules is sufficient to activate a T cell with the appropriate V_β region in its receptor (fewer than 10 molecules per T cell). Activation results in a massive secretion of T-cell cytokines, which include IL-2, IFN-γ, TNF-α, and lymphotoxin. In addition, superantigens can directly activate monocytes and dendritic cells by cross-linking their surface MHC class II molecules. Cross-linking is effected by superantigens bound to T-cell receptor β chains and/or because a number of superantigens, including TSST-1, have two distinct binding sites for MHC class II molecules. Cross-linking of MHC class II molecules causes a rapid and massive release of cytokines such as IL-1, TNF-α, IL-6, CXCL8, and IL-12. This is associated with the upregulation of B7 co-stimulatory molecules on these cells, which, together with cytokine action, further amplifies T-cell activation by superantigen. Thus, minute amounts of superantigen are sufficient to rapidly activate a large number of T cells and monocytes/macrophages, resulting in an

amplification loop and in a massive outpouring of cytokines, which leads to the rapid appearance of clinical symptoms.

Answer 3

Liver injury may occur as a result of decreased organ perfusion during hypotension. However, immunologic mechanisms may also contribute to injury. Hepatocytes express Fas, a cell-surface molecule crucial for the induction of apoptosis (programmed cell death). T-cell activation by superantigens and the massive release of cytokines results in the upregulation of the natural ligand for Fas—FasL—on the surface of circulating lymphocytes. Cross-linking of Fas on hepatocytes by FasL on circulating lymphocytes results in the triggering of apoptosis in hepatocytes. In addition, circulating cytokines such as TNF-α are also capable of triggering cell death and can result in liver injury.

Answer 4

Protection against toxic shock is conferred by antibodies against the superantigen, which neutralize it before it can cause disease. To stimulate an antibody response, the superantigen must be recognized, internalized, and processed by superantigen-specific B cells, which then present the antigenic peptides to antigen-specific T cells. These are activated to become helper T cells that can in turn stimulate the production of superantigen-specific antibodies on reexposure to the superantigen. Antibodies against other antigens that cross-react with the superantigen may also confer protection.

In humans, there is evidence that during and after TSST-associated illness, $V_\beta2$ T cells become anergic and thus cannot provide help to superantigen-specific B cells. Patients with TSS therefore fail to develop TSST-1-specific antibody. So Claire is, unfortunately, likely to be at risk of another episode of TSS. Hopefully, she will eventually develop anti-TSST-1 antibodies.

Case 48

Answer 1

The absence of delayed-type hypersensitivity to a wide range of antigens unrelated to *M. leprae* is called anergy. This should not be confused with T-cell or B-cell anergy, although it might operate by similar mechanisms. In tuberculoid leprosy, there is a strong delayed-type hypersensitivity to *M. leprae* and no anergy. The existence of anergy in the lepromatous form of leprosy but not in the tuberculoid form is most probably due to the presence of regulatory CD8 T cells in lepromatous leprosy. The CD8 T cells secrete the cytokines IL-10 and LT and thereby suppress antigen presentation by macrophages. These cytokines not only influence the T_H1 versus T_H2 phenotype, as discussed in the Case, but can also suppress T-cell responses to other unrelated antigens. IL-10 and LT suppress not only the *M. leprae*-specific T cells but also neighboring T cells, leading to global hyporesponsiveness, which was manifested in Ursula's case as anergy to candida and mumps antigens. However, Ursula's case is somewhat atypical; in many patients with lepromatous leprosy the unresponsiveness is confined to *M. leprae*, and responses are made to other antigens. Other pathogens use the IL-10 pathway to produce anergy; the Epstein–Barr virus, for example, produces a viral protein, vIL-10, that is homologous to human IL-10. Measles virus induces anergy by binding to CD46 on monocytes and inhibiting their production of IL-12.

Answer 2

The immune response in patients with lepromatous leprosy is skewed toward the T_H2 phenotype, leading to a disseminated infection. Because tuberculoid leprosy involves a T_H1 response and significantly reduced symptoms, we would like to switch the response to the T_H1 phenotype. Cytokines with the potential to inhibit T_H2 and induce a T_H1 response are IL-2, IFN-γ, and IL-12. Local injection of IFN-γ has been shown to lead to partial reversal of anergy and reduction of lesions. IFN-γ has also been shown to be effective in the treatment of similar diseases, such as leishmaniasis. In the visceral form of leishmaniasis, the T-cell response is also skewed to the T_H2 phenotype. This is in contrast to the cutaneous form of leishmaniasis, which is accompanied by a T_H1 response. IL-12 might also be beneficial, because it can induce T_H1 cells and does not activate T_H2 cells.

Answer 3

In lepromatous leprosy, a humoral immune response driven by T_H2 cells predominates, with vigorous antibody production, leading to hypergammaglobulinemia as observed in Ursula. The cytokines produced in T_H2 responses lead to enhanced immunoglobulin production. IL-4 induces isotype switching to IgE and increased production of IgG4 and IgE. IL-10 stimulates the production of IgG1 and IgG3, whereas IL-5 stimulates immunoglobulin production globally. It is therefore not uncommon to find hypergammaglobulinemia in patients who are producing a vigorous T_H2 response to an antigen.

Answer 4

Ursula's T_H2 response to *M. leprae* leads to increased production of IL-4 and IL-10. When she encounters a new antigen, her immune system will be awash with IL-4, triggering a T_H2 response to that antigen. This T_H2 response with its associated IL-4 production leads to an IgE response. Because asthma and other atopic diseases are T_H2-driven diseases involving IgE production, Ursula has a higher risk of developing asthma.

Case 49

Answer 1

John's hoarseness resulted from angioedema of the vocal cords. His wheezing was due to forced expiration of air through bronchi that had become constricted. In this case, constriction resulted from the release by activated mast cells of histamine and leukotrienes that caused the smooth muscles of the bronchial tubes to constrict.

Answer 2

John's parents were instructed to avoid feeding him any food containing peanuts and to read the labels of packaged foods scrupulously to avoid anything containing peanuts. They were advised to inquire in restaurants about food containing peanuts. Because green peas, also a legume, contain an antigen that cross-reacts with peanuts and might also incite an anaphylactic reaction, peas were withdrawn from John's diet. A Medi-Alert bracelet, indicating his anaphylactic reaction to peanuts, was ordered for John. The parents were also given an Epi-Pen syringe pre-filled with epinephrine to keep at home or while traveling, in case John developed another anaphylactic reaction.

Answer 3

Epinephrine acts at β_2-adrenergic receptors in smooth muscle surrounding blood vessels and bronchi. It has opposing effects on the two types of muscle. It contracts the muscle surrounding the small blood vessels, thereby constricting them, stopping vascular leakage, and raising the blood pressure. It relaxes that of the bronchi, making breathing easier.

Answer 4
Histamine and tryptase are released by activated mast cells; high levels in the blood indicate the massive release from the mast cells that occurs during an anaphylactic reaction.

Answer 5
Immediately after a systemic anaphylactic reaction the patient is unresponsive in a skin test owing to the massive depletion of mast-cell granules and failure of the blood vessels to respond to mediators. This is called tachyphylaxis and lasts for 72–96 hours after the anaphylactic reaction. For this reason, John had to come back to the Allergy Clinic a few days later for his tests.

Answer 6
Increased incidence of peanut allergy has been related to the increasing topical use of peanut-oil-based creams to treat dry skin in infants. Although peanut oil is not the culpable allergen, it is often contaminated by allergenic peanut proteins.

Answer 7
It has been shown that depletion of IgE by the administration of a humanized mouse anti-human IgE antibody that binds circulating IgE, but not mast-cell-bound IgE, results in protection from peanut anaphylaxis. This therapy works because it results in the eventual depletion of mast-cell-bound IgE, which is in equilibrium with serum IgE. It is safe because the anti-IgE antibody does not trigger mast-cell activation.

Case 50

Answer 1
During inspiration, the negative pressure on the airways causes their diameter to increase, allowing an inflow of air. During expiration, the positive expiratory pressure tends to narrow the airways. This narrowing is exaggerated when the airway is inflamed and bronchial smooth muscle is constricted, as in asthma. This causes air to be trapped in the lungs, with an increase in residual lung volume at the end of expiration. Breathing at high residual lung volume means more work for the muscles and increased expenditure of energy; this results in the sensation of tightness in the chest. The high residual lung volume is also the cause of the hyperinflated chest observed on the chest radiograph. The peribronchial inflammation in asthma causes bronchial marking around the airways.

Answer 2
Chronic allergic asthma is not simply due to constriction of the smooth muscles that surround the airway: it is largely due to the inflammatory reaction in the airway, which consists of cellular infiltration, increased secretion of mucus, and swelling of the bronchial tissues. This explains the failure of bronchodilators, which dilate smooth muscles, to maintain an open airway and their failure to completely reverse the decreased air flow during Frank's acute attacks. Steroids are therefore given to combat the inflammatory reaction of the late-phase response.

Answer 3
Allergic individuals have a tendency to respond to allergens with an immune response skewed to the production of T_H2 cells rather than T_H1. The cells produce the interleukins IL-4 and IL-13, cytokines that induce IgE production in humans. T_H2 cells also make IL-5, which is essential for eosinophil maturation.

Furthermore, activated T cells and bronchial epithelial cells secrete CCL11 (formerly known as eotaxin), which attracts eosinophils in the airways. The production of IL-4 and IL-5 by T_H2 cells responding to allergens in atopic individuals explains the frequent association of IgE antibody response and eosinophilia in these patients.

Answer 4

IgE-mediated hypersensitivity to an allergen is tested for by injecting a small amount of the allergen intradermally. In allergic individuals, this is followed within 10–20 minutes by a wheal-and-flare reaction at the site of injection (see Fig. 50.5), which subsides within an hour. The wheal-and-flare reaction is due mainly to the release of histamine by mast cells in the skin. This increases the permeability of blood vessels and the leakage of their contents into the tissues, resulting in the swollen wheal; dilation of the fine blood vessels around the area produces the diffuse red 'flare' seen around the wheal. This reaction is almost completely inhibited by antagonists of the histamine type 1 receptor, the major histamine receptor expressed in the skin.

Answer 5

The recurrence of the redness and swelling at the site of previous immediate allergic reactions represents the late-phase response characterized by a cellular infiltrate.

Answer 6

Nonsteroidal anti-inflammatory drugs (NSAIDs) such as aspirin and ibuprofen can induce wheezing in certain patients. This is classically seen in patients with Sampter's triad: asthma, nasal polyps, and NSAID sensitivity. NSAIDs inhibit the enzyme cyclooxygenase (COX). Normally, the actions of COX lead to the synthesis of prostaglandins from arachidonic acid. COX inhibition leads to shunting of the arachidonic acid precursor away from prostaglandin synthesis and into the leukotriene synthesis pathway (see Fig. 50.7). The increased leukotriene biosynthesis leads to bronchial smooth muscle constriction and cell proliferation, plasma leakage, mucus hypersecretion, and eosinophil migration, culminating in symptoms of wheezing and asthma exacerbation. Leukotriene E4 levels can be measured in the urine. In patients with aspirin sensitivity, E4 levels are higher at baseline and rise an additional fivefold after aspirin ingestion before returning to baseline as the aspirin-induced wheezing resolves.

Answer 7

Repeated administration of relatively high doses of allergen by subcutaneous injection is thought to favor antigen presentation by antigen-presenting cells that produce IL-12. This results in the induction of T_H1 cells rather than T_H2 cells. The presence of T_H1 cells tends to lead to an IgG antibody response rather than an IgE response because the T_H1 cells produce IFN-γ, which prevents further isotype switching to IgE. The IgG antibody competes with the IgE antibody for antigen. Furthermore, IgG bound to allergen inhibits mast-cell activation (via FcεRI) and B-cell activation (via surface immunoglobulin) by allergen because of inhibitory signals delivered subsequent to the binding of Fcγ receptors on these cells. This is thought to be one mechanism damping down the allergic response. Another is no further boosting of IgE production because IL-4 and IL-13 are not secreted. Existing IgE levels themselves may not fall by much, because IFN-γ does not affect B cells that have already switched to IgE production.

Answer 8

Most human allergy is caused by a limited number of inhaled protein allergens that elicit a T_H2 response in genetically predisposed individuals. These allergens are relatively small, highly soluble protein molecules that are presented to the immune system by the mucosal route at very low doses. It has been estimated that the maximum exposure to ragweed pollen allergens is less than 1 μg per year. It seems that transmucosal presentation of very low doses of allergens favors the activation of IL-4-producing T_H2 cells and is particularly efficient at inducing IgE responses. The dominant antigen-presenting cell type in the respiratory mucosa expresses high levels of co-stimulatory B7.2 molecules. Expression of B7.2 on antigen-presenting cells is thought to favor the development of T_H2 cells. In contrast, injection of antigen subcutaneously in large doses, as occurs on vaccination, results in antigen uptake in the local lymph nodes by a variety of antigen-presenting cells and favors the development of T_H1 cells, which inhibit antibody switching to IgE.

Case 51

Answer 1

Corticosteroids bind to steroid receptors in inflammatory cells such as T cells and eosinophils. The steroid:receptor complex is translocated into the nucleus, where it can control gene expression, including the expression of cytokine genes, by binding to control elements in the DNA. In addition, corticosteroids increase the synthesis of the inhibitor of the transcription factor NFκB, which controls the expression of multiple cytokine genes. One effect is to inhibit the synthesis of cytokines and the release of preformed mediators and arachidonic acid metabolites. Although topical steroids are very effective, excessive or prolonged use of powerful steroids can lead to local skin atrophy.

Answer 2

The immunosuppressant cyclosporin A acts primarily on T cells and interferes with the transcription of cytokine genes. The drug binds to an intracellular protein, cyclophilin, and this complex in turn inhibits calcineurin, which normally dephosphorylates nuclear factor of activated T cells (NFAT), a major cytokine gene transcription factor. FK506, or tacrolimus, is another immunosuppressant with a spectrum of activity similar to that of cyclosporin. Tacrolimus binds to the cytoplasmic protein FK506-binding protein, and this complex also inhibits calcineurin. Tacrolimus has a smaller molecular size and higher potency than cyclosporin A and, perhaps because of these features, it seems to be effective as a topical formulation.

Answer 3

Patients with atopic dermatitis have defective local innate cell-mediated immunity, which is required for the control of herpesvirus and vaccinia virus infections: T_H2 cytokine expression in the affected skin inhibits the production of antimicrobial peptides by keratinocytes. Cell-mediated adaptive immune responses involve T_H1 CD4 cells and CD8 cytotoxic cells; patients with atopic dermatitis have selective activation of T_H2 rather than T_H1 cells, as shown by their reduced delayed-type hypersensitivity skin reactions. They also have decreased numbers and function of CD8 cytotoxic T cells. Furthermore, monocytes from patients with atopic dermatitis secrete increased amounts of IL-10 and prostaglandin E_2 (PGE_2). Both IL-10 and PGE_2 inhibit the production of the T_H1 cytokine IFN-γ, and IL-10 also inhibits T-cell-mediated reactions.

Answer 4

Scratching causes tissue damage that stimulates the keratinocytes to secrete cytokines and chemokines (IL-1, IL-6, CXCL8, GM-CSF, and TNF-α). IL-1 and TNF-α induce the expression of adhesion molecules such as E-selectin, ICAM-1, and VCAM-1 on endothelial cells, which attract lymphocytes, macrophages, and eosinophils into the skin. These infiltrating cells secrete cytokines and inflammatory mediators that perpetuate keratinocyte activation and cutaneous inflammation.

Answer 5

The skin of more than 90% of patients with atopic dermatitis is colonized by *Staphylococcus aureus*. Recent studies suggest that *S. aureus* can exacerbate or maintain skin inflammation in atopic dermatitis by secreting a group of toxins known as superantigens, which cause polyclonal stimulation of T cells and macrophages (see Case 47). T cells from patients with atopic dermatitis preferentially express T-cell receptor β chains $V_\beta 3$, 8, and 12, which can be stimulated by staphylococcal superantigens, resulting in T-cell proliferation and increased IL-5 production. Staphylococcal superantigens can also induce expression of the skin homing receptor (CLA) in T cells, which is mediated by IL-12. In addition, nearly half of patients with atopic dermatitis produce IgE directed against staphylococcal superantigens, particularly SEA, SEB, and toxic shock syndrome toxin-1 (TSST-1). Basophils from patients with atopic dermatitis who produce antitoxin IgE release histamine on exposure to the relevant toxin. These findings suggest that local production of staphylococcal exotoxins at the skin surface could cause IgE-mediated histamine release and thereby trigger the itch–scratch cycle that exacerbates the eczema.

Answer 6

A mouse model of atopic dermatitis suggests that sensitization directly through the skin can result in allergen-induced asthma. In this model, patch application of allergen to the shaved skin of a normal mouse results in an eczematous dermatitis and subsequent allergen-specific airway hypersensitivity such that exposure to allergen by inhalation causes airway hyperresponsiveness typical of the asthmatic state.

There is epidemiologic evidence that sensitization of infants to food allergens through the skin may predispose to food allergy, and that unlike oral exposure to food allergens it does not induce tolerance, but rather results in IgE antibody formation that can cause anaphylaxis.

Case 52

Answer 1

The symptoms were caused by the activation of complement-generated C3a, which releases histamine from mast cells and causes hives. The swelling around the mouth and eyelids is a form of angioedema. There is a more complete discussion of the role of the complement and the kinin systems in the pathogenesis of angioedema in Case 31.

Answer 2

Gregory almost certainly had developed vasculitis in the small blood vessels of his brain, and this compromised oxygen delivery to his brain.

Answer 3

He had red cells and albumin in his urine, which indicated an inflammation of the small blood vessels in his kidney glomeruli. He also developed purpura in his feet and ankles. Purpura (which is the Latin word for purple) indicates hemorrhage from small blood vessels in the skin that are inflamed and have become plugged with clots. A skin biopsy of one lesion showed the deposition of IgG and C3 around the small blood vessels, suggesting that an immune reaction was taking place.

Answer 4
You would expect to see massive follicular hyperplasia, polyclonal B-cell activation, and many mature plasma cells in the medulla. The massive B-cell activation in the lymph nodes leads to an overflow of plasma cells from the medulla of the nodes into the efferent lymph. It is otherwise very, very rare to find plasma cells in the blood, as were found in Gregory's blood. They find their way to the bloodstream via the thoracic duct. The enlargement of the spleen was almost certainly due to hyperplasia of the white pulp. Some plasma cells probably enter the blood from the hyperplastic follicles in the spleen.

Answer 5
The acute-phase reaction is caused by interleukin (IL)-1 and to a greater extent by IL-6, which are released from monocytes that have been activated by the uptake of immune complexes. The acute-phase response consists of marked changes in protein synthesis by the liver. The synthesis of albumin drops sharply, as does the synthesis of transferrin. The synthesis of fibrinogen, C-reactive protein, amyloid A, and several glycoproteins is rapidly upregulated. The precise advantage to the host of the acute-phase reaction is not well understood, but it is presumably a part of innate immunity, which aids host resistance to pathogens before the adaptive immune system becomes engaged.

Answer 6
His serum C1q level was decreased. This almost always indicates complement consumption by immune complexes via the classical pathway. (In hereditary angioedema (see Case 31) the C1q level is normal; in this disease, complement is activated because of a defect in an inhibitor and not by the formation of immune complexes.) The level of C3 in Gregory's serum was also lowered, a further indication of complement consumption (see Case 32).

Answer 7
No! The skin test is positive when there are IgE antibodies bound to the mast cells in the skin. Gregory did not have IgE antibodies against penicillin, as confirmed by the negative RAST test. Serum sickness is caused by complement-fixing IgG antibodies.

Case 53

Answer 1
Pentadecacatechol can be transferred from the initial point of contact to other areas of the skin by the fingernails after scratching the itchy lesion at the primary site of hapten introduction. This is why it is essential to cut the fingernails short and thoroughly wash off the skin and scalp to remove the chemical and prevent further spread.

Answer 2
The half-life of some of the proteins haptenated by pentadecacatechol can be quite long. CD4 memory T cells will continue to be activated as long as the haptenated peptides are being generated. In Paul's case this went beyond the third week after contact with poison ivy.

Answer 3
Once an individual has been sensitized, the reaction often becomes worse with each exposure, as each reexposure not only produces the hypersensitivity reaction but generates more effector and memory T cells. Memory T cells that mediate delayed hypersensitivity reactions, such as contact sensitivity

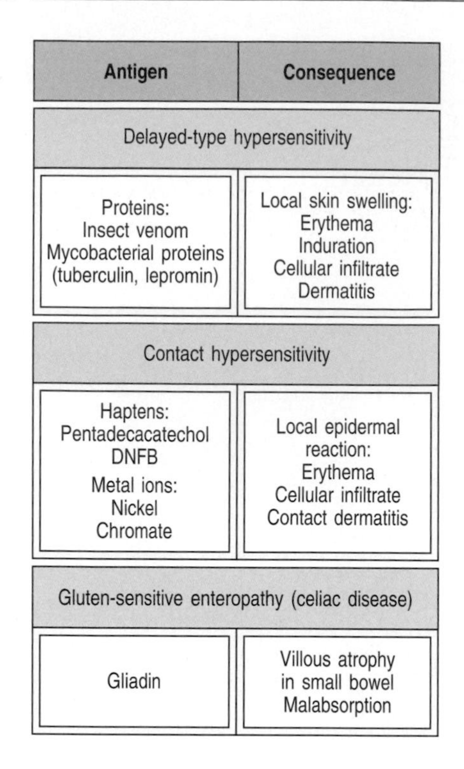

Fig. A53.7 Some type IV hypersensitivity reactions. Depending on the source of antigen and its route of introduction, these clinical conditions have different names and consequences. DNFB, dinitrofluorobenzene.

to poison ivy and the tuberculin test, can persist for most of the life of the individual.

Answer 4

You could perform a patch test. In this test a patch of material impregnated with the hapten is applied to the skin under seal for 48 hours. The area is then examined for redness, swelling, and vesicle formation. Alternatively, peripheral blood mononuclear cells can be incubated with the hapten and T-cell proliferation assessed 6–9 days later.

Answer 5

The risk of Brian's developing poison ivy sensitivity is at least as high as that for a normal child. This is because antibody plays no discernible role in the genesis of delayed hypersensitivity reactions, and T-cell function is normal in X-linked agammaglobulinemia. In fact, clinical observations suggest that boys with X-linked agammaglobulinemia may develop more severe forms of poison ivy sensitivity. It has been suggested that, in the absence of antibody, more hapten is available for conjugation with self proteins and that, in the absence of antigen presentation by β cells, the T-cell response is skewed more towards T_H1 cells.

Answer 6

The artificially induced tuberculin reaction is a good model of a delayed hypersensitivity reaction. This skin test detects infection with the bacterium *Mycobacterium tuberculosis*, or previous immunization against tuberculosis with the live attenuated vaccine BCG. Small amounts of tuberculin, a protein derived from *M. tuberculosis*, are injected subcutaneously; a day or two later, a sensitized person develops a small, red, raised area of skin at the site of injection. In countries where BCG is administered routinely to babies, the tuberculin test can be used to test for T-cell function. This is because antigen-specific memory T cells are long-lived, and the sensitivity to tuberculin will persist throughout life. In the USA, children are not immunized with BCG. However, they all receive a full course of diphtheria and tetanus vaccines, which in each case contain purified protein toxoids as the antigen. Contact sensitivity to these two antigens can be used to test T-cell function. Alternatively, antigen derived from the yeast-like fungus *Candida albicans*, which is a normal inhabitant of the body flora, can be used to induce a delayed hypersensitivity reaction in the skin.

Answer 7

Some of the commoner environmental causes of delayed hypersensitivity reactions are insect bites or stings, which introduce insect venom proteins under the skin, and skin contact with chemicals in the leaves of some plants, or with metals such as nickel, beryllium, and chromium (Fig. A53.7). Nickel sensitivity is quite common and often occurs at the site of contact with nickel-containing jewelry. Contact sensitivity to beryllium has been well documented in factory workers engaged in manufacturing fluorescent light bulbs. Celiac disease (see Case 44) is a type of delayed sensitivity reaction seen in people who are allergic to the protein gliadin, a constituent of wheat grains and flour. Patients with celiac disease therefore have to avoid all food products containing wheat flour.

Figure Acknowledgments

Case 4
Fig. 4.1 from Castigli, E., Geha, R.S.: Molecular basis of common variable immunodeficiency. *Journal of Allergy and Clinical Immunology* 2006, **117**:740–747. Copyright © 2006, with permission from the American Academy of Allergy, Asthma, and Immunology.
Fig. 4.3 from Cunningham-Rundles, C., Bodian, C.: Common variable immunodeficiency: Clinical and immunological features of 248 patients. *Clinical Immunology* **92**:34–48, Copyright © 1999, with permission from Elsevier.
Fig.4.4 reprinted by permission from Macmillan Publishers Ltd: *Nature Genetics* 2005, **37**:829–834, © 2005.

Case 5
Fig. 5.2 bottom left panel from *Diagnostic Immunopathology* (2nd edition), Eds. R.B. Colvin, A.K. Bahn, and R.T. McCluskey. New York, Raven Press 1994, 246–247. © 1995, Raven Press.
Fig. 5.6 reprinted with permission from *Immunology* (6th edition), by Roitt, I., et al., p. 306. © Harcourt Publishers Limited 2001.
Fig. 5.8 from *Janeway's Immunobiology*, 8th edition, by Kenneth Murphy. © 2012 by Garland Science. Used by permission of Garland Science.

Case 7
Fig. 7.3 from *Janeway's Immunobiology*, 8th edition, by Kenneth Murphy. © 2012 by Garland Science. Used by permission of Garland Science.

Case 13
Fig. 13.2 reprinted by permission from *Nature* 1998, Fig. 4b, **395**:462–469. © 1998, Macmillan Magazines Ltd.
Fig. 13.3 from *Janeway's Immunobiology*, 8th edition, by Kenneth Murphy. © 2012 by Garland Science. Used by permission of Garland Science.

Case 15
Fig. 15.1 from *Janeway's Immunobiology*, 7th edition, by Kenneth Murphy, Paul Travers, and Mark Walport. © 2008 by Garland Science. Used by permission of Garland Science.
Fig. 15.2 reprinted with permission from Macmillan Publishers Ltd: Nagle, D.L., et al.: Identification and mutation analysis of the complete gene for Chediak-Higashi syndrome. *Nature Genetics* 1996; **14**:307–311. Copyright © 1996.
Fig. 15.3b and 15.5 reprinted with permission from Huizing, M., Helip-Wooley, A., Westbroek, W., Gunay-Aygun, M., Gahl, W.A.: Disorders of lysosome-related organelle biogenesis: clinical and molecular genetics. *Annual Review of Genomics and Human Genetics* 2008; **9**:359–386. Permission from Annual Reviews conveyed through Copyright Clearance Center, Inc.
Fig. 15.4 used with permission from The Jackson Laboratory.

Case 16
Fig. 16.4 from *Blood* 1986, **68**:1329–1332. © 1986 with permission of the American Society of Hematology.
Fig. 16.7 from *Immunity* 1998, **9**:81–91. © 1998, Cell Press.

Case 17
Fig. 17.1 from Wack, A., et al.: Direct visualization of thymocyte apoptosis in neglect, acute and steady-state negative selection. *International Immunology* 1996, **8**:1537–1548. By permission of Oxford University Press.

Case 18
Fig. 18.1 from Kamradt, T., Mitchison, N.A.: Advances in immunology: Tolerance and autoimmunity. *The New England Journal of Medicine*. 2001; **344**:655–664. © 2001 Massachusetts Medical Society. All rights reserved.

Case 19
Fig. 19.3 from "Inherited Disorders with Autoimmunity and Defective Lymphocyte Regulation" by Jennifer Puck, et al., in *Primary Immunodeficiency Disease* (1998) edited by Ochs, H.D., et al. Used by permission of Oxford University Press, Inc.

Case 20
Fig. 20.3 © The Rockefeller University Press. *The Journal of Experimental Medicine*, 2008, **205**:1551–1557. doi:10.1084/jem.20080218.
Fig. 20.4 from VCU Health System Pediatric Radiology (http://www.pedsradiology.com/) with permission from Dr. Lakshmana D. Narla.
Fig. 20.5 from Grimbacher, B., et al.: Hyper-IgE syndrome with recurrent infections – An autosomal dominant multisystem disorder. *The New England Journal of Medicine*. 1999; **340**:692–702. © 1999 Massachusetts Medical Society. All rights reserved.

Case 21
Fig. 21.3 reprinted with permission from *Color Atlas of Pediatric Dermatology*, 4th edition, by Weinberg, S., Prose, N., Kristal, L. Copyright © 2008 by The McGraw-Hill Companies. All rights reserved.

Case 25
Fig. 25.1 from *The Immune System*, 3rd edition, by Peter Parham. © 2009 by Garland Science. Used by permission of Garland Science.

Case 29
Fig. 29.3 from McDonald, D.R., Brown, D., Bonilla, F.A., Geha, R.: Interleukin receptor-associated kinase 4 deficiency impairs toll-like receptor dependent innate anti-viral immune responses. *Journal of Allergy and Clinical Immunology* 2006, **118**(6):1357–1362. Copyright © 2006, with permission from the American Academy of Allergy, Asthma, and Immunology.

Case 30
Fig. 30.3 from Hamdan, J., University of Illinois at Chicago College of Medicine (http://www.ezhemeonc.com/index.php/hematological-disorders/).

Case 31
Fig. 31.5 from Williams, K.: Hereditary angioneurotic edema. *Clinical Immunology* 1975, **4**:174–188. Copyright © 1975, with permission from Elsevier.

Case 33
Fig. 33.2 from *Blut* 1990, **60**:309–318. © Springer-Verlag.

Case 34
Fig. 34.1 from *The Immune System*, 3rd edition, by Peter Parham. © 2009 by Garland Science. Used by permission of Garland Science.
Fig. 34.3 from Petrilli, V., Papin, S., Tschopp, J.: The inflammasome. *Current Biology* 2005, **15**:R581. Copyright © 2005, with permission from Elsevier.
Fig. 34.4 from *Janeway's Immunobiology*, 8th edition, by Kenneth Murphy. © 2012 by Garland Science. Used by permission of Garland Science.
Fig. 34.5 from Boschan, C.: Neonatal-onset multisystem inflammatory disease (NOMID) due to a novel S331R mutation of the CIAS1 gene and response to interleukin-1 receptor antagonist treatment. *American Journal of Medical Genetics Part A* 2006, **140A**:883–886.
Fig. 34.6 from Notarangelo, L., et al.: Primary immunodeficiency diseases: an update from the International Union of Immunological Societies Primary Immunodeficiency Diseases Classification Committee Meeting in Budapest, 2005. *Journal of Allergy and Clinical Immunology* 2006, **117**:883–896. Copyright © 2006, with permission from the American Academy of Allergy, Asthma, and Immunology.

Case 35
Fig. 35.1 from *Janeway's Immunobiology*, 8th edition, by Kenneth Murphy. © 2012 by Garland Science. Used by permission of Garland Science.

Case 36
Fig. 36.4 from *Janeway's Immunobiology*, 8th edition, by Kenneth Murphy. © 2012 by Garland Science. Used by permission of Garland Science.

Case 38
Fig. 38.2 from Shakil, A.O., Di Bisceglie, A.M.: Vasculitis and cryoglobulinemia related to hepatitis C. *The New England Journal of Medicine*. 1994; **331**:1624. © 1994 Massachusetts Medical Society. All rights reserved.

Case 39
Fig. 39.2 from *Janeway's Immunobiology*, 8th edition, by Kenneth Murphy. © 2012 by Garland Science. Used by permission of Garland Science.
Fig. 39.3 from Peppercorn, M., Moss, A., Cheifetz, A. Crohn's Disease: Cheifetz, A.S., Moss, A.C., Peppercorn, M.A. [internet]. Version 11. Knol. 2008 Oct 20. Available from http://knol.google.com/k/mark-peppercorn-md/crohn-s-disease/jlMpZGOT/pR6xwg.
Fig. 39.4 reprinted by permission from Macmillan Publishers Ltd: Xavier, R.J., Podolsky, D.K.: Unraveling the pathogenesis of inflammatory bowel disease. *Nature* 2007, **448**:427–434. Copyright © 2007.

Case 40
Figs. 40.2 and **40.3** reprinted from Weiner, H.L., et al.: Serial magnetic resonance imaging in multiple sclerosis: correlation with attacks, disability, and disease stage. *Journal of Neuroimmunology* 2000, **104**:164–173. Copyright © 2000, with permission from Elsevier.
Fig. 40.4 from *Janeway's Immunobiology*, 8th edition, by Kenneth Murphy. © 2012 by Garland Science. Used by permission of Garland Science.

Case 43
Fig. 43.4 from *The Immune System*, 3rd edition, by Peter Parham. © 2009 by Garland Science. Used by permission of Garland Science.

Case 44
Fig. 44.3 from Farrell, R.J., Kelly, C.P.: Celiac sprue. *The New England Journal of Medicine* 2002, **346**(3):180–188.
Fig 44.4 reprinted by permission from Macmillan Publishers Ltd*: Nature Reviews in Immunology* 2002, **2**(9):647–655, © 2002.

Case 45
Fig. 45.3a, b from *Second International Workshop on Cell Mediated Cytotoxicity,* Eds. Henkart & Martz, 1985, pp 99–119. © 1985, Kluwer/Plenum Publishers.

Case 47
Fig. 47.4 © The Rockefeller University Press. *The Journal of Experimental Medicine*. 1990, **172**:981–984. doi:10.1084/jem.172.3.981.

Case 48
Fig. 48.4 from Kaplan, G.: *International Review of Experimental Pathology* 1986, **28**:45–78. © 1986, with permission from Elsevier.

Case 51
Fig. 51.4 from the *Journal of Allergy and Clinical Immunology* 1983, **71**:47–56. © 1983, Mosby.

Case 53
Fig. 53.3 from http://www.lib.uiowa.edu/hardin/md/pictures22/dermnet/contact_dermatitis_rhus34.jpg. Copyright © 2007 Interactive Medical Media LLC. All rights reserved.

Index

Note: Figures in the Answer section are labeled by Case number, in the form **Fig. A30.1,** and those for Case numbers 1 to 53 are labeled in the form **Fig. 1.1**, **Fig 2.1** etc.

A

D

E

F

I

R